T0134667

Communications in Computer and Information Science   1637

More information about this series at https://link.springer.com/bookseries/7899

Haijun Zhang · Yuehui Chen · Xianghua Chu ·
Zhao Zhang · Tianyong Hao · Zhou Wu ·
Yimin Yang (Eds.)

Neural Computing for Advanced Applications

Third International Conference, NCAA 2022
Jinan, China, July 8–10, 2022
Proceedings, Part I

 Springer

Editors
Haijun Zhang (iD)
Harbin Institute of Technology
Shenzhen, China

Yuehui Chen (iD)
University of Jinan
Jinan, China

Xianghua Chu (iD)
Shenzhen University
Shenzhen, China

Zhao Zhang (iD)
Hefei University of Technology
Hefei, China

Tianyong Hao (iD)
South China Normal University
Guangzhou, China

Zhou Wu (iD)
Chongqing University
Chongqing, China

Yimin Yang (iD)
Western University
London, ON, Canada

ISSN 1865-0929   ISSN 1865-0937 (electronic)
Communications in Computer and Information Science
ISBN 978-981-19-6141-0   ISBN 978-981-19-6142-7 (eBook)
https://doi.org/10.1007/978-981-19-6142-7

This Springer imprint is published by the registered company Springer Nature Singapore Pte Ltd.
The registered company address is: 152 Beach Road, #21-01/04 Gateway East, Singapore 189721, Singapore

Preface

Neural computing and artificial intelligence (AI) have become hot topics in recent years. To promote the multi-disciplinary development and application of neural computing, a series of NCAA conferences was initiated on the theme of "make the academic more practical", providing an open platform of academic discussions, industrial showcases, and basic training tutorials. This volume contains the papers accepted for this year's International Conference on Neural Computing for Advanced Applications (NCAA 2022). NCAA 2022 was organized by the University of Jinan, Shandong Jianzhu University, South China Normal University, Harbin Institute of Technology, Chongqing University, and Hefei University of Technology, and it was supported by Springer. Due to the effects of COVID-19, the mainstream part of NCAA 2022 was turned into a hybrid event, with both online and offline participants, in which people could freely connect to live broadcasts of keynote speeches and presentations.

NCAA 2022 received 205 submissions, of which 77 high-quality papers were selected for publication in this volume after double-blind peer review, leading to an acceptance rate of just under 38%. These papers have been categorized into 10 technical tracks: neural network theory and cognitive sciences; machine learning, data mining, data security and privacy protection, and data-driven applications; computational intelligence, nature-inspired optimizers, and their engineering applications; cloud/edge/fog computing, the Internet of Things/Vehicles (IoT/IoV), and their system optimization; control systems, network synchronization, system integration, and industrial artificial intelligence; fuzzy logic, neuro-fuzzy systems, decision making, and their applications in management sciences; computer vision, image processing, and their industrial applications; natural language processing, machine translation, knowledge graphs, and their applications; neural computing-based fault diagnosis, fault forecasting, prognostic management, and system modeling; and spreading dynamics, forecasting, and other intelligent techniques against coronavirus disease (COVID-19).

The authors of each paper in this volume have reported their novel results of computing theory or application. The volume cannot cover all aspects of neural computing and advanced applications, but may still inspire insightful thoughts for the readers. We hope that more secrets of AI will be unveiled, and that academics will drive more practical developments and solutions.

June 2022

Haijun Zhang
Yuehui Chen
Xianghua Chu
Zhao Zhang
Tianyong Hao
Zhou Wu
Yimin Yang

Organization

Honorary Chairs

John MacIntyre   University of Sunderland, UK
Tommy W. S. Chow   City University of Hong Kong, Hong Kong

General Co-chairs

Haijun Zhang   Harbin Institute of Technology, China
Yuehui Chen   University of Jinan, China
Zhao Zhang   Hefei University of Technology, China

Program Co-chairs

Tianyong Hao   South China Normal University, China
Zhou Wu   Chongqing University, China
Yimin Yang   University of Western Ontario, Canada

Organizing Committee Co-chairs

Yongfeng Zhang   University of Jinan, China
Yunchu Zhang   Shandong Jianzhu University, China
Choujun Zhan   Nanfang College Guangzhou, China
Mingbo Zhao   Donghua University, China

Local Arrangement Co-chairs

Menghua Zhang   University of Jinan, China
Weijie Huang   University of Jinan, China
Ming Wang   Shandong Jianzhu University, China
Chengdong Li   Shandong Jianzhu University, China
Xiangping Zhai   Nanjing University of Aeronautics and
     Astronautics, China

Registration Co-chairs

Yaqing Hou   Dalian University of Technology, China
Jing Zhu   Macau University of Science and Technology,
     China

Shuqiang Wang	Chinese Academy of Sciences, China
Weiwei Wu	Southeast University, China
Zhili Zhou	Nanjing University of Information Science and Technology, China

Publication Co-chairs

Kai Liu	Chongqing University, China
Yu Wang	Xi'an Jiaotong University, China
Yi Zhang	Fuzhou University, China
Bo Wang	Huazhong University of Science and Technology, China
Xianghua Chu	Shenzhen University, China

Publicity Co-chairs

Liang Feng	Chongqing University, China
Penglin Dai	Southwest Jiaotong University, China
Dong Yang	University of California, Merced, USA
Shi Cheng	Shaanxi Normal University, China
Reza Maleklan	Malmö University, Sweden

Sponsorship Co-chairs

Wangpeng He	Xidian University, China
Bingyi Liu	Wuhan University of Technology, China
Cuili Yang	Beijing University of Technology, China
Jicong Fan	Chinese University of Hong Kong, Shenzhen, China

NCAA Steering Committee Liaison

| Jingjing Cao | Wuhan University of Technology, China |

Web Chair

| Xinrui Yu | Harbin Institute of Technology, China |

Program Committee

Dong Yang	City University of Hong Kong, Hong Kong
Sheng Li	University of Georgia, USA
Jie Qin	Swiss Federal Institute of Technology (ETH), Switzerland

Xiaojie Jin	Bytedance AI Lab, USA
Zhao Kang	University of Electronic Science and Technology, China
Xiangyuan Lan	Hong Kong Baptist University, Hong Kong
Peng Zhou	Anhui University, China
Chang Tang	China University of Geosciences, China
Dan Guo	Hefei University of Technology, China
Li Zhang	Soochow University, China
Xiaohang Jin	Zhejiang University of Technology, China
Wei Huang	Zhejiang University of Technology, China
Chao Chen	Chongqing University, China
Jing Zhu	Macau University of Science and Technology, China
Weizhi Meng	Technical University of Denmark, Denmark
Wei Wang	Dalian Ocean University, China
Jian Tang	Beijing University of Technology, China
Heng Yue	Northeastern University, China
Yimin Yang	University of Western Ontario, Canada
Jianghong Ma	City University of Hong Kong, Hong Kong
Jicong Fan	Chinese University of Hong Kong, Shenzhen, China
Xin Zhang	Tianjing Normal University, China
Xiaolei Lu	City University of Hong Kong, Hong Kong
Penglin Dai	Southwest Jiaotong University, China
Liang Feng	Chongqing University, China
Xiao Zhang	South Central University for Nationalities, China
Bingyi Liu	Wuhan University of Technology, China
Cheng Zhan	Southwest University, China
Qiaolin Pu	Chongqing University of Posts and Telecommunications, China
Hao Li	Hong Kong Baptist University, Hong Kong
Junhua Wang	Nanjing University of Aeronautics and Astronautics, China
Yu Wang	Xi'an Jiaotong University, China
Binqiang Chen	Xiamen University, China
Wangpeng He	Xidian University, China
Jing Yuan	University of Shanghai for Science and Technology, China
Huiming Jiang	University of Shanghai for Science and Technology, China
Yizhen Peng	Chongqing University, China
Jiayi Ma	Wuhan University, China
Yuan Gao	Tencent AI Lab, China

Xuesong Tang	Donghua University, China
Weijian Kong	Donghua University, China
Zhili Zhou	Nanjing University of Information Science and Technology, China
Yang Lou	City University of Hong Kong, Hong Kong
Chao Zhang	Shanxi University, China
Yanhui Zhai	Shanxi University, China
Wenxi Liu	Fuzhou University, China
Kan Yang	University of Memphis, USA
Fei Guo	Tianjin University, China
Wenjuan Cui	Chinese Academy of Sciences, China
Wenjun Shen	Shantou University, China
Mengying Zhao	Shandong University, China
Shuqiang Wang	Chinese Academy of Sciences, China
Yanyan Shen	Chinese Academy of Sciences, China
Haitao Wang	China National Institute of Standardization, China
Yuheng Jia	City University of Hong Kong, Hong Kong
Chengrun Yang	Cornell University, USA
Lijun Ding	Cornell University, USA
Zenghui Wang	University of South Africa, South Africa
Xianming Ye	University of Pretoria, South Africa
Reza Maleklan	Malmö University, Sweden
Xiaozhi Gao	University of Eastern Finland, Finland
Jerry Lin	Western Norway University of Applied Sciences, Norway
Xin Huang	Hong Kong Baptist University, Hong Kong
Xiaowen Chu	Hong Kong Baptist University, Hong Kong
Hongtian Chen	University of Alberta, Canada
Gautam Srivastava	Brandon University, Canada
Bay Vo	Ho Chi Minh City University of Technology, Vietnam
Xiuli Zhu	University of Alberta, Canada
Rage Uday Kiran	University of Toyko, Japan
Matin Pirouz Nia	California State University, Fresno, USA
Vicente Garcia Diaz	University of Oviedo, Spain
Youcef Djenouri	Norwegian University of Science and Technology, Norway
Jonathan Wu	University of Windsor, Canada
Yihua Hu	University of York, UK
Saptarshi Sengupta	Murray State University, USA
Wenxiu Xie	City University of Hong Kong, Hong Kong
Christine Ji	University of Sydney, Australia

Jun Yan	Yiducloud, China
Jian Hu	Yiducloud, China
Alessandro Bile	Sapienza University of Rome, Italy
Jingjing Cao	Wuhan University of Technology, China
Shi Cheng	Shaanxi Normal University, China
Xianghua Chu	Shenzhen University, China
Valentina Colla	Scuola Superiore Sant'Anna, Italy
Mohammad Hosein Fazaeli	Amirkabir University of Technology, Iran
Vikas Gupta	LNM Institute of Information Technology, India
Tianyong Hao	South China Normal University, China
Hongdou He	Yanshan University, China
Wangpeng He	Xidian University, China
Yaqing Hou	Dalian University of Technology, China
Essam Halim Houssein	Minia University, Egypt
Wenkai Hu	China University of Geosciences, China
Lei Huang	Ocean University of China, China
Weijie Huang	University of Jinan, China
Zhong Ji	Tianjin University, China
Qiang Jia	Jiangsu University, China
Yang Kai	Yunnan Minzu University, China
Andreas Kanavos	Ionian University, Greece
Zhao Kang	Southern Illinois University Carbondale, USA
Zouaidia Khouloud	Badji Mokhtar Annaba University, Algeria
Chunshan Li	Harbin Institute of Technology, China
Dongyu Li	Beihang University, China
Kai Liu	Chongqing University, China
Xiaofan Liu	City University of Hong Kong, Hong Kong
Javier Parra Arnau	Karlsruhe Institute of Technology, Germany
Santwana Sagnika	Kalinga Institute of Industrial Technology, India
Atriya Sen	Rensselaer Polytechnic Institute, USA
Ning Sun	Nankai University, China
Shaoxin Sun	Chongqing University, China
Ankit Thakkar	Nirma University of Science and Technology, India
Ye Wang	Chongqing University of Posts and Telecommunications, China
Yong Wang	Sun Yat-sen University, China
Zhanshan Wang	Northeastern University, China
Quanwang Wu	Chongqing University, China
Xiangjun Wu	Henan University, China
Xingtang Wu	Beihang University, China
Zhou Wu	Chonqing University, China

Wun-She Yap   Universiti Tunku Abdul Rahman, Malaysia
Rocco Zaccagnino   University of Salerno, Italy
Kamal Z. Zamli   Universiti Malaysia Pahang, Malaysia
Choujun Zhan   South China Normal University, China
Haijun Zhang   Harbin Institute of Technology, Shenzhen, China
Menghua Zhang   University of Jinan, China
Zhao Zhang   Hefei University of Technology, China
Mingbo Zhao   Donghua University, China
Dongliang Zhou   Harbin Institute of Technology, China
Mirsaeid Hosseini Shirvani   Islamic Azad University, Iran

Contents – Part I

Contents – Part II

TE-BiLSTM: Improved Transformer and BiLSTM on Fraudulent Phone Text Recognition

Hongkui Xu[1,2](✉), Junjie Zhou[1], Tongtong Jiang[1], Jiangkun Lu[1], and Zifeng Zhang[1]

[1] Shandong Jianzhu University, Jinan 250101, China
xhkui2009@163.com
[2] Shandong Key Laboratory of Intelligent Buildngs Technology, Jinan 250101, China

Abstract. The number of mobile phone users become extremely large, at the same time telephone fraud cases occur frequently throughout the world, which makes people's property and security destroyed. For this reason, a method to identify fraud phone text based on text classification is proposed in this paper. A hybrid neural network TE-BiLSTM (Transformer-Encoder-Bidirectional Long Short-Term Memory) based on improved Transformer combined with BiLSTM is proposed. Besides, a fraud phone text dataset with multiple types of fraud terms is built. Word2Vec is used to represent the fraud phone text. Multi-head attention mechanism can extract deep semantic information from text in different subspaces and BiLSTM (Bidirectional Long short-term Memory) can exploit the long-distance dependency of text. TE-BiLSTM is used to extract multi-level information, and Softmax layer is used to classify and recognize fraud phone text. Experiments show that our model has the highest experimental results compared to other comparison models. The recognition accuracy of fraud phone text based on TE-BiLSTM is improved by 5.25% compared with Transformer, and increased by 5.58%, 3.92% and 3.67% respectively compared with other traditional models - LSTM, BiLSTM and BiLSTM-ATT (BiLSTM-attention).

Keywords: Fraudulent phone text classification · Word2Vec · Transformer · BiLSTM

1 Introduction

With intelligent devices and Internet users increasing, the telecom fraud crime is escalating, not only bringing huge losses to the country and the people,but also in the society caused extremely bad influence.

Phone fraud, one of the main ways of telecom fraud, refers to criminals using phone to scam. Fraud in the phone network also have negative impact on the security of online services due to the combination of phone and Internet [1]. Traditional phone fraud governance model by analyzing the structure of the fraud telephone number to build blacklist library, so as to realize the interception of suspected numbers. However, this method has a certain lag, and is just a post-mortem. In particular, traditional anti-phone fraud ways have been unable to adapt to the current situation along with the fraud tricks appearing specialization trends.

In recent years, with the development of machine learning, there are some researches that use machine learning and big data analysis to combat phone fraud. Wang and Qu [2] proposed a phone fraud management technology analyzing big data of massive call signaling to model. This way can output the suspected phone number within 3–5 min after the end of call, and the victim can be returned timely by phone or Message, so as to prevent in advance. Cheng et al. [3] proposed a method to build a phone fraud prevention model based on data mining and fraud number analysis, discriminating and intercepting fraud phones in the pre-mid-post stages. The above crackdown means are mostly joint governance between the public security and telecom operators. Studies have shown that phone numbers and call behaviors generated by phones reveal the types of calls, and fraud phones can be identified by analyzing phone number rules and call behavior patterns. Xing et al. [4] used deep learning-based algorithms such as CNN (Convolutional Neural Network), LSTM (Long Short-Term Memory) and feedforward neural network to identify fraud phones. Due to differences in data distribution and the impact of spending a lot of manual labor on data annotation, Zhou et al. [5] based on BERT transfer learning model, a new monitoring and early warning method of network telecom crime platform is proposed. Ying et al. [6] proposed an efficient fraud call detection framework based on parallel graph mining, it can automatically label fraud phone numbers and generate trust values for phone numbers. This paper adopts the method of deep learning to identify fraud phone at telephone terminals. When the user connected the phone, speech can be converted to text using speech recognition technology firstly. Then fraud phone is identified using NLP (Nature Language Processing) technology according to the phone text. The research of this paper is mainly about text classification for phone text and does not involve speech recognition. The first stage of this research is data collection. The phone text is preprocessed next and features are extracted using improved Transformer combined with BiLSTM model. At last, the probability of this phone being a fraud phone is output through Softmax layer.

To take full advantage of Transformer and BiLSTM, we proposed a fraud phone recognition model based on improved Transformer and BiLSTM. The main contributions of this work are listed as follows:

1. A fraud phone dataset is built;
2. The Transformer structure is improved to make it suitable for text classification;
3. Improved Transformer is combined with BiLSTM to improve classification performance.

The rest of this paper is organized as follows. The related work of Transformer and BiLSTM is presented in Sect. 2. Section 3 introduces the validity of proposed TE-BiLSTM model in detail and the principle of Transformer as well as BiLSTM. Section 4 demonstrates the datasets, baselines, experiments and analysis. In Sect. 5, the main conclusions are discussed.

2 Related Work

Word representation is extremely important in the classification task. How to represent the feature information of the word in an appropriate vector is related to the feature extraction process of the neural network, it will affect the final output results of the model directly. Early word representation is one-hot representation, which is too sparse and vector representations between different words are independent; TF-IDF [7] is a keyword extraction method, which has improved a lot compared with one-hot, but still has problems such as high dimensionality and inaccurate semantic representation, which also exist in discrete representation methods such as bag-of-words and N-gram. Word embedding is a distributed representation based on neural network, represented by Word2Vec [8] and GloVe [9]. In 2013, Mikolov proposed the Word2Vec including two training ways: CBOW (Continuous Bag-of-words) and Skip-Gram. CBOW uses the context of a word to predict the word and Skip-Gram uses a word to predict the context of the word. This method can represent the similarity relationship between words. What's more, it avoids the problem of dimensional explosion by using low vector dimensions. This paper chooses Word2Vec as the word embedding model as shown in Fig. 1, and the vector dimension is 300d.

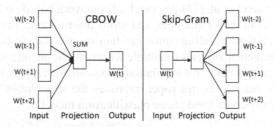

Fig. 1. Two training methods for Word2Vec

Early feature extraction techniques generally use machine learning methods, such as support vector machines [10], naive Bayes, decision trees and so on. In recent years, deep learning has become the mainstream method. Compared with traditional machine learning algorithms, deep learning has stronger data processing capabilities, and does not require cumbersome feature engineering. In particular, it has good generalization capability and strong transferability.

RNN (Recurrent Neural Network) is a recursive model that has the advantage of learning and remembering long-term information of sequences, and has often been used for time series prediction. However, when the length of the processed data is too long, there will be serious gradient disappearance and gradient explosion and etc. To solve these problems, Hochreiter et al. [11] proposed LSTM (Long Short-term Memory Neural Network), and its unique three "gate" structure can selectively control information transmission. Li et al. [12] used it for news text classification. Jakub et al. [13] modified it and used it for short text classification and sentiment classification. Subsequently, Cho et al. [14] proposed a gated recurrent unit (GRU) with a simpler structure replacing three "gates" with two "gates", which greatly reduced the parameters, improved the model's performance and training speed. In the microblog sentiment classification task on the topic of Coronavirus disease, Zhang et al. [15] proposed an interactive multi-task learning method, using the BiLSTM + attention + CRF model for classification. Although the recurrent neural network model has a good performance in processing time series data, it cannot capture key information. Therefore, attention mechanism is proposed.

Attention mechanism is widely used in image processing, natural language processing, speech recognition and other fields because of its simple application and strong flexibility. DeRose et al. [16] proposed a visual analysis method to understand how the attention mechanism changes during NLP task training, describing the deepest classification-based attention mechanism and how the attention mechanisms of the previous layers are flow in the entered words. Bahdanau et al. [17] first used it in machine translation tasks to improve significantly the translation performance of the model; Liu et al. [18] and Sharaf Al-deen et al. [19] combined the attention mechanism with neural networks to construct a classification model. Leng et al. [20] combined multi-head attention mechanism and RNN to construct a decoder with enhanced multi-head attention mechanism and applied it to emotion analysis tasks. Xu et al. [21] converted social network texts into word vectors and used BiLSTM hybrid network combined with self-attention mechanism to deal with spam problems.

In 2017, Google proposed Transformer [22], maximizing the advantages of the attention mechanism. Graterol et al. [23] proposed a framework based on Transformer that enables social robots to detect emotions and store information. Multi-head attention mechanism, a special form of self-attention mechanism, has a multi-head structure that can make better comprehensive use of multiple features. The encoder-decoder structure of Transformer is very suitable for translation tasks. However, it is not good at classification tasks. For this reason, this paper improves the structure of Transformer and combines BiLSTM to build a fraud phone classification model.

3   Model Architecture

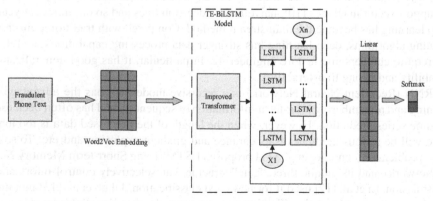

Fig. 2. The overall process of text classification of fraud phone

This part introduces the overall process of fraud phone's recognition in detail. The semantics of fraud statements are generally related to the relevant content of the context and some key words. The semantic content of the fraud phone text is complex and more difficult to distinguish than the general text. It is necessary to extract the deep semantic information of the fraud text. Therefore, we proposed the model TE-BiLSTM. We adopted Word2Vec as the Word embedding model. After passing through the word embedding layer, the fraud phone text forms a 300-dimensional word embedding vector. TE-BiLSTM conducts encoding, dimension mapping, feature extraction and other operations on word embedding vector, then sends it to Softmax layer to achieve classification, the flow chart is shown in Fig. 2.

3.1 TE - BiLSTM Model

The proposed TE-BiLSTM hybrid network model is shown in Fig. 3. The overall structure of this model is based on improved Transformer and BiLSTM, composed of 5 parts, Transformer encoder, vector splicing, improved Transformer decoder, BiLSTM and classification output.

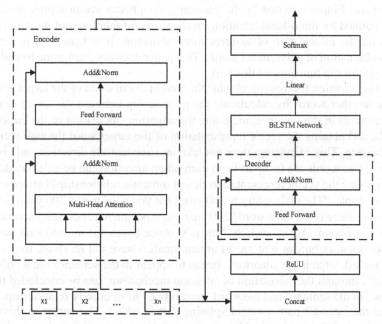

Fig. 3. TE-BiLSTM (Transformer-encoder-bidirectional long short-term memory) structure

(1) Transformer encoder: This part can realize multi-layer stacking. In particular, every layer is composed of multi-head self-attention mechanism, residual connection, normalization and feedforward neural network.

(2) Vector splicing: The word embedding vector is simply spliced with the vector processed by Encoder. In this way, the vector space has both the word embedding vector features and the vector features processed by Encoder, this can enrich the feature composition. Meanwhile, the ReLU activation function is introduced.

(3) Improved Transformer decoder: This part is the decoder calculation module, composed of residual connection, normalized operation layer and feedforward neural network layer. It is very different from the decoder part of original Transformer and this part completes feature mapping of the splicing vector.

(4) BiLSTM: This part uses BiLSTM network to extract long-distance features further from the feature vectors of the upper part.

(5) Classification output: Linear layer reduces the dimension of vectors, then Softmax layer outputs the results of classification.

3.2 Improved Transformer

Transformer has a lot of advantages in model structure, this paper makes the following improvements to Transformer, so that it can have a good performance in fraud phone text recognition.

(1) The part of input is coded by the encoder part, a better vector representation can be obtained by multi-head attention mechanism and feedforward neural network, making the information of features more abundant. It is beneficial to semantic disambiguation of polysemous words. This paper does not change the Encoder part and retains the functions of this part.

   The attention mechanism obtains the weight distribution of the target word relative to other words by calculating the relationship between the target word and other words in the sentence, including the attention calculation of the target word itself, and obtains the vector representation of the target word through weighting processing. This calculation does not take into account the direction and distance, even if two words are far apart from each other, attention can be calculated, which can capture the vector representation rich in semantic relationships between words. For example, "The daily salary is 200 yuan, but you need to pay 200 yuan deposit", although there is the same word "200 yuan" in this sentence, but their vector expression is different; Again, such as "just post office contacted me and said you what public security bureau sent me an urgent email, I have not received, they call me call me ask what is the situation", began to appear in the sentence the words "post office", through the calculation of attention mechanism, can be concluded that the "they" in the sentence and the word "post office" most closely relationship.

(2) After the Encoder part, a vector splicing part is added. The vector in this part is composed of the word embedding vector and the vector splicing after the Encoder part encoding operation. In this way, the splicing vector contains both the features of the word embedding vector and the features after the Encoder part processing, which enrich the feature representation of the vector. After splicing, the nonlinear relationship between neural network layers is increased through activation function processing.

$$L = \mathrm{Cat}\,(L_W, L_E) \tag{1}$$

where L_W represents the word embedding vector, L_E represents the vector after Encoder representation.

(3) The Decoder part in Transformer is mainly based on the word information encoded by the Encoder part and the decoding operation combined with the context, which is unnecessary in the classification task. Compared with the original Decoder part of Transformer, this paper removes all the attention mechanism, leaving only the residual connection & normalization layer and the feedforward neural network layer, because the feature information between the feature vectors formed by the vector splicing part is not well fused. After the improved Decoder part of the feedforward neural network layer on the vector mapping processing, making the vector has a good fusion in combination. The first layer of the feedforward neural network is the activation function Relu, and the second layer is the linear activation function, whose formula is as follows:

$$FFN = \max(0, LW_1 + b_1)W_2 + b_2 \tag{2}$$

where W_1, W_2, b_1 and b_2 are training parameters, L represents the spliced vector. The purpose of retaining residual connection is to consider that Decoder module can be stacked with multiple layers, and residual connection can ensure that initial information is transmitted to deeper modules.

3.3 BiLSTM Network Layer

LSTM can mine the time series variation rules of relatively long intervals and delays in time series [24] and control the transmission of historical information through gate functions, thus possessing certain time series processing and prediction capabilities.

BiLSTM [25] adds a reverse structure on the basis of LSTM, consisting of forward and backward LSTM. At any time, the output is determined by the forward and backward states together, that is, context-based judgment, which is important for extracting contextual information from text.

At moment i, the final output of the network is obtained by bitwise addition of the forward and backward feature vectors, as shown in Eq. (3).

$$h_i = \vec{h}_i \oplus \overleftarrow{h}_i \tag{3}$$

where h_i represent the output vector of hidden layer at moment i.

After word embedding and vector representation of TE-BiLSTM, fraud phone text forms feature vectors with abundant features. The Linear layer reduces its dimension.

$$O = W^T H + b_i \tag{4}$$

where W^T is the training parameter matrix, b_i is the bias parameter, H represents the output vector of BiLSTM. Then we used Softmax function to achieve the classification of fraud phone text.

4 Experiment and Analysis

4.1 Dataset

The content of dataset covers almost all types of fraud, including finance, education, postal, banking, dating, brush list, winning the lottery, posing as police officers type of fraud and so on, occurring frequently in our dairy life [26]. The number of training sets, validation sets and test sets is 6000, 3000 and 1200 respectively. The details of dataset are shown in the following table:

4.2 Evaluation Criteria

Accuracy, Precision, Recall and F1 measures are used to evaluate the model in this paper, and the formula is shown as follows:

$$Accuracy = \frac{TP + TN}{TP + TN + FP + FN} \tag{5}$$

$$Precession = \frac{TP}{TP + FP} \tag{6}$$

$$Recall = \frac{TP}{TP + FN} \tag{7}$$

$$F1 = \frac{2*Precession*Recall}{Precessiom + Recall} \tag{8}$$

where TP represents the fraud sample is predicted to be fraud, TN represents the normal sample is predicted to be normal, FP represents the normal sample is predicted to be fraud, and FN represents the fraud sample is predicted to be normal.

4.3 Parameter Settings

Word2Vec is used as the word embedding, and TE-BiLSTM is used for feature extraction. Word2Vec has 300 dimensions, LR = 1e–5, BATCH_SIZE = 128, Dropout = 0.5. The number of BiLSTM hidden layers is set to 128. In order to obtain the optimal experimental parameters, the number of heads of multiple attention mechanism in Encoder part of Transformer is fixed at first: head = 6. PAD_SIZE is the maximum length of sentence processing, set to 32, 64 and 128 respectively, and the number of Encoder and Decoder layers set to 1–10. The following is the experiment of Encoder/Decoder layers under different PAD_SIZE.

From the Fig. 4, we can see that the average accuracy of experimental results is the highest when parameter PAD_SIZE = 64, and the highest accuracy is achieved when the number of Encoder/Decoder layers is 9 and PAD_SIZE = 64.

From Fig. 5, we can see that when Head = 6, the model has the highest accuracy (Table 1).

In the above experiments, parameters of the model TE-BiLSTM are determined in this paper. The summary of parameters of the model is shown in the Table2.

Table 1. Statistical table of fraud phone text dataset

Dataset	Training set	Validation set	Test set
Fraud	3000	1500	601
Normal	3000	1500	599
Total	6000	3000	1200

Table 2. Experimental parameter setting

Parameter	Value or Type
Optimizer	AdamOptimizer
BATCH_SIZE	128
LEARNING_RATE(LR)	1e–5
Dropout	0.5
Embedding_Dim	300
Number of BiLSTM hidden layers	128
PAD_SIZE	64
Number of encoder/Decoder layers	9
Head	6

4.4 Experimental Results and Discussion

After a series of parameters of the model are determined, multi-class comparative experiments are carried out. Four different models are used to compare with TE-BiLSTM proposed in this paper. Numerical comparison of experimental results shows the superiority of TE-BiLSTM model in text recognition of fraud calls. Table 3 shows the classification results on the fraud phone text dataset.

In this experiment, word embedding is used as the input for text classification, Word2Vec is used as the word embedding model to represent the fraudulent phone text, and then it is transmitted to different neural network models for feature extraction. The following is an introduction to the neural network model used in this paper:

(1) LSTM [11], short and long time memory network is used to extract the features of fraud phone text.
(2) BiLSTM [27], BiLSTM network is used to extract the features of scam phone text, and then BiLSTM output data is sent to Softmax layer for classification.
(3) BiLSTM-ATT [28], has the same structure as the model in (2). The difference is that attention mechanism is added at the back of BiLSTM network to extract richer features.
(4) Transformer [22], encoding operation through Encoder part of Transformer, and finally sent into Softmax layer for classification.

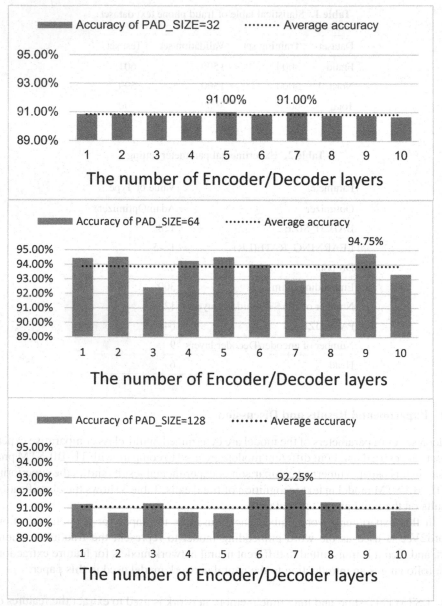

Fig. 4. Experimental results of fraud data set with different maximum length of sentence processing and different Encoder/Decoder layers selected

Figure 6 shows Confusion matrix on the fraud phone text dataset. Figure 7 is the experimental results of the fraud phone dataset and the bar chart of evaluation indexes of each model.

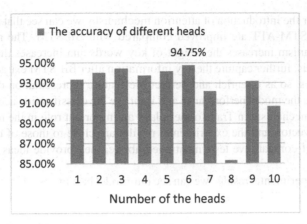

Fig. 5. Experimental accuracy of different head numbers in multiple attention mechanism

Table 3. Classification results on the fraud phone text dataset

Model	Accuracy	Precision	Recall	F1-score
LSTM	0.8917	0.9011	0.8917	0.8911
BiLSTM	0.9083	0.9146	0.9084	0.9080
BiLSTM-ATT	0.9108	0.9157	0.9109	0.9106
Transformer	0.8950	0.9092	0.8952	0.8941
TE-BiLSTM	0.9475	0.9475	0.9475	0.9475

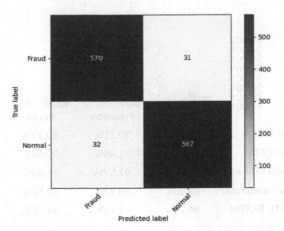

Fig. 6. Confusion matrix on the fraud phone text dataset

BiLSTM adopts bidirectional structure for feature extraction, and simultaneously extracts the feature information above and below. Compared with LSTM, BiLSTM has richer feature representation, so it is higher than LSTM in accuracy and other evaluation indicators. After the introduction of attention mechanism, we can see that all evaluation indexes of BiLSTM-ATT are improved compared with BiLSTM. The reason is that attention mechanism increases the weight of key words and increases the attention to key words, that is, further capture the key information after BiLSTM extraction of long-distance features, so as to enrich and perfect the feature representation of text vector, so as to improve the model performance and improve the classification accuracy. Multi-head attention mechanism in Transformer plays an important role in the representation of text feature vector, and the experimental results are close to those of other models. After a series of comparative tests mentioned above, they provide ideas for proposed model.

From the experimental results, we can see our model has improved in terms of Accuracy, Precision, Recall, F1 score and other indicators compared with other comparison models. The accuracy of our model is improved by 5.25% compared with Transformer, and by 5.58%, 3.92% and 3.67% respectively compared with other traditional models LSTM, BiLSTM and BILSTM-ATT (BiLSTM-attention), showing TE- BiLSTM model has superior performance on the fraud phone text dataset. By analyzing the experimental results, compared with other models, model TE-BiLSTM combines the attention mechanism with BiLSTM, and the data can be presented with more abundant features. Because the fraud phone text is represented by Word2Vec, input TE-BiLSTM model, encoded by Transformer Encoder part, integrated with the feature information after the function of multi-head attention mechanism, so that the whole sentence semantics of the data can be better represented. Then the encoding output vector and Word2Vec input vector are sent together into the improved Decoder part after splicing. After the Decoder part mapping the spliced vector, the features of the data are better fused together. BiLSTM part is used for bidirectional feature extraction of sentence timing sequence, and finally the vector representation with abundant features is obtained.

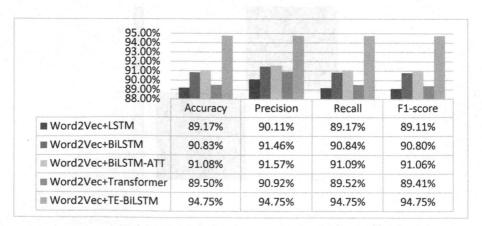

Fig. 7. Histogram of classification results on the fraud phone text dataset

5 Conclusions

This paper studies the problem of fraud phone calls in the society. A text data set of fraud phone with various types of fraud terms is constructed. We proposed a fraud phone text recognition model TE-BiLSTM, which is based on improved Transformer and BiLSTM. We used Word2Vec as the word embedding to represent the fraud phone text, through a series of comparative experiments, we verified the feasibility of the proposed model, which provides a reference for dealing with the problem of fraud phone calls. Word2Vec is a static word embedding model, and it cannot automatically adjust its word embedding representation according to the context information. In the future, we will try dynamic word embedding model such as BERT. The types of fraud phone call cases have been updated. In the future we will continue to pay attention to the cases of fraud phone calls, update the fraud routines in the data set, and enrich the dataset.

References

1. Sahin, M., Francillon, A., Gupta, P., Ahamad, M.: Sok: fraud in telephony networks. In: Proceedings of the 2017 IEEE European Symposium on Security and Privacy (EuroS&P), pp. 235–250 (2017)
2. Wang, Z., Qu, J.: Research on anti telecommunications fraud technology based on big data. Telecom Eng. Tech. Stand. 30, 86–89 (2017). https://doi.org/10.13992/j.cnki.tetas.2017. 04.025
3. Cheng, J., Xiao, Y., Fang, Y., Li, S.: Research on telephone fraud prevention architecture based on big data. Telecom World 27, 13–15 (2020)
4. Xing, J., Yu, M., Wang, S., Zhang, Y., Ding, Y.: Automated fraudulent phone call recognition through deep learning. Wire. Commun. Mob. Comput. (2020)
5. Zhou, S., Wang, X., Yang, Z.: Monitoring and early warning of new cyber-telecom crime platform based on BERT migration learning. China Commun. 17, 140–148 (2020)
6. Ying, J.J.-C., Zhang, J., Huang, C.-W., Chen, K.-T., Tseng, V.S.: PFrauDetector: a parallelized graph mining approach for efficient fraudulent phone call detection. In: Proceedings of the 2016 IEEE 22nd International Conference on Parallel and Distributed Systems (ICPADS), pp. 1059–1066 (2016)
7. Dang, N.C., Moreno-García, M.N., De la Prieta, F.: Sentiment analysis based on deep learning: a comparative study. Electronics 9, 483 (2020)
8. Mikolov, T., Chen, K., Corrado, G., Dean, J. Efficient estimation of word representations in vector space. arXiv preprint arXiv:1301.3781 (2013)
9. Pennington, J., Socher, R., Manning, C.D.: Glove: global vectors for word representation. In: Proceedings of the Proceedings of the 2014 conference on empirical methods in natural language processing (EMNLP), pp. 1532–1543 (2014)
10. Tadesse, M.M., Lin, H., Xu, B., Yang, L.: Detection of depression-related posts in reddit social media forum. IEEE Access 7, 44883–44893 (2019)
11. Hochreiter, S., Schmidhuber, J.: Long short-term memory. Neural Comput. 9, 1735–1780 (1997)
12. Li, C., Zhan, G., Li, Z.: News text classification based on improved Bi-LSTM-CNN. In: Proceedings of the 2018 9th International Conference on Information Technology in Medicine and Education (ITME), pp. 890–893 (2018)
13. Nowak, J., Taspinar, A., Scherer, R.: LSTM recurrent neural networks for short text and sentiment classification. In: Proceedings of the International Conference on Artificial Intelligence and Soft Computing, pp. 553–562 (2017)

14. Cho, K.: Learning phrase representations using RNN encoder-decoder for statistical machine translation. arXiv preprint arXiv:1406.1078 (2014)
15. Zhang, H., Sun, S., Hu, Y., Liu, J., Guo, Y.: Sentiment classification for chinese text based on interactive multitask learning. IEEE Access **8**, 129626–129635 (2020)
16. DeRose, J.F., Wang, J., Berger, M.: Attention flows: analyzing and comparing attention mechanisms in language models. IEEE Trans. Visual Comput. Graph. **27**, 1160–1170 (2020)
17. Bahdanau, D., Cho, K., Bengio, Y.: Neural machine translation by jointly learning to align and translate. arXiv preprint arXiv:1409.0473 (2014)
18. Liu, Y., Xu, Q.: Short Text classification model based on multi-attention. In: Proceedings of the 2020 13th International Symposium on Computational Intelligence and Design (ISCID), pp. 225–229 (2020)
19. Sharaf Al-deen, H.S., Zeng, Z., Al-sabri, R., Hekmat, A.: An improved model for analyzing textual sentiment based on a deep neural network using multi-head attention mechanism. Appl. Syst. Innov. **4**, 85 (2021)
20. Leng, X.-L., Miao, X.-A., Liu, T.: Using recurrent neural network structure with enhanced multi-head self-attention for sentiment analysis. Multimedia Tools Appl. **80**(8), 12581–12600 (2021). https://doi.org/10.1007/s11042-020-10336-3
21. Xu, G., Zhou, D., Liu, J.: Social network spam detection based on ALBERT and combination of Bi-LSTM with self-attention. Secur. Commun. Netw. 2021 (2021)
22. Vaswani, A., et al.: Attention is all you need. In: Proceedings of the Advances in Neural Information Processing Systems, pp. 5998–6008 (2017)
23. Graterol, W., Diaz-Amado, J., Cardinale, Y., Dongo, I., Lopes-Silva, E., Santos-Libarino, C.: Emotion detection for social robots based on NLP transformers and an emotion ontology. Sensors **21**, 1322 (2021)
24. Jang, B., Kim, M., Harerimana, G., Kang, S.-U., Kim, J.W.: Bi-LSTM model to increase accuracy in text classification: combining Word2vec CNN and attention mechanism. Appl. Sci. **10**, 5841 (2020)
25. Graves, A., Schmidhuber, J.: Framewise phoneme classification with bidirectional LSTM and other neural network architectures. Neural Netw. **18**, 602–610 (2005)
26. Xie, L.: Research on information investigation of telecom network fraud. J. People's Pub. Secur. Univ. China (Sci. Technol.) **26**, 85–93 (2020)
27. Nguyen, T.H., Cho, K., Grishman, R.: Joint event extraction via recurrent neural networks. In: Proceedings of the Proceedings of the 2016 Conference of the North American Chapter of the Association for Computational Linguistics: Human Language Technologies, pp. 300–309 (2016)
28. Zhou, P., et al.: Attention-based bidirectional long short-term memory networks for relation classification. In: Proceedings of the Proceedings of the 54th annual meeting of the association for computational linguistics (volume 2: Short papers), pp. 207–212 (2016)

Cross Elitist Learning Multifactorial Evolutionary Algorithm

Wei Li[1], Haonan Luo[1(✉)], and Lei Wang[2]

[1] School of Computer Science and Engineering, Xi'an University of Technology, Xi'an 710048, China
liwei@xaut.edu.cn
[2] Shaanxi Key Laboratory for Network Computing and Security Technology, Xi'an 710048, China

Abstract. Multifactorial Evolutionary Algorithm (MFEA) is a popular optimization algorithm in recent years. It has implicit parallelism and can solve different problems at the same time in the same search space. However, premature convergence is a significant shortcoming of MFEA. The main reason for this phenomenon is the lack of population diversity and the inability to migrate effectively between tasks. To address this shortcoming, this paper proposes a cross elitist learning multifactorial evolutionary algorithm (CEL-MFEA). Firstly, the cross elitist strategy guides the individuals to learn from the elite individual, which can improve the convergence of the proposed CEL-MFEA. Secondly, the Nelder-Mead algorithm is used to provide an effective knowledge transfer between different tasks. Finally, the opposition learning mechanism is employed to ensure population diversity. The comprehensive ability of the proposed CEL-MFEA is evaluated by 9 classical multifactorial optimization problem test sets. Experimental results show that the CEL-MFEA is more competitive than MFEA algorithm and several popular multifactorial optimization algorithms.

Keywords: Multifactorial evolutionary algorithm · Cross elitist learning · Opposition learning

1 Introduction

In recent years, evolutionary algorithms have been widely used to solve optimization problems. Inspired by Darwin's theory of evolution, it takes the population as the unit, and the individuals in the population exchange information through reproduction and mutation. Evolutionary algorithm follows the rules survival of the fittest [1], and can continuously derive better individuals than their parents, so as to find the optimal solution. Since evolutionary algorithm was proposed, many engineering problems can be solved with it due to its robustness [2], such as parameter optimization, industrial scheduling [3].

With the increasing complexity of engineering problems, evolutionary algorithms have three main research directions. The first research point is the single-objective optimization [4], that is, the classical algorithm to solve simple optimization problems.

H. Zhang et al. (Eds.): NCAA 2022, CCIS 1637, pp. 15–29, 2022.
https://doi.org/10.1007/978-981-19-6142-7_2

The second research point is multi-objective optimization [5], which pursues the best compromise solutions of multiple optimization problems. The third research point is multifactorial optimization [6], which aims at finding the optimal solutions of multiple optimization problems at the same time. It is worth noting that there are many optimization problems belong to multi-objective optimization and multifactorial optimization. The difference is that multi-objective optimization mainly finds a set of compromise solutions that can satisfy multiple conflicting objectives. The characteristics of multifactorial optimization is that multiple tasks are optimized in parallel in the same space. The tasks are not limited to single-objective or multi-objective problems. The algorithm finds the optimal solution of different tasks in one run.

Multifactorial evolutionary algorithm (MFEA) has been verified to be effective by many practical and engineering problems since it was proposed. Yi *et al.* [7] proposed IMFEA to optimize the path planning of robot problem. On the basis of multigeomorphic path planning [8], different obstacles are set. Binh *et al.* [9] conducted many experiments on the cluster shortest path tree problem and verified the good performance of MFEA. Recently, some developed MFEA algorithms have been proposed to strengthen the comprehensive ability of the algorithm. Yin *et al.* [10] proposed a MFEA with cross-task search direction (MFEA-DV), where the offspring are generated by the sum of the difference vectors between the elite individuals of one task and the individuals of another task. In the MFEA-DV algorithm, elite individuals are used to accelerate convergence, while the sum of difference vectors can ensure the population diversity. Bali *et al.* [11] proposed a linearized domain adaptive method, where the search space of simple tasks is transformed into a search space that is similar to complex tasks. Feng *et al.* [12] proposed the multifactorial particle swarm optimization (MFPSO) and the multifactorial differential evolution algorithm (MFDE), and discussed the universality of population search mechanism.

The main characteristics of MFEA and its variants are unified search space, coding and decoding, knowledge transfer between tasks and selection strategy based on scalar fitness value. However, it is noteworthy that the correlation between tasks will affect the transfer efficiency between tasks [13]. If there is little correlation between two tasks, it will lead to negative transfer. Many methods have been proposed to reduce the negative transfer between tasks, such as MFEA-AKT [14] and GFMFDE [15]. This paper proposes a cross elitist learning MFEA algorithm (CEL-MFEA). First, ordinary individuals in one task learns from elite individuals in another task. This method is conducive to promote convergence. Moreover, the operators of Nelder-Mead algorithm is used to improve the inter-task knowledge transfer and strengthen information sharing among different tasks. Finally, the opposition learning mechanism is applied to ensure the population diversity.

The main contributions of this paper are as follows.

1) A cross elitist learning method is developed. The optimal solution of a task is called the elitist solution of the task. In the process of evolution, each individual of a task learns from the elite individual of another task, which can effectively promote the individuals move closer to the optimal solution region and enhance the convergence of the algorithm.

2) A novel inter task interaction strategy is proposed. The operators in Nelder-Mead algorithm [16] are used for information transmission between tasks to strengthen the interaction between tasks.
3) Population diversity is enhanced by opposition learning [17] mechanism. To prevent convergence to local optimum, the individuals with poor ranking according to the scalar fitness value are selected, and the individual is replaced with the new individual generated by the eliminated individuals through opposite learning, which will help to maintain the population diversity.

2 Preliminaries

2.1 Multifactorial Evolutionary Algorithm

In MFEA [6], knowledge transfer between tasks is realized by chromosome exchange of genetic material. Suppose there are K tasks $\{T_1, T_2, ..., T_K\}$, and the dimension of each task is recorded as D_k ($k = 1, 2, ..., K$). To facilitate coding individuals, a unified search space is expressed as $D_{multitask} = \max\{D_k\}$. N individuals are generated randomly and recorded as the population P. The core components of the algorithm are defined as follows.

1) *Factorial Cost* (f_{ij}). F_{ij} is the objective value of the i-th individual on the task T_j.
2) *Factorial Rank* (r_{ij}). R_{ij} is the ranking of the i-th individual according to f_{ij} on the task T_j.
3) *Scalar Fitness* (φ_i). $\varphi_i = 1/\min_{j \in \{1, ..., K\}}\{r_{ij}\}$.
4) *Skill Factor* (τ_i). $\tau_i = \mathrm{argmin}_{j \in \{1, ..., K\}}\{r_{ij}\}$.

With the above definition, the framework of MFEA is described as follows. N individuals are generated randomly. Then, f_{ij}, r_{ij}, φ_i and τ_i of each individual are evaluated. When the stop conditions are not met, repeat the following steps. The offspring is generated through the crossover and mutation. The parent and child individuals are combined to $2N$ individuals, and their scalar fitness and skill factors are updated. Finally, N individuals are selected as the next generation by the scalar fitness. The structure of MFEA is described in Algorithm 1.

2.2 Nelder-Mead Algorithm

The Nelder-Mead Algorithm (NMA) proposed by caponio *et al.* is a direct search method with optimal guidance. Research shows that NMA can carry out local search more efficiently [18]. NMA selects $M + 1$ points $\{p_1, p_2, ..., p_{M+1}\}$ and calculates their centroid points through Eq. (1).

$$\bar{p} = \left(\Sigma_{i=1}^{M} p_i\right)/M \tag{1}$$

Four operators are mainly used in the algorithm, which are reflection point (p_r), expansion point (p_e), outside contraction point (p_{oc}) and inside contraction point (p_{ic}). Given by Eq. (2–5).

$$p_r = \bar{p} + \rho(\bar{p} - p_{M+1}) \tag{2}$$

$$p_e = \bar{p} + \chi(p_r - \bar{p}) \tag{3}$$

$$p_{oc} = \bar{p} + \gamma(p_r - \bar{p}) \tag{4}$$

$$p_{ic} = \bar{p} - \gamma(\bar{p} - p_{M+1}) \tag{5}$$

where $\rho = 1$, $\chi = 2$, $\gamma = 1/2$.

Algorithm 1 The Framework of MFEA

1: Initialize the population P_0

2: Evaluate each individual and calculate the skill factor

3: **while** the stop conditions are not reached **do**

4:     Initialize the offspring population P_c

5:     **while** the number of offspring population is less than N **do**

6:         $\{p_i, p_j\} \leftarrow$ Two individuals were randomly selected from P_t

7:         **if** $\tau_i == \tau_j$ **then**

8:             $\{p_a, p_b\} \leftarrow$ crossover(p_i, p_j)

9:             Assign τ_i to p_a and p_b

10:         **else if** *rand* $\leq rmp$ **then**

11:             $\{p_a, p_b\} \leftarrow$ crossover(p_i, p_j)

12:             Randomly assign τ_i and τ_j to p_a and p_b

13:         **else**

14:             $p_a \leftarrow$ mutation(p_i) and assign τ_i to p_a

15:             $p_b \leftarrow$ mutation(p_j) and assign τ_j to p_b

16:         **end if**

17:         $P_c \leftarrow P_c \cup \{p_a, p_b\}$

18:     **end while**

19:     Evaluate each individual

20:     $Q \leftarrow P_t \cup P_c$ and compute the scalar fitness of each individual

21:     $P_{t+1} \leftarrow$ Select top N individuals from Q

22: **end while**

2.3 Opposition Learning

Opposition learning can expand the search range of the current population. More especially, the opposition learning mechanism enlarges the possibility of the algorithm jumping out of the local optimum. In the D-dimensional space in range $[a, b]$, the opposition point p' of point p is given by Eq. (6).

$$p'^j = a + b - p^j \qquad (6)$$

where j represents the dimension, $j = 1, 2,...,D$.

3 Proposed Method

3.1 Cross Elitist Learning Strategy

In general, the elitist solution is preserved to accelerate the population convergence, which can guide the population to evolve along the promising direction. The proposed cross elitist learning strategy guides the individuals to learn from the elite individual with different tasks. The detailed operation is described as follows. Suppose there are two tasks T_1 and T_2. The individuals of T_1 learn from the elite individual of T_2, and the individuals of T_2 learn from the elite individual of T_1. This can accelerate the convergence of the algorithm for different tasks. Algorithm 2 shows the cross elitist learning strategy.

Algorithm 2 Cross elitist learning

1: **Input:** p_1, p_2, $pbest_{T1}$, $pbest_{T2}$

2: **Output:** p_a, p_b

3: **if** $\tau_1 == \tau_{pbestT1}$ and $\tau_2 == \tau_{pbestT2}$ **then**

4:     $p_a \leftarrow$ crossover(p_1, $pbest_{T2}$)

5:     $p_b \leftarrow$ crossover(p_2, $pbest_{T1}$)

6: **else**

7:     $p_b \leftarrow$ crossover(p_1, $pbest_{T1}$)

8:     $p_a \leftarrow$ crossover(p_2, $pbest_{T2}$)

9: **end if**

10: Assign skill factors $\tau_{pbestT2}$ to p_a and $\tau_{pbestT1}$ to p_b

---

3.2 Knowledge Transfer Based on Nelder-Mead Method

Information exchange between tasks is a crucial step in multifactorial evolutionary algorithm. Suppose there are two tasks T_1 and T_2. In MFEA, when the skill factors of two randomly selected individuals are different, the transfer between different tasks is controlled by the parameter *rmp* (random mating probability). Two offspring individuals are generated by the crossover operator, and then they are randomly assigned skill factors p_a or p_b.

In the multifactorial evolutionary algorithm, knowledge transfer between different tasks with high similarity can promote each other. However, knowledge transfer between different tasks with low similarity may bring the so-called negative migration phenomenon. Here, the Nelder-Mead algorithm is used to make better communication between tasks. First, $M + 1$ individuals are randomly selected from the parent population. Next, the centroid points of the first M individuals, their reflection points, expansion points, outside contraction points and inside contraction points are calculated in turn. Finally, two points are randomly selected from the four points, and they are randomly assigned skill factors as the generated offspring individuals. Algorithm 3 illustrates the process of the knowledge transfer based on Nelder-Mead method.

Algorithm 3 Knowledge transfer based on Nelder-Mead method

1: **Input:** $p_1, p_2, ..., p_{M+1}$

2: **Output:** p_a, p_b

3: Compute the centroid points $\bar{p}$ by eq.(1)

4: Compute the points p_r, p_e, p_{oc}, p_{ic} by eq.(2)- eq.(5)

5: $\{p_a, p_b\} \leftarrow$ Random selected two points from $\{p_r, p_e, p_{oc}, p_{ic}\}$

3.3 Opposition Learning

In the multifactorial optimization framework, the range of vectors is between [0, 1]. Therefore, the solution of its opposition learning can be given by Eq. (7).

$$p_i^j = 1 - p_i^j \tag{7}$$

where i represents the i-th solution and $j = 1,2,...,D_{multitask}$.

To further prevent premature convergence to the local optimum, this paper employs the opposition learning with a certain probability in each generation. If the opposition learning is triggered, H individuals are randomly selected from the eliminated individuals. The new H individuals are obtained by the opposition learning according to Eq. (7). Then the H solutions with poor ranking in the new population are replaced by these new generated solutions.

3.4 The Framework of CEL-MFEA

The method proposed in this paper is given by Algorithm 4.

3.5 The Computional Time Analysis

Suppose there are T optimization tasks, the population size is N, and the unified search space dimension is D. We need to evaluate N D-dimensional individuals on each task, and the time complexity is $O(T \cdot N \cdot D)$. The knowledge transfer, cross elitist learning and opposition learning are linear and executed under a certain probability. If the execution times are r_1, r_2 and r_3 respectively, the time complexity of the three strategies is $O(r_1 \cdot 4 \cdot N \cdot D)$, $O(r_2 \cdot N \cdot D)$ and $O(r_3 \cdot N \cdot D)$. Therefore, the total time complexity is $O(N \cdot D)$.

Algorithm 4 The Framework of CEL-MFEA

1: Initialize the population P_0
2: Evaluate each individual and calculate the skill factor
3: **while** the stop condition are not reached **do**
4:      Initialize the offspring population P_c
5:      **while** the number of offspring population is less than N **do**
6:          $\{p_i, p_j\}$ ←randomly selected two individuals from P_t
7:          **if** $\tau_i == \tau_j$ **then**
8:              $\{p_a, p_b\}$ ← crossover(p_i, p_j)
9:              Assignτ_i to p_a and p_b
10:         **else if** $rand \leq rmp$ **then**
11:             $\{p_a, p_b\}$ ←Knowledge transfer by algorithm 3
12:         **else**
13:             **if** $rand \leq pmu$ **then**
14:                 p_a ← mutation(p_i) and assign skill factor τ_i to p_a
15:                 p_b ← mutation(p_j) and assign skill factor τ_j to p_b
16:             **else**
17:                 $\{p_a, p_b\}$ ← CEL($p_i, p_j, pbest_{T1}, pbest_{T2}$) Cross elitist learning by algorithm 2
18:             **end if**
19:         **end if**
20:         $P_c \leftarrow P_c \cup \{p_a, p_b\}$
21:     **end while**
22:     Evaluate each individual and calculate the skill factor
23:     $Q \leftarrow P_t \cup P_c$ and compute the scalar fitness of each individual
24:     P_{t+1} ← Select top N individuals from Q
25:     **if** $rand \leq pol$ **then**
26:         Compute the opposition points by eq.(7)
27:         Replace the poorly ranked H solutions in P_{t+1} with these solutions
28:     **end if**
29: **end while**

---

4 Comparative Studies of Experiments

4.1 General Experimental Setting and Parameter Setting

To effectively evaluate the performance of CEL-MFEA, the classic single-objective Multifactorial Optimization (MFO) test set proposed by *Da* et al. was used in this experiment. It contains 9 sets of continuous single-objective MFO benchmark problem sets. Each benchmark consists of two classical optimization problems. These problems consider different optimal solution intersection degree and fitness landscape. P1(CI + HS), P2(CI

+ MS) and P3(CI + LS) are three complete intersection problems. P4(PI + HS), P5(PI + MS) and P6(PI + LS) are three partial intersection problems. P7(NI + HS), P8(NI + MS) and P9(NI + LS) are three no intersection problems. The symbol 'H', 'M' and 'L' indicates high similarity, medium similarity and low similarity of tasks, respectively. The detail of these problems is given in Table 1.

Table 1. Properties of MFO benchmark problems

Intersection degree	Problem	Landscape	
		Task 1	Task 2
Complete	P1	Multimodal, Nonseparable	Multimodal, Nonseparable
	P2	Multimodal, Nonseparable	Multimodal, Nonseparable
	P3	Multimodal, Nonseparable	Multimodal, Separable
Partial	P4	Multimodal, Nonseparable	Unimodal, Separable
	P5	Multimodal, Nonseparable	Multimodal, Nonseparable
	P6	Multimodal, Nonseparable	Multimodal, Nonseparable
No	P7	Multimodal, Nonseparable	Multimodal, Nonseparable
	P8	Multimodal, Nonseparable	Multimodal, Nonseparable
	P9	Multimodal, Nonseparable	Multimodal, Separable

4.2 Comparative Experiments

In this part, CEL-MFEA is compared with MFEA, MFEA-b, MFDE and MFPSO. MFEA-b is a variant of MFEA and modifies some parameter values of MFEA. MFDE uses DE/rand/1/bin operation in DE for inter-task communication. MFPSO is proposed to solve multitasking problems with particle swarm optimization algorithm. The population size $N = 100$, and the evaluation times $FES = 200000$, that is, each task is evaluated 100000 times respectively. Each algorithm is run independently for 30 times. The parameters of the compared algorithm are in accordance with the parameters proposed in the paper. The parameters of CEL-MEFA are as follows: $rmp = 0.3$, $M = 3$, $pmu = 0.8$, $pol = 0.2$, $H = 20$.

Table 2 shows the statistical results of Task1 obtained by 5 algorithms for solving MFO benchmark problem sets. The optimal values are displayed in bold. From the Table 2, it can be seen that CEL-MFEA wins MFEA, MFEA-b, MFDE and MFPSO in 7 of the 9 problems. For the P5 problem and P8 problem, MFDE ranks the first, while CEL-MFEA ranks the second.

The convergence curves obtained by 5 algorithms are shown in Fig. 1 and Fig. 2. The logarithmic scale is taken to show the convergence effect more clearly. It can be seen from figure that the proposed CEL-MFEA algorithm performs best on Task1 including the problem sets of P1, P2, P3, P4, P6, P7, P9.

Table 2. The average optimal values of Task1 obtained from MFEA, MFEA-b, MFDE, MFPSO and CEL-MFEA

Problem	MFEA	MFEA-b	MFDE	MFPSO	CEL-MFEA
P1	1.07E + 00	6.21E−01	1.02E−03	9.10E−01	**0.00E + 00**
P2	7.82E + 00	2.91E + 00	8.39E-02	7.69E ǀ 00	**2.82E−10**
P3	2.11E + 01	2.12E + 01	2.12E + 01	2.13E + 01	**2.00E + 01**
P4	7.68E + 02	3.75E + 02	8.13E + 01	9.91E + 02	**5.98E-05**
P5	7.06E + 00	2.64E + 00	**1.54E−03**	6.13E + 00	1.61E + 00
P6	2.10E + 01	4.48E + 00	4.58E−01	1.30E + 01	**3.30E−03**
P7	7.03E + 04	3.37E + 03	9.93E + 01	5.87E + 05	**4.69E + 01**
P8	1.02E + 00	9.93E−01	**3.76E−03**	1.11E + 00	4.27E−02
P9	7.48E + 02	3.85E + 02	9.97E + 01	3.21E + 03	**3.07E−04**

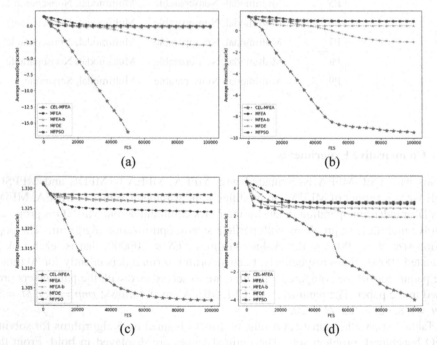

Fig. 1. The convergence curve of Task1 from MFEA, MFEA-b, MFDE, MFPSO and CEL-MFEA on MFO test suite. (a) P1. (b) P2. (c) P3. (d) P4.

Similarly, Table 3 shows the statistical results of Task2 obtained by 5 algorithms for solving MFO benchmark problem sets. The optimal values are displayed in bold. Table 3 shows that CEL-MFEA performs better than MFEA, MFEA-b, MFDE and MFPSO in 7 of the 9 problems. For the P3 problem, MFEA-b ranks the first, while CEL-MFEA

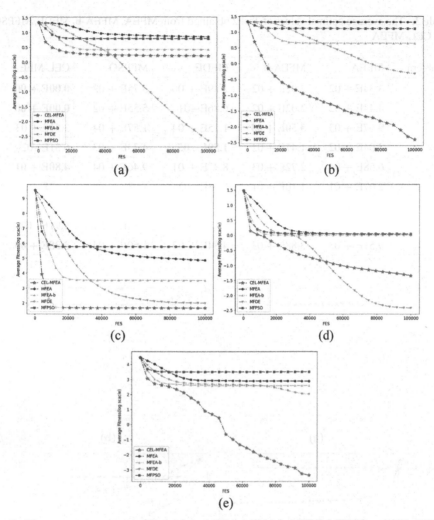

Fig. 2. The convergence curve of Task1 from MFEA, MFEA-b, MFDE, MFPSO and CEL-MFEA on MFO test suite. (a) P5. (b) P6. (c) P7. (d) P8. (e) P9.

ranks the second. For the P4 problem, MFDE ranks the first, while CEL-MFEA ranks the second.

The convergence curves obtained by 5 algorithms are shown in Fig. 3 and Fig. 4. It can be seen from figure that the proposed CEL-MFEA algorithm performs best on Task2 including the problem sets of P1, P2, P5, P6, P7, P8, P9.

The two problems in P5 and P8 are multimodal problems with medium similarity. One task may lead another task to deviate from the search direction, and the result of one task is not optimal. The convergence result of P3 problem on Task2 is almost equal to MFEA-b, and we can think it is the best. Task2 of P4 problem is a unimodal problem. The multimodal problem in Task1 will also make the optimal solution deviate from the direction of Task2.

Table 3. The average optimal values of Task2 obtained from MFEA, MFEA-b, MFDE, MFPSO and CEL-MFEA

Problem	MFEA	MFEA-b	MFDE	MFPSO	CEL-MFEA
P1	3.34E + 02	2.02E + 02	2.89E + 00	3.75E + 02	**0.00E + 00**
P2	4.44E + 02	2.02E + 02	6.86E−01	5.55E + 02	**0.00E + 00**
P3	9.67E + 03	**3.50E + 03**	1.15E + 04	1.57E + 04	3.51E + 03
P4	2.04E + 02	1.47E + 02	**1.89E−05**	4.73E + 03	6.31E−02
P5	6.58E + 04	2.72E + 03	8.37E + 01	9.46E + 04	**4.80E + 01**
P6	2.07E + 01	1.10E + 01	8.33E−02	1.08E + 01	**1.57E−03**
P7	4.45E + 02	2.51E + 02	2.50E + 01	6.05E + 02	**1.16E−05**
P8	2.74E + 01	2.02E + 01	3.10E + 00	3.18E + 01	**5.91E−02**
P9	9.51E + 03	3.88E + 03	4.34E + 03	1.60E + 04	**3.38E + 03**

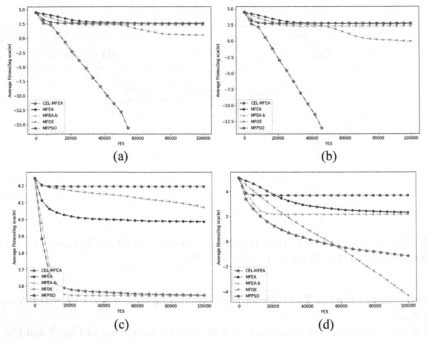

(a) (b)

(c) (d)

Fig. 3. The convergence curve of Task2 from MFEA, MFEA-b, MFDE, MFPSO, CEL-MFEA on MFO test suite. (a) P1. (b) P2. (c) P3. (d) P4.

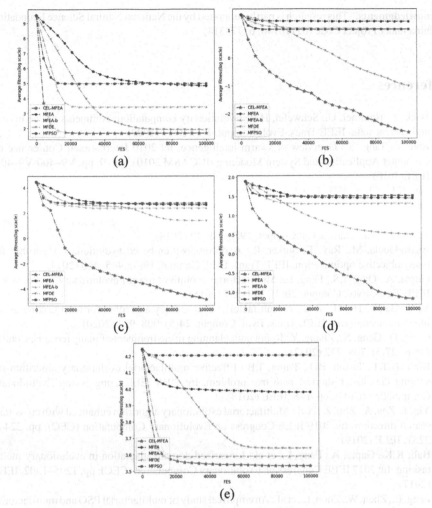

Fig. 4. The convergence curve of Task2 from MFEA, MFEA-b, MFDE, MFPSO, CEL-MFEA on MFO test suite. (a) P5. (b) P6. (c) P7. (d) P8. (e) P9.

5   Conclusions

To avoid the premature convergence of MFEA, a cross elitist learning MFEA (CEL-MFEA) is proposed in this paper. The cross elitist learning strategy can improve the convergence speed of the population. The inter task transfer strategy based on Nelder-Mead operators can reduce the risk aroused by the negative transfer between tasks. The opposition learning strategy improves the diversity of the population and enables the population to find better solutions in the promising area. Finally, 9 classical multifactorial problem test units are used to evaluate the comprehensive ability of CEL-MFEA. The experimental results show that CEL-MFEA performs better than other algorithms on most optimization problems and has strong competitiveness.

Acknowledgments. This research is partly supported by the National Natural Science Foundation of China under Project Code (62176146, 61773314).

References

1. Back, T., Hammel, U., Schwefel, H.P.: Evolutionary computation: comments on the history and current state. IEEE Trans. Evol. Comput. **1**(1), 3–17 (1997)
2. Zhu, Y., Tang, X.: Overview of swarm intelligence. In: 2010 International Conference on Computer Application and System Modeling (ICCASM 2010), vol. 9, pp. V9-400-V9-403. IEEE(2010)
3. Tan, K.C., Chew, Y.H., Lee, L.H.: A hybrid multiobjective evolutionary algorithm for solving vehicle routing problem with time windows. Comput. Optim. Appl. **34**(1), 115–151 (2006)
4. Guo, S.M., Yang, C.C.: Enhancing differential evolution utilizing eigenvector-based crossover operator. IEEE Trans. Evol. Comput. **19**(1), 31–49 (2014)
5. Asafuddoula, M., Ray, T., Sarker, R.: A decomposition-based evolutionary algorithm for many objective optimization. IEEE Trans. Evol. Comput. **19**(3), 445–460 (2014)
6. Gupta, A., Ong, Y.S., Feng, L.: Multifactorial evolution: toward evolutionary multitasking. IEEE Trans. Evol. Comput. **20**(3), 343–357 (2015)
7. Yi, J., Bai, J., He, H., et al.: A multifactorial evolutionary algorithm for multitasking under interval uncertainties. IEEE Trans. Evol. Comput. **24**(5), 908–922 (2020)
8. Gong, D., Geng, N., Zhang, Y.: Robot path planning in environment of many terrains. Control Decis. **27**(5), 708–712 (2012)
9. Binh, H.T.T., Thanh, P.D., Trung, T.B.: Effective multifactorial evolutionary algorithm for solving the cluster shortest path tree problem. In: 2018 IEEE Congress on Evolutionary Computation (CEC), pp. 1–8. IEEE (2018)
10. Yin, J., Zhu, A., Zhu, Z., et al.: Multifactorial evolutionary algorithm enhanced with cross-task search direction. In: 2019 IEEE Congress on Evolutionary Computation (CEC), pp. 2244–2251. IEEE (2019)
11. Bali, K.K., Gupta, A., Feng, L., et al.: Linearized domain adaptation in evolutionary multitasking. In: 2017 IEEE Congress on Evolutionary Computation (CEC), pp. 1295–1302. IEEE (2017)
12. Feng, L., Zhou, W., Zhou, L., et al.: An empirical study of multifactorial PSO and multifactorial DE. In: 2017 IEEE Congress on Evolutionary Computation (CEC), pp. 921–928. IEEE (2017)
13. Da, B., Ong, Y.S., Feng, L., et al.: Evolutionary multitasking for single-objective continuous optimization: Benchmark problems, performance metric, and baseline results. arXiv preprint arXiv:1706.03470 (2017)
14. Zhou, L., Feng, L., Tan, K.C., et al.: Toward adaptive knowledge transfer in multifactorial evolutionary computation. IEEE Trans. Cybern. **51**(5), 2563–2576 (2020)
15. Tang, Z., Gong, M., Wu, Y., et al.: A multifactorial optimization framework based on adaptive intertask coordinate system. IEEE Trans. Cybern. (2021)
16. Lagarias, J.C., Reeds, J.A., Wright, M.H., et al.: Convergence properties of the Nelder-Mead simplex method in low dimensions. SIAM J. Optim. **9**(1), 112–147 (1998)

17. Tizhoosh, H.R.: Opposition-based learning: a new scheme for machine intelligence. In: International Conference on Computational Intelligence for Modelling, Control and Automation and International Conference on Intelligent Agents, Web Technologies and Internet Commerce (CIMCA-IAWTIC 2006), vol. 1, pp. 695–701(2005)
18. Piotrowski, A.P.: Adaptive memetic differential evolution with global and local neighborhood-based mutation operators. Inf. Sci. **241**, 164–194 (2013)

Heterogeneous Adaptive Denoising Networks for Recommendation

Sichen Jin, Yijia Zhang$^{(\boxtimes)}$, and Mingyu Lu

School of Information Science and Technology, Dalian Maritime University,
116024 Liaoning, China
{sichen_jin,zhangyijia,lumingyu}@dlmu.edu.cn

Abstract. Due to the complexity and diversity of real-world relationships, recommender systems are better suited to represent complex data using heterogeneous information networks (HINs), called HIN-based recommendations. However, it is a challenge to efficiently obtain the embedding and remove the noise from the dataset. In our work, we innovatively propose a recommendation model called Heterogeneous Graph Convolutional Network Recommendation with Adaptive Denoising Training (HGCRD). Our model uses a random walk strategy based on meta-path to obtain a valid sequence of nodes. Then for the generated node networks, we use graph convolutional networks (GCNs) to learn the node embeddings. Also, to eliminate the noise in the dataset, we incorporate an adaptive denoising training (ADT) strategy in the training. Experimental results on three public datasets show that HGCRD performs significantly better than the competitive baseline.

Keywords: Recommendation system · Denoising training · Heterogeneous information networks · Graph convolutional networks

1 Introduction

In recent years, recommendation systems have played an increasingly important role in various services. It can recommend content of interest to users. From the traditional recommendation algorithm [1] to the latest recommendation algorithm based on neural network [2–4], most of them are based on homogeneous network. Homogeneous networks contain only one type of edges and nodes, it is not easy to model the complex real world with homogeneous networks. In contrast, heterogeneous information networks (HINs) can contain many types of edges and nodes, so heterogeneous information networks are proposed as a powerful modeling approach [5–7]. Therefore, it is a challenge to design an effective recommendation system for HINs.

In Fig. 1, we give an example of HIN. This example takes the Douban book dataset as an example. The graph contains multiple types of nodes and edges. The edge of type 1 indicates that a user belongs to a reading group. Bob and LiMing both have an edge connected to Group1, indicating that these two people

H. Zhang et al. (Eds.): NCAA 2022, CCIS 1637, pp. 30–43, 2022.
https://doi.org/10.1007/978-981-19-6142-7_3

belong to the same group. An edge of type 2 indicates a connection between users, which means that even two users may have the same interest. The edge of Type 3 connects users to the same book, indicating that the user has read a certain book. Also, the book has three attributes: the edges of types 4, 5 and 6, which indicate the year of publication, the author and the publisher of the book.

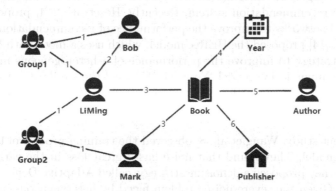

Fig. 1. An illustrative example of HIN.

Secondly, the homogenous information network based on the corresponding semantics obtained through the meta-path lacks practical training. Traditional Deepwalk [9], Metapath2vec [10], and other methods have more or fewer problems, and it is not easy to accurately learn the node embeddings in the network. In response to this, we adopted the latest graph neural networks(GNNs) technology in our model. We passed the homogenous information networks obtained into a neural network containing three convolutional layers for the training. GCN can effectively encode the local structure of the node and spread the local neighbor information of the node. Through our constructed multi-layer GCNs, all edge and node information can be encoded into the embeddings.

The main contributions of this work are as follows:

- Propose an effective recommendation model, called Heterogeneous Graph Convolution networks for Recommendation with adaptive Denoising training (HGCRD). HGCRD uses the latest graph neural network technology to encode the semantic and topological information of HIN.
- Creatively add the Adaptive Denoising Training(ADT) strategy to the training process of heterogeneous information networks. This strategy does not require additional data in model training, significantly improves the model performance.

2   Related Work

In this section, we discuss four aspects of our work: recommender systems, heterogeneous information networks, denoising strategies, and graph neural networks.

The early research mainly adopted the collaborative filtering (CF) method, which uses historical interaction to make recommendations. Ma et al. [11] incorporated social relationships into a matrix decomposition model. Ling et al. [12] considered rating and comment information and proposed a unified model based on the combination of content filtering and collaborative filtering. Shi et al. [14] proposed heterogeneous information networks to solve cold start and node embeddings in the recommendation system. Recently, He et al. [2–4] proposed using graph neural networks to improve the performance of recommendation models.

Shi et al. [14] proposed the HERec model, which uses a meta-path based random walk strategy to improve the performance of a heterogeneous information network recommendation model. Yu et al. [15] used the similarity of meta-path as a regularization term in the matrix decomposition. Luo et al. [16] proposed a collaborative filtering social recommendation method based on heterogeneous relationships.

In a recent study, Wang et al. [8] observed the training process of the recommendation model. They found that noise has a great loss in the early stage of training. So they proposed a denoising strategy, called Adaptive Denoising Training (ADT). Given the severe noise problem faced by heterogeneous information networks, we creatively integrate ADT strategy into the training of heterogeneous information networks, which has not been studied before. We experimentally demonstrated the effectiveness of introducing ADT into the training of HIN.

The NeuMF [2] model proposed by He et al. in 2017 combined the neural network and the recommendation system for the first time. More and more academics are trying to add various neural network structures to the training of recommendation systems, such as NGCF [3] and LightGCN [4]. Unlike the models above, recent graph-based recommendation systems use GNNs technology to learn to generate embeddings from user-item bipartite graphs. In order to learn better graph embeddings, HGCRD uses three-layer GCNs to training the embedding. This structure can effectively learn the embedding of heterogeneous graph networks, which is of great help to improve the model's performance.

3   Model Architecture

3.1   Input and Embedding Module

Inspired by the recent NGCF and LightGCN, we use representational learning methods to extract information from the HIN. Given a heterogeneous information network, we can use a set of meta-paths to transform the heterogeneous information network into a homogeneous information network. Then, we can use graph embedding methods to obtain node embeddings.

In recommendation system, many heterogeneous information networks are processed in a meta-path way. Through meta-path, we can transform a heterogeneous information network into a homogeneous information network. For example, using the UBU meta-path, by randomly walking on the heterogeneous graph, we can extract a path based on the semantics of this meta-path, such as $u1 \rightarrow b1 \rightarrow u2 \rightarrow b2 \rightarrow u4 \rightarrow b3$. We delete other nodes that are different

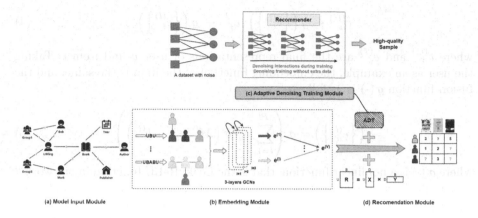

Fig. 2. The schematic diagram of the HGCRD model.

from the first node type, and you can get a path $u1 \rightarrow u2 \rightarrow u4$ that contains only one type of node. Then this path contains the semantic information and implicit feedback contained in the meta-path of UBU: users of the same book are likely to have the same hobbies. Next, it is shown the process from sequence to representation.

The graph Laplacian method is an effective technique for learning node embedding from graph structure [3, 32]. The embedding of nodes is learned from the homogeneous graph by this method. Firstly, calculating the graph Laplacian matrix L^c:

$$L^c = D^{-\frac{1}{2}} H D^{\frac{1}{2}} \tag{1}$$

where D is the diagonal degree matrix of homogenous graph H.

After obtaining the graph Laplacian matrix L^c, the embedding update function can be derived as follows:

$$E^{(m)} = \sigma\left((L^c + I)\, E^{(m-1)} W_o^{(m)} + L^c E^{(m-1)} \odot E^{(m-1)} W_n^{(m)} \right) \tag{2}$$

where $E^{(m)}$ is the embedding obtained after m-th graph convolutional layer. $\sigma(\cdot)$ is the LeakyReLU activation function often used in recommendation systems [3]. I is the identity matrix. $W_o^{(m)}$ and $W_n^{(m)}$ are trainable weight matrices, which retain not only the original information of each node, but also the information of its neighbors.

With the help of multilayer GCN, the embedding of each layer can be obtained: $E^{(1)}$, $E^{(2)}$, and $E^{(3)}$. By combining the embeddings of each layer, we can get the final embedding $e^{(l)}$:

$$e^{(l)} = \begin{bmatrix} E^{(1)} & E^{(2)} & E^{(3)} \end{bmatrix} \tag{3}$$

The usage of fusion function can directly affect the performance of the final model. We use the function $g(\cdot)$ to fuse the node embedding.:

$$e_u^{(U)} \leftarrow g\left(\left\{e_u^{(l)}\right\}\right), e_i^{(I)} \leftarrow g\left(\left\{e_i^{(l)}\right\}\right) \tag{4}$$

where $e_u^{(U)}$ and $e_i^{(I)}$ are the final representations of user u and item i. Taking the user as an example, the embedding function of the item is the same, and the fusion function $g\left(\cdot\right)$ is as follows:

$$g\left(\left\{e_u^{(l)}\right\}\right) = \sigma\left(\sum_{l=1}^{|P|} w_u^{(l)} \sigma\left(M^{(l)} e_u^{(l)} + b^{(l)}\right)\right) \tag{5}$$

where $\sigma\left(\cdot\right)$ is a nonlinear function, that is the LeakyReLU function in our work. l represents different meta-paths. P is the number of meta-paths. $M^{(l)}$ and $b^{(l)}$ are the transformation matrix and bias vector of the l-th meta-path. $w_u^{(l)}$ is the preference weight of user u for the l-th meta-path. Although we only use two nonlinear transformations, it can be flexibly extended to multiple nonlinear layers.

3.2   Recommendation and ADT Module

Combining the fusion function, ADT strategy and matrix decomposition framework to learn the parameters of the proposed model can effectively improve the accuracy of the recommender system. The ADT strategy is a very effective noise removal strategy. The experiments prove that this strategy can effectively remove the noise in the dataset. During the experiments, the authors found that the noisy data generally have a large loss. Therefore, the noise can be selectively removed from the dataset by this feature. The authors propose a truncated cross-entropy loss (T-Loss for short), which discards a larger portion of the total loss. It is defined as follows:

$$L_{T-Loss}(u, i) = \begin{cases} 0 & Loss(u, i) > \tau \\ Loss(u, i) & otherwise \end{cases} \tag{6}$$

where τ is a pre-defined threshold. u and i represent user-item pairs composed of user u and user i. Although easy to explain and implement, fixed thresholds may ignore important nodes during the training process. This is because the loss in training becomes smaller and smaller. To accommodate the overall trend of training losses, the fixed threshold is replaced with a dynamic threshold function $\tau\left(T\right)$. The threshold is dynamically changed by the number of iterations T. In addition, since the loss value varies with different datasets, it is more flexible to design it as a drop rate function $\epsilon\left(T\right)$. The drop rate function is as follows:

$$\epsilon\left(T\right) = min\left(\alpha T, \epsilon_{max}\right) \tag{7}$$

where ϵ_{max} is the upper bound of the drop rate, which is the maximum drop rate. T is the number of iterations. α is a hyperparameter that has been adjusted to reach the maximum drop rate. We do not use more complex functions such as polynomial or logarithmic functions. Although these functions are more expressive, they will inevitably increase the number of hyperparameters and increase the cost of model tuning.

Finally, since we are focusing on implicit feedback and combining the fusion function with the matrix factorization method, the loss function is as follows:

$$Loss(u, i) = \sum_{\langle u, i, r_{u,i} \in R \rangle} (r_{u,i} - \widehat{r_{u,i}})^2 + \lambda \sum_u \left(\left\| \Theta^{(U)} \right\|_2 + \left\| \Theta^{(I)} \right\|_2 \right) \qquad (8)$$

where $\widehat{r_{u,i}}$ is the predicted score of user u for item i. λ is a regularization parameter. $\Theta^{(U)}$ and $\Theta^{(I)}$ are all parameters in the fusion function $g(\cdot)$ of users and items, respectively. In this work, SGD is used to optimize the loss function..

4 Experiments

In this section, we demonstrate the effectiveness of HGCRD through a series of experiments. Based on summing up experience, we first compared with popular models to prove the model's effectiveness in this article. Then, we perform an ablation experiment to verify the effectiveness of the added components. And the hyperparameter experiment considers the influence of hyper-parameter on the model.

4.1 Evaluation Datasets

We evaluate model performance using three datasets commonly found in the recommender system domain: Yelp, Douban-movie, and Douban-book. The detailed information is provided in Table 1. And in Table 2, we show the seleted meta-paths in the experiments.

- **Yelp:** Yelp is a famous review site in the United States. It contains user ratings of various restaurants, hotels, services and other content.
- **Douban-movie:** This is a dataset from the movie review website Douban. It contains data on user ratings of various movies.
- **Douban-book:** This is a dataset from the Douban book domain. It contains user rating data for various books.

4.2 Overall Performance

We consider the following methods for comparison:

- **PMF:** PMF (Probabilistic Matrix Factorization) is optimized on the basis of Regularized Matrix Factorization. It can solve large-scale, sparse and unbalanced data.
- **SoMF:** It takes into account the diversity of data and incorporates social information into the matrix decomposition model.
- **FMHIN:** It is a factorization machine that uses various auxiliary information such as context.
- **HeteMF:** It is a recommendation model combining meta-path algorithm and matrix decomposition algorithm.

Table 1. Details of the three datasets.

Dataset	Relations (A-B)	Number of A	Number of B	Number of (A-B)	Ave. degrees of A	Ave. degrees of B
Yelp	User-Business	16,239	14,284	198,397	12.2	13.9
	User-User	10,580	10,580	158,590	15.0	15.0
	User-Compliment	14,411	11	76,875	5.3	6988.6
	Business-City	14,267	47	14,267	1.0	303.6
	Business-Category	14,180	511	40,009	2.8	78.3
Douban-movie	User-Movie	13,367	12,677	1,068,278	79.9	84.3
	User-User	2,440	2,294	4,085	1.7	1.8
	User-Group	13,337	2,753	570,047	42.7	207.1
	Movie-Director	10,179	2,449	11,276	1.1	4.6
	Movie-Actor	11,718	6,311	33,587	2.9	5.3
	Movie-Type	12,678	38	27,668	2.2	728.1
Douban-book	User-Book	13,024	22,347	792,026	60.8	35.4
	User-User	12,748	12,748	169,150	13.3	13.3
	Book-Author	21,907	10,805	21,905	1.0	2.0
	Book-Publisher	21,773	1,815	21,773	1.0	11.9
	Book-Year	21,192	64	21,192	1.0	331.1

Table 2. The meta-path used in HGCRD.

Dataset	Meta-paths
Yelp	UBU, UBCiBU, UBCaBU
	BUB, BCiB, BCaB
Douban-movie	UMU, UMDMU, UMAMU, UMTMU
	MUM, MAM, MDM, MTM
Douban-book	UBU, UBABU, UBPBU, UBYBU
	BUB, BPB, BYB

- **SemRec:** It is a recommendation model that combines a weighted meta-path algorithm with a collaborative filtering algorithm.
- **DSR:** It differs from other models in that dual similarity is considered in MF. Items with high and low similarity are regularized.
- **HERec:** It is a recommendation framework that combines meta-paths and random walk.
- **NGCF:** It adds neural networks to the recommendation model. The embedding propagation is used to get the topological information.
- **HGCRD:** It is the proposed recommendation model based on the heterogeneous graph neural network. Creatively incorporates an adaptive denoising strategy. In the following experiments, we will verify the effectiveness of each component of the model.

Our chosen baselines cover baselines of all common types of recommender systems. We set the following experimental parameters: the number of

Table 3. Validity experiment results on three datasets.

Dataset	MAE/RMSE	PMF	SoMF	FMHIN	HeteMF	SemRec	DSR	HERec	NGCF	HGCRD
Yelp	MAE	1.0412	1.0095	0.9013	0.9487	0.9043	0.9054	0.8395	0.7852	**0.7502**
	RMSE	1.4268	1.3392	1.1417	1.2549	1.1637	1.1186	1.0907	0.9941	**0.9601**
Douban-movie	MAE	0.5741	0.5817	0.5696	0.0.5750	0.5695	0.5681	0.5519	0.5385	**0.4933**
	RMSE	0.7641	0.768	0.7248	0.7556	0.7399	0.7225	0.7053	0.6722	**0.6487**
Douban-book	MAE	0.5774	0.5756	0.5716	0.574	0.5675	0.574	0.5502	0.5416	**0.5091**
	RMSE	0.7414	0.7302	0.7199	0.736	0.7283	0.7206	0.6811	0.6637	**0.6195**

embedding sizes $d = 64$, after [9], we set the number of latent factors to 10. Considering the characteristics of our dataset, the value of ϵ_{max} is set to 0.05. Default settings are used for all other baselines. The experiment uses evaluation metrics commonly used in the field of recommendation systems: MAE and RMSE. We divided the dataset into a training set and a test set, which accounted for 80% and 20%, respectively. The results of the experiment are shown in Table 3. We have the following conclusions:

(1) The HIN-based methods (HERec, HeteMF, SemRec, FMHIN, and DSR) outperformed the traditional MF-based methods (PMF and SoMF), indicating the effectiveness of heterogeneous information. This is because HIN captures information about the attributes of the nodes, which in turn improves the recommendation performance.

(2) The proposed HGCRD approach always outperforms all baselines. Compared with other HIN-based methods, HGCRD uses a more logical way to capture the information contained in the data. This is due to the new HIN embedding method and denoising strategy adopted by HGCRD.

(3) Compared with the best baseline, HGCRD has an improvement of about 10% on all datasets, which shows that our proposed model has excellent recommendation performance. It can better simulate the complex relationships of the real world. It proves that our HGCRD model performs better after combining the three-layer GCNs and ADT strategy.

4.3   Ablation Experiments

In this section, we verified the effectiveness of the added components. Among them, $HGCRD_{GCN}$ removes the three-layer graph convolutional network in the model and replaces it with the ordinary graph embedding model Deepwalk. $HGCRD_{ADT}$ is the model obtained after removing the denoising strategy. $HGCRD_{NULL}$ is a model obtained by removing the three-layer graph convolution and denoising strategy simultaneously. The experimental results are shown in Table 4.

The following conclusions can be drawn from Table 3. $HGCRD_{NULL}$ has the worst effect. When the GCN is replaced and the denoising module is removed, the model's performance drops significantly. It can be seen that the two modules added have a more significant contribution to the model. Compared with $HGCRD_{NULL}$, $HGCRD_{GCN}$ and $HGCRD_{ADT}$ have improved performance,

Table 4. The influence of component on the model.

Model	Yelp		Douban-movie		Douban-book	
	MAE	RMSE	MAE	RMSE	MAE	RMSE
HGCRD$_{NULL}$	0.8477	1.0716	0.5643	0.6882	0.5931	0.7171
HGCRD$_{GCN}$	0.7952	1.0271	0.5396	0.6354	0.5563	0.6552
HGCRD$_{ADT}$	0.7831	1.0163	0.5244	0.6491	0.5540	0.6637
HGCRD	**0.7502**	**0.9601**	**0.4933**	**0.6487**	**0.5091**	**0.6195**

indicating that the two added components play an essential role in the performance of the model. HGCRD works best, indicating that the two components can play the most significant role when they exist simultaneously.

4.4   Detailed Analysis of HGCRD

In this section, we consider the effect of hyperparameters. Firstly, we verify the influence of GCN's layers m on the model. Then we consider the sensitivity of ϵ_{max} on the model. Finally, the effect of the embedding dimension d is not negligible either.

Impact of GCN's Layers. Considering the influence of GCN's layers m on the model. We set five value of m as in $\{1, 2, 3, 4, 5\}$. The results of the experiments are shown in Fig. 3. The abscissa represents the number of m. The ordinate represents the improvement rate of the MAE relative to the best baseline (NGCF). Only the results of MAE is reported. The same trend appears on RMSE.

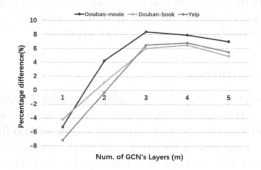

Fig. 3. The influence of GCN's layers on the model.

Based on the results in Fig. 3, the following conclusions are drawn:

(1) Increasing the depth of GCN can better increase the learning ability of the recommendation model. The value of m gradually increases from 1 to 3, and

the model's performance is constantly improving. When there is only one convolutional layer, since the model only considers first-order proximity, it is difficult to learn higher-order structures;

(2) When we further increase the depth of GCN, reaching 4 or even 5 layers, the effect of the model will basically unchanged or decrease to a certain extent. We believe that the excessive depth of GCN causes the over-fitting. Moreover, because the number of layers is too deep, additional noise may be introduced. Further, affect the performance of the model. And it can be seen that the three convolutional layers are enough to capture high-order structures, and the number of layers of GCN will continue to be superimposed, which leads to a higher complexity of the model;

(3) The performance of $HGCRD_{GCN-3}$ is better than other variants. Increasing the GCN's layers, the performance is almost unchanged or even decreased. And greatly increasing the calculation time. Therefore, the best number of GCN's layers is 3. Moreover, experiments show that the superposition of three layers of GCN is more reasonable.

Sensitivity of Hyper-parameter in ADT. Considering the influence of ϵ_{max} on the model. ϵ_{max} adjusts the upper limit of the drop rate. We set seven value of ϵ_{max} as in $\{0, 0.01, 0.03, 0.05, 0.1, 0.15, 0.2\}$. The results of the experiments are shown in Table 5.

Table 5. The influence of ϵ_{max} on the model.

ϵ_{max}	Yelp		Douban-movie		Douban-book	
	MAE	RMSE	MAE	RMSE	MAE	RMSE
0	0.7831	0.9940	0.5361	0.6832	0.5671	0.6607
0.01	0.7663	0.9833	0.5233	0.6741	0.5395	0.6452
0.03	0.7608	0.9715	0.5102	0.6534	0.5243	0.6304
0.05	**0.7502**	**0.9601**	**0.4933**	**0.6487**	**0.5091**	**0.6195**
0.1	0.7850	0.9877	0.5127	0.6679	0.5182	0.6387
0.15	0.8131	1.0223	0.5339	0.6952	0.5464	0.6793
0.2	0.8566	1.0459	0.5784	0.7381	0.6077	0.7048

We considered the performance of HGCRD under different drop rates. The value of ϵ_{max} varies from 0 to 0.2. From Table 4 we can get the following conclusions:

(1) The ADT strategy effectively removes the noise in the model and dramatically reduces the influence of noise interaction on the model;

(2) The best drop rate is 0.05. When it is lower than 0.05, the performance of the model decreases. One possible reason is that the drop rate is too low, and the lost data is too small, which makes it difficult to remove the noise data effectively.

(3) When the drop rate is greater than 0.05, the effect of the model also decreases. It may be that too much data is discarded, causing many normal data to be discarded as noise data. Therefore it is difficult to learn effective parameters.

Impact of Embedding Dimensions. Considering the influence of embedding dimensions d on the model. We set five value of d as in $\{16, 32, 64, 128, 256\}$. The results of the experiments are shown in Fig. 4. The abscissa represents the number of d. The ordinate represents the improvement rate of the MAE relative to the best baseline (NGCF). Only the results of MAE is reported. The same trend appears on RMSE.

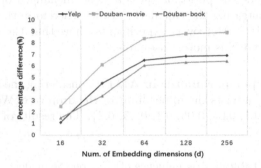

Fig. 4. The influence of embedding dimensions on the model.

Based on the results in Fig. 4, the following conclusions are drawn:

(1) Increasing the embedding dimensions can better learn the embedding of nodes. The value of d gradually increases from 16 to 256, and the model's performance continues to improve. It shows that increasing the embedding dimensions is valid.
(2) When the value of d gradually increases from 16 to 64, the effect of the model is greatly improved. When the embedding dimension is 16, the improvement rate of the model is low. We believe that the embedding dimension is too low to learn the embedding of nodes.
(3) When the value of d gradually increases from 64 to 256, the effects of the model are improved. However, the rate of improvement is relatively low. Moreover, because the embedding dimension is too large, it will significantly increase the training time. Therefore, we believe that the best embedding dimension is 64.

4.5 Limitation

Table 2 shows the selected meta-paths in our model. In fact, the choice of meta-paths depends on specific professional knowledge. Only by choosing the correct meta-path can the semantic information be accurately reflected. If the selected meta-path is not appropriate, it will cause the embedding of users and items to be unable to be extracted correctly. In turn, the recommendation performance of the model decreases.

The symmetric meta-path will limit the extraction of information at the semantic level. Recommendation models based on heterogeneous information networks also face the problem of cold-start. Our model also faces some difficulties in dealing with this problem. Therefore, our model has certain limitations.

5 Conclusion and Future Work

In this paper, we propose a novel recommendation model for heterogeneous information networks (i.e., HGCRD). We present some issues in the study of heterogeneous information networks. To solve these problems, we propose to use ADT policy denoising and GCNs to solve the middle layer problem of unified networks. The recommendation effectiveness is effectively improved. Using the HGCRD recommendation model, the learned embedding information is further integrated into the extended matrix decomposition model by the fusion function. Finally, the recommendation model is optimized by adaptive denoising training. The validity of HGCRD was also verified in the experiment. We also verified the validity of each component of the model and the reasonableness of the speculation parameters.

In the future, we will consider the issue of bias in datasets. Moreover, extending the proposed model to embeddings of arbitrary nodes with arbitrary meta-paths is an exciting work. How to improve the interpretability of recommendation methods based on meta-path semantics is also an important topic in recommendation systems research.

Acknowledgment. This work is supported by a grant from the Natural Science Foundation of China 62072070 and Social and Science Foundation of Liaoning Province (No. L20BTQ008).

References

1. Koren, Y., Bell, R., Volinsky, C.: Matrix factorization techniques for recommender systems. Computer **42**(8), 30–37 (2009)
2. He, X., et al.: Neural collaborative filtering. In: Proceedings of the 26th International Conference on World Wide Web. Perth, Australia, pp. 173–182 (2017)
3. Wang, X., et al.: Neural graph collaborative filtering. In: Proceedings of the 42nd International ACM SIGIR Conference on Research and Development in Information Retrieval. Paris, France, pp. 165–174 (2019)
4. He, X., et al.: LightGCN: simplifying and powering graph convolution network for recommendation. In: Proceedings of the 43rd International ACM SIGIR Conference on Research and Development in Information Retrieval. Xi'an, China, pp. 639–648 (2020)

5. Sun, Y., et al.: PathSim: meta path-based top-k similarity search in heterogeneous information networks. Proc. VLDB Endow. **4**(11), 992–1003 (2011)
6. Shi, C., et al.: A survey of heterogeneous information network analysis. IEEE Trans. Knowl. Data Eng. **29**(1), 17–37 (2017)
7. Shi, C., huan, Yu Philip, S.: Heterogeneous Information Network Analysis and Applications. Springer, Cham, pp. 1–227 (2017). ISBN 978-3-319-56211-7. https://doi.org/10.1007/978-3-319-56212-4
8. Wang, W., et al.: Denoising implicit feedback for recommendation. In: Proceedings of the 14th ACM International Conference on Web Search and Data Mining, pp. 373–381 (2021)
9. Perozzi, B., Al-Rfou, R., Skiena, S.: DeepWalk: online learning of social representations. In: Proceedings of the 20th ACM SIGKDD International Conference on Knowledge Discovery and Data Mining. USA, pp. 701–710 (2014)
10. Dong, Y., Chawla, N.V., Swami, A.: Metapath2vec: scalable representation learning for heterogeneous networks. In: Proceedings of the 23rd ACM SIGKDD International Conference on Knowledge Discovery and Data Mining. Halifax, Canada, pp. 135–144 (2017)
11. Ma, H., et al.: Recommender systems with social regularization. In: Proceedings of the fourth ACM International Conference on Web Search and Data Mining, pp. 287–296 (2011)
12. Ling, G., Lyu, M.R., King, I.: Ratings meet reviews, a combined approach to recommend. In: Proceedings of the 8th ACM Conference on Recommender Systems. California, USA, pp. 105–112 (2014)
13. Ye, M., et al.: Exploiting geographical influence for collaborative point-of-interest recommendation. In: Proceedings of the 34th International ACM SIGIR Conference on Research and Development in Information Retrieval. Beijing, China, pp. 325–334 (2011)
14. Shi, C., et al.: Heterogeneous information network embedding for recommendation. IEEE Trans. Knowl. Data Eng. **31**(2), 357–370 (2019)
15. Yu, X., et al.: Personalized entity recommendation: a heterogeneous information network approach. In: Proceedings of the 7th ACM International Conference on Web Search and Data Mining. New York, USA, pp. 283–292 (2014)
16. Luo, C., et al.: Hete-CF: Social-based collaborative filtering recommendation using heterogeneous relations. In: IEEE International Conference on Data Mining, pp. 917–922. IEEE, Shenzhen, China (2014)
17. Shi, C., et al.: RHINE: relation structure-aware heterogeneous information network embedding. IEEE Trans. Knowl. Data Eng. **34**(1), 433–447 (2022)
18. Wang, X., et al.: Heterogeneous graph attention network. In: The World Wide Web Conference. San Francisco, CA, USA, pp. 2022–2032 (2019)
19. Lao, N., Cohen, W.W.: Relational retrieval using a combination of path-constrained random walks. Mach. Learn. **81**(1): 53–67 (2010)
20. Shi, C., et al.: HeteSim: a general framework for relevance measure in heterogeneous networks. IEEE Trans. Knowl. Data Eng. **26**(10), 2479–2492 (2014)
21. Yan, S., et al.: Graph embedding and extensions: a general framework for dimensionality reduction. IEEE Trans. Pattern Analy. Mach. Intell. **29**(1), 40–51 (2007)
22. Tu, C., et al.: Max-margin DeepWalk: discriminative learning of network representation. In: IJCAI, pp. 3889–3895 (2016)
23. Wei, X., et al.: Cross view link prediction by learning noise-resilient representation consensus. In: Proceedings of the 26th International Conference on World Wide Web. Perth, Australia, pp. 1611–1619 (2017)

24. Cao, S., Lu, W., Xu, Q.: Deep neural networks for learning graph representations. In: Proceedings of the AAAI Conference on Artificial Intelligence. Phoenix, Arizona, pp. 1145–1152 (2016)
25. Liang, D., et al.: Factorization meets the item embedding: regularizing matrix factorization with item co-occurrence. In: Proceedings of the 10th ACM Conference on Recommender Systems. Boston, MA, USA, pp. 59–66 (2016)
26. Wang, D., Cui, P., Zhu, W.: Structural deep network embedding. In: Proceedings of the 22nd ACM SIGKDD International Conference on Knowledge Discovery and Data Mining. San Francisco, CA, USA, pp. 1225–1234 (2016)
27. Cao, S., Lu, W., Xu, Q.: GraRep: learning graph representations with global structural information. In: Proceedings of the 24th ACM International On Conference on Information and Knowledge Management. Melbourne, VIC, Australia, pp. 891–900 (2015)
28. Chang, S., et al.: Heterogeneous network embedding via deep architectures. In: Proceedings of the 21th ACM SIGKDD International Conference on Knowledge Discovery and Data Mining. Sydney, NSW, Australia, pp. 119–128 (2015)
29. Tang, J., Qu, M., Mei, Q.: PTE: predictive text embedding through large-scale heterogeneous text networks. In: Proceedings of the 21th ACM SIGKDD International Conference on Knowledge Discovery and Data Mining. Sydney, NSW, Australia, pp. 1165–1174 (2015)
30. Xu, L., et al.: Embedding of embedding (EOE) joint embedding for coupled heterogeneous networks. In: Proceedings of the Tenth ACM International Conference on Web Search and Data Mining. Cambridge, UK, pp. 741–749 (2017)
31. Chen, T., Sun, Y.: Task-guided and path-augmented heterogeneous network embedding for author identification. In: Proceedings of the Tenth ACM International Conference on Web Search and Data Mining. Cambridge, UK, pp. 295–304 (2017)
32. Meng, Z., et al.: Jointly learning representations of nodes and attributes for attributed networks. ACM Trans. Inf. Syst. 38(2): 16:1–16:32 (2020)
33. Rendle, S.: Factorization machines with LIBFM. ACM Trans. Intell. Syst. Technol. 3(3), 57:1–57:22 (2012)

Formation Control Optimization via Leader Selection for Rotor Unmanned Aerial Vehicles

Bu Liu[✉], Xiangping Bryce Zhai[✉], Bin Du[✉], and Jing Zhu

Nanjing University of Aeronautics and Astronautics, Nanjing, China
{waitingbuliu,blueicezhaixp,iniesdu,drzhujing}@nuaa.edu.cn

Abstract. Unmanned aerial vehicles (UAV) are becoming more and more widely used and playing an increasingly important role in lots of military and civilian tasks. This paper presents a leader selection algorithm of formation based on rotor UAVs. We assume that a certain UAV is the leader, and use it as a benchmark to convert the state of multiple UAVs into the relative motion state of the leader in this algorithm. The purpose is to select a suitable leader for the formation of UAVs and minimize the time of forming a target formation. In order to solve these problems, we use the linear quadratic regulator (LQR) algorithm to realize the formation control of the UAVs. We also solve the combined mapping problem between the UAVs and target formations which is a sub-problem of leader selection. Finally, the simulation experiments show the processes of UAVs formation. The numerical results further verify the effectiveness of this algorithm.

Keywords: UAV · Leader selection · Formation control · Linear quadratic regulator (LQR) · Genetic algorithm (GA)

1 Introduction

After decades of development, the technology of unmanned aerial vehicles (UAV) has gradually improved and even matured. At the same time, unmanned aerial vehicles are becoming more and more widely used, playing an increasingly important role in many military and civilian tasks such as battlefield surveillance and investigation, border patrol, forest fire detection, and public safety monitoring. Compared with a single UAV, multi-UAVs formation flight has more advantages in large-scale surveillance, combat distance, attack capability, etc. Moreover, the

Supported by National Defense Science and Technology Innovation Zone Foundation under Grant No. 19-163-16-ZD-022-001-01, the Fund of Prospective Layout of Scientific Research for NUAA (Nanjing University of Aeronautics and Astronautics), the Foundation of Key Laboratory of Safety-Critical Software (Nanjing University of Aeronautics and Astronautics), Ministry of Industry and Information Technology (No. NJ2020022), and also supported by National Natural Science Foundation of China (No. 61701231).

success rate of executing UAV missions is also higher. Multi-UAVs formations can be flexibly deployed according to mission requirements, forming different formations according to different terrains or missions. Therefore, how to efficiently generate UAVs target formations within a limited time or shorter distance has attracted the attention of many researchers.

For UAVs formations control and optimization, many scholars at home and abroad have also made many excellent results. Zheng et al. [1] proposed a quadrotor UAV flight formation reconstruction method with collision avoidance function, and calculated the probability route map(PRM) navigation method of a nonlinear quadrotor helicopter model in a dynamic environment. In [2], Q-learning reinforcement learning method was used to adjust parameters, and according to the relative motion relationship of formation flying, the UAV longitudinal and horizontal fuzzy controllers are designed to solve the problem of multi-UAVs formation control. The UAV cluster formation flight and obstacle avoidance control based on the backbone network are studied in [3], and a master-slave strategy is proposed to make the backbone UAVs in the subnet only obtain virtual master-slave information. Besides, the authors of [4] proposed a new model of autonomous search, detection and mission execution of UAVs distributed formation. In [5], a proportional integral derivative (PID) control method was used to study the formation maintenance of UAVs. However, in previous studies, most of the attention was focused on the UAVs formation control and formation maintenance phase. In fact, the formation generation phase of UAVs can also be studied emphatically.

In this paper, we consider the case where the UAV is abstracted as a mass point in a two-dimensional coordinate system to build a model and describe its motion process. It is well known that calculating the state information of each UAV is necessary to form a target formation finally in the process of multi-UAVs formation. In the scenario considered in the leader selection algorithm, choosing leader as the anchor point, the multiple follower UAVs states are converted into the relative motion state of leader, as shown in Fig. 1. Then, the formation task can be completed only by studying the relative motion state. Therefore, a reasonable choice of leader in multiple UAVs has also become a key point to improve the performance of UAVs formation. In order to make the UAVs formation generation fast and efficient, while minimizing the time and energy consumption of UAVs formation, we propose a leader selection algorithm based on it.

The main contributions of this paper can be summarized as follows:

- We construct a formation model based on leader, which converts the formation control problem of the UAV formation system into a positional movement following the leader.
- Based on the LQR control algorithm and proposed model, we propose a corresponding algorithm for the leader selection problem of the UAVs formation.
- We apply the genetic algorithm (GA) to solve the mapping problem between the UAVs and target formation.

The rest of this paper is organized as follows. In Sect. 2, we introduce the system model of leader-based relative UAVs formation. GA and LQR algorithm in Sect. 3 are applied to solve the UAVs formation's leader selection problem. Section 4 shows the experimental results to validate effectiveness of the proposed algorithm. Finally, we conclude our work in Sect. 5.

2   System Model

Consider an leader-based formation system containing M UAVs denoted by the set of $\mathcal{M} = \{1, \ldots, M\}$, where one is selected as the leader, and the other UAVs are regarded as followers. In our scenario, we prioritize the shape of the target formation rather than where it is located. And we assume that the target formation (TF) shape is pre-determined, and the position of TF is determined by the currently selected leader position (as shown in Fig. 1). Our goal is to select one of the M UAVs as leader, and the other UAVs as followers to minimize the generation time of UAVs formations. In the system, the state of each UAV can be expressed as:

$$\begin{cases} \dot{\boldsymbol{p}}_i = \boldsymbol{v}_i, \\ \dot{\boldsymbol{v}}_i = \boldsymbol{u}_i, \\ \boldsymbol{s}_i = [\boldsymbol{p}_i, \boldsymbol{v}_i], \quad i = 1, 2, \ldots, M \end{cases} \tag{1}$$

where $\boldsymbol{p}_i = [x_i, y_i]$, $\boldsymbol{v}_i = [v_{xi}, v_{yi}]$ and $\boldsymbol{u}_i = [u_{xi}, u_{yi}] \in \mathbb{R}^2$ denote the position vector, velocity vector and control input of ith UAV, and $\boldsymbol{s}_i = [x_i, y_i, v_{xi}, v_{yi}]$ denotes motion state of ith UAV. For simplicity, we use the $\boldsymbol{p}_i$, $\boldsymbol{v}_i$ and $\boldsymbol{u}_i$ vector to represent the state of the ith UAV, $i \in \mathcal{M}$.

Without loss of generality, we construct a 2-D Cartesian coordinate system where the initial $\boldsymbol{p}_i$ and $\boldsymbol{v}_i$ of all UAVs in the system are randomly generated, and the initial control input $\boldsymbol{u}_i$ is $[0, 0]$, $i \in \mathcal{M}$. Besides, we assume that the initial positions and TF positions of all UAVs are in the first quadrant. In our model, the final velocity of the formation tends to be consistent with the velocity of the selected leader. Obviously, the leader can be regarded as a stationary point or an anchor point relative to other followers. We can see the conversion process in Fig. 1, that only control the UAVs formation by calculating the relative motion state of leader and other followers to generate a target formation.

We assume that when the lth UAV is the leader ($l \in \mathcal{M}$), the state of leader is represented as $\boldsymbol{p}_l$, $\boldsymbol{v}_l$ and $\boldsymbol{s}_l$ according to (1). Thus, the total position and velocity matrix of all followers is expressed as:

$$\begin{cases} \mathcal{P}^l = [\boldsymbol{p}_1^l, \boldsymbol{p}_2^l, \ldots, \boldsymbol{p}_{M-1}^l]^{\mathrm{T}}, \\ \mathcal{V}^l = [\boldsymbol{v}_1^l, \boldsymbol{v}_2^l, \ldots, \boldsymbol{v}_{M-1}^l]^{\mathrm{T}}, \end{cases} \tag{2}$$

where $\boldsymbol{p}_j^l$ represents the position of the jth follower when the lth UAV is the leader, $1 \leq j \leq M - 1$, $l \in \mathcal{M}$.

Besides, the position of target formation is affected by the selected leader and the ranking position of the leader. When the leader's position in the target formation is ranked as k ($1 \leq k \leq M$), the total position set of TF corresponding

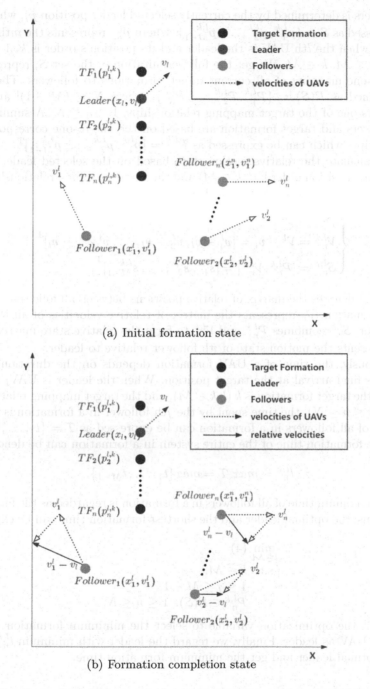

(a) Initial formation state

(b) Formation completion state

Fig. 1. The illustration of relative motion state transition process, where (a) shows the initial state of the leader selection algorithm, and the black arrow in (b) is the relative state of followers and leader.

to followers is determined by the currently selected leader position $\boldsymbol{p}_l$, which can be expressed as $\mathcal{S} = \{\boldsymbol{p}_1^{l,k}, \boldsymbol{p}_2^{l,k}, \ldots, \boldsymbol{p}_{M-1}^{l,k}\}$, where $\boldsymbol{p}_m^{l,k}$ represents the mth target position when the lth UAV is the leader and its position's order is k, $1 \leq m \leq M-1$, $l \in \mathcal{M}$, $k \in \mathcal{M}$. We get the full permutation of the set $\mathcal{S}$, representing a one-to-one mapping of all current target formations to followers. Therefore, it is defined as $P(\mathcal{S}) = \{\mathcal{P}_1^{l,k}, \mathcal{P}_2^{l,k}, \ldots, \mathcal{P}_N^{l,k}\}$, where $N = (M-1)!$ and $\mathcal{P}_n^{l,k}$ represents one of the target mapping relationships, $1 \leq n \leq N$. Assuming that the followers and target formation are based on the one-to-one correspondence of subscripts which can be expressed as $\mathcal{P}_1^{l,k} = [\boldsymbol{p}_1^{l,k}, \boldsymbol{p}_2^{l,k}, \ldots, \boldsymbol{p}_{M-1}^{l,k}]^{\mathrm{T}}$.

We calculate the relative motion state based on the selected leader UAV$_l$, the leader's position order k ($l, k \in \mathcal{M}$) and the target mapping relationship $\mathcal{P}_n^{l,k}$ ($1 \leq n \leq N$), which can be expressed as:

$$\begin{cases} \mathcal{P}_r^{l,k} = \mathcal{P}^l - \mathcal{P}_n^{l,k}, \\ \mathcal{V}_r^{l,k} = \mathcal{V}^l - \boldsymbol{v}_l = [\boldsymbol{v}_1^l - \boldsymbol{v}_l, \boldsymbol{v}_2^l - \boldsymbol{v}_l, \ldots, \boldsymbol{v}_{M-1}^l - \boldsymbol{v}_l]^{\mathrm{T}}, \\ \mathcal{S}_r^{l,k} = [\mathcal{P}_r^{l,k}, \mathcal{V}_r^{l,k}] = [\boldsymbol{s}_1^{l,k}, \boldsymbol{s}_2^{l,k}, \ldots, \boldsymbol{s}_{M-1}^{l,k}]^{\mathrm{T}}, \end{cases} \tag{3}$$

where $\mathcal{P}_r^{l,k}$ denotes the matrix of relative positions between all followers and the target formation, $\mathcal{V}_r$ represents the matrix of relative velocities of all followers and leader, $\mathcal{S}_r^{l,k}$ combines $\mathcal{P}_r^{l,k}$ and $\mathcal{V}_r^{l,k}$ to form the relative state matrix where $\boldsymbol{s}_i^{l,k}$ represents the motion state of ith follower relative to leader.

Obviously, the time of a UAV formation depends on the duration of the follower's final arrival at the target position. When the leader is UAV$_l$ and its order in the target formation is k ($l, k \in \mathcal{M}$), and the target mapping relationship is $\mathcal{P}_n^{l,k}$ ($1 \leq n \leq N$), the time spent by the jth follower in a formation is t_j. The time set of all followers in a formation can be expressed as $\mathcal{T} = \{t_1, \ldots, t_{M-1}\}$. Then the formation time of the entire system in a formation can be denoted as:

$$t_n^{l,k} = max\ \mathcal{T} = max\{t_1, \ldots, t_{M-1}\}, \tag{4}$$

i.e., the maximum time of all followers in a formation is regarded as the formation time. Thus, the optimal leader and the shortest formation time can be expressed as:

$$\begin{aligned} &\min_{l \in \mathcal{M}}\ (4) \\ &\text{s.t.}\ \ k \in \mathcal{M}, \\ &\qquad 1 \leq j \leq M-1, \\ &\qquad \mathcal{P}_n^{l,k} \in P(\mathcal{S}),\ 1 \leq n \leq N, \end{aligned} \tag{5}$$

where l is the optimization variable to select the minimum formation time of different UAV as leader. Finally, we regard the leader with minimum $t_n^{l,k}$ in (5) as the optimal leader and get the minimum formation time.

3   Leader Selection Algorithm

In the proposed leader selection algorithm, it mainly includes two sub-problems: UAV formation control and the target mapping problem mentioned above. The

UAV formation control is based on LQR which mainly focuses on the control of velocity and position; the target formation mapping problem is to solve the one-to-one mapping between the current followers and the target formation position.

LQR is based on state space technology to design an optimized dynamic controller. The system model is a linear system given in the form of state space, and its objective function is a quadratic function of state and control input. The system state space expression is composed of two parts: state equation and output equation. Thus, the vector matrix form of the system state space can be expressed as:

$$\begin{cases} \dot{x} = Ax + Bu, \\ y = Cx + Du, \end{cases} \tag{6}$$

where A is the state matrix of the system, B is the output matrix, C is the input matrix, D is the direct transfer matrix of the input, x,y and u denote the state vector, the output vector and the input vector respectively.

In this system, assuming that the UAV$_l$ ($l \in \mathcal{M}$) is the leader and its position's order is k ($k \in \mathcal{M}$). The state vector x of the jth follower contains four state variables, i.e., the position and velocity in a two-dimensional coordinate system, and the input variable u contains two variables as control inputs. Two conditions are required for the formation of multiple UAVs to reach a stable state: the UAVs reach target formation positions and the velocities of followers and leader tend to be consistent. Both conditions need to be met at the same time. Supposing the jth follower corresponds to the mth target formation position $p_m^{l,k}$, its expected output is $Y_j = [p_m^{l,k}, v_l]^T$ which meets the conditions of reaching the position of TF and keeping the velocity consistent with leader. The initial state is $X_j = [p_j^l, v_j]^T$ and the expected input is $U_j = [0\ 0]^T$. However, we propose to convert the state of multiple UAVs into the relative motion state of the leader. Accordingly, both the initial state vector and the output state vector in the state equation must be changed. Derived from (3), $X_j = [p_j^l - p_m^{l,k}, v_j^l - v_l]^T = s_j^{l,k}$, $Y_j = [0\ 0\ 0\ 0]^T$, $U_j = [0\ 0]^T$.

Obviously, Y_j and U_j need to satisfy the physical model. Thus, the state space description of the jth follower can be obtained from (1), where

$$A = \begin{bmatrix} 0 & 0 & 1 & 0 \\ 0 & 0 & 0 & 1 \\ 0 & 0 & 0 & 0 \\ 0 & 0 & 0 & 0 \end{bmatrix}, \quad B = \begin{bmatrix} 0 & 0 \\ 0 & 0 \\ 1 & 0 \\ 0 & 1 \end{bmatrix}, \quad C = \begin{bmatrix} 1 & 0 & 0 & 0 \\ 0 & 1 & 0 & 0 \\ 0 & 0 & 1 & 0 \\ 0 & 0 & 0 & 1 \end{bmatrix}, \quad D = 0. \tag{7}$$

On the one hand, we hope that the system can reach a stable state with minimal deviation. Besides, we want to pay a small cost to achieve our goal. For this reason, the cost function that comes with LQR is defined as:

$$J = \frac{1}{2} \int_0^\infty x^T Q x + u^T R u\ dt, \tag{8}$$

where Q is the state weight and positive semi-definite matrix, and R denotes the control weight and positive definite matrix.

LQR defines the state feedback controller u and the state feedback matrix K as:

$$u = -Kx, \tag{9}$$

and bring u into the previous system state equation, which can be expressed as:

$$\dot{x} = (A - BK)x = A_c x. \tag{10}$$

Then we calculate the value of the feedback matrix K by solving the Riccati equation and set the initial state and the simulation time interval to simulate the system with the K feedback matrix. Finally, through the simulation of the system, the total state of each UAV and the formation time can be obtained under the current leader and target formation mapping.

In the second sub-problem of the leader selection algorithm, the mapping problem between followers and target formation has a time complexity of $O(n!)$. In the case of a small number of UAVs, the optimal solution can be obtained by enumeration. However, with the increase in the size and number of UAV formations, the search space of combinatorial optimization problems also increases exponentially. Sometimes it is difficult to find the optimal solution by enumeration method in calculation, and the problem of combinatorial explosion will occur. This problem is essentially an NP problem of determining the one-to-one mapping between the current UAV reaching the target position, because it is a mapping combinatorial optimization problem from one sequence to another. For such complex problems, genetic algorithm (GA) is one of the best tools to find satisfactory solutions. Practice has proved that genetic algorithm is very effective for NP problems in combinatorial optimization.

GA uses chromosomes to represent solutions to optimization problems, with each chromosome corresponding to a solution. It maintains a certain number of chromosomes to solve optimization problems, and this series of chromosomes is called a population. The first generation population is randomly generated. The fitness function is used to evaluate each individual in the population, and it is extracted from the optimization problem (5). Applied to our model, a solution represents a mapping (i.e., $\mathcal{P}_n^{l,k}, 1 \leq n \leq N$), and the population $P(\mathcal{G})$ is randomly selected from the set $P(\mathcal{S})$ according to the population size G. The fitness function is the formation time $t_n^{l,k}$ corresponding to the mapping in the case where the leader UAV_l and the order k of the leader are determined. Since GA is equivalent to narrowing the range of the set $P(\mathcal{S})$, the constraints of problem (5) will also change. So we reformulate the problem (5) into the following form:

$$\min_{l \in \mathcal{M}} (4)$$
$$\text{s.t.} \ \ k \in \mathcal{M}, \tag{11}$$
$$1 \leq j \leq M - 1,$$
$$\mathcal{P}_n^{l,k} \in P(\mathcal{G}), \ 1 \leq n \leq N,$$

where $P(\mathcal{G})$ is constantly changing with the number of iterations and genetic operator operations.

The main process of GA is mainly divided into the following steps. It first calculates the individual fitness value $t_n^{l,k}$ according to the mapping $\mathcal{P}_n^{l,k}, 1 \leq$

$n \leq N$ when the leader UAV_l and the order k of the leader is known, and then performs genetic operations such as selection, crossover and mutation according to the fitness value (i.e., the formation duration) to continuously update the population until the iteration is completed.

Based on LQR and GA, we propose a leader selection algorithm for UAV formation. The pseudo code is shown in Algorithm 1:

Algorithm 1. Leader Selection for Formation

Input:

    Each UAV's position p_i, velocity v_i and state $s_i, \forall i \in \mathcal{M}$,

    Simulation time interval Δt,

    Parameter matrix A, B, C, Q and R for LQR algorithm,

    Population size G, iterations I, selection, crossover and mutation

probability p_s, p_c, p_m for GA.

Output:

    Minimum formation time T and optimal leader UAV_l.

1: $T = \infty$, leader=1

2: **for** $i = 1; i < M; i++$ **do**

3:     // Let the current ith UAV as leader

4:     $p_l = p_i; v_l = v_i$;

5:     **for** $k = 1; k < M - 1; k++$ **do**

6:         Obtain the relative velocities $\mathcal{V}_r^{l,k}$ of all followers from (2);

7:         **while** $I > 0$ **do**

8:             **for** $m = 1; m < G; m++$ **do**

9:                 Obtain the target formation positions $\mathcal{P}_n^{l,k}, \mathcal{P}_n^{l,k} \in P(\mathcal{G})$;

10:                $\mathcal{P}_r^{l,k} = \mathcal{P}^l - \mathcal{P}_n^{l,k}$;

11:                time_temp = -∞;

12:                **for** $j = 1; j < M - 1; j++$ **do**

13:                     time_cost,state= LQR($s_j^{l,k}$);

14:                     **if** time_cost>time_temp **then**

15:                         time_temp = time_cost;

16:                         state_temp = state;

17:                     **end if**

18:                **end for**

19:                **if** T >time_temp **then**

20:                    T = time_temp;

21:                    state_final = state_temp;

22:                    leader=i;

23:                **end if**

24:             **end for**

25:             $I--$;

26:             Perform genetic operator to update the population $P(\mathcal{G})$;

27:         **end while**

28:     **end for**

29: **end for**

30: **return** $T, leader$;

The pseudo codes 9–13 of Algorithm 1 take the calculated relative state as a parameter into the LQR algorithm to obtain the formation duration and simulated motion state data of the followers. In pseudo codes 7–27 of this algorithm, the process of GA finding the optimal solution and updating the population is described.

4   Simulation Results

In this section, we show the simulation results and the performance of the proposed leader selection algorithm. Table 1 gives a summary of the simulation parameters for this algorithm.

Table 1. The simulation parameters.

Parameter	Value
Simulation time interval Δt	0.05s
The parameter matrix of LQR Q	$\mathbf{I_4}$
The parameter matrix of LQR R	$\mathbf{I_2}$
Population size G	100
Number of iterations	100
Selection probability p_s	0.9
Crossover probability p_c	0.8
Mutation probability p_m	0.06

We choose five UAVs as the experimental simulation objects, and make them all in the first quadrant. Then, the proposed leader selection algorithm for rotor UAVs is used to obtain the optimal leader and a group of motion states for the UAVs. Table 2 gives a summary of each UAV's initial values. Next, we show the experimental processes and results in detail.

Table 2. The simulation parameters of UAVs.

UAV_i	p_i	v_i
UAV_1	$[50\ 60]^T$	$[10\ 10]^T$
UAV_2	$[70\ 20]^T$	$[-10\ 20]^T$
UAV_3	$[40\ 20]^T$	$[30\ 10]^T$
UAV_4	$[80\ 50]^T$	$[20\ 20]^T$
UAV_5	$[60\ 60]^T$	$[-20\ 10]^T$

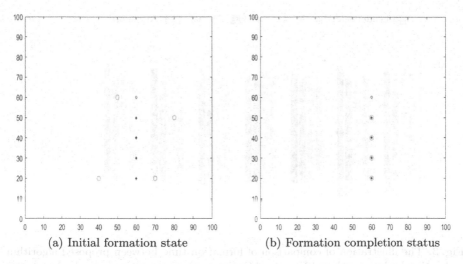

(a) Initial formation state   (b) Formation completion status

Fig. 2. Illustration of the UAVs motion state designed by leader selection algorithm, where blue dots indicate TF, and the green circles represent UAVs.

Based on the parameters of each UAV in Table 2, we can get the minimum formation duration set {268,273,274,274,267} when each UAV is regarded as the leader. Therefore, the optimal leader can be obtained as UAV_5. Then let the leader selected by this algorithm lead the followers to reach the target formation. The formation process led by the optimal leader is shown in Fig. 2. We assume that the leader is stationary and generate the UAV formation in the relative state. It can be clearly seen that in the process of changing from the random initial state to the target vertical formation.

We compare the formation duration of our algorithm with the optimal solution's duration, as shown in Fig. 3. We found that these cases are basically similar to the optimal solution. It shows that our proposed algorithm can obtain approximately optimal duration. In addition, we also compare the formation duration obtained by our proposed algorithm with the average formation duration. Our proposed algorithm can greatly reduce the formation generation time and form target formations.

In addition, we apply the GA algorithm to calculate an approximate solution to solve the combinatorial explosion problem. It can be seen from Fig. 4 that in small-scale UAV formations, our proposed algorithm executes slowly, which is due to the population size and iterations of the GA algorithm. However, as the size of the UAV increases, it has little effect on the execution time of our algorithm. Instead, the optimal solution has high time complexity, and the execution time grows exponentially. When the number of UAVs reaches more than 10, the optimal solution can no longer be executed on MATLAB because it occupies too much memory. It can be seen that the leader selection algorithm can solve this problem very well.

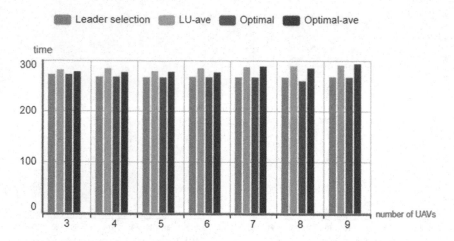

Fig. 3. The illustration of comparison of formation time between proposed algorithm and optimal solution, where LU-ave and Optimal-ave represent respectively the average formation duration of the two scenarios.

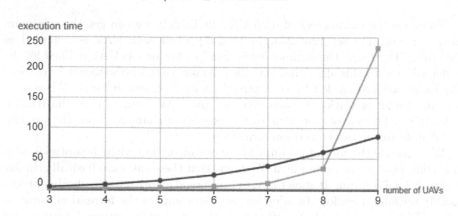

Fig. 4. The illustration of comparison of the execution time between the optimal solution and the leader selection algorithm with the increase of UAV's size

5   Conclusion

In this paper, we studied the leader selection problems of formation based on rotor UAVs. We considered the case where UAVs were regarded as mass points, and controled the motion states to realize the formation. We designed a leader-based relative motion system model, which could transform the complex motion of multiple UAVs into the motion of a single individual, and reduced the amount of computation. We proposed the leader selection algorithm to choose best leader and described the process of formation. The experimental results showed that this algorithm could effectively shorten the formation time and solved the combined explosion problem in the UAV formation mapping problem. In addition, we will study UAVs formation problems of virtual leader selection in our future works.

References

1. Farooq, M.U., Ziyang, Z., Ejaz, M.: Quadrotor UAVs flying formation reconfiguration with collision avoidance using probabilistic roadmap algorithm. In: International Conference on Computer Systems, Electronics and Control (ICCSEC), pp. 866–870 (2017)
2. Rui, P.: Multi-UAV formation maneuvering control based on Q-Learning fuzzy controller. In: 2nd International Conference on Advanced Computer Control, pp. 252–257 (2010)
3. Fu, X., Zhang, J., Chen, J., Wang, S.: formation flying and obstacle avoidance control of UAV cluster based on backbone network. In: IEEE 16th International Conference on Control & Automation (ICCA), pp. 859–863 (2020)
4. Sial, M.B., Wang, S., Wang, X., Wyrwa, J., Liao, Z., Ding, W.: Mission oriented flocking and distributed formation control of UAVs. In: IEEE 16th Conference on Industrial Electronics and Applications (ICIEA), pp. 1507–1512 (2021)
5. Luo, D., Xu, W., Wu, S., Ma, Y.:UAV formation flight control and formation switch strategy. In: 8th International Conference on Computer Science & Education, pp. 264–269 (2013)
6. Olfati-Saber, R.: Flocking for multi-agent dynamic systems: algorithms and theory. IEEE Trans. Autom. Control **51**(3), 401–420 (2006)
7. Saber, R.O., Murray, R.M.: Flocking with obstacle avoidance: cooperation with limited communication in mobile networks. In: Proceedings of the 42nd IEEE Conference on Decision and Control, vol. 2, pp. 2022–2028, December 2003
8. Fu, X., Zhang, J., Chen, J., Wang, S.: Formation flying and obstacle avoidance control of UAV cluster based on backbone network. In: IEEE 16th International Conference on Control & Automation (ICCA), pp. 859–863 (2020)
9. Sial, M.B., Wang, S., Wang, X., Wyrwa, J., Liao, Z., Ding, W.: Mission oriented flocking and distributed formation control of UAVs. In: IEEE 16th Conference on Industrial Electronics and Applications (ICIEA), pp. 1507–1512 (2021)
10. Godsil, C., Royle, G.: Algebraic Graph Theory, Vol. 207 of Graduate Texts in Mathematics. Springer-Verlag, New York (2001). https://doi.org/10.1007/978-1-4613-0163-9

Dynamic Monitoring Method Based on Comparative Study of Power and Environmental Protection Indicators

Xiaojiao Liang(✉), Chunling Ma, Chuanguo Ma, and Shaofei Xin

State Grid Shandong Electric Power Company Dongying Power Supply Company,
Dongying, China
bendideren@163.com

Abstract. In order to promote my country's comprehensive green development, this paper conducts a research on the correlation mapping between electricity consumption and environmental protection data. Based on the existing short-term electricity load forecasting model, the electricity load forecast data is divided into active electricity and reactive electricity, so as to realize the comprehensive tracking of the operation status of the enterprise's electrical equipment. The comparative learning algorithm is used to obtain the association mapping of the electricity consumption-environmental data distribution, so that the Encoder data of the pollution discharge data can be obtained based on the electricity consumption. Subsequently, the generator generates enterprise short-term emission forecast data based on the coded data, and combines the Wasserstein distance with the mean square error of similar samples to construct a loss function to improve the quality of the generated data. Analysis and comparison of calculation examples through simulation experiments has show the effectiveness and feasibility of the proposed.

Keywords: Electricity load · Environmental protection index · Comparative study · GAN · Monitoring warning

1 Introduction

At present, the environmental protection testing index data of enterprises has the characteristics of many points, wide area, and long line. Relying on the inspection and supervision of the national law enforcement department can often only make up for it. Moreover, the single monitoring method inevitably has the occurrence of fraudulent monitoring data and destruction of monitoring equipment. As an essential energy in the production activities of enterprises [1], which has mirror image and high coverage, which can timely, accurately and comprehensively reflect the production status of enterprises and the use of environmental protection equipment [2]. Accurate short-term pollution forecast can assist environmental protection departments to formulate response strategies in advance to avoid the occurrence of severe events such as smog and acid rain. Due to the inevitable correlation between input energy and production, this paper analyzes

H. Zhang et al. (Eds.): NCAA 2022, CCIS 1637, pp. 56–70, 2022.
https://doi.org/10.1007/978-981-19-6142-7_5

the correlation between electricity consumption data and environmental protection indicators of key pollutant discharge monitoring enterprises. Through comparison mapping and data fitting, the short-term forecast of enterprise pollutant discharge can be realized.

At present, relevant institutions have begun to pay attention to the prediction of enterprise pollution discharge based on enterprise electricity consumption. However, the research content is relatively small. Paper [3] uses regression analysis method to build an environmental impact prediction model based on industry electricity consumption, and realizes the monitoring and early warning of air pollution prevention and control for key enterprises. In addition, Paper [4] studies air pollution prevention and audit methods based on big data visualization technology, and analyzes the relationship between enterprise electricity consumption data and air pollution; Paper [5] analyzes the impact of air pollution on short-term electricity consumption on the demand side. Paper [6] adjusts the power generation control strategy according to the real-time monitoring data of pollution emissions, thereby reducing the amount of air pollution emissions. The above literatures all show that there is a correlation between electricity consumption data and environmental protection data. However, no effective mapping relationship between electricity consumption and environmental protection has been established. At present, relatively accurate short-term forecasts of electricity consumption data can be achieved, and short-term forecasts of pollutant emissions can be completely based on this. Based on the existing research foundation and technology, the article carries out short-term forecast of enterprise pollutant discharge. The main improvements are as follows:

(1) Industrial electricity involves rotating high-power electrical appliances such as motors, so production equipment will generate reactive power. Even with compensating devices, there will be power fluctuations [7]. Therefore, in order to measure the operation status and production volume of production equipment, the electricity load is divided into reactive power and active power.
(2) In order to maximize the correlation analysis of electricity consumption and environmental protection characteristics, the article adopts comparative learning [8] to realize the correlation mapping between electricity consumption and environmental protection features. By zooming in similar instances in the projected space and pushing away dissimilar instances, the distance distribution of mapping codes preserves the differentiated pollutant discharge information of enterprises. Even when enterprises lack sample data, they can make reasonable predictions based on similar samples and their own characteristics.
(3) In order to improve the global similarity and consistency between the predicted data and the actual data, the Wassersein distance is used when the predicted data is generated based on the mapped coded data combined with the mean square error of similar samples, the loss function of the generator is constructed, and the data distribution of the generated curve is further drawn closer to the real data, effectively avoiding the distortion of the predicted data.

2 Short-term Emissions Forecasting Process

This paper obtains electricity consumption and environmental protection monitoring data based on related systems, and uses web crawling technology to obtain data on weather, holidays and emission reduction policies. After all kinds of data are preprocessed, similar enterprises are classified [9]. The overall data processing flow as follows (Fig. 1):

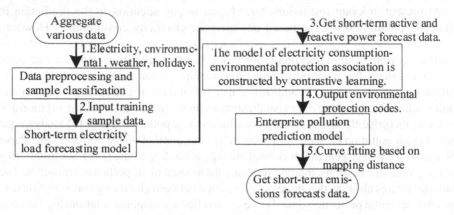

Fig. 1. Short-term emission forecast

First, in order to obtain the short-term active and reactive power forecast data, the short-term electricity load prediction model based on two-layer LSTM [10] processes the input data which includes active power, reactive power, meteorological data, and holiday data. Subsequently, this paper uses the comparative learning algorithm to realize the correlation mapping between electricity load and environmental protection index data, so as to construct a similar mapping model of electricity consumption-environmental protection correlation. Contrastive learning can achieve differentiated feature extraction based on data differences between enterprises. After the above prediction data is input into the model, the encoder of the pollutant discharge prediction data is output. Finally, the encoder and the recent actual sewage monitoring data of the enterprise are input into the enterprise sewage prediction model, which uses the combination of Wasserstein distance and similar historical sample data as the loss function. The model can predict the short-term pollutant discharge data of enterprises based on historical electricity consumption data and pollutant discharge data.

2.1 Building a Short-Term Electricity Load Forecasting Model

In this paper, the LSTM algorithm is used to build a short-term electricity load prediction model. In order to effectively track the operation status of enterprise equipment, the historical electricity load data of the enterprise is divided into active power and reactive power data. The model training data set contains enterprise history active power, reactive power, data, the temperature data, holiday data, The output is the predicted value of short-term active and reactive power. Take the company's active power and reactive power data at time t, combined with temperature data, holiday data, and emission reduction policy data (such as corporate pollution exceeding the standard and being restricted from running) as the data input of the LSTM model, and the input layer is 100 memory neurons.

The predicted output value of the active power and reactive power of the previous neural unit is h_{t-1}, and each neuron input has a corresponding weight. After sigmoid activation, a value of 0–1 is obtained, that is, three gate values. The calculation formula as follows:

$$q_t = \sigma(W_q x_t' + U_q h_{t-1})$$
$$r_t = \sigma(W_r x_t' + U_r h_{t-1}) \qquad (1)$$
$$o_t = \sigma(W_o x_t' + U_o h_{t-1})$$

where the output value of the input gate is q_t, the output value of the forget gate is r_t, and the output value of the output gate is o_t, W_q, U_q, W_r, U_r, W_o, U_o are the corresponding parameters of each gate.

The input value x_t' and the output h_{t-1} of the previous unit have corresponding weights W_c and U_c, and the tanh activation function is used. The output value is equivalent to obtaining the new memory value $\tilde{c}_t$ obtained based on the input value at time t. The calculation formula is as follows:

$$\tilde{c}_t = \tanh(W_c x_t' + U_c h_{t-1}) \qquad (2)$$

$$c_t = r_t \circ c_{t-1} + q_t \circ \tilde{c}_t \qquad (3)$$

where r_t is the output value of the forget gate at time t, q_t is the output value of the input gate at time t, and c_{t-1} is the final memory value of the previous neuron at time t-1. The final output value of h_t is

$$h_t = o_t \circ \tanh(c_t) \qquad (4)$$

The hidden layer adopts the same single-layer structure as the input layer, the output layer is a two-dimensional output, and the output data is the predicted value of the short-term enterprise active power and reactive power $\hat{x}_t$.

2.2 Constructing a Similarity Mapping Model for Electricity Consumption and Environmental Protection

The historical active power, reactive power data and historical sewage monitoring data of the enterprise constitute training data set τ, and M sample data of enterprises are randomly selected from the training data set to form a branch training set A, which generates a total of two branches that act on the upper and lower branches. The two training sets are the historical sewage monitoring data set A_1, the historical active power data and the historical reactive power data A_2. Each branch contains the sample data of M enterprises, the sample data of the upper branch is fixed, while the data of the lower branch is dynamically updated. Randomly extract a certain type of historical pollutant discharge monitoring data $P^t = [p_1^t, p_2^t, \cdots, p_n^t]$ of a certain enterprise and input it into the upper branch, and input the active power and reactive power data at the corresponding time into the lower branch.

$$X^t = \begin{bmatrix} x_1^t & x_2^t & \cdots & x_n^t \end{bmatrix} = \begin{bmatrix} x_{A1}^t, x_{A2}^t, \cdots, x_{An}^t \\ x_{R1}^t, x_{R2}^t, \cdots, x_{Rn}^t \end{bmatrix} \qquad (5)$$

where t represents the daily sampling time point p_1^t, x_{A1}^t and x_{R1}^t are the same dimension column vector of the sampling sequence, n represents the daily monitoring sampling number of a certain enterprise, p_i^t and x_i^t are positive examples of each other, $i, j \in (1, 2, \cdots, n)$. During training, any other input sample data p_j and x_j of the same enterprise or different enterprises in the upper branch and the lower branch are negative examples of p_i^t and x_i^t.

Subsequently, a representation learning architecture is constructed by which the training data is projected into a hyperplane representation space. In order to shorten the distance of positive examples and push the distance of negative examples farther, the upper and lower branches adopt a two-tower asymmetric structure, and the internal levels and the number of neurons are different. Historical sewage monitoring data set A_1, and the data enters the upper branch, through the feature Encoder (using the Transform encoder Encoder structure), the Encoder module has a total of 4 layers, thereby mapping the input sample data to a hyperplane Represents the vector z_p in the space. The architecture of the first part of the lower branch is the same as that of the upper branch, but the number of neurons in the middle layers is different, and the parameters are not shared. The output is represented as $h_x = f_{\theta'}(x)$, and then input the nonlinear transformation structure Projector, the output is $z_x = g_{\theta'}(h_x)$. The matrix internal structure of vectors z_x and z_p is consistent with the quantity (Fig. 2).

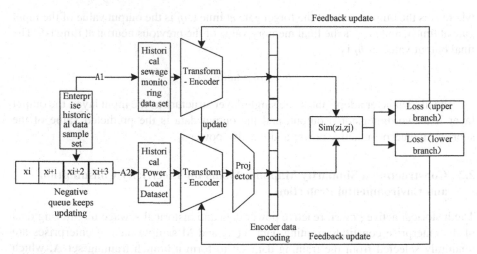

Fig. 2. Training process of electricity-environmental protection correlation mapping.

In the representation space, in order to realize the correlation analysis between the electricity data and the environmental protection indicators, it is necessary to make the representation vector z after the mapping of the positive electricity data and the pollutant discharge monitoring data overlap or be as close as possible, and the distance from the negative sample is farther away. Therefore, L2 is used as the distance metric to represent the similarity between space vectors $S(z_i, z_j)$:

$$S(z_i, z_j) = \frac{z_i^T z_j}{\|z_i\|_2 \|z_j\|_2} \tag{6}$$

In order to make the distance between positive examples close and the distance between any negative examples far in the representation space, the following loss function is used. Upper branch loss function:

$$loss = -\log \frac{e^{S(z_{xi}, z_{pj})/\tau}}{\sum_{j=1, i \neq j}^{M \times n} e^{S(z_{pi}, z_{pj})/\tau}} \tag{7}$$

Lower branch loss function:

$$loss = -\log \frac{e^{S(z_{xi}, z_{pj})/\tau}}{\sum_{j=1, i \neq j}^{M \times n} e^{S(z_{xi}, z_{xj})/\tau}} \tag{8}$$

where τ is the temperature hyperparameter, which represents the difficulty recognition penalty coefficient, which is used to widen the distance between positive samples and negative samples with high similarity, so as to avoid model collapse due to parameter adjustment based on high-difficulty negative samples. $S\left(z_{x_i}, z_{p_i}\right)$ represents the similarity between positive samples, that is, the similarity between the electricity data samples and the pollutant discharge monitoring data samples of an enterprise in the same day, $S\left(z_{P_i}, z_{p_j}\right)$ represents the similarity between positive and negative samples of sewage monitoring, $S(z_{x_i}, z_{xj})$ represents the similarity between positive and negative samples of electricity data, that is, the positive electricity data samples and other negative electricity data samples of an enterprise in the same time period, or The similarity between the positive and other negative pollution monitoring data samples. The smaller the objective loss function, the better. Therefore, the larger the numerator, the greater the similarity between positive examples. While sample data of different enterprises or different time periods are similar in spatial mapping. so as to maintain the individual differences of the sample data while realizing the association analysis.

2.3 Construction of Enterprise Pollution Prediction Model

In the training stage of the enterprise pollutant discharge prediction model, if the similarity between the Encoder data of the input sample and the Encoder data of the training sample is within a certain threshold range, it means that the similarity between the two is high, that is, the similar sample data exists in the historical pollutant discharge data sample. Therefore, in the process of data fitting, the influence of the weight of similar samples increases, and the pollutant discharge monitoring data is more inclined to fit the pollutant discharge forecast data with similar samples; if the similarity exceeds the threshold, it means that the existing enterprise pollutant discharge samples are all different from the input samples, so the loss is The function is more inclined to generate pollution prediction data through the generator based on the encoder output encoding of electricity. The training process for this part is as follows (Fig. 3):

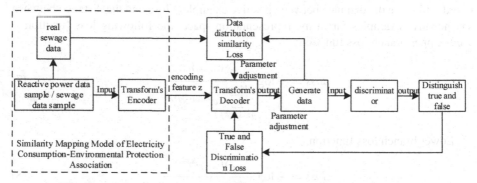

Fig. 3. Training process for generating pollutant emission prediction data.

In order to measure the distance between the prediction and the real data, this paper uses the Wasserstein distance as a component of the loss function. Wasserstein distance is used to measure the distance between two probability distributions. That is, the shortest distance to move date from distribution 1 to distribution 2. The advantage over KL divergence and JS divergence is that the distance can still be measured when the two data distributions have no or negligible overlap. At this time, the JS divergence is constant, and the KL divergence is meaningless. The Wasserstein distance is calculated the loss function as follows:

$$loss = E\left(\underset{Encoder}{\mathrm{var}}\,(p_1, p_2, \cdots, p_h, \tilde{p})\right)$$
$$+ \ell \underset{\tilde{p} \sim P_g, p \sim P_r}{E}\left[D(p, \tilde{p})\right]$$
$$+ \lambda \underset{\hat{p} \sim P_{\hat{p}}}{E}\left[\left(\left\|\nabla_{\hat{p}} D(p, \hat{p})\right\|_2 - 1\right)^2\right] \tag{9}$$

where P_g is the set of generated data, P_r is the set of actual data, $\tilde{p}$ is the generated data sample, p is the actual data sample, and $p_1, p_2, \cdots, p_h$ is the similar sample encoded by the Encoder of the actual data sample p, and represents the sample cluster set with high similarity of historical pollutant discharge data. It can be determined by setting the threshold. $\underset{Encoder}{\mathrm{var}}\,(p_1, p_2, \cdots, p_h, \tilde{p})$ represents the mean square error between the generated data and the historical pollutant discharge data of similar enterprises. By minimizing this value, the approximate data fitting generation of the pollutant discharge data of similar enterprises is realized. ℓ is the weight data generated based on the data volume b of similar enterprises. $\ell = a/e^b$, a is a custom parameter. When there is a lot of similar enterprise data, the value of ℓ decreases. At this time, the generated data is more inclined to rely on the pollutant discharge data of similar enterprises. When there is little or no similar enterprise data, the value of ℓ increases, and the generated data depends on the generator. $\lambda \underset{\hat{p} \sim P_{\hat{p}}}{E} \left[\left(\left\| \nabla_{\hat{p}} D(p, \hat{p}) \right\|_2 - 1 \right)^2 \right]$ is the weight gradient penalty, which is used to ensure that the generated data is close to the real sample data, but does not exceed the real sample data, $\left\| \nabla_{\hat{p}} D(p, \hat{p}) \right\|_2$ is the gradient penalty coefficient, and $\hat{p}$ is the sampling composition data between the generated data and the real data. Gradient penalty to stabilize the gradient change, $\hat{p} = \varepsilon p + (1 - \varepsilon)\tilde{p}$, λ is the weight; $\underset{\tilde{p} \sim P_g, p \sim P_\gamma}{E} \left[D(p, \tilde{p}) \right]$ is the average Wasserstein distance between all generated data in the set and its corresponding actual data samples, $D(p, \tilde{p})$ is the Wasserstein distance between the generated samples and the actual samples, the calculation formula is as follows:

$$D(p, \tilde{p}) = \min_{\sigma \in \sum_N} \left(\sum_i \left\| p_i - \tilde{p}_{\sigma(i)} \right\|^2 \right) \tag{10}$$

where p_l represents the actual sewage monitoring data at time t, $\tilde{p}$ is the generated data, $\tilde{p}_{\sigma(t)}$ is the real data at time t mapped to the generated sewage monitoring data at time $\sigma(t)$ when the data was generated, since the actual data and the generated data have the same number of elements, and the data measures are the same, so the Wasserstein distance is measured by the quadratic difference, and $\sum_N$ represents the set formed by all the sorted arrangements of N elements.

The training process is as follows:

(1) Initialize generator parameters θ_g and discriminators θ_d;
(2) In the iterative training process, the generator is trained once, and the discriminator is trained several times. The process is as follows:

Train the discriminator, fix the generator parameters, and update the maximizing discriminator $loss_d$:

$$loss_d = E\left(\underset{Encoder}{var} (p_1, p_2, \cdots, p_h, \tilde{p}) \right)$$

$$+ \ell \underset{\tilde{p} \sim P_g, p \sim P_r}{E} \left[D(p, \tilde{p}) \right]$$

$$+ \lambda \underset{\hat{p} \sim P_{\hat{p}}}{E} \left[\left(\left\| \nabla_{\hat{p}} D(p, \hat{p}) \right\|_2 - 1 \right)^2 \right] \tag{11}$$

$$\theta_d \leftarrow \theta_d + \eta \nabla loss_d \tag{12}$$

where η is the learning rate.I In order to realize the curve fitting process based on similar sample data, the mean square error discrimination condition of historical similar samples is added. Train the generator, fix the discriminator parameters, and update the minimized generator $loss_g$:

$$loss_g = \underset{\tilde{p} \sim P_g, p \sim P_r}{E} \left[D(p, \tilde{p}) \right] \tag{13}$$

$$\theta_g \leftarrow \theta_g - \eta \nabla loss_g \tag{14}$$

After the training is completed, the electricity consumption-environmental protection correlation similarity mapping model inputs the electricity data and some pollutant discharge sampling point data, outputs the Encoder code. These codes into the enterprise pollutant discharge prediction model, and outputs the enterprise pollutant discharge forecast data.

3   Case Analysis

3.1   Sample Data Preprocessing

The data collection frequency of enterprise power consumption data and environmental protection data is 24 points/day. The power collection data is the historical active power data and reactive power data obtained based on the multi-function energy meter. Reactive power is the power consumption generated by converting electrical energy into capacitance or inductance during the generation of enterprise operating equipment, including forward reactive + reverse reactive power. The environmental monitoring indicator is the nitrogen oxide concentration in the air pollution indicator. Data preprocessing is performed on the acquired data. The 2385 key sewage monitoring enterprises in the province are classified and modeled based on the industry, and the historical data from January 2019 to December 2020 is selected as the model training sample set, and the historical data from January 2021 to May 2021 is used as the model. To verify the sample set, in order to avoid the imbalance of sample categories, only 2/5 of the historical samples of the normal production of the enterprise are selected for training and verification, and the historical data in mid-June 2021 is used as the model test data sample set to realize the enterprise within a short period of one week. Evaluating the effect of pollutant discharge forecasting.

3.2   Data Spatial Distribution Visualization

In the training phase, after the active, reactive power and NOx data are input into the electricity consumption-environmental protection correlation similarity mapping model, the Encoder code of the enterprise pollution prediction model generator is obtained from the upper and lower branches. In order to visualize the spatial distribution law of NOx data extracted by the comparative learning algorithm, that is, on the premise of retaining the detailed information of enterprise sewage, the clustering and classification of sample data are realized, and the representation space is relatively uniformly distributed, and electricity data is used as an enhanced sample data, through comparative learning to achieve the fitting of the distribution law of nitrogen oxide emission monitoring data in the same period.

The article uses a nonlinear dimensionality reduction algorithm, that is, t-distributed random neighbor embedding to reduce the 6-dimensional Encoder data to three-dimensional space, and uses matlab to correlate the Encoder codes generated by the NOx data and electricity consumption data in the three-dimensional space, as shown in the following figure:

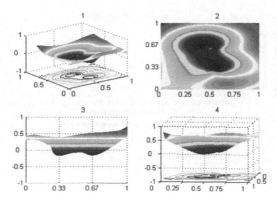

Fig. 4. 3D view of encoder data.

In Fig. 4, view 1 is the azimuth $-37.5°$, the elevation $30°$. View 2 is the azimuth $0°$, the elevation $90°$. View 3 is the azimuth $90°$, the elevation $0°$. View 4 is the azimuth $-7°$, elevation $-10°$. The upper branch obtains the Encoder of the enterprise pollutant discharge prediction model generator, which is located in the upper area of view 2 and the left area of view 3. The lower branch obtains the Encoder of the enterprise pollutant discharge prediction model generator based on the active and reactive power data. The encoding is located in the lower area of view 2 and the right area of view 3. Figure 4 shows the symmetry and correlation of the Encoder code generated by the "twin-tower" structure from different perspectives. Since only 829 key pollutant discharge enterprises can realize daily monitoring of nitrogen oxides, the remaining 1,556 key pollutant discharge monitoring enterprises have upper branches. The NOx data samples are missing, limited by the limited amount of NOx data, which leads to deviations in the code of the lower branch fitting the sewage data based on the electricity consumption data, as shown in the yellow and red data areas in Fig. 4, this part of the area of the average nitrogen oxide emission index per unit of electricity of enterprises is lower than 5.89, and most of them are manufacturing enterprises. Therefore, the daily monitoring data is insufficient, so without corresponding pollution data mapping. Figure 4 shows that through comparative learning, the Encoder data distribution of the pollutant discharge index data can be extracted based on the power consumption data of the enterprise, so as to provide the recurring code of the pollutant discharge data for the enterprise pollutant discharge prediction model generator.

3.3 Comparative Analysis of Results

In the process of model data processing, the article mainly improves from the following three points:

(1) Divide the power load data into active and reactive power data.
(2) The input code of the enterprise pollutant discharge prediction model generator is obtained by comparing the electricity consumption data with NOx data.
(3) The loss function uses the Wasserstein distance and combines the dynamic weights to adjust the influence of similar samples to achieve curve fitting.

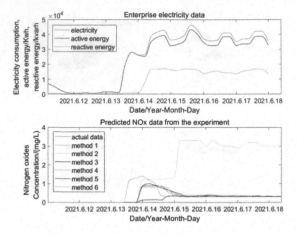

Fig. 5. Comparison chart of experimental results.

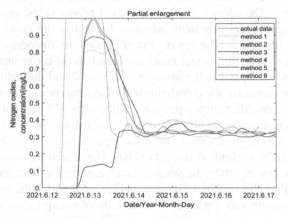

Fig. 6. Partial enlarged view of forecast data.

As shown in Fig. 5 and Fig. 6, after the electricity consumption-environmental protection correlation similarity mapping model and enterprise pollution prediction model are trained, the electricity consumption data is input into the electricity consumption-environmental protection correlation similarity mapping model, and the Encoder data is output, and then the Encoder data is input. Enterprise pollutant discharge forecast model, output forecast data of enterprise nitrogen oxide discharge. The example enterprise is a steel product processing enterprise, so it is less affected by external weather and holidays. The enterprise is based on the production plan. Therefore, the short-term electricity consumption forecast data is basically consistent with the actual electricity consumption data, so the actual electricity consumption is not shown in the experiment. Minor deviation of data from predicted electricity consumption data. The comparison experiment method is as follows:

Method 1: without comparative learning. Under the condition of sufficient NOx training data, the active and reactive power data as model input, and the experimental model uses the VAE-GAN method;

Method 2: Unlike method 1, NOx sample is insufficient;

Method 3: The NOx sample is sufficient, electricity load as the input data;

Method 4: The NOx sample is sufficient, the method in the text is adopted;

Method 5: The NOx sample is insufficient, the method in the text is adopted;

Method 6: The NOx sample is sufficient, Wassersein distance replaced by kl.

As shown in Fig. 5 and Fig. 6, the method 4 and method 5 using the method in this paper have a high similarity to the curve fitting of the actual monitoring data, especially when the method 4 has sufficient training data for nitrogen oxide monitoring, the fitting best effect; Method 1 does not use comparative learning, but adopts the VAE-GAN method. The active and reactive electricity data is used as the model input, and the NOx data is used as the learning and reproduction data for training. Under the condition of sufficient training samples, the effect is also relatively Ideal, but in the case of insufficient NOx training sample data, as shown in Method 2, fitting prediction will be made based on the emission data of enterprises with similar power consumption scales, rather than the comparison of electricity consumption and environmental protection indicators in the article Learning to obtain the similarity code, generating a large deviation in the prediction results based on the similarity of electricity consumption only, and failing to capture the unique information of the enterprise's own sewage; Method 3 is to use the electricity load instead of the active and reactive power data in the text as the input of the model, as shown in the Fig. 5, at 8:00 on June 13, 2021, the enterprise enters the production and operation state, and the simulation demonstrates the decontamination The equipment is not turned on, causing the actual nitrogen oxide index to exceed the standard. The reactive power data can better reflect this imagination, while the overall power load data cannot achieve effective prediction based on the operation of the equipment; Method 6 uses the kl divergence method for the loss function. Although the kl divergence can reflect the distribution of the data well, it is difficult to track the lag characteristics of the sewage data, and the model training process takes a long time and the parameter adjustment is difficult, so it is easy to generate The phenomenon that the gradient is zero, as shown in Fig. 6, the fitting effect of the model after the training of method 6 has a certain deviation, and because the KL divergence distance is difficult to measure the lag of the data, the KL divergence data in Table 1 not counted.

This paper selects 295 key polluting enterprises for 7-day nitrogen oxide emission concentration prediction, uses RMSE (Root Mean Squared Error) root mean square error (normalized mean), and evaluates the effect of emission concentration prediction on the results of 12 experiments. The KL divergence is used to measure the information loss of a predicted data distribution compared to the actual data distribution. The smaller the data, the higher the curve fitting degree, and the fitting curve is integrated to obtain the forecast of the cumulative pollutant discharge of the enterprise. The prediction accuracy of the predicted value and the actual pollutant discharge is evaluated by the mean absolute percentage error (MAPE, Mean Absolute Percentage Error), as shown in the following Table 1:

Table 1. Curve similarity evaluation table

Method name	RMSE	KL divergence	MAPE
Method 1	3.08%	$2.34*10-2$	10.65%
Method 2	68.30%	1.58	462.65%
Method 3	10.68%	1.43	56.33%
Method 4	1.03%	$7.8*10-3$	4.56%
Method 5	1.74%	$9.88*10-3$	6.73%
Method 6	15.17%	/	78.82%

According to the RMSE and KL divergence, the curve fitting of the predicted value obtained by the method 4 and method 5 described in this paper is the best. A MAPE of 0% indicates a perfect model, and a MAPE greater than 100% indicates an inferior model. Table 1 shows that Method 4 and Method 5 can predict relatively accurate corporate pollutant emissions. In summary, the method in this paper can effectively realize the prediction of pollutant discharge monitoring data based on electricity consumption data.

In order to further verify the impact of the similar enterprise sample data in the comparative learning and loss function on the prediction model. In this paper, under the assumption that an enterprise lacks NOx samples, the effect of the number of similar enterprises on generating forecast data is tested. The L1 distance and mean error are obtained by comparing with the real data, as shown in the following figure (Fig. 7):

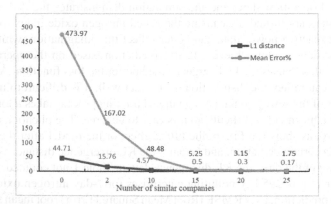

Fig. 7. Similar enterprise data affects the generation of forecast data.

The above figure shows that in the absence of a company's NOx samples and no similar companies, that is, only based on electricity consumption data as input and based on GAN to generate pollution forecast data, the L1 distance and mean error are both bad. However, when the number of similar enterprises is 2, both the L1 distance and the mean error decrease rapidly. With the increase of similar enterprises, the mapping between the sample data in the comparative learning process is more dense, so that the encoded data can be generated based on the nearest similar samples. The sample weights of similar enterprises in the loss function can assist the model to generate data on the basis of similar enterprise data, and further ensure the stability of the data generated by GAN.

4 Conclusion

Through the comparative learning algorithm, the correlation mapping between electricity consumption data and environmental protection indicators is realized., the lower branch electric power data of the positive example enhanced sample of the upper branch environmental protection data fits the distribution of the corresponding positive example mapping data, and realizes that the positive examples attract and the negative examples repel each other, and there is no corresponding upper branch positive example or the electricity data samples with few positive examples of sewage discharge in the upper branch are relatively evenly distributed based on the similarity and distance of environmental protection sampling data and electricity data between enterprises, so as to ensure that when the training samples of pollution discharge data monitoring are missing, similar enterprise data can be obtained based on the electricity consumption-environmental protection correlation mapping, and the pollutant discharge data can be fitted by the generated data and the similar enterprise data to generate the short-term pollutant discharge forecast data of the enterprise.

Since it is still in the initial stage of exploration to predict the pollutant discharge of enterprises based on electricity consumption data. And some companies have the problem of missing environmental protection data, which affects the stability of the generated data by GAN. In the future, with the increase of environmental monitoring data of various enterprises, the distribution of similar sample data will be more dense, which can further improve the quality of generated data. In addition, this paper only predicts the nitrogen oxide indicator data. With the increase of point source emission monitoring data, the prediction of multiple indicators can be explored, so as to provide technical research reference for environmental monitoring.

References

1. Du, Y., Yang, X., Guo, L., et al.: Coupling evaluation of green development and safety and efficiency of distribution network incontext of double carbon. Sci. Technol. Eng. **21**(30), 12973–12981 (2021)
2. Matindife, L., Sun, Y., Wang, Z.: Disaggregated power system signal recognition using capsule network. In: Zhang, H., Zhang, Z., Wu, Z., Hao, T. (eds.) NCAA 2020. CCIS, vol. 1265, pp. 345–356. Springer, Singapore (2020). https://doi.org/10.1007/978-981-15-7670-6_29
3. Sun, K., Chen, Z., Fan, M., et al.: Design of special action plan for pollution prevention and control of key enterprises based on electric power big data mining. Distrib. Utilization **38**(04), 28–36 (2021)
4. Chen, W., Gao, J.: Research on air pollution prevention audit method based on big data visualization analysis technology. Finan. Acc. **04**(07), 65–68 (2019)
5. He, Z., Zhang, T., Hu, Y., et al.: Short-term electric load forecasting model considering the influence of air pollution prevention and control policy. Smart Power **47**(05), 1–9 (2019)

6. An, J., Chen, Q., Dai, F., et al.: Power green dispatch strategies for air pollution prevention and control. Power Syst. Technol. **45**(02), 605–612 (2021)
7. Babuta, A., Gupta, B., Kumar, A., et al.: Power and energy measurement devices: a review, comparison, discussion, and the future of research. Measurement **172**, 108961 (2021)
8. Ting, C., Simon, K., Mohammad, N., et al: A simple framework for contrastive learning of visual representations. In: Proceeding of the 37th International Conference on Machine Learning. Cambridge MA: JMLR, pp. 1597–1607 (2020)
9. Huang, S., Kang, Z., Xu, Z.: Deep K-Means: a simple and effective method for data clustering. In: Neural Computing for Advanced Applications. NCAA 2020. Communications in Computer and Information Science, vol.1265, pp. 272–283 (2020) https://doi.org/10.1007/978-981-15-7670-6_23
10. Xu, Y., Chhim, L., Zheng, B., Nojima, Y.: Stacked deep learning structure with bidirectional long-short term memory for stock market prediction. In: Zhang, H., Zhang, Z., Wu, Z., Hao, T. (eds.) NCAA 2020. CCIS, vol. 1265, pp. 447–460. Springer, Singapore (2020). https://doi.org/10.1007/978-981-15-7670-6_37

Combustion State Recognition Method in Municipal Solid Waste Incineration Processes Based on Improved Deep Forest

Xiaotong Pan[1,2] ⓘ, Jian Tang[1,2(✉)] ⓘ, Heng Xia[1,2] ⓘ, Weitao Li[3] ⓘ, and Haitao Guo[1,2] ⓘ

[1] Faculty of Information Technology, Beijing University of Technology, Beijing 100124, China
freeflytang@bjut.edu.cn
[2] Beijing Laboratory of Smart Environmental Protection, Beijing 100124, China
[3] School of Electrical Engineering and Automation, Hefei University of Technology, Anhui Province 230009, China

Abstract. It is important to accurately identify the combustion state of the municipal solid waste incineration (MSWI) processes. Stable state not only can greatly improve the combustion efficiency, but also can ensure safety of the MSWI processes. What's more, the pollution emission concentration would be greatly reduced. Aiming at the situation that domain experts identify the combustion state in terms of self-experience in the actual MSWI processes, this study proposes an efficient method based on improved deep forest (IDF). First, the image preprocessing methods such as defogging and denoising, were used to preprocess the combustion flame image to obtain a clear one. Then, the multi-source features (brightness, flame and color) were extracted. Finally, the multi-source features were used as the input of cascade forest module in terms of substituting multi-grained scanning module. Therefore, a combustion state recognition model of MSWI processes based on IDF was established. Based on actual flame images of industrial processes, many experiments has been done. The results showed that the constructed model can reach a recognition accuracy of 95.28%.

Keywords: Municipal solid waste incineration · Combustion state recognition · Features extraction · Improved deep forest

1 Introduction

The generation rate of municipal solid wastes (MSWs) increases year by year [1]. The MSWs which weren't treated in time have caused serious environmental pollution. Facing the increasingly serious global environmental pollution [2], countries all over the world began to vigorously promote MSWs incineration (MSWI) technology to replace the original composting one [3]. However, when the MSWI technology that imported from developed countries is applied to China, there are a series of "endemic" phenomena [4, 5]. That is to say, at the actual plant, the MSWI processes is generally controlled by domain experts manually. In order to ensure the sufficiency, stability and safety of

MSWI processes, it is necessary to accurately identify the combustion state. At present, the partition of combustion state of MSWI processes is mainly based on the location of the flame combustion line. In view of the complex and changeable, it is of great practical significance to take effective and feasible methods to recognize the combustion state.

The combustion flame images collected by acquisition equipment generally contain some noise and interference. Therefore, it is generally necessary to preprocess flame image. By this way, it can restore the real combustion scene as much as possible. To solve this problem, some experts and scholars have carried out detailed researches. A fast image defogging algorithm for a single input image was proposed [6], which can restore a fog-free combustion image. In order to effectively separate the flame and black smoke in the vent flare, reference [7] used different parameters to normalize the torch combustion image. The results shown that when the mean and variance values of three channels were set to 0.5, and the pixel values of the image were scaled to [−1, 1], the flame can be effectively separated from the background area. In order to eliminate the influence of fringe noise on THz image, notch filter was used to eliminate fringe noise [8]. Further, reference [9] verified that the median filter algorithm can effectively remove the noise of image.

Flame image features that are used to reflect the combustion state are diverse. In order to identify the combustion state, feature extraction is necessary. Zhang [10] et al. studied a recognition method of cement rotary kiln combustion state based on flame image. First, the flame image was divided into several target regions. Then, the average gray value, average brightness value and color feature of the target area were extracted. The combustion state identification model is constructed based on these extracted features. As the abnormal working state is easy to occur in the smelting processes of electric melting magnesium furnace, Liu [11] et al. proposed a diagnosis method by using dynamic image of furnace body. First, the spatial features of local sub-block image were extracted. Then, the monitoring indicators of the defined area was combined. Finally, the level of abnormal working state was obtained. Wu et al. [12] proposed a diagnosis method of abnormal working state, in which the spatial and temporal features of images were extracted based on neural network. Although this study realized the automatic labeling based on weighted median filter, the identification method based on flame image in industrial processes is relatively fixed. However, the coupling between different abnormal regions is serious in the MSWI processes.

There are some researches on the identification of combustion state for MSWI processes. Reference [13] extracted 12 features of MSWI flame image, in which the recognition model of combustion state was constructed based on BP neural network. Consistent with the disadvantages of traditional artificial neural network, this method needs lots of training data to achieve a better result [14, 15]. Qiao et al. [16] proposed a combustion state recognition method based on the color moment feature of flame image for the MSWI processes, in which the color moment feature of flame image was extracted by using sliding window and the identification model was built by using the least square support vector machine (LS-SVM). These models either have a slow convergence or have a poor recognition accuracy.

In recent years, Zhou et al. [17] proposed the deep forest (DF) model for classification problem. Compared with other deep learning algorithms based on neural network, DF has a lot of advantages. It can automatically adjust model scale and maintain a good representation learning ability [18, 19]. Subsequently, a large number of researchers have carried out many researches in the field of computer vision. Tang et al. [20] constructed a waste mobile phone recognition model using DF algorithm, which achieved a good recognition accuracy. The classification of hyperspectral images based on DF was studied [21], whose effectiveness was verified based on common hyperspectral dataset with character of sample insufficient. To solve the small sample problem, the DF method has great advantages in accuracy and anti-parameter sensitivity [22]. The above research results show that DF has good modeling effect on small samples and high dimensional dataset. However, the multi-grained module in DF has great computational consumption. This is an urgent problem to be solved.

Motivated by the above problems, this paper proposes a combustion state identification method for the MSWI processes based on improved deep forest (IDF). First, image preprocessing is used to restore fog-free images. Then, the multi-source features such as brightness, flame and color of combustion flame image are extracted. Finally, the extracted multi-source features are used to replace the multi-grained scanning module as the input of cascade forest (CF) module. The efficiency of the model has been verified based on actual flame images of a MSWI plant in Beijing.

2 Processes Description and Modeling Strategy

The MSWs are transported to the plant by the feeder truck. Then, the operator controls the grab throw the MSWs into incinerator. The combustion processes is generally divided into three stages, i.e., drying, combustion and burnout stages. During MSWI processes, the domain experts judge the combustion state by observing the flame image in furnace. Meanwhile, they continuously adjust the grate speed and air volume to keep the combustion state as stable as possible.

Based on the above analysis, this paper proposes a modeling strategy composed of image preprocessing module and improved deep forest (IDF) module, which is shown in Fig. 1. In which, $\{I_n(u)\}_{n=1}^{N}$ represents the flame images, $I_n(u)$ represents the nth image, N is the number of image samples and u represents the uth pixel in the image, $\{L_n^{median}(u)\}_{n=1}^{N}$ represents the images after preprocessing, $\{X_n\}_{n=1}^{N}$ represents the extracted feature sets, and $\hat{y}$ represents the identification results.

The functions of the above modules are shown as follows:

① Image preprocessing module: It is used to eliminate the noise introduced by environment and transmission and to separate the flame from the furnace background.

② IDF module: First, multi-source features such as brightness, flame and color of combustion flame images are extracted. Then, they are used as the input of recognition model based on CF module, in which random forest (RF) and completely random forest (CRF) are used as the basic learners to identify the combustion state. Finally, the prediction results are gotten by using the simple weighting average algorithm.

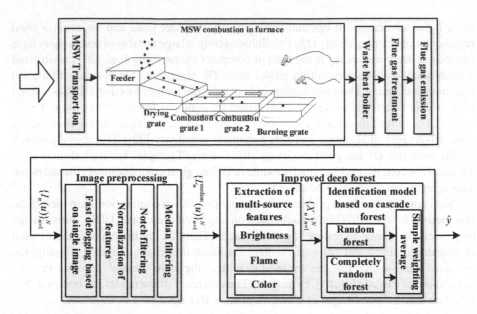

Fig. 1. Modeling strategy of the combustion state identification based on IDF

3 Algorithm Implementation

3.1 Image Preprocessing Module

In MSWI processes, fly ash and smoke are generated ceaselessly. RGB image $\{I_n(u)\}_{n=1}^{N}$ inevitably contains signal interference and other physical noise. Therefore, before to make feature extraction, image preprocessing is employed to restore the flame image $\{L_n^{\text{median}}(u)\}_{n=1}^{N}$.

Fast Defogging Algorithm Based on Single Image

First, the minimum values of the R, G and B channels from original image $I_n(u)$ are taken to obtain $H_n(u)$. By considering the relationship between transmittance $l_n(u)$ and image $H_n(u)$, we can get.

$$l_n(u) \geq 1 - I_n(u) \times (A_n)^{-1} = 1 - H_n(u) \times (A_n)^{-1}, \tag{1}$$

where A_n represents global atmospheric light. Then, a rough estimation of transmittance $l_n(u)$ can be obtained,

$$l_n(u) = 1 - H_n^{\text{ave}}(u) \times (A_n)^{-1} + \Psi H_n^{\text{ave}}(u) \times (A_n)^{-1}, \tag{2}$$

where $\Psi \in [0, 1]$. By set $\eta = 1 - \Psi$, the following result is obtained,

$$l_n(u) = 1 - \eta H_n^{\text{ave}}(u) \times (A_n)^{-1}. \tag{3}$$

To prevent the image being too dark or too bright after defogging, we set up $\tilde{\eta} = \rho h_n^{av}$. In which, ρ is an adjustable parameter with range $0 \leq \rho \leq 1/(h_n^{av})^{-1}$, and h_n^{av} is the mean of all elements of $H_n(u)$.

To ensure that Ψ is positive value, the upper limit is set to be 0.9. Then, we have the relation,

$$\eta = \min(\rho h_n^{av}, 0.9). \tag{4}$$

Thus, $l_n(u)$ is obtained as.

$$l_n(u) = \max(1 - \eta H_n^{ave}(u) \times (A_n)^{-1}, 1 - H_n(u) \times (A_n)^{-1}), \tag{5}$$

where $1 - \eta H_n^{ave}(u) \times (A_n)^{-1}$ is a rough estimate of $l_n(u)$ and $1 - H_n(u) \times (A_n)^{-1}$ is the lower limit of $l_n(u)$.

Then, ambient light $Z_n(u)$ can be obtained,

$$Z_n(u) = \min(\eta H_n^{ave}(u), H_n(u)). \tag{6}$$

The range of A_n is $(\max(H_n^{ave}(u)), \max(\max_{c \in r,g,b}(I_n^c(u))))$. The mean value of A_n is directly taken as follows,

$$A_n = \frac{1}{2}(\max(H_n^{ave}(u)) + \max(\max c \in r, g, b(I_n^c(u)))). \tag{7}$$

Finally, the image after defogging is obtained,

$$F_n(u) = \frac{I_n(u) - l_n(u)}{1 - (A_n)^{-1}l_n(u)}. \tag{8}$$

Features Normalization

After features normalization, the flame can be separated from the background. Here, zero mean normalization method is adopted,

$$W_n(u) = \frac{F_n(u) - \mu_n}{\sigma_n}, \tag{9}$$

where μ_n is the mean value, σ_n is the standard deviation, and $W_n(u)$ is the normalized image.

Notch Filtering

Noise needs to be eliminated in the frequency domain. The notch filter has a very narrow stop band, whose ideal frequency response is expressed as follow,

$$|H(e^{jw})| = \begin{cases} 1, w \neq w_0 \\ 0, w = w_0 \end{cases}. \tag{10}$$

After notch filtering, the image $\{V_n(u)\}_{n=1}^N$ can be obtained,

$$V_n(u) = \text{notch}\{ W_n(u)\} \; n = 1, 2, \ldots \ldots N. \tag{11}$$

Median Filtering

The isolated noise points caused by fly ash need to be eliminated by median filter. The median filtered image is obtained as follow,

$$L_n^{\text{median}}(u) = \underset{(u,v) \in \varpi}{\text{median}}\{V_n(u)\} n = 1, 2, \cdots \cdots N \tag{12}$$

3.2 Improved Deep Forest Module

Traditional DNN algorithms usually need large dataset to ensure training results. At the same time, the structure of the trained model is very complex. So it has great limitations in practical application in terms of limit hardware resource. DF is a forest ensemble algorithm composed of multiple forests, which maintains high performance and trainability on small-scale datasets. The structure of traditional DF model includes multi-Grained scanning and CF modules, as shown in Fig. 2.

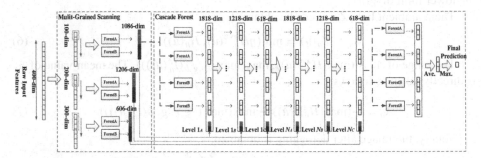

Fig. 2. Structure of the traditional DF model

In Fig. 2, it is assumed that the dimension of the input eigenvector is 400. The purpose of multi-grained scanning module is to deal with the relationship between data and features. So it is used to enhance the performance of CF module. In this paper, it is considered that the combustion state identification needs a real-time performance. To meet this requirement, the multi-grained scanning module is replaced with multi-source feature extraction module.

Extraction of multi-Source Features Sub-module
The flame image usually contains lots of information. It is considered to extract different flame features (size, brightness, color) to express the combustion state.
Brightness Feature
For the flame image of MSWI processes, its brightness feature can be described from different views.

(1) Average value of gray image

First, the original color image is converted into gray image [6], which is shown as follow,

$$G_n(u) = 0.11 \times (L_R^{\text{median}})_n(u) + 0.59 \times (L_G^{\text{median}})_n(u) + 0.3 \times (L_B^{\text{median}})_n(u), \quad (13)$$

where $(L_R^{\text{median}})_n(u)$, $(L_G^{\text{median}})_n(u)$ and $(L_B^{\text{median}})_n(u)$ represent the color components of R, G and B channels at uth pixel of the nth image, respectively.

The average value of gray image is calculated as follow,

$$\text{Gray\_ave}_n = \frac{1}{U} \sum_{u=1}^{U} G_n(u). \tag{14}$$

(2) Variance of gray image
    The variance of gray image is calculated as follow,

$$\text{Gray\_var}_n = \frac{1}{U} \sum_{u=1}^{U} [G_n(u) - \text{Gray\_ave}_n]^2. \tag{15}$$

The former two features mainly describe the image brightness in terms of computer vision. Image in HSV color space is closer to the human vision. Therefore, the flame image is transferred from RGB space to HSV space firstly. Then, the brightness feature is extracted from V channel.

(3) Average value of brightness
    Here, the image converted to HSV space is represented as $\{L_n^{HSV}(u)\}_{n=1}^{N}$. The average value of V-channel $(L_V^{HSV})_n(u)$ is calculated as the brightness feature,
(4) Variance of brightness

$$\text{Bright\_ave}_n = \frac{1}{U} \sum_{u=1}^{U} (L_V^{HSV})_n(u). \tag{16}$$

The variance of brightness is calculated as follow,

$$\text{Bright\_var}_n = \frac{1}{U} \sum_{u=1}^{U} [(L_V^{HSV})_n(u) - \text{Bright\_ave}_n]^2. \tag{17}$$

Flame Feature
The calculation of flame area is based on V channel.

(1) Effective area of flame
    The effective area of flame is defined as the total number of pixels in the image with whose brightness value is greater than the specified threshold θ_{th}. It is calculated as follow,

$$A\_v_n = \sum_{u=1}^{U} \Theta[(L_V^{HSV})_n(u) - \theta_{th}], \tag{18}$$

where $\Theta(\cdot)$ is the unit step function.

(2) High temperature area of flame

The high temperature area is defined as the total number of pixels in the image with whose brightness value is greater than the specified threshold ω_{th}. It is calculated as follow,

$$G\_v_n = \sum_{u-1}^{U} \Theta[(L_V^{HSV})_n(u) - \omega_{th}]. \tag{19}$$

Color Feature

Color moment is a simple and effective color feature. The first-order, second-order and third-order moments are used to express the color information in this study.

The first-order moment v_n^{color} is calculated as follow,

$$v_n^{color} = \frac{1}{U} \sum_{u=1}^{U} L_n^{median}(u). \tag{20}$$

The second-order moment σ_n^{color} is calculated as follow,

$$\sigma_n^{color} = (\frac{1}{U} \sum_{u=1}^{U} (L_n^{median}(u) - v_n^{color})^2)^{\frac{1}{2}}. \tag{21}$$

The third-order moment δ_n^{color} is calculated as follow,

$$\delta_n^{color} = (\frac{1}{U} \sum_{u=1}^{U} (L_n^{median}(u) - v_n^{color})^3)^{\frac{1}{3}}. \tag{22}$$

Finally, the extracted features are combined to obtain the final multi-source feature set, i.e.,

$$X_n = [Gray\_ave_n, Gray\_var_n, Bright\_ave_n, Bright\_var_n, A\_v_n, G\_v_n, Color\_T_n].$$

Identification Model Based on Cascade Forest Sub-module

In each layer of CF, there are several forests composed of decision trees. The CF module used in this study contains 2 RFs and 2 CRFs in each layer.

Random Forest Algorithm

RF is a machine learning algorithm in terms of ensemble learning [23]. Bootstrap is used to randomly sample the training sets $D = \{ (x_i, y_i), i = 1, 2, \cdots b\} \in R^{B \times M}$, whose process can be described as.

$$\{ (x^{j,M^j}, y^j)_1^b\}_{b=1}^{B} = f_{RSM}(f_{Bootstrap}(D, P), M^j), \tag{23}$$

where $\{ (x^{j,M^j}, y^j)_1^b\}_{b=1}^{B}$ represents the jth training subset; $f_{RSM}(\cdot)$ represents the bootstrap function, $m = 1, \cdots, M^j$, M^j represents the number of features selected by the jth training subset in the forest and $M^j \ll M$.

The above function was used J times. Then, the RF training sets can be obtained. The above process is shown as follow,

$$\left.\begin{array}{c} D \\ J \end{array}\right\} \Rightarrow \left\{ \begin{array}{c} \{(\mathbf{x}^{1,M^1}, y^1)_1^b\}_{b=1}^B \\ \cdots \\ \{(\mathbf{x}^{j,M^j}, y^j)_1^b\}_{b=1}^B \\ \cdots \\ \{(\mathbf{x}^{1,M^J}, y^J)_1^b\}_{b=1}^B \end{array} \right., \tag{24}$$

where J is the number of bootstrap times and the number of DTs in RF. Further, J DTs in RF model are built with the above training subsets. The construction strategy is shown in Fig. 3.

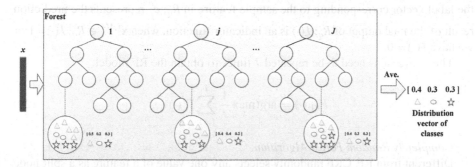

Fig. 3. Modeling strategy of RF

We take the jth training subset as an example. The number of best segmentation feature M_{sel}^j and syncopation point s need to be searched based on Gini index criterion, which can be represented to solve the following optimization problem.

$$(M_{\text{sel}}^j, s) = \arg\min[\frac{y_{C_{Left}}^j}{y^j}\text{Gini}(y_{C_{Left}}^j) + \frac{y_{C_{Right}}^j}{y^j}\text{Gini}(y_{C_{Right}}^j)], \tag{25}$$

$$\text{Gini}(\cdot) = \sum_{k=1}^K p_k(1-p_k) = 1 - \sum_{k=1}^K p_k^2, \tag{26}$$

$$s.t. \begin{cases} C_{Left} > \theta_{\text{Forest}} \\ C_{Right} > \theta_{\text{Forest}} \\ \text{Gini}(y_{C_{Left}}^j) > 0 \\ \text{Gini}(y_{C_{Right}}^j) > 0 \end{cases}$$

where k represents the kth class in labels, and p_k indicates the proportion of kth class in the total number of tags, and θ_{Forest} indicates the threshold of the number of samples contained in the leaf node.

Based on the above criteria, the optimal variable number and the value of cut-off point are found by traversing all input features. The input feature space is divided into left and right regions. Then, the above process for each region need to be repeated until the leaf node number of samples contained is less than the threshold, or the Gini index is 0. Finally, the input feature space is divided into Q regions. To build the model of classification tree, the following functions are defined,

$$\Gamma^j(\cdot) = \sum_{q=1}^{Q} c_j^q I(\mathbf{x}^{j,M^j} \in R_q). \tag{27}$$

$$c_j^q = [p_1, \cdots, p_k, \cdots, p_K]^T (\mathbf{y}_{N_{R_q}}^j \in R_q, N_{R_q} \leq \theta_{Forest}), \tag{28}$$

where N_{R_q} represents the number of training samples contained in R_q; $\mathbf{y}_{N_{R_q}}^j$ represents the label vector corresponding to the sample feature in R_q; c_j^q represents the prediction result of the final output of R_q; $I(\cdot)$ is an indicator function, when $\mathbf{x}^{j,M^j} \in R_q$, $I(\cdot)=1$; or we have $I(\cdot)= 0$.

The above steps need to be repeated J times to obtain the RF model,

$$F_{RF}(\cdot) = \arg(\max_k \frac{1}{J} \sum_{j=1}^{J} \Gamma^j(\cdot)). \tag{29}$$

Completely Random Forest Algorithm
Different from RF, CRF randomly selects any one value of a feature as a split node in the feature space. The obtained CRF model is represented by $F_{CRF}(\cdot)$.
Simple Weighting Average
Each layer of CF module uses two $F_{RF}(\cdot)$ and two $F_{CRF}(\cdot)$ as base learners for cascade learning. The model is constructed by using the idea of stack. For the input samples X_n, the last layer of CF outputs the $4K$-dim class distribution vector $R = [r_1^{RF}, r_2^{RF}, r_1^{CRF}, r_2^{CRF}]$. We take the average and maximum value to obtain the final predicted value.

$$\hat{y} = \max[\frac{1}{4} \times (r_1^{RF} + r_2^{RF} + r_1^{CRF} + r_2^{CRF})]. \tag{30}$$

4 Experimental Verification

4.1 Description of Data

The combustion images used in this experiment come from an MSWI plant in Beijing. According to the different positions of combustion line, those images are divided into three combustion states, i.e., forward moving, normal moving and backward moving of the combustion line. The calibration criterion of flame image is shown in Fig. 4. The size of image is 718×512. The numbers of samples is 571. The number of samples corresponding to three state are 184, 213 and 174, respectively. The training set, verification set and testing set are divided in a ratio of 2:1:1.

Fig. 4. Calibration of flame image

4.2 Results of Experiment

Results of Image Preprocessing

For the studied flame image, the large amount of smoke contained in the image should be removed firstly. The defogging effect is shown in Fig. 5 (b). Thus, the amount of smoke in image is significantly reduced after defogging.

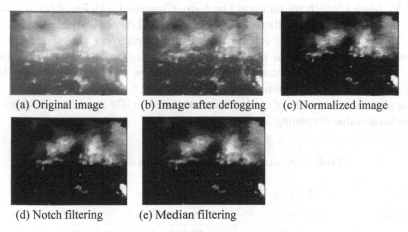

(a) Original image   (b) Image after defogging   (c) Normalized image

(d) Notch filtering   (e) Median filtering

Fig. 5. Results of image preprocessing

In order to improve the computational efficiency, it is necessary to normalize the image. Figure 5 (c) shows the normalized image. It can be seen that the normalized image achieves the purpose of effectively separating the flame from background. However, there are still serious fringe interference in the image. It will be eliminated by notch filtering, as shown in Fig. 5 (d). Thus, the fringe noise in the filtered image is obviously suppressed. The median filter is used to eliminate small noise. It is shown in Fig. 5 (e).

Identification Results of IDF

The preprocessed image is converted from color space to gray space, which is shown in Fig. 6 (a).

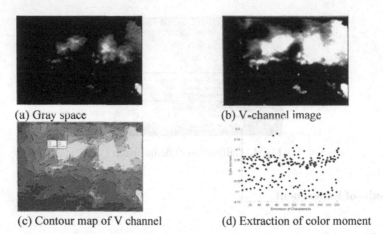

(a) Gray space

(b) V-channel image

(c) Contour map of V channel

(d) Extraction of color moment

Fig. 6. Results of multi-source features extraction

The preprocessed flame image is transformed from RGB space to HSV space. The threshold values of flame effective zone and high temperature zone are selected based on the V-channel, which are shown in Fig. 6 (b). Compared with Fig. 6 (c), it is determined that the threshold of flame effective area is 0.226 and the threshold of high temperature area is 0.941. The experimental results of reference [16] showed that the recognition performance is the best one when the sliding window scale is 1/5 of the image. By experimental verification, the sliding window scale consistent with [16] is finally selected. The extraction results are shown in Fig. 6 (d).

The identification results of combustion state based on IDF are shown in Tables 1–2. It is the mean value of running 20 times.

Table 1. Accuracy of identification results based on IDF

Features	Training set	Validation set	Testing set
Brightness	95.58%	60.66%	66.01%
Flame	90.96%	57.55%	59.44%
Color	99.98%	94.76%	94.93%
Combination	99.98%	94.68%	95.28%

The results show that the color features have the highest contribution to the effect of recognition. The contribution of flame feature is the lowest one. Compared with the color features, the training results of combined features keep the same effects on the training set. However, combined features have a better effect on the testing set, which shows that it has a stronger generalization.

Comparison of Different Methods

The comparative results with the existing methods are shown in Table 3.

Table 2. F1 score of identification results based on IDF

Features	Training set	Validation set	Testing set
Brightness	95.70%	61.28%	66.56%
Flame	91.05%	57.97%	59.33%
Color	99.98%	95.01%	95.08%
Combination	99.98%	94.92%	95.43%

Table 3. Comparative results of different methods

Methods	Recognition accuracy on testing set
Color moment + LS_SVM [16]	75%
Convolutional neural network	75.56%
Color moment + RF	94.81%
Our method	**95.28%**

The results show that our method has the highest recognition accuracy. The reasons include: (1) The extracted features have a good expressive ability for the combustion state information contained in the flame image. (2) The tree structure can exert more powerful learning and classification capabilities for features with physical meaning. (3) The transfer of supervised information between cascade layers greatly enhances the learning ability of the model. These results prove the efficiency of our proposed method.

5 Conclusion

A combustion state identification method on MSWI processes is proposed in this paper. The main contributions are shown as follows: (1) A modeling strategy based on image preprocessing and IDF is proposed. (2) Multi-source features with physical meaning are extracted from the flame image to replace the multi-grained scanning module in traditional DF, which reduces the computational consumption. Based on the actual running data, the accuracy on testing set of the proposed method is more than 95%. The future study is to extract the powerful and interesting local features in the flame image.

Acknowledgments. The work was supported by Beijing Natural Science Foundation (No. 4212032), National Natural Science Foundation of China (No. 62073006).

References

1. Lu, J., Jin, Q.: A study on methods of municipal solid waste treatment and resource utilization. Res. Conserv. Environ. Protect. **05**, 121–122 (2021)

2. Niu, Y., Chen, F., Li, Y., Ren, B.: Trends and sources of heavy metal pollution in global river and lake sediments from 1970 to 2018. J. Rev. Environ. Contamin. Toxicol. 1-35 (2020)

3. Bai, B.: A discussion on treatment and utilization of urban waste. J. Intell. Build. Smart City, **12**, 120–121 (2021)

4. Saikia, N., et al.: Assessment of Pb-slag, MSWI bottom ash and boiler and fly ash for using as a fine aggregate in cement mortar. J. Hazard. Mater. **154**(1–3), 766–777 (2008)

5. Korai, M., Mafar, R., Uqaili, M.: The feasibility of municipal solid waste for energy generation and its existing management practices in Pakistan. J. Renew. Sustain. Energy Rev. **72**, 338–353 (2017)

6. Huang, L.: Fast single image defogging algorithm. J. Optoelectr. Laser. **22**(11), 1735–1736 (2011)

7. Xie, X., Gu, K., Qiao, J.: A new anomaly detector of flare pilot based on siamese network. In: 20th Chinese Automation Congress, pp. 2278–2238 (2020)

8. Zou, Y.: A Research on Fringe Noise Elimination Method of THz Image. Dissertation of Capital Normal University (2009)

9. Khatri, S., Kasturiwale, H.: Quality assessment of median filtering techniques for impulse noise removal from digital images. In: International Conference on Advanced Computing & Communication Systems, pp. 1–4. IEEE(2016)

10. Zhang, R., Lu, S., Yu, H., Wang, X.: Recognition method of cement rotary kiln burning state based on Otsu-Kmeans flame image segmentation and SVM. Optik, **243**, 167418 (2021)

11. Liu, Q., Kong, D., Lang, Z.: Diagnosis of abnormal working state of electric melting magnesium furnace based on multistage dynamic principal component analysis. J. Autom. **47**(11), 1–8 (2021)

12. Wu, G., Liu, Q., Chai, T., Qin, S.: Diagnosis of abnormal working state of electric melting magnesium furnace based on time series image and deep learning. J. Autom. **45**(8), 1475–1485 (2019)

13. Zhou, Z.: A Research on Combustion State Diagnosis of Waste Incinerator based on Image Processing and Artificial Intelligence. Dissertation of Southeast University (2015)

14. Fang, Z.: A high-efficient hybrid physics-informed neural networks based on convolutional neural network. IEEE Trans. Neural Netw. Learn. Syst. **99**, 1–13 (2021)

15. Rubio, J.: Stability analysis of the modified levenberg-marquardt algorithm for the artificial neural network training. IEEE Trans. Neural Netw. Learn. Syst. **99**, 1–15 (2021)

16. Qiao, J., Duan, H., Tang, J.: A recognition of combustion state in municipal solid waste incineration process based on color feature extraction of flame image. In: 30th China Process Control Conference, pp. 1–9 (2019)

17. Zhou, Z., Feng, J.: Deep forest. arXiv preprint, 1702.08835 (2017)

18. Molaei, S., Havvaei, A., Zare, H., Jalili, M.: Collaborative deep forest learning for recommender systems. IEEE Access **99**, 1 (2021)

19. Shao, L., Zhang, D., Du, H., Fu, D.: Deep forest in ADHD data classification. IEEE Access **99**, 1(2019)

20. Tang, J., Wang, Z., Xia, H., Xu,Z.H.: Han: Deep forest recognition model based on multi scale features of waste mobile phones for intelligent recycling equipment. In: 31th China Process Control Conference, pp. 10–19 (2020)

21. Li, M., Ning, Z., Pan, B., Xie, S., Xi, W., Shi, Z.: Hyperspectral Image Classification Based on Deep Forest and Spectral-Spatial Cooperative Feature. Springer, Cham (2017)

22. Yin, X., Wang, R., Liu, X., Cai, Y: Deep forest-based classification of hyperspectral images. In: 37th Chinese Control Conference, pp. 10367–10372. IEEE (2018)

23. Bbeiman, L., Quinlan, R.: Bagging predictors. Mach. Learn. **26**(2), 123–140 (1996)

RPCA-Induced Graph Tensor Learning
for Incomplete Multi-view Inferring
and Clustering

Xingfeng Li[1], Yinghui Sun[1], Zhenwen Ren[1,2(✉)], and Quansen Sun[1(✉)]

[1] School of Computer Science and Engineering, Nanjing University of Science and
Technology, Nanjing 210094, China
{yinghuisun,rzw,sunquansen}@njust.du.cn

[2] Department of National Defence Science and Technology, Southwest University of Science
and Technology, Mianyang 621010, China

Abstract. The existing incomplete multi-view graph clustering (IMGC) methods mainly focus on leveraging available samples among different views to
explore the weak local structure information, resulting in inexact or unreliable
affinity graphs for clustering. More importantly, they fail to exploit the spatial
structure information among graphs. To graciously address both, in this paper, we
propose a novel IMGC method, claimed as robust principle component analysis
(RPCA)-induced graph tensor learning (RPCA-IGTL) for incomplete multi-view
inferring and clustering. This model can synchronously perform missing part filling, complete data inferring, and diagonalized graph tensor learning to obtain the
more exact and reliable affinity graphs for clustering. Especially, the proposed
method first designs a RPCA-induced local manifold learning framework, which
bridges complete data inferring in feature space and the diagonalized graph learning in graph semantic space. Both objects can boost each other to fully exploit the
underlying local structure information among incomplete view data. Besides, a t-
SVD based tensor low-rank constraint introduces to exploit the spatial structure
of graph tensor and complementary information of diagonalized affinity graphs.
Extensive experiments have demonstrated the effectiveness of our method compared to the previous state-of-the-art methods.

Keywords: Incomplete multi-view inferring and clustering · Robust principle
component analysis · Low-rank graph tensor

1 Introduction

Multi-view clustering (MVC) has become an increasingly pervasive research topic due
to its powerful ability of handling multi-view data for improving clustering performance
during the past few decades [4, 19]. The implementation of these methods quite depends
on an assumption that views among samples are visible. However, this assumption will
not hold anymore owing to the absence of multi-view data, especially in certain practical applications [9, 22]. This tremendously limits the aforementioned MVC methods to
handle incomplete MVC (IMVC) tasks.

© The Author(s), under exclusive license to Springer Nature Singapore Pte Ltd. 2022
H. Zhang et al. (Eds.): NCAA 2022, CCIS 1637, pp. 85–99, 2022.
https://doi.org/10.1007/978-981-19-6142-7_7

Recently, lots of IMVC methods have been proposed to alleviate the absence problem of data and can be roughly divided into three main categories: kernel based IMVC, non-negative matrix factorization (NMF) based IMVC (NMF-IMVC) [3], and graph based IMVC, also known as incomplete multi-view graph clustering (IMGC) [26,27]. For first category, [16] first proposes to complete kernel Gram matrix to deal with incomplete kernel problem for IMVC. Unfortunately, it is merely appropriate for the incomplete case that at least one base kernel is complete, which greatly prevents its application in practice. To this end, many scholars propose absent multiple kernel learning models for incomplete clustering. Typically, multiple kernel k-means with incomplete kernels clustering (MKK-IKC) [10] proposes to learn the kernel coefficients and the clustering partition matrix in a joint objection to fill the incomplete base kernels and clustering. However, it is great difficult for kernel based IMVC methods to tune the suitable kernel parameters for predefined base kernels.

For second category, most of them directly fuse multiple complementary representations into a low-dimensional consensus representation for clustering [5]. Another methods use mean values or zeros to fill the incomplete parts and then couple a weighted strategy [2,18,21,23]. These methods can directly obtain a consensus representation for reducing the impact of the incomplete samples of different views. Despite its effectiveness, NMF-IMVC methods always obtain an uncompacted representation since they neglect the intrinsic (such as inter-view) structure of multi-view data. For instance, incomplete multi-view clustering with flexible locality structure diffusion (IMVC-FLSD) [23] integrates the consensus representation learning and the objection weighted learning into a unified model to explore the certain paired similarity in different views.

Compared to most NMF-IMVC methods, IMGC methods can better preserve the geometric structure of data, especially for the local geometric structure among data in graph learning process, which has been proved to very significant for clustering [20,22,25]. Incomplete multi-view spectral clustering with adaptive graph learning (IMVSC-AGL) [20] directly uses available multi-view data to generate a consensus representation for clustering. To sum up, the above mentioned IMVC methods always focus on adequately exploring the effective information of available samples to obtain preferable clustering, while lacking of the ability of recovering missing views.

To this end, a unified embedding alignment framework (UEAF) [22] attempts to address IMVC problem by recovering the missing views rather than only leveraging available views data. However, it is seriously sensitive to noise and outliers, resulting in the inferior affinity graph and clustering results.

To address above problems, we propose a novel IMGC method, namely Robust Principle Component Analysis Induced Graph Tensor Learning (RPCA-IGTL), to synergistically recover missing views and infer complete data without noise for improve the IMVC. Particularly, we first extend the traditional RPCA model to manifold regularized RPCA framework for handling the challenging incomplete multi-view recovering problem. And then, we couple the adaptive neighbors graph learning with manifold regularized RPCA framework into a unified objective function. Besides, an enhanced block diagonal constraint and t-SVD based tensor low-rank constraint are simultaneously imposed on the learning of affinity graph, so as to guide the missing data recovering and complete data inferring. In summary, the contributions of this paper mainly include the following:

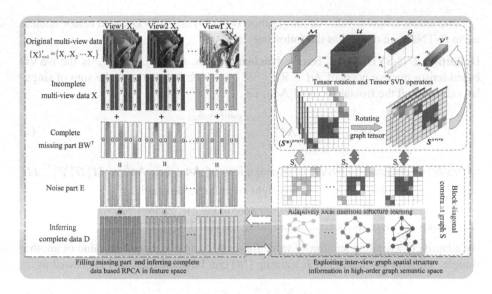

Fig. 1. Framework of the proposed RPCA-IGTL.

- To address the original incomplete and noise data, we design an elegant RPCA-induced manifold learning framework. It integrates missing part completing, low-rank complete data inferring, and local manifold learning into a unified objective function, such that the more exact and reliable graphs can be learned for improving the clustering performance.
- Instead of suboptimal block diagonal regularizer with extra parameter, a more flexible block diagonal constraint with parameter free is considered to precisely control the number of diagonal blocks of graphs in framework, which is greatly significant for clustering. Further, with the t-SVD based tensor low-rank constraint on the learned diagonal graphs, the spatial structure and complementary information among incomplete graphs can be exploited as possible.
- Compared to two base line and several state-of-the-art IMVC and IMGC methods, our method achieves important improvement on various scale datasets.

2   Related Work

2.1   Notation Summary

Through this paper, we denote 3-order tensor, matrix and vector as $\mathcal{M} \in \mathbb{R}^{n_1 \times n_2 \times n_3}$, $\mathrm{M}^{n_1 \times n_2}$, M, respectively. $\mathrm{M} \succeq 0$ is the positive semi-definite matrix. The $\mathcal{M}_{i,j,k}$ is the element of $\mathcal{M}$, where the fiber is denoted as $\mathcal{M}(:, j, k)$, $\mathcal{M}(i, :, k)$, and $\mathcal{M}(i, j, :)$, as well as the slice is denoted as horizontal slice $\mathcal{M}(i, :, :)$, lateral slice $\mathcal{M}(:, j, :)$, and frontal slice $\mathcal{M}(:, :, k)$. For convenience, $\mathcal{M}(:, :, k)$ is simplified as M^k or $\mathcal{M}^k$. $\mathcal{M}_f = \mathtt{fft}(\mathcal{M}, [\,], 3)$ and $\mathcal{M} = \mathtt{ifft}(\mathcal{M}_f, [\,], 3)$ are the fast Fourier transformation (FFT) and inverse FFT along the third direction of tensor $\mathcal{M}$, respectively.

The tensor singular value decomposition (t-SVD) and t-SVD based tensor nuclear norm (t-TNN) are defined as the following

Definition 2 (t**-SVD-Based Tensor Nuclear Norm,** t**-TNN**). $\|\mathcal{M}\|_{\circledast}$ is the t-SVD based tensor nuclear norm of $\mathcal{M} \in \mathbb{R}^{n_1 \times n_2 \times n_3}$, which is denoted as the sum of singular values of all the frontal slices of $\mathcal{M}_f$, i.e.,

$$\|\mathcal{M}\|_{\circledast} = \sum_{k=1}^{n_3} \left\|\mathcal{M}_f^k\right\|_* = \sum_{i=1}^{\min(n_1,n_2)} \sum_{k=1}^{n_3} \left|\mathcal{S}_f^k(i,i)\right| \qquad (1)$$

where $\mathcal{S}_f^k$ is obtained by t-SVD of frontal slices of $\mathcal{M}_f$, i.e., $\mathcal{M}_f^k = \mathcal{U}_f^k \mathcal{S}_f^k \mathcal{V}_f^{k\mathsf{T}}$. $\mathcal{U}_f^k$ and $\mathcal{V}_f^k$ are the corresponding left and right singular value matrices.

2.2 Adaptive Neighbors Graph Learning (ANGL)

By adaptively assigning an affinity value for each sample as neighbor value of another sample, adaptive neighbors graph learning [14] can be mathematically expressed as

$$\min_{\mathbf{S}^v} \sum_{i=1}^{n} \sum_{j=1}^{n} \mathbb{D}(\mathbf{x}_i^v, \mathbf{x}_j^v) s_{ij}^v \qquad (2)$$

$\mathbb{D}$ is the distance criterion between any two samples, and the larger distance indicates the smaller affinity value s_{ij}^v corresponding to v-th view. (2) can effectively learn an affinity graph with local geometrical structure preserving.

2.3 Robust Principle Component Analysis (RPCA)

The model of RPCA takes the form of

$$\min_{\mathbf{D},\mathbf{E}} \|\mathbf{D}\|_* + \alpha\phi(\mathbf{E}) \text{ s.t. } \mathbf{X} = \mathbf{D} + \mathbf{E} \qquad (3)$$

which has a strong recovery guarantees as the first polynomial-time method [1]. $\mathbf{X}$ in (3) denotes the damaged or noisy samples, and is decomposed as the low-rank sample matrix and sparse error matrix, respectively. $\phi(\cdot)$, α and $\|\cdot\|_*$ denote the certain regularizer of noise, regularization parameter and nuclear norm, respectively. Most algorithms have demonstrated that low-rank representation can well characterize the relationship between data. Low-rank representation in (3) $\mathbf{D}$ can be deemed as the clean data recovered from original data $\mathbf{X}$, which has been extensively used in video surveillance and image denoising [11].

3 Proposed Method

Different from partial multi-view learning [18], incomplete multi-view learning is a significantly challenging research topic. Since the incomplete view data (missing parts) will destroy the original affinity between multi-view data, resulting in deceptive affinity values. Further, the unreliable affinity graphs constructed from these data will cause the terrible clustering results. Thus, we develop a RPCA inferring framework (RPCA-IGTL) to simultaneously infer the missing view data and diagonalized graphs learning for clustering. Intuitively, Fig. 1 has given a clear pipeline of RPCA-IGTL.

3.1 Model of RPCA-IGTL

In complete partial multi-view graph clustering, adaptive neighbors graph learning (ANGL) becomes increasingly prevalent and has achieved the promising results recently [2]. Based on ANGL of (2), multi-view graph clustering (MVGC) can be formulated as

$$\min_{\mathbf{S}^v} \sum_{v=1}^{r} \sum_{i,j=1}^{n} (\|x_i^v - x_j^v\|_2^2 s_{ij}^v + \gamma s_{ij}^{v\,2}) \tag{4}$$

$$\text{s.t. } \forall i, s_i^\top \mathbf{1} = 1, 0 \leq s_i \leq 1, \text{rank}(\mathbf{L}_S^v) = n - c$$

where smaller distance between samples $\mathbf{x}_i^v$ and $\mathbf{x}_j^v$ automatically learns a larger affinity value s_{ij}^v to construct the affinity graph $\mathbf{S}^v$, and vice versa. γ is a parameter to control the neighbors between samples and r represents the number of view. Here, $\mathbf{L}_S^v = \mathbf{P}^v - (\mathbf{S}^v + (\mathbf{S}^v)^\top)/2$ and c are a Lagrangian matrix and the number of cluster, respectively. The degree matrix $\mathbf{P}^v$ can be computed via $p_{ii}^v = \sum_j s_{ij}^v$. $\text{rank}(\mathbf{L}_S^v) = n - c$ is a widely used Lagrangian rank constraint to pursue a graph with exact connected components for improving clustering performance. Owing to Lagrangian rank constraint and the capacity of local manifold structure, the complete MVGC methods have received lots of attention recently. However, (4) needs all views data to be complete. To address this challenging problem, it urgently requires a way to infer or recover the missing view data. To this end, we consider that RPCA has a strong recovery ability as the first polynomial-time method in (3). Here, one may utilize clean data to substitute $\mathbf{X}$ in (4). Apparently, there is only available samples information without missing samples information in $\mathbf{D}$, such that the missing view information is ignored. Inspired by (3), we first design an inferring framework as mentioned in the left of Fig. 1, and this framework can be mathematically fulfilled as

$$\min_{\mathbf{D},\mathbf{E}} \|\mathbf{D}\|_* + \alpha\phi(\mathbf{E}) \text{ s.t. } \mathbf{X} + \mathbf{BW}^\top = \mathbf{D} + \mathbf{E} \tag{5}$$

where $\mathbf{BW}^\top$ is the complete missing part, and $\mathbf{W}$ is a index matrix to explicitly enforces the entries to be zeros corresponding to the missing samples from $\mathbf{B}$. $\mathbf{W}$ can be construct via

$$W_{i,j}^v = \begin{cases} 1, & \text{if the } i\text{-th missing sample is } x_j^v \\ 0, & \text{otherwise} \end{cases} \tag{6}$$

Not here that we can take full advantage of different incomplete prior information to construct corresponding index matrix $\mathbf{W}$, so as to deal with various of incomplete cases. By extending (5) to multi-view model, and we seamlessly integrate this model and (4) into a unified objection as

$$\min_{\substack{\mathbf{B}^v, \mathbf{D}^v, \mathbf{E}^v, \\ \mathbf{S}^v, \mathbf{Z}^v}} \sum_{v=1}^{r} \|\mathbf{D}^v\|_* + \alpha\|\mathbf{E}^v\|_1 + \beta\text{Tr}(\mathbf{D}^v \mathbf{L}_S^v (\mathbf{D}^v)^\top) + \gamma\|\mathbf{S}^v\|_F^2 \tag{7}$$

$$\text{s.t. } \forall v, \mathbf{X}^v + \mathbf{B}^v(\mathbf{W}^v)^\top = \mathbf{D}^v + \mathbf{E}^v, \text{rank}(\mathbf{L}_S^v) = n - c$$

where β is hyper-parameter. $\phi(\cdot) = \|\cdot\|_1$ on $\mathbf{E}^v$ is l_1 norm, which can alleviate error in the available samples of $\mathbf{X}^v$ and make the most of useful information in $\mathbf{X}^v$ transfer to $\mathbf{D}^v$. Frobenius norm is imposed on $\mathbf{S}^v$ to avoid trivial solution.

Although (7) has the power of inferring the information of missing views, it still exists the following limits: First, it neglects the complementary information among different views and spatial structure of graphs. Second, 1) n is usually much larger than c, so that it is illogical to find a high-rank $\mathbf{L}_S^v$ by minimizing $\text{rank}(\mathbf{L}_S^v)$; 2) more importantly, $\text{rank}(\mathbf{L}_S^v) = n - c$ is not able to control the targeted number of blocks; 3) the solution of $\text{rank}(\mathbf{L}_S^v) = n - c$ always introduces an extra hyper-parameter. To address these problems, we propose a novel IMGC method, named RPCA-induced graph tensor learning (RPCA-IGTL) for incomplete multi-view inferring and clustering, as follows

$$\min_{\substack{\mathbf{B}^v, \mathbf{D}^v, \mathbf{E}^v, \\ \mathbf{Z}^v, \mathcal{S}}} \sum_{v=1}^{r} \|\mathbf{D}^v\|_* + \alpha\|\mathbf{E}^v\|_1 + \beta\text{Tr}(\mathbf{D}^v\mathbf{L}_S^v\mathbf{D}^{v\top}) + \gamma\|\mathcal{S}\|_\circledast \tag{8}$$

$$\text{s.t. } \forall v, \mathbf{X}^v + \mathbf{B}^v(\mathbf{W}^v)^\top = \mathbf{D}^v + \mathbf{E}^v, \text{Tr}(\mathbf{S}^v) = c, 0 \preccurlyeq \mathbf{S}_{ij}^v \preccurlyeq 1 (\mathbf{S}^v)^\top = \mathbf{S}^v$$

where $\text{Tr}(\cdot)$ is a trace norm. To address the first problem, we stack the r affinity graphs into a tensor $\mathcal{S}^* \in \mathbb{R}^{n \times n \times r}$ to well capture the complementary information among different views and spatial structure information of graphs. And then $\mathcal{S}^*$ is rotated to $\mathcal{S}^{n \times r \times n}$ so as to better exploit inter-view information, which can simultaneously reduce the computational complexity from $\mathcal{O}(rn^2\log(r) + rn^3)$ to $\mathcal{O}(rn^2\log(n) + r^2n^2)$. Further, a t-SVD based tensor nuclear norm is imposed on $\mathcal{S}$ inspired by following two facts: 1) the views of different $\mathbf{S}^v$, originating from the same dataset source, possess some consensus structure information; 2) the learned tensor $\mathcal{S}$ should enjoy the low-rank property since the number of clusters c is always much less than the number of samples n. For another problem, the more exactly block diagonal constraint $\text{Tr}(\mathbf{S}^v) = c, 0 \preccurlyeq \mathbf{S}_{ij}^v \preccurlyeq 1, (\mathbf{S}^v)^\top = \mathbf{S}^v$ not only directly encourages affinity graphs to be block diagonal rather than inducing an extra hyper-parameter to solve $\text{rank}(\mathbf{L}_S^v) = n - c$, but also can control the number of blocks. As mentioned in the right part of Fig. 1, the imposed block diagonal constraint can control the graphs with intra-cluster dense and inter-cluster sparse as the input of tensor, which will vastly benefit to exploiting inter-view graph spatial structure information in high-order graph semantic space.

3.2 Optimization

By introducing an auxiliary variable $\mathcal{G}$ and $\{\mathbf{Z}\}_{v=1}^r$, a seven-step alternating direction method of multipliers (ADMM) is developed to optimize non-convex problem (8) as follows

$$\min_{\substack{\mathbf{B}^v, \mathbf{D}^v, \mathbf{E}^v, \\ \mathbf{S}^v, \mathbf{Z}^v, \mathcal{G}}} \sum_{v=1}^{r} \|\mathbf{D}^v\|_* + \alpha\|\mathbf{E}^v\|_1 + \beta\text{Tr}(\mathbf{Z}^v\mathbf{L}_S^v(\mathbf{Z}^v)^\top) + \gamma\|\mathcal{G}\|_\circledast + \frac{\mu}{2}\|\mathcal{G} - \mathcal{S} + \frac{\mathcal{Y}_3}{\mu}\|_F^2$$

$$+ \frac{\mu}{2}\|\mathbf{X}^v + \mathbf{B}^v(\mathbf{W}^v)^\top - \mathbf{D}^v - \mathbf{E}^v + \frac{\mathbf{Y}_1^v}{\mu}\| + \frac{\mu}{2}\|\mathbf{Z}^v - \mathbf{D}^v + \frac{\mathbf{Y}_2^v}{\mu}\|_F^2 \tag{9}$$

$$\text{s.t. } \forall v, \mathbf{Z}^v = \mathbf{D}^v, \mathcal{G} = \mathcal{S}, \text{Tr}(\mathbf{S}^v) = c, 0 \preccurlyeq \mathbf{S}_{ij}^v \preccurlyeq 1, (\mathbf{S}^v)^\top = \mathbf{S}^v$$

where μ is a penalty parameter. $\mathbf{Y}_1^v$, $\mathbf{Y}_2^v$ and $\mathcal{Y}_3^v$ are Lagrangian multipliers.

▶ **Step 1. D-subproblem:** Fixing the other variables except the recovering $\mathbf{D}$, we update $\mathbf{D}$ via

$$\min_{\mathbf{D}^{(v}} (\|\mathbf{D}^v\|_* + \frac{\mu}{2}\|\mathbf{D}^v - \mathbf{H}^v\|_F^2) \tag{10}$$

where $\mathbf{H}^v = (\mathbf{X}^v + \mathbf{B}^v(\mathbf{W}^v)^\top + \mathbf{Z}^v - \mathbf{E}^v + (\mathbf{Y}_1^v + \mathbf{Y}_2^v)/\mu)/2$, and (10) can obtain a closed-form solution via by the singular value thresholding operator [24].

▶ **Step 2. Z-subproblem:** By keeping the other variables unchanged except $\mathbf{Z}$, the subproblem of $\mathbf{Z}$ becomes

$$\min_{\mathbf{Z}^v} \&(\beta \mathrm{Tr}(\mathbf{Z}^v \mathbf{L}_S^v (\mathbf{Z}^{(v)})^\top) + \frac{\mu}{2} \|\mathbf{Z}^v - \mathbf{D}^v + \frac{\mathbf{Y}_2^v}{\mu}\|_F^2) \qquad (11)$$

By taking the derivative of $\mathbf{Z}$ and setting it to be zeros, we then have $\mathbf{Z}^{(v)^*} = (\mu \mathbf{D}^v - \mathbf{Y}_2^v)(2\beta \mathbf{L}_S^v + \mu \mathbf{I})^{-1}$.

▶ **Step 3. E-subproblem:** The other variables are fixed except $\mathbf{E}$, the $\mathbf{E}$ subproblem of problem (9) reduces to

$$\min_{\mathbf{E}^v} \& \alpha \|\mathbf{E}^v\|_1 + \frac{\mu}{2} \|\mathbf{E}^v - \mathbf{C}^v\|_F^2 \qquad (12)$$

where $\mathbf{C}^v = \mathbf{X}^v + \mathbf{B}^v(\mathbf{W}^v)^\top - \mathbf{D}^v + \frac{\mathbf{Y}_1^v}{\mu}$. Such subproblem can first be written in vector form, and efficiently solved by Lemma 3.2 in [8].

▶ **Step 4. B-subproblem:** The inferring matrix $\mathbf{B}$ can be obtained by fixing the other variables except $\mathbf{B}$ as follows

$$\min_{\mathbf{B}^v} \frac{\mu}{2} \|\mathbf{X}^v + \mathbf{B}^v(\mathbf{W}^v)^\top - \mathbf{D}^v - \mathbf{E}^v + \frac{\mathbf{Y}_1^v}{\mu}\|_F^2 \qquad (13)$$

By setting the first-order derivative of $\mathbf{B}$ to zeros, we can obtain the closed-form solution as $\mathbf{B}^v = -(\mathbf{X}^v - \mathbf{D}^v - \mathbf{E}^v + \mathbf{Y}_1^v/\mu) * \mathbf{W}^v * ((\mathbf{W}^v)^\top \mathbf{W}^v)^{-1}$.

▶ **Step 5. S-subproblem:** The other variables remain unchanged except $\mathbf{S}$, subproblem of $\mathbf{S}$ can be reformulated as

$$\min_{\mathbf{S}^v} \beta \sum_{v=1}^{r} \mathrm{Tr}(\mathbf{Z}^v \mathbf{L}_S^v (\mathbf{Z}^v)^\top) + \frac{\mu}{2} \|\mathbf{G}^v - \mathbf{S}^v + \mathbf{Y}_3^v/\mu\|_F^2$$
$$\Rightarrow \min_{\mathbf{S}^v} \|\mathbf{S}^v - \mathbf{A}^v\| \qquad (14)$$
$$\text{s.t.} \quad \forall v, \mathrm{Tr}(\mathbf{S}^v) = c, 0 \preccurlyeq \mathbf{S}_{ij}^v \preccurlyeq 1, (\mathbf{S}^v)^\top = \mathbf{S}^v$$

where $\mathbf{A}^v = \mathbf{G}^v + \mathbf{Y}_3^v/\mu - \frac{\beta}{\mu} * \mathbf{Q}^v$, and $q_{ij}^v = \|z_i^v - z_j^v\|_2^2$. q_{ij}^v is the ij-th element of $\mathbf{Q}^v$. For each $\mathbf{S}^v$, the optimal solution can be obtained by Theorem 1.

Theorem 1 *For a symmetric affinity matrix $\mathbf{S} \in \mathbb{R}^{n \times n}$, the spectral decomposition of $\mathbf{S}$ is denoted as $\mathbf{A} = \mathbf{U} \mathrm{Diag}(\delta)\mathbf{U}^\top$. The following problem*

$$\min_{\mathbf{S}} \|\mathbf{S} - \mathbf{A}\|_F^2 \ s.t. \ \mathrm{Tr}(\mathbf{S}) = c, \mathbf{S}^\top = \mathbf{S}, 0 \preceq \mathbf{S} \preceq 1 \qquad (15)$$

has optimal solution given by $\mathbf{S}^* = \mathbf{U} \mathrm{Diag}(\rho^*)\mathbf{U}^\top$, where ρ^* is the solution to

$$\min_{\rho} \|\rho - \delta\|_2^2, \ s.t. \ 0 \leq \rho \leq 1, \rho^\top 1 = c. \qquad (16)$$

Proof. For two symmetric matrices $\mathbf{S} \in \mathbb{R}^{n \times n}$ and $\mathbf{A} \in \mathbb{R}^{n \times n}$, and let $\rho_1 \geq \rho_2 \geq \cdots \geq \rho_n$ and $\sigma_1 \geq \sigma_2 \geq \cdots \geq \sigma_n$ be the ordered eigenvalues of $\mathbf{S}$ and $\mathbf{A}$, respectively. Due to the fact that $\text{Tr}(\mathbf{S}^\top \mathbf{A}) \leq \sum_{i=1}^{n} \rho_i \sigma_i$ shown in [13], we obtain

$$
\begin{aligned}
\|\mathbf{S} - \mathbf{A}\|_F^2 &= \text{Tr}\left(\mathbf{S}^\top \mathbf{S}\right) + \text{Tr}\left(\mathbf{A}^\top \mathbf{A}\right) - 2\text{Tr}\left(\mathbf{S}^\top \mathbf{A}\right) \\
&= \sum_{i=1}^{n} \rho_i^2 + \sum_{i=1}^{n} \delta_i^2 - 2\text{Tr}\left(\mathbf{S}^\top \mathbf{A}\right) \\
&\geq \sum_{i=1}^{n} \left(\rho_i^2 + \delta_i^2 - 2\rho_i \delta_i\right) \\
&= \|\rho - \delta\|_2^2
\end{aligned}
\tag{17}
$$

Note here that the above equality holds when $\mathbf{S}$ admits the spectral decomposition $\mathbf{A} = \mathbf{U}\text{Diag}(\sigma)\mathbf{U}^\top$. Addionally, the constraints $0 \preceq \mathbf{S} \preceq 1, \text{Tr}(\mathbf{S}) = c$ are equicalent to $0 \leq \rho \leq 1, \rho^\top 1 = c$, respectively. Thus, $\mathbf{S}^* = \mathbf{U}\text{Diag}(\sigma)\mathbf{U}^\top$ is optimal to problem (16) with σ^* being optimal to problem (17). After that, we can obtain the final solution, $\mathbf{S}^* = (\mathbf{S}^* + (\mathbf{S}^*)^\top)/2$, to satisfy the constraint $\mathbf{A}^\top = \mathbf{A}$. The proof is completed. ∎

Finally, an efficient iterative algorithm in [14] can be employed to solve (17).

▶ **Step 6. $\mathcal{G}$-subproblem:** The subproblem of $\mathcal{G}$ can be transformed as follows

$$
\min_{\mathcal{G}} \gamma \|\mathcal{G}\|_{\circledast} + \frac{\mu}{2} \|\mathcal{G} - \mathcal{S} + \mathcal{Y}_3/\mu\|_F^2
\tag{18}
$$

Let $\mathcal{M} = \mathcal{S} - \mathcal{Y}_3/\mu$ and according to the following Theorem 2, we can apply the tensor tubal-shrinkage of $\mathcal{M}$ to solve the problem (18).

Theorem 2 [28] *For a scalar $\tau > 0$ and two three-order tensors $\mathcal{T} \in \mathbb{R}^{n_1 \times n_2 \times n_3}$, $\mathcal{M} \in \mathbb{R}^{n_1 \times n_2 \times n_3}$, the global optimal solution of the following problem*

$$
\min_{\mathcal{T}} \tau \|\mathcal{T}\|_{\circledast} + \frac{1}{2} \|\mathcal{T} - \mathcal{M}\|_F^2
\tag{19}
$$

where $\tau = \frac{\gamma}{\mu}$. And (19) can be computed by tensor tubal-shrinkage operator as follows

$$
\mathcal{T} = \mathcal{C}_{n_3\tau}(\mathcal{M}) = \mathcal{U} * \mathcal{C}_{n_3\tau}(\mathcal{G}) * \mathcal{V}^\top,
\tag{20}
$$

where $\mathcal{M} = \mathcal{U} * \mathcal{G} * \mathcal{V}^\top$ and $\mathcal{C}_{n_3\tau} = \mathcal{G} * \mathcal{Q}$. $\mathcal{Q} \in \mathbb{R}^{n_1 \times n_2 \times n_3}$ denotes a f-diagonal tensor and each diagonal element of $\mathcal{Q}$ is defined as $\mathcal{Q}_f(i, i, j) = \left(1 - \frac{n_3\tau}{\mathcal{G}(i,i,j)}\right)_+$.

▶ **Step 7. ADMM variables-subproblem:** We update variables of ADMM via

$$
\begin{aligned}
\mathbf{Y}_1^v &= \mathbf{Y}_1^v + \mu(\mathbf{X}^v + \mathbf{B}^v(\mathbf{W}^v)^\top - \mathbf{D}^v - \mathbf{E}^v) \\
\mathbf{Y}_2^v &= \mathbf{Y}_2^v + \mu(\mathbf{Z}^v - \mathbf{D}^v) \\
\mathcal{Y}_3 &= \mathcal{Y}_3 + \mu(\mathcal{G} - \mathcal{S}) \\
\mu &= \min(\eta \mu, \mu_{\max})
\end{aligned}
\tag{21}
$$

where both η and $\mu_{\max}$ are the scalars of ADMM. The pseudo-code is depicted as in Algorithm 1, whose convergence condition is $\max\{|\text{obj}^{t+1} - \text{obj}^t|, \|\mathcal{S}^{t+1} - \mathcal{S}^t\|_F^2\} \leq$

Algorithm 1 Algorithm to the proposed IMGC method.

Require: Multiple incomplete data $\{\mathbf{X}^v\}_{v=1}^r, \alpha, \beta$ and $\gamma, \mu = 0.01, \mu_0 = 10^{10}, \eta = 2$.
Ensure: Graphs $\mathbf{S}^v$, inferring data $\mathbf{B}^v$, recovering data $\mathbf{D}^v$.
1: Initialize $\{\mathbf{D}\}_{v=1}^r, \{\mathbf{B}\}_{v=1}^r, \{\mathbf{E}\}_{v=1}^r, \{\mathbf{Z}\}_{v=1}^r, \{\mathbf{G}\}_{v=1}^r, \{\mathbf{Y}_1\}_{v=1}^r, \{\mathbf{Y}_2\}_{v=1}^r, \{\mathbf{Y}_3\}_{v=1}^r$ to
   be zeros. $\{\mathbf{S}\}_{v=1}^r$ are initialized by constructing k-nearest neighbor graphs from $\{\mathbf{X}^v\}_{v=1}^r$.
2: **repeat**
3:   Update the $\mathbf{D}$ via (10);
4:   Update the $\mathbf{Z}$ via (11);
5:   Update the $\mathbf{E}$ via (12);
6:   Update the $\mathbf{B}$ via (13);
7:   Update candidate graphs $\{\mathbf{S}^v\}_{v=1}^r$ via (14);
8:   Construct $\mathcal{S}$ via bvfold and rotate on $\{\mathbf{S}^v\}_{v=1}^r$;
9:   Update the tensor $\mathcal{A}$ via (19);
10:   Update the ADMM variables via (21);
11: **until** $\max\{|\text{obj}^{t+1} - \text{obj}^t|, \|\mathcal{S}^{t+1} - \mathcal{S}^t\|_F^2\} \le \epsilon$;
12: **Output** $\widehat{\mathbf{S}} = (\sum_{v=1}^r \mathbf{S}^v)/r$.

Table 1. Benchmark datasets.

Database	Class	View	Objective	Samples
BBCSport	5	4	Documents	116
Handwritten	10	2	Digit images	2000
Caltech7	7	6	Generic object	1474
MNIST	10	3	Digit images	10000

ϵ. Here, obj, t and $\epsilon = 10^{-5}$ are the objection value of function, iteration number and threshold value, respectively.

Computational Complexity: The computational complexity of (9) involves the following main subproblems, including (10), (11), (12), (13), (15), and (19). And the major computational cost involves (10), (11), (15), and (19). (10) needs to compute matrix SVD with the complexity $\mathcal{O}(rn^2)$ by leveraging an approximation rank technique like PROPACK package [6]. (11) requires the complexity of $\mathcal{O}(rn^3)$ due to the matrix inversion. (15) involves the skinny SVD operation to compute the c largest eigenvalue-eigenvector pair of graphs with $\mathcal{O}(rn^2)$. (19) involves fft and ifft operators on the $n \times r \times n$ tensor, the complexity is $\mathcal{O}(rn^2 \log(n))$, then it also requires to compute the SVD of each frontal slice with size $n \times r$ in the Fourier domain, thus the complexity is $\mathcal{O}(rn^2 \log(n)) + \mathcal{O}(r^2 n^2)$ in total. Here, we ignore (12), (13) and the variables of ADMM since they are basic matrix operations. The computational complexity of Algorithm 1 can be summarized as $\mathcal{O}(n^3)$ since t and r are far less than n.

4 Experiment

Evaluation and Datasets: As shown in Table 2, 4 popular datasets from various applications, cluster-numbers, view-numbers and sample-numbers are employed to evaluate

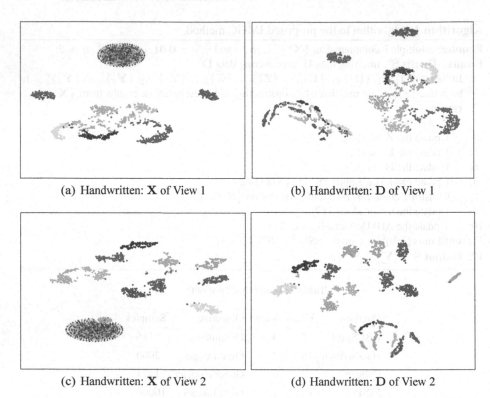

(a) Handwritten: **X** of View 1   (b) Handwritten: **D** of View 1

(c) Handwritten: **X** of View 2   (d) Handwritten: **D** of View 2

Fig. 2. The visualization via t-SNE [12] for original data and inferring complete data with missing rate of 30% on Handwritten dataset.

the compared methods with three criterions: accuracy (ACC), normalized mutual information (NMI), and Purity. Here, the bigger values of criterions indicate a better clustering performance. (1) **Handwritten** contains 2000 samples and averagely distributes in 10 classes, we merely choose pixel average features and Fourier coefficient features to construct two views. (2) **Caltech7** is a prevalent 7-object dataset with 1474 samples, which is consisted of six features corresponding to 6 views, including LBP, GIST, Gabor, wavelet moments, CENTRIST and HOG. (3) **BBCSport** has 116 samples from 5 classes, whose 4 views contains different feature dimensions with 2158, 2113, 2063 and 1991 in this paper. (4) **MNIST** consists of 70000 samples of 28-by-28 pixel size, where we randomly select the 10000 samples, and then following [29] to construct 3 different views.

Compared Methods: We compare our approach with two base line methods, including best single view (BSV) and Concat, as well as several state-of-the-art IMGC methods, including PVC [7], GPMVC [15], IMG [26], MIC [18], OMVC [17], DAIMC [3], UEAF [22], and GIMC-FLSD [23], APGLF [27].

Incomplete Data Construction: Following the incomplete data construction from [22], 10%, 30%, and 50% samples of all evaluated datasets (except MNIST) are randomly

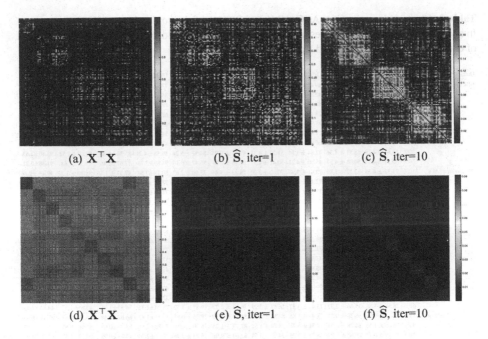

(a) $\mathbf{X}^\top \mathbf{X}$ (b) $\widehat{\mathbf{S}}$, iter=1   (c) $\widehat{\mathbf{S}}$, iter=10

(d) $\mathbf{X}^\top \mathbf{X}$ (e) $\widehat{\mathbf{S}}$, iter=1   (f) $\widehat{\mathbf{S}}$, iter=10

Fig. 3. The graph recovering with missing rate of 50%.

moved for each view to construct the incomplete multi-view datasets with corresponding missing rates, where each sample owns at least one view. Further, 10%, 30%, 50% and 70% samples are randomly selected as the paired samples for MNIST dataset. Then, we remove half of the remaining samples for one view, and follow the previous strategy to remove the next view until the last view.

Inferring Complete Data and Affinity Graph Recovery: Before performing Algorithm 1, the dense clusters of ellipse in Fig. 2 (a) and (c) are caused by the incomplete information of original data $\mathbf{X}$, where incomplete part is filled with 0. After performing our Algorithm 1, the dense clusters of ellipse in Fig. 2 (b) and (d) have a good discriminant, where incomplete information are inferred and filled with the valid information value. This intuitively proves the validity of inferring complete data $\mathbf{D}$.

As mentioned in Fig. 3, we have shown the affinity graph of our method on BBC-Sport and Handwritten datasets. It is easy to find that: 1) compared to original affinity graph (here, we adopt inner product to produce graph $\mathbf{X}^\top \mathbf{X}$.), the edges of graphs corresponding to the missing part can be gradually completed well with the increasing of iterations; 2) even in missing rate with 50%, both BBCSport and Handwritten have clear block diagonal structure, *i.e.*, blocks of graphs are equivalent to the class of original data, this plays a critical role for clustering.

Experimental Results: Following [23], we repeatedly perform 5 times independent experiments of all the comparison methods, and then the average clustering performance and standard deviation are presented in Table (2). From the experimental results of Table (2), we can obtain the following observations: 1) our RPCA-IGTL

Table 2. Clustering results (*i.e.*, ACC, NMI and Purity) of the comparison methods in terms of five varying scale datasets. The best results are highlighted in bold. For compared methods, most of experimental results are copied from [23].

Dataset	Metrics	MR	BSV	Concat	GPMVC	MIC	DAIMC	OMVC	UEAF	GIMC-FLSD	APGLF	our
BBCSport	ACC	10%	67.35±3.86	69.32±4.56	49.48±6.46	50.51±2.10	62.53±8.16	51.45±5.67	77.13±5.21	79.02±4.47	**91.38±4.21**	88.90±3.26
		30%	54.17±3.91	60.21±3.85	42.89±4.66	47.05±3.96	60.66±9.71	44.28±5.01	85.21±4.61	76.29±1.91	81.03±1.86	**84.48±1.26**
		50%	49.37±2.96	50.82±3.02	40.13±4.77	45.52±1.87	54.51±9.18	49.57±4.37	68.27±4.01	69.71±4.20	75.00±4.16	**78.45±3.81**
	NMI	10%	65.35±3.09	62.80±3.61	28.14±6.93	30.47±2.91	49.24±8.20	39.37±6.46	69.52±3.89	71.81±3.60	79.37±1.66	**81.09±1.83**
		30%	50.39±4.68	40.31±5.28	18.94±4.84	26.30±4.57	46.99±9.23	40.32±5.12	67.21±3.21	64.64±2.54	71.53±2.11	**80.31±3.68**
		50%	36.81±3.95	27.13±5.21	17.82±5.18	24.54±3.19	36.54±9.40	42.65±4.91	54.81±3.08	52.28±5.91	55.27±5.21	**73.07±2.96**
	Purity	10%	72.57±4.98	79.12±4.26	56.83±6.16	56.19±2.13	71.49±7.03	52.56±5.78	87.21±4.92	88.45±2.70	91.38±3.26	**93.97±3.61**
		30%	55.86±4.28	68.51±4.36	47.84±4.16	51.22±4.38	69.71±9.91	54.52±5.24	86.38±4.21	85.34±1.68	87.07±2.52	**93.10±2.63**
		50%	42.21±4.34	50.08±4.37	45.60±4.21	50.41±2.91	62.90±4.88	56.47±5.16	76.89±3.76	77.67±5.16	79.31±4.88	**91.38±3.21**
Handwritten	ACC	10%	71.32±4.35	82.12±1.89	79.61±3.96	70.62±3.20	78.23±3.49	73.90±2.63	80.18±2.56	87.27±1.98	88.10±1.72	**99.45±2.36**
		30%	55.31±4.21	78.31±1.27	77.15±3.28	63.07±2.59	75.15±2.01	68.19±3.78	77.12±2.38	84.29±1.62	87.45±2.11	**98.35±2.34**
		50%	41.43±5.89	66.91±1.14	71.28±2.49	55.17±3.29	64.28±3.92	59.12±3.50	71.26±2.89	77.27±2.28	81.55±2.69	**98.15±1.67**
	NMI	10%	58.85±1.48	78.20±2.90	72.19±2.16	63.27±2.11	64.16±2.17	61.16±3.01	70.13±1.58	79.16±1.06	89.04±1.21	**98.53±2.37**
		30%	52.03±1.20	65.34±2.13	69.16±2.31	55.21±1.98	61.16±2.37	58.22±3.38	65.23±1.21	75.27±1.95	85.22±1.31	**96.02±1.08**
		50%	45.52±1.16	56.31±2.66	67.26±3.03	50.15±2.16	52.15±2.11	51.15±2.16	61.31±1.27	69.83±1.21	82.85±1.29	**95.63±1.27**
	Purity	10%	66.15±1.98	83.01±1.56	81.15±2.84	73.11±2.46	78.26±2.17	73.16±2.27	83.72±2.67	87.93±1.62	88.10±1.74	**99.45±0.91**
		30%	58.86±1.35	70.23±2.09	78.22±2.38	64.27±2.27	74.27±2.53	67.27±2.83	80.12±2.98	83.16±1.15	85.50±1.13	**98.35±1.21**
		50%	52.29±1.69	61.35±2.95	72.27±3.93	58.72±2.27	65.82±3.82	60.27±3.28	73.21±3.52	76.87±2.27	82.85±76.33	**98.15±1.09**
Catech7	ACC	10%	61.32±2.78	45.48±2.67	43.34±3.30	41.77±3.64	42.26±4.03	38.89±2.64	51.63±4.23	48.20±1.55	66.82±1.35	**77.82±1.73**
		30%	45.23±3.15	44.31±1.54	40.81±4.90	40.47±3.36	41.16±3.49	37.77±3.91	43.62±1.02	47.14±1.12	48.14±2.65	**50.14±2.83**
		50%	44.24±1.87	40.61±3.01	34.25±3.93	38.88±4.95	38.35±2.78	36.50±2.79	37.11±3.16	44.08±2.41	46.31±1.88	**49.32±1.94**
	NMI	10%	45.83±1.29	45.71±1.04	29.08±3.00	35.52±1.72	42.71±2.65	27.74±1.57	40.13±2.36	44.84±0.85	50.34±2.06	**60.55±2.13**
		30%	36.25±3.12	41.74±2.45	21.24±7.05	31.10±1.95	40.29±2.44	23.76±3.47	32.21±2.08	42.32±1.05	45.16±3.84	**48.04±2.95**
		50%	28.58±1.56	35.66±2.31	13.41±3.31	26.67±3.29	36.22±2.27	19.87±3.32	25.24±1.51	36.53±1.91	42.65±1.37	**44.82±1.25**
	Purity	10%	84.17±0.94	86.28±0.39	78.41±1.86	80.86±1.12	84.63±1.37	78.95±0.88	82.31±1.87	86.62±0.34	86.58±0.87	**87.11±0.64**
		30%	76.28±1.36	84.28±0.91	74.31±6.28	78.24±1.62	83.79±1.25	76.90±2.19	79.88±2.56	85.63±0.65	85.96±0.97	**86.16±0.87**
		50%	72.19±0.77	79.68±0.84	68.71±3.75	74.99±3.58	82.60±0.96	75.23±2.46	77.61±2.30	83.33±0.96	84.27±1.86	**85.96±1.97**

overwhelmingly surpasses the compared methods, especially in Handwritten and BBC-Sports datasets, our method has a larger improvement with the increasing of missing rate, the reason may be that our method can well capture the graph spatial information of graph semantic space; and 2) the clustering performance decreases on all datasets for each method with the increasing of missing rate, and the great decline on BBCSport, and Handwritten datasets for BSV and Concat methods can demonstrate the validity of exploring inter-view information; 3) compared to UEAF and GIMC-FLSD, *w.r.t.* 50% high missing rate, our method improves 34.32% and 25.80% in terms of Purity, respectively. Compared to recent proposed APGLF [27], our method has also achieved the satisfied clustering performance. This profits from the learned inferring complete data and high-quality affinity graphs.

Note here that the datasets of Table (2) are the widely used to deal with the IMVC methods. To better evaluate the performance of the proposed method, we also employ the large scale MNIST to perform incomplete clustering and report the results in Table (3). As can be seen, our method still has a superior clustering performance in large scale dataset.

Parameter Sensitivity Analysis: Our RPCA-IGTL involves three parameters α, β, and γ required to be set properly, which can control the noise term $\mathbf{E}^v$, manifold regularized term $\mathrm{Tr}(\mathbf{D}^v \mathbf{L}_S^v) \mathbf{D}^{v\top}$, and the effect of spatial graph tensor $\mathcal{S}$, respectively. As shown in Fig.(4), it is insensitive for our parameters to work well for a wide range of α, β. Although it seems to be a little sensitive to γ, this indicates effectiveness of high-order spatial structure information. In fact, we always find satisfied clustering performance in a large range (*i.e.*, $\alpha \in [10^3, 10^{-4}]$, $\beta \in [10^{-5}, 10^{-1}]$, and $\gamma \in [10^{-3}, 10^{-1}]$).

Table 3. Experimental results of UEAF, GIMC-FLSD, APGLF and our RPCA-IGTL on the MNIST dataset with 10,000 samples.

Methods	Metrics	Missing rate	UEAF	GIMC-FLSD	APGLF	Proposed
MNIST	ACC	10%	80.28	86.18	89.31	**97.19**
		30%	77.39	83.06	87.69	**95.89**
		50%	69.28	71.68	84.46	**93.61**
		70%	64.72	68.82	80.29	**92.20**
	NMI	10%	69.86	74.25	88.45	**94.64**
		30%	67.71	70.41	86.12	**93.27**
		50%	58.36	61.32	73.96	**91.05**
		70%	49.16	56.26	70.45	**84.31**
	Purity	10%	80.28	86.18	88.46	**97.19**
		30%	78.31	83.06	85.46	**95.89**
		50%	70.03	72.21	74.16	**94.21**
		70%	66.26	69.18	71.79	**92.93**

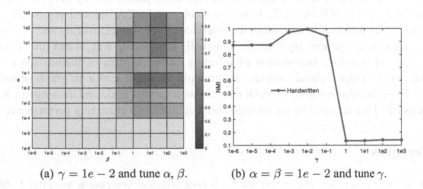

(a) $\gamma = 1e - 2$ and tune α, β.   (b) $\alpha = \beta = 1e - 2$ and tune γ.

Fig. 4. The NMI in terms of α, β and γ on the Handwritten dataset with a missing rate of 30%.

Convergence Analysis: To demonstrate the convergence of Algorithm (1), we experimentally record the objective values and NMI in terms of BBCSport and Handwritten datasets with missing rate 10%, 30%, 50% at each iteration. As illustrated in Fig. 5 , we find that: 1) the residual curves for different missing rate converge rapidly and consistently till to the stable point, where the objective values of each iteration are calculated via $\max\{|obj^{t+1} - obj^t|, \|\mathcal{S}^{t+1} - \mathcal{S}^t\|_F^2\} \leq \epsilon$; and 2) NMI *w.r.t.* iterator of RPCA-IGTL consistently and gradually increases until objective values become stable. This can prove the fast and stable convergence property of Algorithm (1).

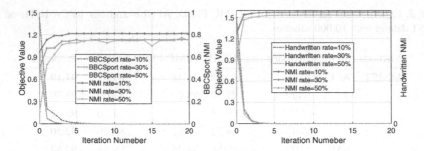

Fig. 5. The convergence curves and NMI of the proposed RPCA-IGTL method on BBCSport dataset in terms of 10%, 30%, and 50% missing rate.

5 Conclusion

Although the existing ANGL-based IMGC methods have the power of handling incomplete multi-view data, while they cannot take full use of the structure information hidden in the incomplete view data. To address these problems, we propose a novel IMGC method, *i.e.*, RPCA-IGTL. It designs an elegant RPCA-induced manifold learning framework and jointly introduces the graph tensor low-rank constrain and enhanced block diagonal constraint. By leverage a mutually reinforcing way, it can fully exploit the manifold structure information of inferring complete data in feature space and the inter-view graph spatial structure information of graph tensor in graph semantic space. Experimental results on several datasets with various scale and missing rate have demonstrated the superiority for inferred complete data and clustering performance.

References

1. Candès, E.J., Li, X., Ma, Y., Wright, J.: Robust principal component analysis? J. ACM (JACM) **58**(3), 1–37 (2011)
2. Hu, M., Chen, S.: Incomplete multi-view clustering. Springer International Publishing (2016)
3. Hu, M., Chen, S.: Doubly aligned incomplete multi-view clustering, pp. 2262–2268
4. Huang, S., Tsang, I., Xu, Z., Lv, J.C.: Measuring diversity in graph learning: a unified framework for structured multi-view clustering. In: IEEE TKDE, pp. 1–1 (2021)
5. Wen, J., Zhang, Z., Xu, Y., Zhong, Z.: Incomplete multi-view clustering via graph regularized matrix factorization. In: Leal-Taixé, L., Roth, S. (eds.) ECCV 2018. LNCS, vol. 11132, pp. 593–608. Springer, Cham (2019). https://doi.org/10.1007/978-3-030-11018-5_47
6. Larsen, R.M.: Propack-software for large and sparse svd calculations 2008-2009 (2004). http://sun.stanford.edu/rmunk/PROPACKpp
7. Li, S.Y., Jiang, Y., Zhou, Z.H.: Partial multi-view clustering. In: AAAI, vol. 28 (2014)
8. Liu, G., Lin, Z., Yan, S., Sun, J., Yu, Y., Ma, Y.: Robust recovery of subspace structures by low-rank representation. IEEE Trans. Pattern Anal. Mach. Intell. **35**(1), 171–184 (2012)
9. Liu, X., et al.: Efficient and effective regularized incomplete multi-view clustering. IEEE Trans. Pattern Anal. Mach. Intell. **43**, 2634–2646 (2020)
10. Liu, X., et al.: Multiple kernel k k-means with incomplete kernels. IEEE Trans. Pattern Anal. Mach. Intell. **42**(5), 1191–1204 (2019)

11. Lu, C., Feng, J., Chen, Y., Liu, W., Lin, Z., Yan, S.: Tensor robust principal component analysis with a new tensor nuclear norm. IEEE Trans. Pattern Anal. Mach. Intell. **42**(4), 925–938 (2020)
12. Van der Maaten, L., Hinton, G.: Visualizing data using t-sne. J. Mach. Learn. Res. **9**(86), 2579–2605 (2008)
13. Marshall, A.W., Olkin, I., Arnold, B.C.: Inequalities: Theory of Majorization and Its Applications. SSS, Springer, New York (2011). https://doi.org/10.1007/978-0-387-68276-1
14. Nie, F., Wang, X., Jordan, M.I., Huang, H.: The constrained laplacian rank algorithm for graph-based clustering. In: Thirtieth AAAI Conference on Artificial Intelligence, pp. 1969–1976 (2016)
15. Rai, N., Negi, S., Chaudhury, S., Deshmukh, O.: Partial multi-view clustering using graph regularized nmf. In: 2016 23rd International Conference on Pattern Recognition (ICPR), pp. 2192–2197. IEEE (2016)
16. Rai, P., Trivedi, A., Daumé III, H., DuVall, S.L.: Multiview clustering with incomplete views. In: NIPS Workshop, pp. 1–4 (2010)
17. Shao, W., He, L., Lu, C.t., Philip, S.Y.: Online multi-view clustering with incomplete views. In: 2016 IEEE International Conference on Big Data (Big Data), pp. 1012–1017 (2016)
18. Shao, W., He, L., Yu, P.S.: Multiple incomplete views clustering via weighted nonnegative matrix factorization with $L_{2,1}$ regularization. In: Appice, A., Rodrigues, P.P., Santos Costa, V., Soares, C., Gama, J., Jorge, A. (eds.) ECML PKDD 2015. LNCS (LNAI), vol. 9284, pp. 318–334. Springer, Cham (2015). https://doi.org/10.1007/978-3-319-23528-8_20
19. Shi, S., Nie, F., Wang, R., Li, X.: Fast multi-view clustering via prototype graph. IEEE Transactions on Knowledge and Data Engineering, pp. 1–1 (2021)
20. Wen, J., Yan, K., Zhang, Z., Xu, Y., Zhang, B.: Adaptive graph completion based incomplete multi-view clustering. IEEE Trans. Multimedia **23**, 2493002504 (2020)
21. Wen, J., Xu, Y., Liu, H.: Incomplete multiview spectral clustering with adaptive graph learning. IEEE Trans. Cybern. **50**(4), 1418–1429 (2018)
22. Wen, J., Zhang, Z., Xu, Y., Zhang, B., Fei, L., Liu, H.: Unified embedding alignment with missing views inferring for incomplete multi-view clustering. In: Proceedings of the AAAI Conference on Artificial Intelligence, vol. 33, pp. 5393–5400 (2019)
23. Wen, J., Zhang, Z., Zhang, Z., Fei, L., Wang, M.: Generalized incomplete multiview clustering with flexible locality structure diffusion. IEEE Trans. Cybern. **51**(1), 101–114 (2020)
24. Zhang, C., Fu, H., Hu, Q., Cao, X., Xie, Y., Tao, D., Xu, D.: Generalized latent multi-view subspace clustering. IEEE Trans. Pattern Anal. Mach. Intell. **42**(1), 86–99 (2018)
25. Zhang, P., et al.: Adaptive weighted graph fusion incomplete multi-view subspace clustering. Sensors **20**(20), 5755 (2020)
26. Zhao, H., Liu, H., Fu, Y.: Incomplete multi-modal visual data grouping. In: IJCAI, pp. 2392–2398 (2016)
27. Zheng, X., Liu, X., Chen, J., Zhu, E.: Adaptive partial graph learning and fusion for incomplete multi-view clustering. Int. J. Intell. Syst. **37**(1), 991–1009 (2022)
28. Zhou, P., Lu, C., Feng, J., Lin, Z., Yan, S.: Tensor low-rank representation for data recovery and clustering. IEEE Trans. Pattern Anal. Mach. Intell. **43**(5), 1718–1732 (2019)
29. Zhou, S., et al.: Multi-view spectral clustering with optimal neighborhood laplacian matrix. In: Proceedings of the AAAI Conference on Artificial Intelligence, vol. 34, pp. 6965–6972 (2020)

TRUST-TECH Assisted GA-SVM Ensembles and Its Applications

Yong-Feng Zhang[1]([✉]), Hsiao-Dong Chiang[2], Yun-Fei Qu[3], and Xiao Zhang[1]

[1] University of Jinan, Jinan 250022, SD, China
cse_zhangyf@ujn.edu.cn
[2] Cornell University, Ithaca, NY 14853, USA
[3] State Grid Baoding Electric Power Supply Company, Baoding 071000, HB, China

Abstract. A framework of Genetic Algorithm-Support Vector Machine (GA-SVM) is proposed for SVM parameters (model) selection, and clustering algorithm is also integrated with the framework to generate multiple optimal models, as well as being condition of convergence for GA. Moreover, an ensemble method on various SVM models assisted by TRUST-TECH methodology is put forward, to enhance the generalization ability of a single SVM model. The performance of GA-SVM and ensemble method is testified by applying them in both classification and regression problems. Results show that, comparing with traditional parameters selection method (such as grid search), the proposed GA-SVM framework and ensemble strategy can solve general classification and regression issues more efficiently and automatically with better performance.

Keywords: Support vector machine · Genetic algorithm · Clustering algorithm · SVM model selection · Ensemble

1 Introduction

Support Vector Machine (SVM) has become a competitive learning algorithm since it was developed by Vapnik and his co-workers in 1995 [1]. For recent decades, abundant literature has proven its effectiveness in solving classification problems in pattern recognition and regression analysis, which could be partially attributed to that SVM is founded on the basis of Structural Risk Minimization (SRM) [2, 3]. Traditional machine learning algorithms, such as Artificial Neural Network (ANN), form the learning rules based on the principle of Empirical Risk Minimization (ERM), i.e., the minimization of training errors. However, SRM balances ERM and VC dimension to ensure better generalization ability. The VC dimension denotes the learning ability of a set of functions, or informally, the higher the VC dimension is, the more complicated the learning machine becomes.

One major advantage of SVM is that the optimal solution is unique when model parameters are fixed, since the constructing/construction of SVM model equals solving a convex quadratic programming problem. However, challenges exist at least in these fields: 1) selection of model parameters are usually empirical; 2) training process, especially for huge data, can be computationally expensive. The Sequential Minimization Optimization (SMO) [4] algorithm is one of the most popular methods in solving the QP problems in SVM and affirms its valid in many aspects and applications. Hence, we focus more on the first challenge in this paper. Selection of model parameters, (or referred as model selection, tuning parameters) aims to find the optimal parameters that minimize the generalization errors of the SVM model, and these errors can often be estimated either by direct testing based on data which have not been used for training or by given bounds in previous studies such as [5–7]. Selection of parameters has always been intractable since multiple parameters should be tuned simultaneously and searching space of parameters could be problems-depended in most cases. Usual applicable strategies include grid search [8], heuristic algorithms [9, 10], and gradient-based meth ods introduced in [11]. The Grid search method is always exhaustive and its accuracy may depend on the steps of the 'grid', while traditional GA or SA may be stuck in a local optimal solution and be time-consuming for convergence.

Another strategy for improving generalization ability is the ensemble of several SVM models. Literature [12] defines the ensemble of classifiers as 'a set of classifiers whose individual decisions are combined in some way (typically by weighted or unweighted voting) to classify new examples'. The literature also illustrates the reason why ensembles could often outperform individual classifiers from views of statistics, computations and representations. The ensemble of SVM actually consists of two tasks: 1) generation of different SVM models; 2) combination of these models in an appropriate manner. Literature [13] describes representative methods of both tasks and discusses their performance based on simulation results.

In this paper, we put forward a novel SVM ensemble method by adopting GA-combined clustering algorithm and TRUST-TECH technologies [14, 15]. GA combined with SVM (or GA-SVM) hybrid algorithm has shown its superior ability compared with conventional SVM applications [16, 17]. Moreover, the concept of clustering is introduced in SVM parameters tuning to ensure the diversity of SVM models as an ensemble, as well as being a criterion to end the evolution of GA. TRUST-TECH was developed to find high-quality solutions for general nonlinear optimization problems. It has been successfully applied to solve machine learning problems including optimal training ANNs, and in this paper it is adopted for optimal ensemble of SVM models.

2 Preliminaries

2.1 Support Vector Machine

SVM was first introduced to solve data classification problems or pattern recognition (Support Vector Classification, SVC), and the maximal margin classifier forms the strategy of the first SVM, namely to find the maximal margin hyperplane in an appropriately chosen kernel-induced feature space [18]. For a given training set of pairs (x_i, y_i), where the instance $x_i \in R^n$ and the label $x_i \in \{1, -1\}$, a typical representation of the

implementation is given as (1)

$$\min_{\omega,b,\xi} \quad \frac{1}{2}\omega^T\omega + C\sum_{i=1}^{l}\xi_i$$
$$s.t. \quad y_i(\omega^T\phi(x_i) + b) \geq 1 - \xi_i, \tag{1}$$
$$\xi_i \geq 0.$$

Here, C is the penalty parameter of the error and is always positive. ξ_i is the slack variable which means allowed errors. Kernel function $K(x_i, x_j) = \phi(x_i)^T\phi(x_j)$ can map non-separable data (input space) into a higher dimension (feature space) where these mapped data would be separable, and usual kernels include radial basis function (RBF) $K(x_i, x_j) = \exp(-\gamma\|x_i - x_j\|^2)$, $\gamma > 0$ linear one $K(x_i, x_j) = \phi(x_i)^T\phi(x_j)$, and polynomial one $K(x_i, x_j) = (\gamma x_i^T x_j + \gamma)^d$, $\gamma > 0$, where γ, d are kernel parameters. Therefore, the free parameters for classification in a given case consist of C and kernel parameters.

SVM for regression analysis (or termed Support Vector Regression, SVR) is developed, for example in ε-SVM to find a function $f(x)$ that allows a deviation less than ε from the actual target y_i for all training data, and meanwhile make it as flat as possible. The expression is given by (2)

$$\min_{\omega,b,\xi} \quad \frac{1}{2}\omega^T\omega + C\sum_{i=1}^{l}(\xi_i + \xi_i^*)$$
$$s.t. \quad y_i - \omega^T\phi(x_i) - b \leq \varepsilon + \xi_i, \tag{2}$$
$$\omega^T\phi(x_i) + b - y_i \leq \varepsilon + \xi_i^*,$$
$$\xi_i, \xi_i^* \geq 0.$$

Compared with (1), the additional parameter ε means that a deviation less then ε would be ignored or could be described as the so-called ε-insensitive loss function $|y_i - f(x_i)|_\varepsilon$ as (3) and Fig. 1 presented.

$$|y_i - f(x_i)|_\varepsilon = \begin{cases} 0 & |y_i - f(x_i)| - \varepsilon \\ |y_i - f(x_i)| - \varepsilon, & Otherwise \end{cases} \tag{3}$$

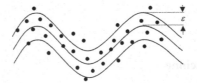

Fig. 1. ε-insensitive loss function

Lots of tools have been developed to solve SVM problems, among which LIBSVM is popular for its high efficiency in SVM classification, regression, probability estimation and other tasks [19]. LIBSVM is also adopted in the following study.

2.2   Genetic Algorithms

Genetic Algorithms (GA), are stochastic methods based on imitating Darwinian natural selection and genetics in biological systems, which have been successfully used for globally researching and optimizing problems [20]. In optimization problems, a set of candidate solutions is termed as population and is coded by specific chromosomes. And all candidate solutions will be evaluated by the objective function (often problems-based) to find the best solutions which can be used to build the next generation of the candidate ones. The successive steps could be described as follows: initialization of the first generation, evaluation of the candidate solutions, selection of the best ones and creation of the next generation using specific genetic operators (crossover, mutation, etc.). The iterative computation will be stopped when it comes to meeting the termination condition [9].

Although it has been applied in many fields, GA still gets its drawbacks: weak ability in local optimal search, premature convergence in certain situations, random solutions, and all these may make this algorithm time-consuming and difficult to guarantee its convergence. Thus, one aim of improvement on GA is how to speed up the convergence to get optimal solutions.

2.3   GA-Combined Clustering

Clustering is a common technique for statistical data analysis, and its main task is to group a set of objectives in a way that they are more similar within the same group compared with objectives in others. General clustering algorithms include K-means, K-modes and their variations.

Iterative Self-organizing Data Analysis Techniques Algorithm (ISODATA) [21] is developed based on K-means and introduces the operation of 'splits' and 'merges' on clusters, so the number of clusters is also variable. Basic processes could be summarized as: 1) Initialization of controlling parameters; 2) Randomly placing the cluster centers, and assigning samples to clusters based on their distance to cluster centers; 3) Calculation of standard deviation within each cluster and the distance between cluster centers, and splitting or merging clusters; 4) A second iteration is performed in the new cluster centers; 5) Termination conditions include the average inter-center distance which falls below the user-defined threshold, the average change in the inter-center distance between iterations that is less than a threshold, or the maximum number of iterations reached.

At the end of the evolution process of GA, a set of optimal solutions are obtained. Evaluation of these optimal solutions utilizing clustering method (ISODATA is preferred in this paper) is to analyze the probability of grouping strategy termed GA-combined Clustering Algorithm (GACA). The purpose of GACA could be addressed as: firstly, this method provides a termination condition for GA, avoiding potential time-consuming evolution process with little effect; secondly, this method describes the distinctions of optimal solutions generated by GA, and is meaningful when multi-local optimal solutions are needed for further work (e.g., ensemble of multiple solutions).

2.4 TRUST-TECH Technology

TRUST-TECH was developed to find high-quality solutions for general nonlinear optimization problems. It has been successfully applied to solve machine learning problems including optimal training ANNs [22, 23], estimating optimal parameters for finite mixture models [24], as well as solving the optimal power flow problem [25].

TRUST-TECH-based methods can escape from a local optimal solution and search for other solutions in a systematic and deterministic way. Another feature of TRUST-TECH is its effective cooperation with existing local and global methods. This cooperation starts with a global method for obtaining promising solutions. Then by working with robust and fast local methods, TRUST-TECH efficiently searches the neighboring subspace of the promising solutions for new local optimal solutions in a tier-by-tier manner. A high-quality optimum can be found from the multiple local optimal solutions.

2.5 Ensemble of SVMs

Ensemble on machine learning has been well studied in [12] and the literature also explains why ensemble can often perform better than any single classifier. Ensemble has also been widely applied in ANN, and can offer an effective way to alleviate the burden of tuning the parameters of a single ANN, moreover, the results show ensemble is effective in improved generalization capability [26, 27]. Factors may affect the ensemble results including accuracy and diversity of member networks [28–30] and the combination strategy used for ensemble [31, 32].

Ensemble of SVMs has also been studied, [13] expects that SVM ensemble can improve classification performance greatly than using a single one, especially in multi-classification cases, and propose to use the SVM ensemble based on the bagging and boosting techniques. Literature [2] uses the boosting technique to train each SVM and take another SVM for combining several SVMs.

In this paper, firstly, a set of SVM individuals would be trained by GA-combined clustering method; secondly, a selection of these SVM individuals would be optimally combined, to form a nonlinear optimization problem; finally, ensemble of SVM individuals will be finished by solving the optimization problem with TRUST-TECH technology.

3 Algorithm Procedures

3.1 SVM Parameters

Given the training data set and methods, SVM parameter p would decide the unique SVM model. Thus, the selection of optimal parameters is actually the process of constructing SVM with best generalization ability. As mentioned in part A, Section II, free parameters in SVC include the penalty parameter C and kernel parameters. If using RBF (Gaussian) kernel $K(x_i, x_j) = \exp(-\gamma \|x_i - x_j\|^2)$, $\gamma > 0$, the SVC parameters to be optimized here will be defined as $p_c = (C, \gamma)$, and meanwhile in SVR the similar definition is $p_r = (C, \varepsilon, \gamma)$ and ε is the tolerable deviation.

The goal of parameters selection (optimization) is to identify proper p_c or p_r that makes accurate classification or regression results on unknown data. One common method, as described in [8], suggests using a grid-search method with cross-validation. One typical cross-validation method is called v-folder cross-validation, which separates all training data with the purpose of forming v subsets with equal size. And then, each subset (validating data) is valid using the SVM trained on the other v-1 subsets (training data). The operation repeats v times so that each subset has been valid and got responding accuracy. The average of all v accuracy could be seen as the accuracy to evaluate the selected parameters. According to [8], cross-validation can also prevent over-fitting problems. One extreme situation of the cross-validation method is the Leave-One-Out (LOO) method, where v is equal to the number of instances of all training data. Thus, all instances will be valid by utilizing the LOO method. Literature [5] proves that LOO can provide an unbiased estimate on the probability of test errors for known data. Yet, the LOO method may be time-consuming, especially when the number of training instances is huge. Grid-search tries p_c or p_r in the potential solution space according to some simple rules, for example, an exponentially growing sequences of C and γ in [8]. This method is straightforward and easy for implementation. However, the solution space in SVR may become three dimensions and its efficiency is worrisome.

3.2 GA-SVM for Parameter Optimization

In this paper, the algorithm of optimizing parameters of SVM automatically by GA is termed GA-SVM. Input of this algorithm is normally an approximate interval of parameters, and output would be a set of optimal parameters. One significant element in the evolution process of GA-SVM is the fitness function, which is a predefined objective function that evaluates all individuals in each generation and then ranks them accordingly. Figure 2 gives the framework of this algorithm.

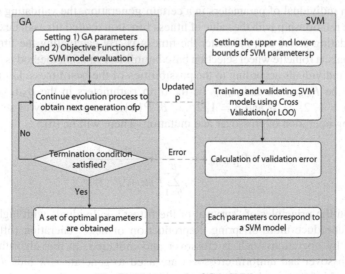

Fig. 2. Framework of GA-SVM

It has to be elaborated in the framework:

1) Upper and lower bounds of parameters p are usually decided by a coarse cross validation process. The simplest case, searching region for p_c is a closed rectangle region and p_r a closed cuboid region.
2) Important GA parameters include size of the population α and evolution generation β. From the view of parameters selection, λ is the number of optimal parameters in each evolution process and β is the number of evolution times of parameters.
3) In the process of υ-folder cross validation, let the number of all known instances is N, and each subset consists of $M = N/\upsilon$ instances where the correctly classified number of instances of the jth validating subset is m_j. Thus, validating error of SVC is defined as (4)

$$\sigma_c = (1 - \frac{1}{\upsilon} \sum_{j=1}^{\upsilon} \frac{m_j}{M}) * 100\% \tag{4}$$

4) Similarly, in SVR, validating error is defined as (5), which is actually the average of Mean Absolute Percentage Error (MAPE). In (5), $f_j(x)$ represents the estimation function of the jth validating subset, and y_i is the actual value of the corresponding x_i.

$$\sigma_r = \frac{1}{\upsilon} \sum_{j=1}^{\upsilon} \sum_{i=1}^{M} \frac{\left| \frac{f_j(x_i) - y_i}{y_i} \right|}{M} * 100\% \tag{5}$$

5) At the beginning of the first loop, i.e., the initialization process, GA will randomly generate λ parameters and then calculate fitness of these parameters.
6) For each individual of parameters in a certain generation, the validating error is of negative relationship with the value of fitness function (called fitness), i.e., the lower the validating errors are, the better the fitness is. By calculating the fitness of all individuals, a roulette wheel selection based nonlinear ranking method is adopted to rank all individuals according to the possibilities of the best fitness. Let the fitness function be $F_{fitness}(x)$, and $p_{\alpha_a \beta_b}$ be the α_ath parameters in the β_bth generation, where $a = 1, 2, ..., \alpha$ and $b = 1, 2, ..., \beta$. $P\{x\}$ is the possibility of selection for following operation of crossover and mutation. There will be (6):

$$P\{p_{\alpha_a \beta_b}\} = \frac{F_{fitness}(p_{\alpha_a \beta_b})}{\sum_{b=1}^{\beta} F_{fitness}(p_{\alpha_a \beta_b})} \tag{6}$$

7) Based on the principles of "survival of the fittest", individuals with higher fitness would reproduce more offspring. Reproduction of next generation (offspring) is realized by operations such as crossover and mutation. In this algorithm, multi-point crossover and uniform crossover are used to generate new individuals with exchanged information between old individuals. Mutation reflects another randomness in the evolution process, which generates new individuals in case that crossover is not effective. Details about the process are presented in [21].

8) In GA-SVM, the evolution process will be stopped until the βth evolution finishes. During the evolution process, parameters in each generation are recorded. At the end, all recorded parameters are compared in order to select one with the lowest validating errors as the best one to build a SVM model.

9) It has to be noted that the best parameter is not always generated at the βth evolution. Numerical studies later in this paper show the evolution process may come to convergence earlier. This explains why a fixed β may be time-consuming with low efficiency.

3.3 Improvement on GA-SVM Using Clustering Method

However, drawbacks of GA-SVM for parameters selection are obvious, especially from the view of ensemble. 1) criteria for convergence is not guaranteed, because a small β may not ensure better evolution results and a big β may be tedious with little efforts; 2) diversity of SVM models for ensemble is not available, for that the evolution process does not consider differences of parameters within one generation and between successive generations.

Clustering method would help in GA-SVM, and in this paper we combine ISODATA with GA (or termed GACA) to optimize SVM parameters. In GACA, the evolution process begins with a much smaller generation $\tilde{\beta}$, and after/when the evolution finishes, a set of optimal parameters are obtained. Then, ISODATA is introduced to analysis the internal relationship of parameters within this set and compared it with the previous one. If the results of clustering satisfy the termination condition, the algorithm finishes; if not, evolution process with generation $\tilde{\beta}$ will continue based at the foundation of the last evolution results and repeat the comparing operation. The algorithm would stop until the termination condition satisfies that predefined repeating times are reached. The procedure is described in Fig. 3.

After clustering on $\{p_1\}$ finishes, GA-SVM evolution will directly generate $\{p_2\}$, i.e., comparison of clustering results happens for the first time only after clustering on $\{p_2\}$ finishes.

In the end, a set of clustered parameters $\{p_{cld}\}$ are acquired, including corresponding clusters. Moreover, in considering of Euclidean distance among parameters, the diversity of SVM models is guaranteed.

3.4 Ensemble on SVM Using TRUST-TECH

Let K be the number of selected clusters of $\{p_{cld}\}$, and p_{cld}^i is a representative parameter (normally the cluster center) selected from the ith cluster, where $i = 1, 2, ..., K$. Usually, p_{cld}^i is the one with the lowest validating error of all parameters in the ith cluster, and all p_{cld}^i should be within a neighborhood of $\min\{p_{cld}\}$. e.g., $p_{cld}^i \leq \eta \min\{p_{cld}\}$, where η is a factor ensuring the low validating error of selected parameters. Hence, parameters with high parameters would not be chosen for the ensemble. The task of finding an optimal ensemble on SVM models is achieved by solving the following optimization problem:

$$\min \ E(v|S, x) = \sum_{n=1}^{Q} (\sum_{m=1}^{K} v_m \circ f_m(x_n) - y_n)^2 \tag{7}$$

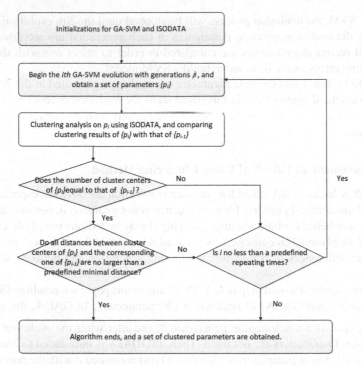

Fig. 3. Framework of Improving GA-SVM using ISODATA

$E(v|S, x)$ is the ensemble function under a given set of SVM models $S = \{s_1, s_2, ...,$ $s_K\}$, v is the rules used for combination, $\{x_n\}$ is the testing data set that is not used in the training and validating process, Q is the number of samples in the testing data set, $f_m(x_n)$ is the estimation function of mth SVM models when input is x_n, and y_n is the desired output. Not that in SVR output of $f_m(x_n)$ is real number, and in SVC (binary classification) only integer 1 or -1.

In this paper, a linear combination of SVM models is considered. Thus $v = (v_1, v_2, ..., v_K)$ represents the weight of each model and an optimization problem turns to a QP problem (8):

$$\min_v \quad E(v) = \frac{1}{2}v^T C v$$
$$s.t. \quad v^T e = 1, \quad v \geq 0 \tag{8}$$

where $e = [1, 1, ..., 1]^T$, C is the correlation matrix describing the relationship between different SVM models. In SVR, C is calculated using (9)

$$C_{ij} = \frac{1}{Q}\sum_{k=1}^{Q}(f_i(x_k) - t_k)(f_j(x_k) - t_k), \quad i, j = 1, 2, ..., K \tag{9}$$

While in SVC, C is calculated using (10)

$$C_{ij} = \frac{1}{Q} \sum_{k=1}^{Q} \sigma_i \sigma_j, \quad i, j = 1, 2, ..., K \tag{10}$$

where σ_i denotes the validating error of the ith SVM tested on a testing data of Q samples.

Since C does not always have positive eigenvalues, the quadratic optimization problem (8) may be non-convex. Therefore, it might have multiple local optimal solutions. Traditional iterative methods (e.g., Interior Point Method, IPM) which are effective in solving the convex quadratic optimization problem may get stuck in a local optimal solution in case of multiple local optimal solutions. TRUST-TECH technology is used to solve (8), and a high-quality solution will be adopted for ensemble.

Using the logarithmic barrier function, the augmented Lagrange function of (8) is

$$L_\mu(v, \theta) = \frac{1}{2} v^T C v + \theta(v^T e - 1) + \mu \sum_{i=1}^{K} \ln v_i \tag{11}$$

where μ is the barrier parameter. Hence, the Karush-Kuhn-Tuker's optimal conditions are

$$\frac{\partial L_\mu}{\partial v} = C v + \theta e - \mu V^{-1} e = 0 \tag{12}$$

$$\frac{\partial L_\mu}{\partial \theta} = v^T e - 1 = 0 \tag{13}$$

where $V = \text{diag}(v_1, v_2, ..., v_K)$. Multiplying both sides of (12) with V, we have

$$H_\mu(v, \theta) = \begin{pmatrix} VCv + \theta Ve - \mu e \\ v^T e - 1 \end{pmatrix} = 0 \tag{14}$$

By solving (12) and (13) with decreasing $\mu \to 0$, IPM will provide a local optimal solution to the original problem (8) where $\mu = 0$.

Actually, IPM plays a role of local solver, and then TRUST-TECH is used to compute multiple local optimal solutions following procedure below:

Step1: Initialization. Set the initial point $v_0 = (1/K, 1/K, ..., 1/K)$ and the set of SEPs $V_s = \emptyset$.

Step2: Calculate the correlation matrix C and compute its eigenvalues $\lambda_1, \lambda_2, ..., \lambda_K$.

Step3: Use v_0 as the initial point, apply the IPM to solve (8) and get an SEP v_{s0}, and update $V_s = \{v_{s0}\}$.

Step4: If $\min_{i=1}^{K} \lambda_i < \rho$ (ρ is a very small positive value), calculate the search direction $\{\vec{d}_1, \vec{d}_2, ...\vec{d}_K\}$.

Step5: For $i = 1:m$, search for an exit point v_e along $\vec{d}_i$ in the generalized gradient system (15),

$$\frac{dx}{dt} = -\nabla H^T(x) H(x) \tag{15}$$

which is defined based on problem (14) with associate energy function

$$E(x) = \frac{1}{2}\|H(x)\|^2 \tag{16}$$

If υ_e along $\vec{d_i}$ is found, step forward along the search direction to the point $\upsilon' = \upsilon_{s0} + \tau(\upsilon_e \quad \upsilon_{s0})$ with τ being a small positive value. υ' will lie in the stability region of the neighboring SEP.

Then, use υ' as an initial point, apply the IPM to get the tier-1 SEP, denoted as υ_{si}, lying in the neighboring stability region, and update V_s as $V_s = V_s \cup \{\upsilon_{si}\}$.

Step6: The optimal combination vector is obtained as $\upsilon^* = \text{argmin}_\upsilon\{E(\upsilon)|\upsilon \in V_s\}$.
End

The final ensemble function of SVR is expressed as (17)

$$Y(x_n) = \sum_{m=1}^{K} v_m f_m(x_n) \qquad m = 1, 2, ...K \tag{17}$$

where $Y(x_n)$ is the final regression results with input x_n.

In the case of SVC (binary classification), a weight-based voting process will determine the final results as (18)

$$Y(x_n) = \begin{cases} 1 & \sum_{m=1}^{K} v_m f_m(x_n) > 0 \\ -1 & \sum_{m=1}^{K} v_m f_m(x_n) < 0 \\ f_{m'}(x_n) & \sum_{m=1}^{K} v_m f_m(x_n) = 0 \end{cases} \tag{18}$$

where $f_{m'}(x)$ is the estimation function corresponding to the maximum $v_m \in \upsilon^*$. For multiple-class problems, a similar voting strategy is adopted, where the basic principle is that the instance is tagged to the most voted classification, and if two or more classes win the same vote, the estimation with the highest weight reaches the final decision.

4   Application of Algorithm in Regression

In this chapter, we will solve a regression problem by using GA-SVM and ensemble. Instead of clustering algorithms, independent repeating of GA-SVM will be applied to generate multiple SVM models, and then all these models are ranked where top ones are selected for ensemble.

4.1   Problem Definition

[33] points out that among all SVR problems, financial data and electric load time series prediction appear to be the worthiest topics. In this thesis, the load forecasting problem in EUNITE Network Competition 2001 is studied. According to [34], the problem is described as follows.

Given Data:

- *Electricity load demand recorded every half hour, from 1997 to 1998.*
- *Average daily temperature, from 1995 to 1998.*
- *Dates of holidays, from 1997 to 1999.*

Task:

- *prediction of maximum daily values of electrical loads for January 1999*

Main evaluation on results:

$$MAPE = 100 * \frac{\sum_{i=1}^{n} \left| \frac{L_i - \hat{L}_i}{L_i} \right|}{n}, n = 31 \qquad (19)$$

To make the final results comparable, raw data are preprocessed similar to [34].

Unlike parameters in SVC, the ones in SVR are a three- dimension $p_r = (C, \varepsilon, \gamma)$ and thus a grid-search method will be more computationally complex (tries of solution increasing from n^2 to n^3, where n is the number of 'grid'). Anyway, GA-SVM deals with this problem similar to SVC.

4.2 Multiple Models Generated by GA-SVM

In this stage, GA-SVM is firstly repeated ρ times and it will get ρ independent optimal parameter. According to [2], SVM models with smaller ω tend to be of higher generalization ability. Then, all ρ parameters would be ranked according to ω in descending orders, and the top ρ' parameter will be used for ensemble.

In the electric load forecasting problems, 50 optimal parameters (or models $S = \{s_1, s_2, ..., s_{50}\}$) are generated by repeating GA-SVM independently and then ranked. Top 10 models $S' = \{s_{39}, s_{49}, s_{47}, s_7, s_{50}, s_{10}, s_{18}, s_{23}, s_{21}, s_{17}\}$ are selected further according to the value of ω, as showing in Table 1. By solving the corresponding optimization problem (8), weight of each model of S' is obtained, denoted as $v' = (0.0365, 0, 0.4845, 0.2358, 0.2432, 0, 0, 0, 0, 0)$. Thus, the final prediction is (20).

$$Y(x_i) = \sum_{S_i \in S'} f_j(x_i) v'_j, i = 1, 2, ...31, j = 1, 2, ...10 \qquad (20)$$

where x_i represents the maximum electric load of the ith date in January, 1999. $f_j(x)$ is the corresponding estimation function of the jth which is the selected model in S' with v'_j, its weight.

To illustrate the performance of ensemble, a comparison between the prediction results of ensemble models, single model (with the best ω) and the actual electric loads is given in Fig. 4. Final results show that with the lowest ω model the *MAPE* is 2.566, while with ensemble the *MAPE* is 1.866, which is better than the result obtained by Lin in EUNITE competition using SVM with the *MAPE* = 1.982.

Table 1. SVM models for ensemble

| Before ranking | | After ranking | | |
Models no	Validating error	Models no	ω (descending order)	Validating error
1	1.833	39	523.4863	2.756
2	1.604	49	4781.672	1.578
3	1.702	47	5349.222	1.557
4	1.818	7	7717.754	1.683
5	1.863	50	9073.668	1.590
6	1.510	10	9114.403	1.665
7	1.683	18	16202.6	1.563
8	1.361	23	16685.5	1.599
9	1.699	21	20775.92	1.568
10	1.665	17	26607.92	1.714
11	1.762	34	30603.84	1.617
12	1.828	2	30905.61	1.604
13	1.825	43	31191.94	1.586
14	1.407	27	55406.93	1.724
15	1.839	16	63099.55	1.610
…	…	…	…	…
50	1.610	42	68794.02	1.754

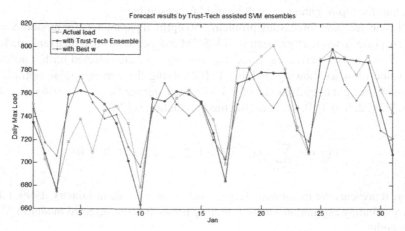

Fig. 4. Comparison among ensemble, single and actual load

5 Conclusion and Future Work

A framework of GA-SVM is proposed in this paper, which will automatically optimize parameters of SVM and generate optimal SVM models. The clustering algorithm is also combined with GA-SVM as an enhancement, providing devious SVM models for future ensemble, as well as being a convergent criterion. This paper also proposes an ensemble strategy with the help of TRUST-TECH. Moreover, experiments in both classification and regression problems show the validity of GA-SVM, the clustering algorithm and the ensemble.

References

1. Cortes, C., Vapnik, V.: Support-vector networks. Mach. Learn. **20**(3), 273–297 (1995)
2. Vapnik, V.: The Nature of Statistical Learning Theory. Springer, New York (2000). https://doi.org/10.1007/978-1-4757-3264-1
3. Vapnik, V.: An overview of statistical learning theory. IEEE Trans. Neural Netw. **10**(5), 988–999 (1999)
4. Platt, J.: Sequential minimal optimization: a fast algorithm for training support vector machines (1998)
5. Vapnik, V., Chapelle, O.: Bounds on error expectation for support vector machines. Neural Comput. **12**(9), 2013–2036 (2000)
6. Chang, M.W., Lin, C.J.: Leave-one-out bounds for support vector regression model selection. Neural Comput. **17**(5), 1188–1222 (2005)
7. Chapelle, O., Vapnik, V.: Model selection for support vector machines. In: Advances in Neural Information Processing Systems, pp. 230–236 (1999)
8. Hsu, C.W., Chang, C.C., Lin, C.J.: A practical guide to support vector classification (2003)
9. Huang, C.L., Wang, C.J.: A GA-based feature selection and parameters optimization for support vector machines. Expert Syst. Appl. **31**(2), 231–240 (2006)
10. Pai, P.F., Hong, W.C.: Support vector machines with simulated annealing algorithms in electricity load forecasting. Energy Convers. Manag. **46**(17), 2669–2688 (2005)
11. Chapelle, O., Vapnik, V., Bousquet, O., Mukherjee, S.: Choosing multiple parameters for support vector machines. Mach. Learn. **46**(1), 131–159 (2002)
12. Dietterich, T.G.: Ensemble methods in machine learning. In: Kittler, J., Roli, F. (eds.) MCS 2000. LNCS, vol. 1857, pp. 1–15. Springer, Heidelberg (2000). https://doi.org/10.1007/3-540-45014-9_1
13. Kim, H.C., Pang, S., Je, H.M., Kim, D., Bang, S.Y.: Constructing support vector machine ensemble. Pattern Recogn. **36**(12), 2757–2767 (2003)
14. Zhang, Y.F., Chiang, H.D.: A novel consensus-based particle swarm optimization-assisted trust-tech methodology for large-scale global optimization. IEEE Trans. Cybern. **47**(9), 2717–2729 (2017)
15. Zhang, Y.F., Chiang, H.D., Wang, T.: A novel TRUST-TECH-enabled trajectory-unified methodology for computing multiple optimal solutions of constrained nonlinear optimization: theory and computation. IEEE Trans. Syst. Man Cybern. Syst. **52**(1), 473–484 (2022)
16. Nguyen, T., Gordon-Brown, L., Wheeler, P., Peterson, J.: GA-SVM based framework for time series forecasting. In: 5th International Conference on Natural Computation, pp. 493–498. IEEE (2009)
17. Wang, L., Xu, G., Wang, J., Yang, S., Guo, L., Yan, W.: GA-SVM based feature selection and parameters optimization for BCI research. In: 7th International Conference on Natural Computation, pp. 580–583. IEEE (2011)

18. Cristianini, N., Shawe-Taylor, J.: An Introduction to Support Vector Machines and Other Kernel-Based Learning Methods. Cambridge University Press, Cambridge (2000)
19. Chang, C.C., Lin, C.J.: LIBSVM: a library for support vector machines. ACM Trans. Intell. Syst. Technol. (TIST) 2(3), 1–27 (2011)
20. Srinivas, M., Patnaik, L.M.: Genetic algorithms: a survey. Computer 27(6), 17–26 (1994)
21. Ball, G. H., Hall, D. J.: ISODATA, a novel method of data analysis and pattern classification. Stanford Research Inst., Menlo Park, CA (1965)
22. Wang, B., Chiang, H.D.: ELITE: ensemble of optimal input-pruned neural networks using TRUST-TECH. IEEE Trans. Neural Netw. 22(1), 96–109 (2007)
23. Zhang, Y.F., Chiang, H.D.: Enhanced ELITE-load: a novel CMPSOATT methodology constructing short-term load forecasting model for industrial applications. IEEE Trans. Ind. Inf. 16(4), 2325–2334 (2020)
24. Reddy, C.K., Chiang, H.D., Rajaratnam, B.: Trust-tech-based expectation maximization for learning finite mixture models. IEEE Trans. Pattern Anal. Mach. Intell. 30(7), 1146–1157 (2008)
25. Chiang, H.D., Wang, B., Jiang, Q.Y.: Applications of TRUST-TECH methodology in optimal power flow of power systems. In: Kallrath, J., Pardalos, P.M., Rebennack, S., Scheidt, M. (eds.) Optimization in the Energy Industry, pp. 297–318. Springer, Heidelberg (2009). https://doi.org/10.1007/978-3-540-88965-6_13
26. Rao, N.S.V.: On fusers that perform better than best sensor. IEEE Trans. Pattern Anal. Mach. Intell. 23(8), 904–909 (2001)
27. Zhou, Z.H., Wu, J., Tang, W.: Ensembling neural networks: many could be better than all. Artif. Intell. 137(1–2), 239–263 (2002)
28. Opitz, D.W., Shavlik, J.W.: Generating accurate and diverse members of a neural-network ensemble. In: Advances in Neural Information Processing Systems, pp. 535–541 (1995)
29. Brown, G.: Diversity in neural network ensembles. University of Birmingham (2004)
30. Windeatt, T.: Accuracy/diversity and ensemble MLP classifier design. IEEE Trans. Neural Netw. 17(5), 1194–1211 (2006)
31. Hashem, S.: Optimal linear combinations of neural networks. Neural Netw. 10(4), 599–614 (1997)
32. Ueda, N.: Optimal linear combination of neural networks for improving classification performance. IEEE Trans. Pattern Anal. Mach. Intell. 22(2), 207–215 (2000)
33. Sapankevych, N.I., Sankar, R.: Time series prediction using support vector machines: a survey. IEEE Comput. Intell. Mag. 4(2), 24–38 (2009)
34. Chen, B.J., Chang, M.W.: Load forecasting using support vector machines: a study on EUNITE competition 2001. IEEE Trans. Power Syst. 19(4), 1821–1830 (2004)

An Early Prediction and Label Smoothing Alignment Strategy for User Intent Classification of Medical Queries

Yuyu Luo[1], Zhenjie Huang[1], Leung-Pun Wong[2], Choujun Zhan[1], Fu Lee Wang[2], and Tianyong Hao[1($\boxtimes$)]

[1] School of Computer Science, South China Normal University, Guangzhou, China
{2020022977,2020022962,haoty}@m.scnu.edu.cn
[2] School of Science and Technology, Hong Kong Metropolitan University, Hong Kong, China
{s1243151,pwang}@hkmu.edu.hk

Abstract. Deep learning models such as RoBERTa and Bi-LSTM are widely utilized in user intention classification tasks. However, in the medical field, there are difficulties in recognizing user intents due to the complexity of medical query representations and medical-specific terms. In this paper, an alignment strategy based on early prediction and label smoothing named EP-LSA is proposed to classify user intents of medical text queries. The EP-LSA strategy uses a Chinese pre-training model RoBERTa to encode sentence features with rich semantic information, predicts the early features of Bi-LSTM in RCNN and aligns them with output features. The early knowledge from early prediction is processed utilizing cross-entropy loss incorporating label smoothing, which enhances random information to the early knowledge and helps the strategy to extract more fine-grained features related to intention labels. Experiment evaluation was performed based on two publicly available datasets KUAKE and CMID. The results demonstrated that the proposed EP-LSA strategy outperformed other baseline methods and demonstrated the effectiveness of the strategy.

Keywords: Intent classification · Early prediction · Label smoothing · Alignment

1 Introduction

User intent classification is an essential component of intelligent question-answering and task-driven dialogue systems. Recently, online health communities provide a convenient way for users to seek medical and health information services [1], where users describe their health conditions and post queries to acquire assistance. As a result, identifying user intents from large amount of query sentences has become a critical task. For example, a query "What tests do diabetics need?" contains a user's confusion about the examination of diabetes disease, and the predefined user intent category of the query in a question-answering system is "medical advice" accordingly.

The information distribution of medical query sentences differs from that of sentences in other fields. The former contains complex and specialized concepts, such as "Charcot triad", "arterial blood gas", and "fasting blood glucose". These are professional terms that are not common in daily life, nevertheless closely related to users' intentions. The same terminology can have different meanings in different contexts, such as "liver" represents an organ of human body, while " liver failing to store blood " represents a state of liver. In addition, ordinary users lack medical expertise and may not express their intentions as accurate and concise as clinicians do. They may have misrepresentations and colloquial expressions. These differences create more challenges for user intent classification in medical domain.

Recently, pre-trained models such as BERT, RoBERTa and ALBERT, have been utilized for user intent classification. All these models have shown potentials and have outperformed traditional classification methods [2–4]. BERT [5] extracted contextually relevant semantic features based on deep bidirectional language representations. Based on the BERT language masking strategy, RoBERTa [6] modified key hyperparameters in the BERT to construct a pre-trained model applicable to Chinese contexts. Many researches combined pre-trained models and neural networks for the classification task. Guo et al. [7] utilized BERT to extract global features between words and sentences and input the features to a capsule network. Liu et al. [8] combined RoBERTa & CNN and utilized CNN to perform convolutional operations on features extracted by RoBERTa to capture important semantic information. All these methods were applicable at obtaining semantic information in sentences. However, they were not adapted to the problem of multiple word meanings in different contexts, as word sense disambiguation. Chen et al. [9] enriched the semantic information of a dataset with an external knowledge base to reduce the ambiguity of words and sparsity of data. However, in the medical domain, medical query sentences contained many complex and variable concepts due to the diversity of disease representations and medical terminology. It was difficult to construct the applicable knowledge base to provide external knowledge for classification tasks. He et al. [10, 11] utilized limited data information and predictions from lower layers of neural network output to improve traditional task representations. These methods took the information obtained from early predictions as deterministic knowledge, but lacked of analysis of the stochastic nature of the information.

Therefore, Sun et al. [12] utilized early prediction to enhance the semantic representation of sentences and achieved state-of-the-art performance. It was proposed to add random information for the information obtained from early prediction, and adjusted the influence degree of the information in downstream task. Motivated by this approach, we apply the idea of early prediction into the user intent classification task and propose a new alignment strategy based on early prediction and label smoothing as EP-LSA, which extracts intent features from medical query sentences from limited data. Firstly, word vectors encoded by the RoBERTa pre-training model are utilized as input. Secondly, we utilize RCNN containing a two-layer Bi-LSTM structure. The EP-LSA strategy obtains the hidden layer output of the first Bi-LSTM layer and inputs it to the classifier for early prediction, as well as aligns it with the features of the second Bi-LSTM layer output. A cross-entropy loss incorporating label smoothing is utilized to process early knowledge from early prediction, adding random information to early knowledge to extract more fine-grained features related to user intent category labels. The publicly available datasets KUAKE and CMID are applied to verify the performance of the strategy. Compared with a list of state-of-the-art baseline methods, the EP-LSA strategy obtains an improvement of 0.34% in F1-score on the KUAKE dataset and an improvement of 3.68% in F1-score on the CMID dataset, respectively.

In summary, the major contribution of the paper lies on: 1) an alignment strategy EP-LSA is proposed for user intent classification of Chinese medical queries, 2) early prediction and label smoothing are applied to obtain early knowledge and provide supervised signals to improve the feature extraction capability of the EP-LSA strategy, 3) The proposed EP-LSA strategy outperforms state-of-the-art baseline methods on standard datasets, verifying its effectiveness.

2   Related Work

The researches on user intent classification are mainly focusing on short query text classification. As an important existing task in natural language processing, text classification has been explored with a number of approaches. In the early stage, traditional machine learning approaches used manually extracted features for text classification [2–4]. However, short sentences conveyed fewer semantic features, which were difficult to extract manually. In addition, manual feature extraction was expensive and time-consuming usually.

In recent years, deep neural networks [13, 14] have been extensively applied to text classification tasks due to their strong capability in automatic extraction of text features. Johnson et al. [15] utilized CNN for text classification and achieved better results compared with traditional methods. Zhang et al. [16] utilized RNN to extract long-term dependencies and key concept information in sentences. Sun et al. [17] demonstrated the effectiveness of an Attention mechanism in text classification by observing whether local attention weights reflected the importance of input text representations. Lai et al. [18] proposed to apply RCNN to text classification by combining the advantages of RNN and CNN, which improved the performance of neural network in acquiring contextual information and essential semantic features. Wang et al. [19] combined CNN and an Attention mechanism to efficiently extract n-gram features of sentences. Wu et al. [20] proposed to extract main semantic features based on a weighted Word2vec-BiLSTM-Attention mechanism. However, all these methods could not provide specific supervised signals for representation learning.

Pre-trained language models have also been widely applied to text classification. They usually obtain better performance compared with non-pre-trained deep neural networks models by fine-tuning parameters during training of downstream tasks. A joint model [21] improved the performance of user intent classification by combining the single-layer CNN with the BERT. To promote Chinese natural language processing, the laboratory of HIT and iFLYTEK Research released the Chinese pre-training RoBERTa [3] model with Whole Word Masking. It was based on the Chinese BERT [23] language mask strategy, and modified key hyperparameters in BERT to provide a more effective pre-trained model for Chinese natural language processing tasks. Due to the domain specificity of sentences in various domains, Chen et al. [9, 24, 25] utilized an external knowledge base to introduce a priori knowledge such as entity relations to enrich the semantic representation of short texts, which further alleviated the problems of word ambiguity and data sparsity.

However, previous researches are mostly based on distributed word embeddings, where words were expressed as identical vectors by matching pre-trained word embeddings, without resolving the problem of multiple meanings of words in different contexts. Even though an external knowledge could be used to enhance contextual knowledge, it was difficult to construct an applicable knowledge base to provide external knowledge for medical classification tasks due to the diversity of disease representations and medical terminology. Therefore, the above methods were limited in their generalization capability. To further solve the problem, Chen et al. [9] constructed early prediction by extracting output features of lower layers of neural network in entity relationship extraction task. It utilized labeling information generated from early prediction to provide a priori knowledge for downstream tasks without the help of external knowledge. The early prediction presented significant impact on performance improvement through experiment evaluations. Therefore, motivated by this idea, this paper introduces early prediction into user intent classification in the medical domain. To further improve the applicability of the early prediction strategy to the user intention classification task, we also apply label smoothing strategy with random probabilities. A cross-entropy loss incorporating label smoothing is applied to process early knowledge from early prediction to provide supervised signal for the early knowledge.

3   The EP-LSA Strategy

We propose an alignment strategy based on early prediction and label smoothing as EP-LSA for extracting semantic features and identify user intents from medical queries. The strategy consists of a RoBERTa and a RCNN with two layers of Bi-LSTM. It treats hidden layer output of the first Bi-LSTM layer as the object of early prediction. After obtaining early knowledge from the early prediction, it aligns early knowledge with features from the second Bi-LSTM layer. Afterwards, cross-entropy loss incorporating label smoothing is utilized to process early knowledge by add random information extract more fine-grained features related to user intent labels. Figure 1(c) demonstrates the architecture of the proposed EP-LSA strategy. As a comparison, Fig. 1(a) presents the architecture of RoBERTa+RCNN model and Fig. 1(b) presents the architecture of RoBERTa+RCNN incorporating early prediction alignment.

3.1   User Intention Feature Extraction

The EP-LSA strategy extracts user intention features from Chinese medical query sentences based on a pre-trained RoBERTa model and RCNN. Taking pre-trained word vectors by RoBERTa as input, the EP-LSA strategy maximizes the capture of contextual information and preserves a large range of sequential information when learning word representations, which is achieved by a Bi-LSTM bidirectional loop structure and a maximum pooling of convolutional layer in RCNN.

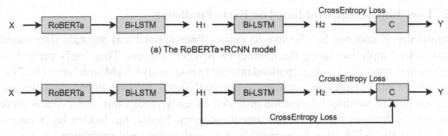

(a) The RoBERTa+RCNN model

(b) The RoBERTa+RCNN model incorporating early prediction alignment

(c) Alignment strategy based on early prediction and label smoothing

Fig. 1. The architecture of (c) the proposed EP-LSA strategy compared to (a) the RoBERTa+RCNN model and (b) the RoBERTa+RCNN model incorporating early predictive alignment.

For a sentence x, the strategy obtains an input character embedding sequence $E = (E_1, E_2, ..., E_{m-1}, E_m)$ using the RoBERTa, where $E_i \in R^{m \times d}$ and m is the maximum length of the sentence. The last layer of encoder generates the contextual embedding sequence $T = (T_1, T_2, ..., T_{m-1}, T_m)$ and $T_i \in R^h$, where h is the dimension of the hidden feature space to represent the sentence. We utilize an RCNN structure with two layers of Bi-LSTM. $c_l(T_i)$ is the left context of word T_i, $c_r(T_i)$ is the right context of word T_i, and $e(T_i)$ is the word embedding of word T_i. We define the word vector representation of the first Bi-LSTM layer and second Bi-LSTM layer in Eq. (1) and Eq. (2). $X_1 = \{x_1^i\}$ is the feature from the output of the first Bi-LSTM layer and $X_2 = \{x_2^i\}$ is the feature from the output of the second Bi-LSTM layer. The maximum pooling operation of the convolution layer is as Eq. (3) and Eq. (4). Finally, the sentence features extracted by the strategy are represented as H_1 and H_2.

$$x_1^i = [c_l(T_i); e(T_i); c_r(T_i)] \tag{1}$$

$$x_2^i = [c_l(x_1^i); e(x_1^i); c_r(x_1^i)] \tag{2}$$

$$H_1 = Max\text{-}pooling(X_1) \tag{3}$$

$$H_2 = Max\text{-}pooling(X_2) \tag{4}$$

3.2   User Intent Alignment Based on Early Prediction

Empirically, it may not be effective to extract finer-grained and accurate user intent features by simply increasing the number of Bi-LSTM layers. Thus, early prediction alignment is expected to be incorporated into the two-layer Bi-LSTM structure of RCNN. This alignment provides an auxiliary supervised signal for user intent classification with the help of labeling information provided by early predictions, and extracts more fine-grained features related to user intent category. Firstly, the hidden layer output of the first Bi-LSTM layer is as input to a classifier for early prediction, and early knowledge containing intention information is obtained. Secondly, the early knowledge is aligned with the features output from the second Bi-LSTM layer to enhance the intrinsic connection between the orderly assignments.

According to the architecture in Fig. 1(a), assuming that x is an intent query sentence, y is a true category label, H_1 and H_2 are contextual representations obtained from a Bi-LSTM, and Y is the prediction label by the classifier. A classification strategy is defined as the map C: $H_2 \rightarrow P(Y)$. The function C utilizes H_2 as input, to obtain the probability distribution over the output space, and H_2 is required to satisfy the Markov relation $y \rightarrow x \rightarrow H_2$. Therefore, the calculation of joint distribution $p(x, y, H_2)$ is shown in Eq. (5).

$$p(x, y, H_2) = p(H_2|x, y)p(x, y) \tag{5}$$

The conditional distribution $p(H_2 \mid x, y) = p(H_2 \mid x)$ satisfies the Markov constraint. There is no semantic information extracted from the label y during user intent feature extraction, and the sentence features H_2 is obtained only from the query sentence x. Without other auxiliary information, H_2 cannot learn any explicit information related to y. Therefore, an early knowledge based on the label y is expected, and the alignment strategy extract more intention features related to y with the help of H_2 and the early knowledge.

The idea of early prediction can be incorporated into RCNN, as shown in Fig. 1(b). The strategy utilizes RCNN structure containing two layers of Bi-LSTM, and hidden layer output H_1 obtained from the first Bi-LSTM layer is utilized as the input to a classifier to generate early knowledge Y'. Y' is an approximation of the final output value Y, containing information related to user intent category. Therefore, in the early prediction of the features of the first Bi-LSTM layer and alignment of the early knowledge with the output features of the second Bi-LSTM layer, Y' can provide information related to the true category label y. It provides supervisory signals for the classifier when processing intention features obtained from the second Bi-LSTM layer. Finally, H_2 is obtained depending on x and y. It is beneficial for the alignment strategy to extract deeper feature representations that are more related to the user intent category.

Afterwards, the loss of early and underlying predictions is calculated using a cross-entropy loss function by Eq. (6) and the result is denoted as CE-Loss. The final total loss is calculated using Eq. (7).

$$CE\_Loss = -\frac{1}{N} \sum_{i} \sum_{k=0}^{K} y_{ic} \log(p_{ic}) \tag{6}$$

$$Final\_Loss = CE\_Loss + \beta \times CE\_Loss \tag{7}$$

K is the number of user intent categories, y_{ic} is the true value of a sample label (0 or 1), and p_{ic} is predicted probability that the sample i belongs to the category c. The label is encoded in a one-hot form with correct categories marked as 1 and incorrect categories marked as 0. $\beta \in [0, 1]$ is a hyperparameter to adjust loss ratio. The small value of β, the less early knowledge from early prediction contributes to the final classification and vice vera.

3.3 Supervised Signals by Label Smoothing

Although early prediction had proven to be effective in some tasks, one fact that has been ignored in previous work [10, 11] is as follows. Early knowledge obtained Y' from early prediction is an approximation to its true category label. Some information from Y' is irrelevant to the true category label, and not all the information contained in Y' is beneficial to the extraction of intent features. Therefore, according to the idea of Sun et al. [12], it is necessary to select relevant information of early knowledge to downstream tasks.

In addition, if the early prediction applies a cross-entropy loss to select the early knowledge, it may enable the lower layer Bi-LSTM learn the fine-grained features of downstream task. However, the layers in a multi-layer neural network are usually ordered, and lower layer Bi-LSTM usually focuses on extracting coarse-grained features. Meanwhile, it is necessary to provide early knowledge with label information to facilitate fine-grained feature extraction. An ideal way to deal with early knowledge is to increase random information, reduce confidence level, and select relevant information. Therefore, the EP-LSA strategy introduces a label smoothing [27] in early prediction. The strategy utilizes label smoothing to add random information to early knowledge from early prediction, and enables lower layer Bi-LSTM to learn coarse-grained semantic features of user intents.

The cross-entropy loss incorporating label smoothing is applied to the EP-LSA strategy to calculate information loss from early prediction. Label smoothing improves the one-hot form of label distribution y to the form of Eq. (8). y' is the smoothed label after using the label smoothing, $\alpha \in [0, 1]$ is a smoothing coefficient, and u is a mean distribution. By this way, the label smoothing brings random information to the early knowledge.

$$y' = (1 - a) \times y + a \times u \tag{8}$$

The calculation of information loss of early knowledge is as Eq. (9) and the loss is denoted as *CELS_Loss*, where u_{ic} is the value obtained by sampling the i-th sample on category c according to the mean distribution u.

$$CELS\_Loss = -\frac{1}{N}\left((1 - a) \sum_{i} \sum_{k=0}^{k} y_{ic} log(p_{ic}) + a \sum_{i} \sum_{k=0}^{k} u_{ic} log(p_{ic}) \right) \tag{9}$$

The calculation of the final total loss is as Eq. (10). $\beta \in [0, 1]$ is a hyperparameter to adjust the loss ratio and the β is the same as in Sect. 3.2.

$$Final\_Loss = CE\_Loss + \beta \times CELS\_Loss \tag{10}$$

4   Experiments and Results

4.1   Datasets

The experiments were conducted based on two publicly available Chinese medical standard datasets KUAKE and CMID. The KUAKE dataset contained 10,800 sentences with predefined labels in 11 user intent categories, such as *diagnosis, cause, method, advice, result, effect*, and *price*. The CMID dataset had 36 categories of user intent labels, including *definition, prevention, infectivity, price, side effect, recovery time, prevention*, etc. Since missing values existed in original data of the CMID dataset and certain categories contained extreme small size of samples, we supplemented the missing values by referring to their contextual values and removed categories containing samples less than 20. Eventually, 13,249 sentences with predefined labels in 31 categories were utilized. The statistics of the two datasets are shown in Table 1.

Table 1. Statistics of the KUAKE and CMID datasets.

Dataset	#Categories	#Training	#Validation	#Test	# Maximum words per sentence
KUAKE	11	6931	1955	1994	60
CMID	31	6093	4078	3078	100

4.2   Evaluation Metrics

Four widely used metrics are applied to evaluate the user intent classification task: Accuracy, Precision, Recall and F1-score. Accuracy is the percentage of correctly predicted samples over total samples. Precision is the proportion of correctly predicted positive samples over all samples that predicted to be positive. Recall is the proportion of correctly predicted positive samples over all relevant samples. F1-score is as a harmonic average of Precision and Recall. The calculations of the metrics are as Eq. (11)–(13).

$$Accuracy = \frac{TP + TN}{TP + FP + TN + FN} \tag{11}$$

$$Precision = \frac{TP}{TP + FP}, \ Recall = \frac{TP}{TP + FN} \tag{12}$$

$$F1-score = \frac{2 \times Precision \times Recall}{Precision + Recall} \tag{13}$$

TP denotes the number of correctly predicted positive samples, TN denotes the number of correctly predicted negative samples, FP denotes the number of incorrectly predicted positive samples, while FN denotes the number of incorrectly predicted negative samples.

4.3 Parameter Settings

The EP-LSA strategy was implemented on the Pytorch deep learning framework including the pre-trained models RoBERTa and RCNN. In RCNN, two-layer structure of Bi-LSTM network was adopted. The dimension of user intent features was 768, the epoch was 10, and the dropout rate was 0.1. The strategy was optimized by Adam for training with the learning rate of $2e^{-5}$ for the RoBERTa and RCNN. The batch size was 16. And the smoothing coefficient α was 0.2, which is selected by set of experiments. Certain parameters were set differently on the two datasets. The maximum sequence length of the KUAKE dataset was limited to 60, while the length of the CMID dataset was limited to 100. Since the dependence of different query texts on the early prediction strategy varied, the hyperparameter β was set to 0.3 for the KUAKE dataset and was set to 0.1 for the CMID dataset.

4.4 Baseline Methods

The EP-LSA strategy was compared with a set of baseline methods on the KUAKE and CMID datasets to explore the effectiveness on user intent classification.

1) Roberta+CNN: The model utilized a Roberta pre-trained model as sentence encoder, combined CNN as feature extractor, as well as used a maximum pooling operation to enhance feature representation capability.
2) Roberta+CNN+Attention: It utilized CNN to obtain vector representation of query texts and combined with an Attention mechanism to extract word information that were important to the sentence.
3) Roberta+Bi-LSTM+Attention: The model utilized Roberta pre-trained word vectors as input, extracted text features by Bi-LSTM, and combined an Attention mechanism to obtain sentence vectors with fused word-level weights.
4) Roberta+RCNN: The joint model combined a Bi-LSTM bidirectional loop structure and a maximum pooling operation to maximize the capture of contextual information, and retained a large range of sequential information when learning word representations.
5) Roberta+RCNN+Focal Loss: The model reduced the weight of easily classified samples with the help of Focal Loss, allowing to focus more on difficulty classified samples during training.

4.5 The Results

The performance of our EP-LSA strategy compared with baseline methods on the KUAKE and CMID datasets are shown in Table 2. Both the Roberta+CNN+Attention and Roberta+Bi-LSTM+Attention utilized an Attention mechanism to extract features from query sentences and achieved an improvement performance compared with the RoBERTa+CNN. The Roberta+CNN+Attention achieved an F1-score of 0.803 and 0.420 on the KUAKE and CMID datasets, respectively. The Roberta+Bi-LSTM+Attention obtained an F1-score of 0.813 on the KUAKE dataset and an F1-score of 0.435 on the CMID dataset, slightly higher than that of the Roberta+CNN+Attention. The

RoBERTa+RCNN acquired the highest F1-score 0.819 on the KUAKE dataset among all the baselines. Our EP-LSA strategy achieved an accuracy of 0.820 and an F1-score of 0.822 on the KUAKE dataset, while achieved an accuracy of 0.534 and an F1-score of 0.451 on the CMID dataset. The EP-LSA strategy had an improvement ratio of 0.3% on F1-score and 0.3% on accuracy compared to the Roberta+RCNN on the KUAKE dataset. In addition, the improvement ratio was more significant on the CMID dataset as 3.7% on F1-score and 0.4% on accuracy compared to the Roberta+RCNN. Due to the misrepresentations and colloquial expressions existed in medical queries, it might be difficult for the Attention mechanism to accurately identify critical words related to user intent. Our EP-LSA strategy generated early knowledge with labeling information in the early prediction stage thus was more effective in feature extraction.

Table 2. The performance comparison on the KUAKE and CMID datasets.

Methods	Datasets	Precision	Recall	F1-score	Accuracy
RoBERTa+CNN	KUAKE	0.778	0.820	0.798	0.798
	CMID	0.420	0.419	0.400	0.484
RoBERTa+CNN+Attention	KUAKE	0.808	0.798	0.803	0.807
	CMID	0.429	0.374	0.420	0.456
RoBERTa+Bi-LSTM+Attention	KUAKE	0.799	0.828	0.813	0.816
	CMID	0.295	0.325	0.435	0.465
RoBERTa+RCNN	KUAKE	0.799	0.839	0.819	0.817
	CMID	0.442	0.390	0.414	0.530
RoBERTa+RCNN+Focal Loss	KUAKE	0.803	0.819	0.817	0.818
	CMID	0.475	0.388	0.427	0.532
EP-LSA	KUAKE	**0.805**	**0.840**	**0.822**	**0.820**
	CMID	**0.490**	**0.420**	**0.451**	**0.534**

Ablation experiments were conducted to verify the contribution of early prediction alignment and label smoothing to intent classification. As shown in Table 3, the EP-LSA strategy (RCNN(2)+Ali(1,2)+label smoothing) improved F1-score by 1.2% compared to the strategy without label smoothing as RCNN(2)+Ali(1,2) on the KUAKE dataset and by 4.9% on the CMID dataset. Compared to the strategy without early prediction alignment and label smoothing as RCNN(2), the EP-LSA strategy improved F1-score by 1.5% on the KUAKE dataset and by 4.5% on the CMID dataset. The result indicated that the early prediction contributed to the extraction of fine-grained intention features, and the label smoothing was able to provide better supervised signals for the early knowledge from early prediction. An additional experiment using RCNN(3) with a three-layer Bi-LSTM structure verified the similar performance. RCNN(3)+Ali(1,3) with the label smoothing, as output features of the first Bi-LSTM layer utilized for early

Table 3. Ablation experiments of early prediction alignment and label smoothing on the KUAKE and CMID datasets. RCNN(n) is the RCNN with n layers of Bi-LSTM, "Ali(1,i)" represents the alignment of output features of the first and the i-th layers in the RCNN.

Methods	Datasets	Precision	Recall	F1-score	Accuracy
RCNN(3)	KUAKE	0.805	0.793	0.801	0.797
	CMID	0.443	0.396	0.414	0.540
RCNN(3)+Ali(1,3)	KUAKE	0.795	0.829	0.812	0.814
	CMID	0.433	0.388	0.409	0.532
RCNN(3)+Ali(1,3)+label smoothing	KUAKE	0.784	0.826	0.805	0.805
	CMID	0.438	0.392	0.414	0.538
RCNN(3)+Ali(2,3)+label smoothing	KUAKE	0.778	0.819	0.798	0.805
	CMID	0.453	0.394	0.422	0.538
RCNN(2)	KUAKE	0.787	0.828	0.807	0.812
	CMID	0.432	0.384	0.406	0.520
RCNN(2)+Ali(1,2)	KUAKE	0.794	0.826	0.810	0.812
	CMID	0.427	0.380	0.402	0.519
RCNN(2)+Ali(1,2)+label smoothing	KUAKE	**0.805**	**0.840**	**0.822**	**0.820**
	CMID	**0.490**	**0.420**	**0.451**	**0.534**

prediction and predicted results aligned with output features of the third layer, exceeded the RCNN(3)+Ali(2,3) with the label smoothing. The result presented that utilizing features of the first Bi-LSTM layer for early prediction was more effective.

Figure 2 shows the performance changes with the increasing value of the smoothing coefficient of the EP-LSA strategy. The performance tended to be unstable when the smoothing coefficient was in the value range of [0.05, 0.3]. It achieved the best F1-score and accuracy on both datasets when the smoothing coefficient was 0.2. In addition, the performance was stable when the value was in the range of [0.3, 0.5]. The EP-LSA strategy utilized the smoothing coefficient to select the random information of the early knowledge. When the smoothing factor was increasing, the random information added to the early knowledge also increased. However, the early knowledge has too much random information and may not help to extract more fine-grained and accurate intent features.

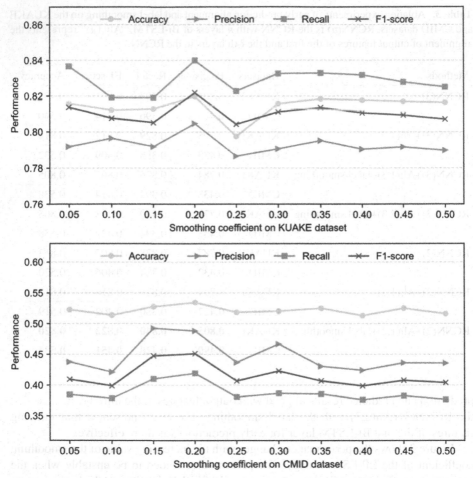

Fig. 2. Influence of the smoothing coefficient of the EP-LSA on the KUAKE and CMID datasets.

5   Conclusions

This paper proposed an EP-LSA strategy for the user intent classification in medical domain. Based on sentence features encoded with RoBERTa, this strategy applied lower layer output features in RCNN for early prediction and aligned early knowledge from early prediction with final output features. Cross entropy loss incorporating label smoothing was also utilized to provide supervised signals for selecting the random information of early knowledge. Experiments on two standard datasets demonstrated the strategy was effective for user intent classification of medical queries compared with a list of state-of-the-art baseline methods.

Acknowledgements. This work was supported by Natural Science Foundation of Guangdong Province (2021A1515011339).

References

1. Cai, R., Zhu, B., Ji, L., Hao, T., Yan, J., Liu, W.: An CNN-LSTM attention approach to understanding user query intent from online health communities. In: 2017 IEEE International Conference on Data Mining Workshops, pp. 430–437 (2017)
2. Hao, T., Xie, W., Wu, Q., et al.: Leveraging question target word features through semantic relation expansion for answer type classification. Knowl. Based Syst. **133**, 43–52 (2017)
3. Xie, W., Gao, D., Hao, T.: A feature extraction and expansion-based approach for question target identification and classification. In: Wen, J., Nie, J., Ruan, T., Liu, Y., Qian, T. (eds.) CCIR 2017. LNCS, vol. 10390, pp. 249–260. Springer, Cham (2017). https://doi.org/10.1007/978-3-319-68699-8_20
4. Shimura, K., Li, J., Fukumoto, F.: Text categorization by learning predominant sense of words as auxiliary task. In: Proceedings of the 57th Annual Meeting of the Association for Computational Linguistics, pp. 1109–1119 (2019)
5. Devlin, J., Chang, M.W., Lee, K., Toutanova, K.: BERT: pre-training of deep bidirectional transformers for language understanding. In: Proceedings of the 2019 Conference of the North American Chapter of the Association for Computational Linguistics, pp. 4171–4186 (2019)
6. Liu, Y., et al.: Roberta: a robustly optimized BERT pretraining approach. arXiv preprint arXiv: 1907.11692 (2019)
7. Guo, H., Liu, T., Liu, F., Li, Y., Hu, W.: Chinese text classification model based on bert and capsule network structure. In: 2021 7th IEEE International Conference on Big Data Security on Cloud, pp. 105–110 (2021)
8. Liu, Y., Liu, H., Wong, L.-P., Lee, L.-K., Zhang, H., Hao, T.: A hybrid neural network RBERT-C based on pre-trained RoBERTa and CNN for user intent classification. In: Zhang, H., Zhang, Z., Wu, Z., Hao, T. (eds.) NCAA 2020. CCIS, vol. 1265, pp. 306–319. Springer, Singapore (2020). https://doi.org/10.1007/978-981-15-7670-6_26
9. Chen, J., Hu, Y., Liu, J., Xiao, Y., Jiang, H.: Deep short text classification with knowledge powered attention. In: Proceedings of the AAAI Conference on Artificial Intelligence, pp. 6252–6259 (2019)
10. He, R., Lee, W.S., Ng, H.T., Dahlmeier, D.: An interactive multi-task learning network for end-to-end aspect-based sentiment analysis. arXiv preprint arXiv:1906.06906 (2019)
11. Zhao, S., Liu, T., Zhao, S., Wang, F.: A neural multi-task learning framework to jointly model medical named entity recognition and normalization. In: Proceedings of the AAAI Conference on Artificial Intelligence, pp. 817–824 (2019)
12. Sun, K., Zhang, R., Mensah, S., Mao, Y., Liu, X.: Progressive multi-task learning with controlled information flow for joint entity and relation extraction. In: Proceedings of the AAAI Conference on Artificial Intelligence, pp. 13851–13859 (2021)
13. Lai, S., Xu, L., Liu, K, et al.: Recurrent convolutional neural networks for text classification. In: Twenty-Ninth AAAI Conference on Artificial Intelligence, pp. 2267–2273. (2015)
14. Zhao, Y., Shen, Y., Yao, J.: Recurrent neural network for text classification with hierarchical multiscale dense connections. In: Proceedings of the 28th International Joint Conference on Artificial Intelligence, pp. 5450–5456 (2019)
15. Johnson, R., Zhang, T.: Deep pyramid convolutional neural networks for text categorization. In: Proceedings of the 55th Annual Meeting of the Association for Computational Linguistics, pp. 562–570 (2017)
16. Zhang, X., Wang, H.: A joint model of intent determination and slot filling for spoken language understanding. In: Proceedings of the 25th International Joint Conference on Artificial Intelligence, pp. 2993–2999 (2016)
17. Sun, X., Lu, W.: Understanding attention for text classification. In: Proceedings of the 58th Annual Meeting of the Association for Computational Linguistics, pp. 3418–3428 (2020)

18. Lai, S., Xu, L., Liu, K., Zhao, J.: Recurrent convolutional neural networks for text classification. In: Twenty-Ninth AAAI Conference on Artificial Intelligence, pp. 2267–2273 (2015)
19. Wang, S., Huang, M., Deng, Z.: Densely connected CNN with multi-scale feature attention for text classification. In: Twenty-Seventh International Joint Conference on Artificial Intelligence, pp. 4468–4474 (2018)
20. Wu, H., He, Z., Zhang, W., Hu, Y., Wu, Y., Yue, Y.: Multi-class text classification model based on weighted word vector and BiLSTM-attention optimization. In: Huang, D.-S., Jo, K.-H., Li, J., Gribova, V., Bevilacqua, V. (eds.) ICIC 2021. LNCS, vol. 12836, pp. 393–400. Springer, Cham (2021). https://doi.org/10.1007/978-3-030-84522-3_32
21. He, C., Chen, S., Huang, S., Zhang, J., Song, X.: Using convolutional neural network with BERT for intent determination. In: 2019 International Conference on Asian Language Processing, pp. 65–70 (2019)
22. Lin, Y., et al.: BertGCN: transductive text classification by combining GCN and BERT. arXiv preprint arXiv:2105.05727 (2021)
23. Cui, Y., et al.: Pre-training with whole word masking for Chinese BERT. IEEE/ACM Trans. Audio Speech Lang. Process. **29**, 3504–3514 (2021)
24. Wang, X., et al.: Learning Intents behind Interactions with Knowledge Graph for Recommendation. In: Proceedings of the Web Conference, pp. 878–887 (2021)
25. Zhong, Y., Zhang, Z., Zhang, W., Zhu, J.: BERT-KG: a short text classification model based on knowledge graph and deep semantics. In: CCF International Conference on Natural Language Processing and Chinese Computing, pp. 721–733 (2021)
26. Chen, N., Su, X., Liu, T., Hao, Q., Wei, M.: A Benchmark dataset and case study for Chinese medical question intent classification. BMC Med. Inform. Decis. Mak. **20**, 1–7 (2020)
27. Müller, R., Kornblith, S., Hinton, G.: When does label smoothing help. arXiv preprint arXiv: 1906.02629 (2019)

An Improved Partition Filter Network for Entity-Relation Joint Extraction

Zhenjie Huang[1], Likeng Liang[1], Xiaozhi Zhu[1], Heng Weng[2], Jun Yan[3], and Tianyong Hao[1(✉)]

[1] School of Computer Science, South China Normal University, Guangzhou, China
{2020022962,lianglikeng,2020022975,haoty}@m.scnu.edu.cn
[2] State Key Laboratory of Dampness, Syndrome of Chinese Medicine, The Second Affiliated Hospital of Guangzhou University of Chinese Medicine, Guangzhou, China
wengh@gzucm.edu.cn
[3] AI Lab, Yidu Cloud (Beijing) Technology Co., Ltd., Beijing, China
Jun.yan@yiducloud.cn

Abstract. The purpose of a joint entity-relation extraction task is to extract entity-relation triples from unstructured text to assist text analysis, knowledge graph construction, etc. The existing sequence-to-sequence or sequence-to-non-sequence models treat the joint extraction task as a triple generation task, sharing the feature space of entity and relation extraction in the same structure. However, fusing the information of both subtasks may cause the problem of feature conflicts and thus decrease model performance. In order to enable each extraction subtask has its own independent feature space to reduce feature conflicts, this paper proposes a dual-decoder to decode entity extraction subtask and relation extraction subtask separately based on an encoder-to-decoder structure. A Dual-Joint-Input-PFN model is proposed by improving the partition filter network as an interaction to capture connection information between two subtasks. The model consists of two Joint-Input-PFNs layers, and each layer accepts two inputs simultaneously and filters the other input according to one of them. The experiments are based on standard datasets WebNLG and NYT, and the effectiveness of the proposed model is verified by comparing with the state-of-the-art baseline methods.

Keywords: Joint entity-relation extraction · Dual-decoder · Information extraction

1 Introduction

Both entity extraction and relation extraction are fundamental and critical tasks for information extraction in natural language processing. The extracted entity-relation triples can be applied to various downstream tasks, such as automatic knowledge graph construction. The early studies [1, 2] employed the pipeline approach, which extracts entities and relations sequentially. In the pipeline manner, as relation extraction depended on entity extraction, errors such as missing or incorrect entities in entity extraction were propagated to relation extraction, and were amplified [3].

In recent years, more and more studies paid attention to joint extraction approaches, which combine entity and relation extraction by multi-task learning and accomplish the two subtasks within one model. Various joint extraction approaches had been proposed. Table filling-based approach utilized a table structure to achieve joint extraction [3–6]. However, this approach required much computational resources during training. Tagging-based approach [7–11] designed novel tagging methods for extracting entities and relations simultaneously. However, elaborately designing a complex and relatively reasonable tagging method required much expertise. Sequence-to-sequence approach [12] treated joint extraction as a triple extraction task. It extracted triples by a sequence generation model, and was beneficial to solve the relation overlap problem. Nevertheless, the construction of the joint extraction task as a sequence generation task leaded to increased exposure bias since there was no order information among triples. In order to avoid the bias issue, some researches employed sequence-to-non-sequence approaches. Zhang et al. [13] constructed a seq2tree method, while Sui et al. [14] developed a seq2set method using a non-autoregressive encoder-decoder. All of these models were able to alleviate negative impact caused by the exposure bias.

However, both sequence-to-sequence and sequence-to-non-sequence approaches employed merely one set of encoder-decoders to construct features of the two subtasks. The parameters that were exclusive to each of the subtasks were generally the last parameters utilized for classification. This structure assumed that the features of the subtasks were mutually compatible and conflict-free. However, Zhong et al. [15] mentioned that there was a high possibility of feature conflicts between the two subtasks, which might significantly limit the performance of models.

Encoder in deep learning models can be shared by both subtasks in encoding phase. This helps the encoder learn to extract more useful information from input, since multi-task learning integrates the losses of both subtasks during training. However, in decoding phase, dual decoders may help to avoid the feature conflict problem. To this end, a novel model named as Dual-Joint-Input-PFN-Decoder is proposed in this paper. The Dual-Joint-Input-PFN-Decoder is based on the seq2set structure of SPN4RE [14] and integrates a Dual-Joint-Input-PFN strategy into dual-decoder. The Dual-Joint-Input-PFN is implemented by two Joint-Input-PFN layers which are proposed based on Partition Filter Network (PFN) [16]. The original PFN is not applicable to be utilized in dual-decoder directly. One reason is that feature structure constructed by dual-decoder is different from that in the original PFN and the other reason is that dual-decoder needs to construct interactions for both features. Thus, Joint-Input-PFN is proposed by improving the original PFN and it receives two features as input and extract favorable interaction from one feature based on another feature. In order to extract the interaction that is beneficial to both subtasks simultaneously, the Dual-Joint-Input-PFN strategy is constructed based on the pairwise Joint-Input-PFN. The strategy captures interaction features from two features that are beneficial to both subtasks and ensure that there is no feature conflict during the construction of the interactions. Based on the Dual-Joint-Input-PFN strategy, this paper decodes the two subtasks separately with a dual-decoder network incorporating the Dual-Joint-Input-PFN strategy for avoiding feature conflicts.

The main contributions of the paper lie on three-fold:

1) A new Dual-Joint-Input-PFN strategy is proposed by incorporating two Joint-Input-PFN strategies improved from Partition Filter Network for construct interactions between entity and relation extractions.

2) A new Dual-Joint-Input-PFN-Decoder model integrates the Dual-Joint-Input-PFN strategy into dual-decoder structure is proposed to utilize interactions of entity and relation extraction for reducing feature conflicts.

3) The proposed model achieves the best performance on two standard datasets compared with state-of-the-art baseline methods, demonstrating its effectiveness.

2 Related Work

The entity-relation extraction task is the fundamental task of many downstream tasks, and the aim of the task is to extract all entity-relation triples from given a sentence. Existing research on joint entity-relation extraction can be divided into categories of pipeline-based models, Table-filling-based models, tagging-based models, seq2seq models, and Multitask learning-based models.

The pipeline-based models [1, 2] were characterized by first extracting entities, and then classifying relations between the entities. However, these models might easily lead to the accumulation of errors. For instance, if a correct entity was missed in entity extraction task, the relations related to this entity could not be extracted correctly in relation extraction. In a backpropagation manner, the information utilized to correct errors can only flow from the relation extraction task to the entity extraction task, not from the entity extraction task to the relation extraction task, resulting in the failure of the models in utilizing connection information between the two subtasks. The Table-filling-based models [3–6] constructed relations between each pair of tokens in given a sentence with the help of table structure, and achieved entity extraction and relation extraction according to relations between tokens. This structure could well solve the problem of triples in overlapping. However, the scale of table structure and the length of sentences were quadratic, so the models often needed to consume a lot of computational resources. The tagging-based models [7–11] elaborated a novel tagging approach to triples extraction. These models could focus on extracting triples with different characteristics by adopting different tagging methods. However, they often required meticulous and complex human involvement consuming a huge amount of time to design an appropriate tagging strategy. Based on the structure of seq2seq models [12, 13, 17], these models adopted a similar encoder-decoder approach to extract entity-relation triples. They could achieve the entity-relation extraction task with excellent performance of existing translation models and could overcome the entity overlapping problem. Although these models had serious exposure bias problem at first, the exposure bias had decreased to be a mainstream problem after continuous improvement. However, there was always a problem of feature conflict with these models. These models tended to employ merely one set of structures to complete the joint extraction task, but the information between the two subtasks was not always beneficial for both, especially the closer to the downstream task, the more the two subtasks have task-specific features. Therefore, the feature conflict brought by mixing two features could limit the performance of the models. Secondly, after separating the two subtasks, an interaction mechanism needed to be constructed to ensure that the association information between the two subtasks were not be lost.

Multitask learning utilized connection information between tasks to integrate multiple tasks into a single model. Joint extraction can be considered as a multi-task learning task. Wang et al. [3] and Sun et al. [18] built interaction mechanisms for the entity extraction and relation extraction through which the model could capture the connection information of the two tasks and thus promote overall model performance. However, these interaction mechanisms did not filter entity and relation features, and direct fusion of two features to construct interaction leaded to the feature conflict issue.

In the joint entity-relation extraction task, encoding sentences to obtain appropriate features could further improve the performance of models. In early research, there were some other networks utilized as encoders for entity and relation extraction tasks, including CNN, LSTM, GRU, GNN, and GCN. With the emergence of large-scale pre-trained language models and performance breakthroughs achieved by them in various NLP tasks, more and more models have begun to utilize these language models as encoders or embedding layers the extraction tasks to better capture semantic information in sentences. BERT [19] was a pre-trained language model that was obtained by training on a large-scale corpus employing a multilayer Transformer encoder [20].

In this paper, the structure of SPN4RE [14] is used as a backbone structure for improvement. In order to solve the problems of feature conflicts, decoder in the original structure improved to be a dual-decoder. In addition, we propose an improved Partition Filter Network strategy into the Dual-Joint-Input-PFN-Decoder model to generate an interaction mechanism for entity extraction and relation extraction to enhance connections during of forward information propagation.

3   Methods

This paper proposes a new Dual-Joint-Input-PFN-Decoder model on by taking the sequence-to-set framework of Sui et al. [14] as a backbone structure for jointly extracting entities and relations in sentences. Our model integrates dual-decoder with Dual-Joint-Input-PFN strategy and Dual-Joint-Input-PFN is implemented by two Joint-Input-PFN strategies for performance improvement. The overall network structure is as shown in Fig. 1. Dual-Joint-Input-PFN-Decoder model needs to generate a fixed size set predictions for each sentence, and the input of the model is initialized by a fixed-size number of learnable embeddings that termed as triple embeddings. After encoding sentences by an encoder, sentence features are extracted and input into the model to transform triple embeddings into output features. Afterwards, the model compares output features with the corresponding labels through bipartite matching and calculates the loss. The improved Dual-Joint-input-PFN-Decoder model receive two inputs at the same time to build interactions between entity extraction and relation extraction and is able to separately decode entity and relation features.

Given a sentence $s = \{x_1, x_2, \cdots, x_n\}$, x_i denotes a token and n denotes the length of the sentence. The model encodes s by pre-trained BERT [19]. The encoding output is denoted as $H^E \in R^{l \times d}$, where l is the length of the encoded sentence containing the three specified symbols [CLS], [SEP], and [PAD], while d is the hidden dimension size of hidden features.

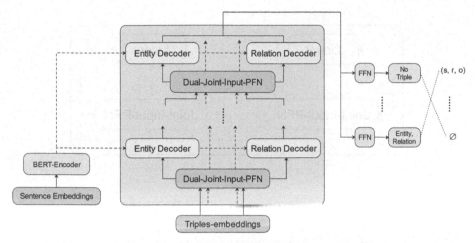

Fig. 1. The architecture of the proposed Dual-Joint-Input-PFN-Decoder model

3.1 The Dual-Joint-Input-PFN Strategy

The original PFN is an interaction strategy built on a multi-task learning framework, in which PFN encodes a feature and generates three types of features: features for entity extraction, features for relation extraction, and features for entity-relation extraction. Particularly, PFN considers that features for entity extraction is irrelevant or even harmful for relation extraction and vice versa. Afterwards, the features for entity extraction/relation extraction and the features for entity-relation extraction are combined to build entity/relation features. However, features constructed by dual-decoder are different from the feature accepted by PFN. Interaction strategy employed in dual-decoder need to be able to receive entity features and relation features simultaneously and select beneficial feature from one type of features based on another type of features. PFN cannot meet the requirements above due to structural limitations.

To improve the PFN, we propose Dual-Joint-Input-PFN strategy which is implemented by two Joint-Input-PFN strategy. The Joint-Input-PFN strategy receives two inputs at the same time and utilize one type of features to partition and filter the other features to obtain beneficial information. However, it is insufficient to utilize merely one Joint-Input-PFN strategy to construct two interactions, since one Joint-Input-PFN cannot construct interactions beneficial to both extraction task. In terms of entity extraction, Joint-Input-PFN utilize entity features to select features from the relation features for keeping useful ones that beneficial for entity extraction. However, it not able to utilize entity feature for selecting useful features from the relation features for relation extraction. Therefore, we proposed Dual-Joint-Input-PFN strategy by employing symmetric Joint-Input-PFN strategy for generating interactions that are beneficial to both entity extraction and relation extraction, presented in Fig. 2. Moreover, interactions are separated constructed, thus it avoids the problem of feature conflicts. For illustration, a Joint-Input-PFN structure for entity extraction is presented as follows since the Dual-Joint-Input-PFN strategy is implemented by two same Joint-Input-PFN strategy, in which one for entity extraction and one for relation extraction.

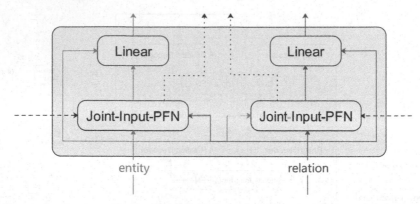

Fig. 2. The Dual-Joint-Input-PFN strategy

The Joint-Input-PFN receives the entity features $H_i^{\text{ent-D}}$ and relation feature $H_i^{\text{rel-D}}$ decoded by the dual-decoder. It also takes the hidden state $H_{i-1}^{\text{Joint-Input-PFN}}$ and cell state c_{t-1} from previous Joint-Input-PFN strategy. The Joint-Input-PFN calculates current cell state $\tilde{c}_i$ utilizing the relation features $H_i^{\text{rel-D}}$ and the hidden state $H_{i-1}^{\text{Joint-Input-PFN}}$, as shown in Eq. (1):

$$\tilde{c}_i = \tanh(\text{Linear}([H_i^{\text{rel-D}}; H_{i-1}^{\text{Joint-Input-PFN}}])) \tag{1}$$

The [;] denotes a connection operation. Afterwards, the master gate [21] is employed to select beneficial features from the current cell state $\tilde{c}_i$. The procedure is as Eq. (2):

$$\tilde{p}_{\tilde{c}_t} = \text{cummax}(\text{Linear}([H_i^{\text{ent-D}}; H_{i-1}^{\text{Joint-Input-PFN}}]))$$
$$\tilde{q}_{\tilde{c}_t} = 1 - \text{cummax}(\text{Linear}([H_i^{\text{ent-D}}; H_{i-1}^{\text{Joint-Input-PFN}}])) \tag{2}$$

The selector contains a $\tilde{p}_{\tilde{c}_t}$ for selecting relation features of current cell state $\tilde{c}_i$ that are beneficial or harmful to the entity extraction, and a $\tilde{q}_{\tilde{c}_t}$ for selecting relation features of $\tilde{c}_i$ that beneficial or irrelevant to the entity extraction. After the selection using the selector, the relation feature of current cell state $\tilde{c}_i$ is split into three parts using the $\rho_{\text{useful},\tilde{c}_t}$ selector, $\rho_{\text{harmful},c_t}$ selector and $\rho_{\text{unrelated},c_t}$ selector, as shown in Eq. (3). The acquired beneficial features, harmful features, and irrelevant features to entity extraction are denoted as Eq. (4):

$$\rho_{\text{useful},\tilde{c}_t} = \tilde{p}_{\tilde{c}_t} \cdot \tilde{q}_{\tilde{c}_t}$$
$$\rho_{\text{harmful},c_t} = \tilde{p}_{\tilde{c}_t} \rho_{\text{useful}}$$
$$\rho_{\text{unrelated},c_t} = \tilde{q}_{\tilde{c}_t} - \rho_{\text{useful}} \tag{3}$$

$$\rho_{\text{ent-useful}} = \rho_{\text{useful},\tilde{c}_t} \cdot \tilde{c}_t + \rho_{s,c_{t-1}} \cdot c_{t-1}$$
$$\rho_{\text{ent-unrelated}} = \rho_{\text{unrelated},\tilde{c}_t} \cdot \tilde{c}_t + \rho_{e,c_{t-1}} \cdot c_{t-1}$$
$$\rho_{\text{ent-harmful}} = \rho_{\text{harmful},\tilde{c}_t} \cdot \tilde{c}_t + \rho_{r,c_{t-1}} \cdot c_{t-1} \tag{4}$$

$\rho_{s,c_{t-1}}, \rho_{e,c_{t-1}}, \rho_{r,c_{t-1}}$ are selector to select useful features from the previous cell states c_{t-1}, calculated in the same way as $\rho_{\text{ent-useful},\tilde{c}_t}, \rho_{\text{ent-unrelated},c_t}, \rho_{\text{ent-harmful},c_t}$. Similarly, for the relation decoder, another Joint-Input-PFN strategy can be constructed to extract the features $\rho_{\text{rel-useful}}$ from the entity feature that are beneficial to the relation feature.

The features $\rho_{\text{ent-useful},\tilde{c}_t}$ are generated by extracting useful feature from relation feature based on entity feature, as interaction features between entity and relation features. The Skipping connection [22] and Linear layer are employed to generate the interactions of features $H_i^{\text{ent-DPFN-D}}$ and $H_i^{\text{rel-DPFN-D}}$, as shown in Eq. (5). Afterwards, the two features with interactions are sent to next layer dual-decoder for further processing.

$$H_i^{\text{ent-DPFN-D}} = \text{Linear}([H_i^{\text{ent-D}}; \rho_{\text{ent-useful}}])$$
$$H_i^{\text{rel-DPFN-D}} = \text{Linear}([H_i^{\text{rel-D}}; \rho_{\text{rel-useful}}]) \tag{5}$$

3.2 The Dual-Joint-Input-PFN-Decoder Model

In existing studies, sequence-to-sequence or sequence-to-non-sequence models have employed one decoder to decode entity extraction and relation extraction together. However, considering that one decoder may cause the problem of feature conflicts during sharing one structure, as mentioned in [15]. We propose to use a Dual-Joint-Input-PFN-Decoder model incorporated with Dual-Joint-Input-PFN strategy to avoid feature conflicts. Based on the transformer-base non-autoregressive decoder [23], our model utilizes dual-transformers fused with our proposed Dual-Joint-Input-PFN strategy for decoding entity and relation features separately.

The Dual-Joint-Input-PFN-Decoder model consists of two identical decoders, each of which is a non-autoregressive Transformer-decoder structure with k layers. Before decoding, the Dual-Joint-Input-PFN-Decoder model takes triples embeddings, denoted as $E \in R^{m \times d}$, as inputs, where m is the maximum number of triples in all sentences. The Transformer-decoder contains a self-attention layer, an inter-attention layer, and a feed forward networks (FFN) layer. The forward propagation process for each layer of the dual-decoder can be formalized as Eq. (6), where i denotes the i-th layer of decoder:

$$H_i^{\text{ent-D}} = \text{Transformer}_{\text{ent}}(H_{i-1}^{\text{ent-D}}, H^E)$$
$$H_i^{\text{rel-D}} = \text{Transformer}_{\text{rel}}(H_{i-1}^{\text{rel-D}}, H^E) \tag{6}$$

However, the forward propagation of the dual-decoder is completely separated without interactions between entity and relation extraction. The Dual-Joint-Input-PFN formally receives two inputs $H_{i-1}^{\text{ent-D}}$ and $H_{i-1}^{\text{rel-D}}$, and generate $H_{i-1}^{\text{ent-DPFN}}$ and $H_{i-1}^{\text{rel-DPFN}}$ with interaction features, as shown in Eq. (7). Thus, the forward propagation process can be revised to Eq. (8) by replacing the inputs with $H_{i-1}^{\text{ent-DPFN}}$ and $H_{i-1}^{\text{rel-DPFN}}$, respectively.

$$(H_{i-1}^{\text{enti-DPFN}}, H_{i-1}^{\text{rel-DPFN}}) = \text{Dual-Joint-Input-PFN}(H_{i-1}^{\text{ent-D}}, H_{i-1}^{\text{rel-D}}) \tag{7}$$

$$H_i^{\text{ent-D}} = \text{Transformer}_{\text{ent}}(H_{i-1}^{\text{ent-DPFN-D}}, H^E)$$
$$H_i^{\text{rel-D}} = \text{Transformer}_{\text{rel}}(H_{i-1}^{\text{rel-DPFN-D}}, H^E) \tag{8}$$

4 Experiment

4.1 Dataset

Our proposed methods are evaluated based on two standard datasets WebNLG [24] and NYT [25], which are widely applied to the joint entity-relation extraction task, e.g., Zeng et al. [26]. WebNLG contains 5019 sentences in training dataset and 703 sentences in test dataset. There are 171 predefined relation types in WebNLG. NYT contains 56196 sentences in training dataset and 5000 sentences in test dataset, with a total of 24 relation types. The experiments were conducted using the Zeng et al. [26] version of the NYT dataset (Table 1).

Table 1. Statistics for WebNLG and NYT

Dataset	#Train	#Valid	#Test	Relation type
WebNLG	5019	500	703	171
NYT	56196	5000	5000	24

4.2 Evaluation Metrics

The evaluation metrics are standard precision, recall and F1-score as follows:

$$Precision = \frac{TP}{TP + FP}$$
$$Recall = \frac{TP}{TP + FN}$$
$$F1 = \frac{2 \times Precision \times Recall}{Precision + Recall} \tag{9}$$

TP denotes the number of positive classes predicted to be positive, FP denotes the number of negative classes predicted to be positive, and FN denotes the number of positive classes predicted to be negative. A triple is regarded as correct if and only if matched the head entity, tail entity, and relation type of the triple are all exactly matched. There are two ways to evaluate whether an entity is correctly matched, namely partial match and exact match. Partial match means that the first word of predicted entity is the same as the first word of the matched label. Exact match means that the whole predicted entity is the same as its matched label. The WebNLG dataset utilizes partial match, and the NYT dataset utilizes exact match for entity matching respectively by following the same strategy in previous work.

4.3 Baselines

Our methods are compared with the following baseline models:

1) NovelTagging [9]: The model proposed a new tagging strategy featured with extracting a triple from sentences instead of extracting entity and relation separately.
2) CopyRE [26]: The model employed an encoder-decoder structure in a sequence-to-sequence manner by utilizing a copy mechanism to identify entities from sentences.
3) GraphRel [27]: The model encoded triples using graph neural networks and employed graph structures to construct interactions between entity and relation extraction.
4) CasRel [7]: The model treated a relation as a mapping from head entities to tail entities, to enable the extraction of entity-relation triples jointly.
5) RIN [18]: Two structures were developed and separately used to implement entity and relation extraction with a Recurrent Interaction Network to construct their connection information.
6) TPLinker [11]: The model proposed a Handshaking tagging strategy to extract overlapping triples.
7) SPN4RE [14]: The model utilized a non-autoregressive model and bipartite matching loss to implement a sequence-to-set framework, to solve the exposure bias issue.

The experiments were implemented using the base version of BERT [19] as encoder. The initial learning rate of BERT was set to 0.00001, while the learning rate of Dual-Joint-Input-PFN-Decoder model was set to 0.00002. The Dual-Joint-Input-PFN-Decoder model was composed of 3-layer non-autoregressive transformers. In addition, a dropout was applied to prevent overfitting, with a rate of 0.1. The maximum gradient was set to 20 to prevent explosive growth of the gradient. The training was performed utilizing the AdamW method and a Layernorm was employed to accelerate training speed. All experiments were conducted on a server with Intel Xeon CPU E5-2609, 96 GB memory, and RTX 2080Ti.

4.4 The Results

The performance comparison of our proposed model with all baseline models are presented in Table 2. Our model achieved the best performance with a precision of 0.932, a recall of 0.929 and a F1-score of 0.931 on WebNLG, and a precision of 0.930, a recall of 0.918 and a F1-score of 0.924 on NYT. Compared with SPN4RE, our model had an improvement of F1-score by 0.2% and 0.1% on the WebNLG and NYT datasets, respectively. Moreover, our model had a clear improvement on prevision over the state-of-the-art SPN4RE by 0.4% and 0.5% on the two datasets, respectively. Comparing the models with interaction mechanism, our model exceeded the RIN by 5.5% and 7.3% in F1-score on the WebNLG and NYT datasets, respectively. Comparing the models with complex tagging strategy, our model exceeded the TPLinker by 1.2% and 0.4% in F1-score on the WebNLG and NYT datasets. The results strongly demonstrate that the Dual-Joint-Input-PFN-Decoder model is able to avoid feature conflicts and build interaction between entity and relation extraction.

To verify the effectiveness on extracting element of relational triples, we conducted a comparison of entity pair extraction and relation type extraction with SPN4RE. Although the performance of entity pairs decreased on F1-score by 0.3%, the performance of relation extraction improved on F1-score by 0.3%, as shown in Table 3. The improvement

Table 2. The results of performance comparison, where * denotes results of reproduced models, 'partial' denotes entities using partial match, and 'exact' denotes entities using exact match.

Models	WebNLG(partial)			NYT(exact)		
	Precision	Recall	F1-score	Precision	Recall	F1-score
NovelTagging	0.525	0.193	0.283	–	–	–
CopyRE-One	0.322	0.289	0.305	–	–	–
CopyRE-Mul	0.377	0.364	0.371	–	–	–
GraphRel-1p	0.423	0.392	0.407	–	–	–
GraphRel-2p	0.447	0.411	0.429	–	–	–
CasRel	**0.934**	0.901	0.918	–	–	–
RIN	0.877	0.879	0.877	0.844	0.860	0.851
TPLinker	0.918	0.920	0.919	0.914	0.926	0.920
SPN4RE	0.928*	**0.929***	0.929*	0.925	**0.922**	0.923
Ours	0.932	**0.929**	**0.931**	**0.930**	0.918	**0.924**

in relation extraction also driven a 0.2% improvement on F1-score of relational triples extraction, which was an indication that the proposed model correctly matched entity pairs and relation properly.

Table 3. The comparison of our model with the SPN4RE on entity pair extraction and relation type extraction on WebNLG

Models	Element	Precision	Recall	F1-score
SPN4RE	(s, o)	**0.952**	**0.952**	**0.952**
	r	0.954	0.955	0.954
	Overall	0.928	0.929	0.929
Ours	(s, o)	0.948	0.945	0.947
	r	**0.958**	0.955	**0.957**
	Overall	**0.932**	0.929	**0.931**

To verify the effectiveness on extraction various numbers of triples contained in a sentence, we conducted an experiment on a sentence with different numbers of triples. The number of triples in a sentence denoted as n. The best performance is achieved at $n = 1$ with a F1-score 0.911. The experimental results are shown in Fig. 3. Although the performance of this model decreases on F1-score by 0.4% and 0.9% compared to SPN4RE at $n = 2$ and $n = 3$, respectively. The performance of the model improves on F1-score by 3% at $n = 1$. Many existing models do not perform well for triple extraction at $n = 1$. Compared with TPLinker and CasRel, the performance of the proposed model at $n = 1$ is improved by 3.1% and 1.8% in F1-score, respectively.

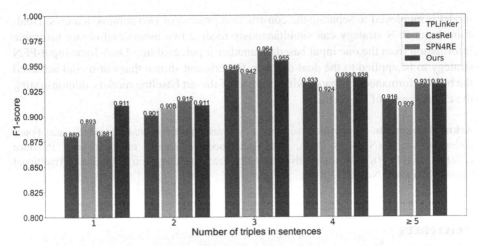

Fig. 3. The comparison of various number of triples in a sentence on WebNLG

To verify the effectiveness of Dual-Joint-Input-PFN-Decoder model, an ablation experiments was conducted: Ours(w/o interaction) denotes our model without interactions and Ours(PFN) denotes the model with PFN, while Ours(Double-Encoder) denotes the model with separate two encoders. The results are shown in Table 4.

Table 4. Results of ablation experiments on WebNLG

Ablation	Precision	Recall	F1-score
Ours(PFN)	0.915	0.908	0.912
Ours(Double-Encoder)	0.918	0.916	0.917
Ours(w/o interaction)	0.927	0.918	0.923
Ours	**0.932**	**0.929**	**0.931**

The performance of the Ours(PFN) decreased on F1-score by 1.9% verifying that the PFN was not suitable in our proposed model. The performance of the Ours(Double-Encoder) decreased by 1.4%, indicating that the utilization of two encoders still lowered the performance. In addition, the performance of the Ours(w/o interaction) decreased by 0.8%, indicating the importance of building interactions between in entity and relation extractions.

5 Conclusion

This paper proposed a Dual-Joint-Input-PFN-Decoder model and Dual-Joint-Input-PFN strategy to solve the problem of feature conflicts in the original model and to better capture the connection information between subtasks. Dual-Joint-Input-PFN-Decoder model is employed to separate the construction process of two subtask features. Dual-Joint-Input-PFN strategy can simultaneously receive two inputs and obtain favorable information from the one input based on another input, enabling Dual-Joint-Input-PFN strategy to be applied to the dual-decoder. Experiment shown that our model achieved the best performance compared with the state-of-the-art baseline models, demonstrating the effectiveness of the method.

Acknowledgements. The work is supported by grants from National Natural Science Foundation of China (No. 61871141), Natural Science Foundation of Guangdong Province (2021A1515011339), and Collaborative Innovation Team of Guangzhou University of Traditional Chinese Medicine (No. 2021XK08).

References

1. Zelenko, D., Aone, C., Richardella, A.: Kernel methods for relation extraction. J. Mach. Learn. Res. **3**, 1083–1106 (2003)
2. Chan, Y.S., Roth, D.: Exploiting syntactico-semantic structures for relation extraction. In: Proceedings of the 49th Annual Meeting of the Association for Computational Linguistics: Human Language Technologies, pp. 551–560 (2011)
3. Wang, J., Lu, W.: Two are better than one: joint entity and relation extraction with table-sequence encoders. In: Proceedings of the 2020 Conference on Empirical Methods in Natural Language Processing (EMNLP), pp. 1706–1721 (2020)
4. Miwa, M., Sasaki, Y.: Modeling joint entity and relation extraction with table representation. In: Proceedings of the 2014 Conference on Empirical Methods in Natural Language Processing (EMNLP), pp. 1858–1869 (2014)
5. Gupta, P., Schütze, H., Andrassy, B.: Table filling multi-task recurrent neural network for joint entity and relation extraction. In: Proceedings of COLING 2016, the 26th International Conference on Computational Linguistics: Technical Papers, pp. 2537–2547 (2016)
6. Ma, Y., Hiraoka, T., Okazaki, N.: Named entity recognition and relation extraction using enhanced table filling by contextualized representations. arXiv preprint arXiv:2010.07522 (2020)
7. Wei, Z., Su, J., Wang, Y., Tian, Y., Chang, Y.: A novel cascade binary tagging framework for relational triple extraction. In: Proceedings of the 58th Annual Meeting of the Association for Computational Linguistics, pp. 1476–1488 (2020)
8. Yu, B., et al.: Joint extraction of entities and relations based on a novel decomposition strategy. In: ECAI 2020, pp. 2282–2289 (2020)
9. Zheng, S., Wang, F., Bao, H., Hao, Y., Zhou, P., Xu, B.: Joint extraction of entities and relations based on a novel tagging scheme. In: Proceedings of the 55th Annual Meeting of the Association for Computational Linguistics (Volume 1: Long Papers), pp. 1227–1236 (2017)
10. Luo, X., Liu, W., Ma, M., Wang, P.: A bidirectional tree tagging scheme for jointly extracting overlapping entities and relations. arXiv e-prints, arXiv-2008 (2020)
11. Wang, Y., Yu, B., Zhang, Y., Liu, T., Zhu, H., Sun, L.: TPLinker: single-stage joint extraction of entities and relations through token pair linking. In: Proceedings of the 28th International Conference on Computational Linguistics, pp. 1572–1582 (2020)
12. Nayak, T., Ng, H.T.: Effective modeling of encoder-decoder architecture for joint entity and relation extraction. In: Proceedings of the AAAI Conference on Artificial Intelligence, pp. 8528–8535 (2020)

13. Zhang, R.H., et al.: Minimize exposure bias of Seq2Seq models in joint entity and relation extraction. In: Proceedings of the 2020 Conference on Empirical Methods in Natural Language Processing: Findings, pp. 236–246 (2020)
14. Sui, D., Chen, Y., Liu, K., Zhao, J., Zeng, X., Liu, S.: Joint entity and relation extraction with set prediction networks. arXiv preprint arXiv:2011.01675 (2020)
15. Zhong, Z., Chen, D.: A frustratingly easy approach for entity and relation extraction. In: Proceedings of the 2021 Conference of the North American Chapter of the Association for Computational Linguistics: Human Language Technologies, pp. 50–61 (2021)
16. Yan, Z., Zhang, C., Fu, J., Zhang, Q., Wei, Z.: A partition filter network for joint entity and relation extraction. In: Proceedings of the 2021 Conference on Empirical Methods in Natural Language Processing, pp. 185–197 (2021)
17. Zeng, X., He, S., Zeng, D., Liu, K., Liu, S., Zhao, J.: Learning the extraction order of multiple relational facts in a sentence with reinforcement learning. In: Proceedings of the 2019 Conference on Empirical Methods in Natural Language Processing and the 9th International Joint Conference on Natural Language Processing (EMNLP-IJCNLP), pp. 367–377 (2019)
18. Sun, K., Zhang, R., Mensah, S., Mao, Y., Liu, X.: Recurrent interaction network for jointly extracting entities and classifying relations. In: Proceedings of the 2020 Conference on Empirical Methods in Natural Language Processing (EMNLP), pp. 3722–3732 (2020)
19. Devlin, J., Chang, M.W., Lee, K., Toutanova, K.: BERT: pre-training of deep bidirectional transformers for language understanding. In: Proceedings of the 2019 Conference of the North American Chapter of the Association for Computational Linguistics: Human Language Technologies, Volume 1 (Long and Short Papers), pp. 4171–4186 (2019)
20. Fu, T.J., Li, P.H., Ma, W.Y.: GraphRel: modeling text as relational graphs for joint entity and relation extraction. In: Proceedings of the 57th Annual Meeting of the Association for Computational Linguistics, pp. 1409–1418 (2019)
21. Shen, Y., Tan, S., Sordoni, A., Courville, A.: Ordered neurons: integrating tree structures into recurrent neural networks. arXiv preprint arXiv:1810.09536 (2018)
22. He, K., Zhang, X., Ren, S., Sun, J.: Deep residual learning for image recognition. In: Proceedings of the IEEE Conference on Computer Vision and Pattern Recognition, pp. 770–778 (2016)
23. Gu, J., Bradbury, J., Xiong, C., Li, V.O., Socher, R.: Non-autoregressive neural machine translation. arXiv preprint arXiv:1711.02281 (2017)
24. Gardent, C., Shimorina, A., Narayan, S., Perez-Beltrachini, L.: Creating training corpora for NLG micro-planners. In: Proceedings of the 55th Annual Meeting of the Association for Computational Linguistics (Volume 1: Long Papers), pp. 179–188 (2017)
25. Riedel, S., Yao, L., McCallum, A.: Modeling relations and their mentions without labeled text. In: Balcázar, J.L., Bonchi, F., Gionis, A., Sebag, M. (eds.) ECML PKDD 2010. LNCS (LNAI), vol. 6323, pp. 148–163. Springer, Heidelberg (2010). https://doi.org/10.1007/978-3-642-15939-8_10
26. Zeng, X., Zeng, D., He, S., Liu, K., Zhao, J.: Extracting relational facts by an end-to-end neural model with copy mechanism. In: Proceedings of the 56th Annual Meeting of the Association for Computational Linguistics (Volume 1: Long Papers), pp. 506–514 (2018)
27. Vaswani, A., et al.: Attention is all you need. In: Advances in Neural Information Processing Systems, pp. 5998–6008 (2017)

Fast Dynamic Response Based on Active Disturbance Rejection Control of Dual Active Bridge DC-DC Converter

Zongfeng Zhang[1], Ao Fu[1], Guofeng Tian[1], Fan Gong[1], Rui Zhang[1], Yu Xing[1], Yue Sun[2(✉)], Yulan Chen[3], and Xingong Cheng[2]

[1] State Grid Rizhao Electric, Rizhao, China
[2] School of Electrical Engineering, University of Jinan, Jinan, China
sunyue_996@163.com, cse_cxg@ujn.edu.cn
[3] City Lighting Service Center, Jinan, China

Abstract. In view of the dynamic characteristics for dual active bridge (DAB) converters in single phase shift control, a control method based on active disturbance rejection control (ADRC) control with output current and input voltage feed-forward is proposed. The basic structure and control method of the linear active disturbance rejection control (LADRC) are introduced and analysis of the dependence of the method on the inductance parameters and the stability of the control system in this paper. Finally, the proposed control method and PI control are compared and validated in a simulation-based platform. Experiments show that the converter responds three times faster than conventional PI control during sudden load changes and is insensitive to inductor parameter deviations.

Keywords: Dual active bridge · Active disturbance rejection control · Dynamic characteristics · Hybrid control · Robustness

1 Introduction

Battery energy storage can smooth out power fluctuations of wind power generation and photovoltaic power generation, improve power quality of grid-connected wind farms and photovoltaic power plants, and promote stable operation of power systems [1,2]. With a rise within the research attention directed toward renewable power systems with storage energy, DAB converter will play a lot of primary role in power conversion system. With symmetrical structure, bidirectional energy flow, electrical isolation, high power density and ease of soft switching control [3].

In addition to the study of the basic characteristics of the converter, improving the dynamic performance of the converter is also a research hot spot of the DAB converter. In the literature [4], a small signal model of DAB converter was established and the dynamic performance of DAB converter was briefly analyzed. The literature [5] proposed a load current feed-forward control method based on

H. Zhang et al. (Eds.): NCAA 2022, CCIS 1637, pp. 142–153, 2022.
https://doi.org/10.1007/978-981-19-6142-7_11

direct power control, which introduces the load current into the control system and significantly improves the load response speed of the converter, but it does not analyze the responsiveness of the converter when the input voltage changes. A hybrid control method with input voltage feed-forward was proposed in the literature [6], and the proposed method was experimentally verified in the literature [7], which essentially uses the output of the proportional-integral (PI) controller as the compensation amount of the phase shift D. Since D takes a small value, its compensation amount is even smaller, which leads to a small PI output, a high demand on PI The inductor parameters are introduced to participate in the control calculation, and the dependence of the inductor parameters is not analyzed, which will reduce the compatibility of the control method.

Traditional PI control is widely utilized in DAB converter applications because of its simple controller design and ability to meet the control needs of most systems [8,9]. However, this simple control method is incapable of resolving the conflict between system velocity and overshot. After changing system parameters, PI control becomes difficult to achieve reasonable control effects, and non-linear phenomena such as bifurcation or chaos may emerge, resulting in excessive voltage and current ripple coefficients, poor anti-disturbance capabilities, and a lack of robustness.

With the explosive development of power electronics, the performance requirements of DAB converters are being redoubled, necessitating a superior velocity, robustness, and adaptability control technique. A variety of control systems, including adaptive control, fuzzy control, neural network control, and active disturbance rejection control, are being researched for DAB converter control. However, adaptive algorithms require a sufficiently accurate model of the control object, making it difficult to settle on controller parameters; using fuzzy control can introduce judder and vibration issues to the circuit, affecting the circuit's stable output; and the complex structure of neural network control makes controller design difficult. As a result, all of the above-mentioned control techniques have certain limitations in real usage.

To improve response velocity, robustness, and flexibility of the system, the control mechanism selected is critical. On the basis of PID control, [10] proposed ADRC, which is a control technique that does not rely on a precise model of the system, with the advantages of robustness, small overshoot, and good dynamic response, compensating for the hysteresis phenomenon in traditional PID control strategies. The control impact of the active disturbance rejection control technique is also superior in the situation of high precision control and visible outside disturbances; however, nonlinear active disturbance rejection control parameters are more difficult to handle. LADRC is developed in [11] to achieve the control aim in an extremely simple and quick manner. It uses a linear function to control the system and simplifies the parameter adjustment process by applying a new bandwidth parameterization technique.

To further explore the control methods in dynamic response of DAB converter, a control method based on ADRC control with output current and input voltage feed-forward is proposed to improve the dynamic characteristics of the

DAB converter. The basic structure and control method of the LADRC controller are introduced. The principle of single-phase shift control of DAB converter is analyzed, and then the steady-state characteristics of its output current are analyzed, based on which, the output current and input voltage feed-forward is used to improve the dynamic characteristics of the DAB converter. Finally, the experimental study comparing the proposed algorithm with the conventional pi control is verified by simulation.

2   Circuit Schematic of DAB

The circuit schematic of the DAB converter is shown in Fig. 1. The DAB converter combines two H-bridges, a high-frequency transformer with ratio n, and two DC capacitors C_1 and C_2 as shown in Fig. 1. L is the sum of the transformer leakage inductance and the external inductance. i_L, i_2 and i_0 are the inductor current, the transformer secondary current and the load current. The phase shift control is used to regulate the phase shift difference φ between the output AC square wave voltages U_{h1} and U_{h2} of the full bridge H1 and H2 to change the magnitude and direction of the inductor current i_L, thus changing the power transmission magnitude and direction of the converter, whose schematic can be simplified to the circuit shown in Fig. 2 [12].

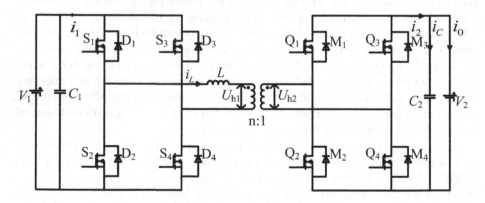

Fig. 1. A circuit schematic of DAB converter

The single phase shift control of DAB converter is adopted as the modulation method, and its timing diagram is shown in Fig. 3. In the Fig. 3, T_{hs} is half a switching period. D is the ratio of the phase shift difference φ between U_{h1} and U_{h2} to half a switching period T_{hs}. The phase shift ratio is

$$D = \frac{\varphi}{\pi} \tag{1}$$

When the converter is in steady state, the operating state of the DAB converter can be divided into four stages according to the single phase shift control operating principle waveform of the DAB converter as shown in Table 1. Two stages $t_0 - t_0'$ and $t_2 - t_2'$, are added to facilitate the analysis of the inductor current commutation process. For ease of representation, the ideal switching function is defined as Eqs. (2) and (3).

$$S_a = \begin{cases} 1, S_1 \& S_3 on, S_2 \& S_4 off \\ -1, S_2 \& S_4 on, S_1 \& S_3 off \end{cases} \tag{2}$$

$$S_b = \begin{cases} 1, S_5 \& S_7 on, S_6 \& S_8 off \\ -1, S_6 \& S_8 on, S_5 \& S_7 off \end{cases} \tag{3}$$

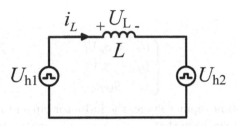

Fig. 2. Simplified circuit of DAB

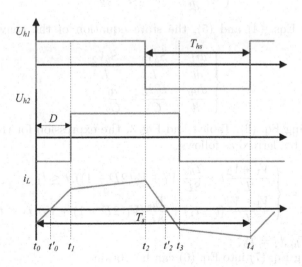

Fig. 3. Voltage and current waveform of single phase shift control

Table 1. Parameters of DABC converter prototype

Time	S_a	S_b
$t_0 - t_1$	1	-1
$t_1 - t_2$	1	1
$t_2 - t_3$	-1	1
$t_3 - t_4$	1	-1

3 Output Steady State Current of the Converter

Neglecting the losses in the converter, U_{h1}, U_{h2} and output current i_2 can be analyzed from Table 1 and literature [12]. Expressions for U_{h1}, U_{h2} and output current i_2 is

$$\begin{cases} U_{h1} = S_a V_1 \\ U_{h2} = S_b V_2 \\ i_2 = S_b i_L \end{cases} \tag{4}$$

Assuming a constant input voltage, the inductor current i_L and output voltage V_2 as state variables have that

$$\begin{cases} L\dfrac{di_L}{dt} = U_L = U_{h1} - U_{h2} \\ C_2\dfrac{du_0}{dt} = i_c = i_2 - i_0 \end{cases} \tag{5}$$

Combining Eqs. (4) and (5), the state equation of the converter can be obtained as

$$\begin{cases} \dfrac{di_L}{dt} = \dfrac{S_a V_1}{L} - \dfrac{S_b V_2}{L} \\ \dfrac{du_0}{dt} = \dfrac{S_b i_L}{C_2} - \dfrac{i_0}{C_2} \end{cases} \tag{6}$$

By combining Eq. (6), Table 1 and Fig. 3, the expression for the inductance current i_L can be derived as follows

$$i_L = \begin{cases} \dfrac{V_1 + V_2}{L}t - \dfrac{T_{hs}}{8L}(V_1 + V_2(2D - 1)), t \le t_1 \\ \dfrac{V_1 - V_2}{L}(t - t_1) + \dfrac{T_{hs}}{8L}(V_1(2D - 1) + V_2), t_1 < t \end{cases} \tag{7}$$

where $t_1 = DT_{hs}, t_2 = T_{hs}$.

Substituting Eq. (7) into Eq. (6) can be obtain

$$\dfrac{du_0}{dt} = \begin{cases} -\dfrac{1}{LC_2}(V_1 + V_2)t + \dfrac{T_{hs}}{8L}(V_1 + V_2(2D - 1)) - \dfrac{i_0}{C_2}, t \le t_1 \\ \dfrac{V_1 - V_2}{LC_2}(t - t_1) + \dfrac{T_{hs}}{8LC_2}(V_1(2D - 1) + V_2) - \dfrac{i_0}{C_2}, t_1 < t \end{cases} \tag{8}$$

Using averaging module methodology of the switch cycle, the average values of V_1, V_2 and i_0 over the average values of $\langle V_1 \rangle_{T_s}$, $\langle V_2 \rangle_{T_s}$ and $\langle i_0 \rangle_{T_s}$ of a switch cycle are used to represent the V_1, V_2 and i_0 with small signal perturbations. With small signal perturbations, and take the average of each phase with the time-weighted average of each phase to obtain the state equation of the converter. The state equation of the converter is

$$\frac{d\langle V_2 \rangle_{T_s}}{dt} = \frac{T_s}{2LC_2} d(1-d)\langle V_1 \rangle_{T_s} - \frac{\langle i_0 \rangle_{T_s}}{C_2} \tag{9}$$

Neglecting the disturbance quantity, the output current steady state quantity of the converter is

$$i_0 = \frac{T_s D(1-D)}{2L} V_1 \tag{10}$$

From Eq. (10), the expression for the phase shift D can be derived

$$D = \frac{1}{2} - \sqrt{\frac{1}{4} - \frac{2i_0 L}{V_1 T_s}} \tag{11}$$

According to Eq. (11), $\sqrt{\frac{1}{4} - \frac{2i_0 L}{V_1 T_s}} > 0$, the output current limiting condition can be obtained as follows

$$i_0 <= \frac{V_1 T_s}{8L} \tag{12}$$

The compensation of the output current Δi is obtained by integrating the output current error, and the control model of the hybrid control method of the output current can be obtained as follows

$$D = \frac{1}{2} - \sqrt{\frac{1}{4} - \frac{2(i_0 + \Delta i)L}{V_1 T_s}} \tag{13}$$

The desired value of the output current is i_{set}, and the actual value of the output current is i_0, can be get from Eq. (13)

$$D = \frac{1}{2} - \sqrt{\frac{1}{4} - \frac{2(i_{set} + \Delta i)L}{V_1 T_s}} \tag{14}$$

Then,

$$\Delta i = K_i \int (i_{(set)} - i_0) \tag{15}$$

where K_i is integral parameter. Combining the condition that Eq. (14) has a real number of solutions yields the integral limit condition of Eq. (15) can be obtain

$$\Delta i =\leq \frac{V_1 T_s}{8L} - i_{set} \tag{16}$$

4   LADRC Model

Building conception that strict mathematical model of system is pronumeral and considering system disturbance f, dynamic system of DAB converter of storage energy can be reformulated within LADRC framework as

$$\begin{cases} \dot{x}_1 = b_0 u + f \\ f = \Delta b u + f_0 x_1 \end{cases} \tag{17}$$

where $x_1 = V_2$ is controlled voltage, and u is control input variable. b_0 represents the known part and Δb is an unknown modeling error. f is denoted as system total disturbance. A logical transition process should be proposed for input signal to accomplish its real-time tracking. Disturbance signal f should be observed and compensated to build up anti-disturbance performance of system.

Let $x_1 = y, x_2 = f, y = V_2$. The enhanced system can be straight denoted from (17) and system can be, clearly, expressed the state-space equation as follows

$$\begin{cases} \dot{x} = Ax + Bu + Eh \\ y = Cx \end{cases} \tag{18}$$

where,

$$x = \begin{bmatrix} x_1 \\ x_2 \end{bmatrix}, A = \begin{bmatrix} 0 & 1 \\ 0 & 0 \end{bmatrix}, B = \begin{bmatrix} b_0 \\ 0 \end{bmatrix}, E = \begin{bmatrix} 0 \\ 1 \end{bmatrix}, C = \begin{bmatrix} 1 & 0 \end{bmatrix}.$$

According to Eq. (18), ESO is designed as follows

$$\begin{cases} \dot{z} = Az + Bu + L(y - \hat{y}) \\ \hat{y} = Cz \end{cases} \tag{19}$$

where $L = [l_1, l_2]^T, z(t) = [z_1(t), z_2(t)]^T$ are observer gain vector and observer state vector. The favored bandwidth parametrization methodology for observer gains is employed [13].

$$s^2 + l_1 s + l_2 = (s + \omega_0)^2 \tag{20}$$

where $\omega_0 > 0$, ω_0 denotes the bandwidth of the second-order LESO, $l_1 = 2\omega_0, l_2 = \omega_0^2$. The ESO of this system is as follows

$$\begin{cases} e_1 = z_1 - y \\ \dot{z}_1 = z_2 - l_1 e_1 + b_0 u \\ \dot{z}_2 = -l_2 e_1 \end{cases} \tag{21}$$

z_1 is the state of tracking output voltage in LESO, and z_2 is the state of tracking system total disturbance of system in LESO [14]. The value of l_1 and l_2 can have an effect on convergence velocity of LESO.

Generally, the control law $u_0(t)$ LADRC is designed as [14]

$$u_0(t) = k_p(r(t) - \hat{y}(t)) \tag{22}$$

The tracking error e(t) of the LADRC can be expressed as

$$e(t) = r(t) - y(t) \tag{23}$$

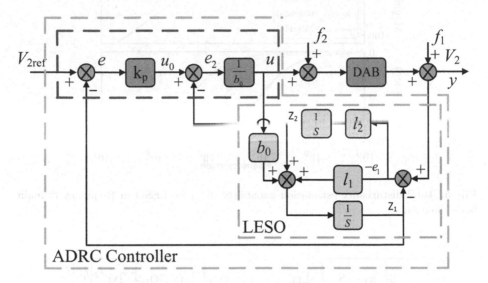

Fig. 4. LADRC control structure of DAB converter

where k_p is the proportionality factor, additionally referred to as the controller bandwidth, structure of the system is shown in Fig. 4. An usual regulation of experience is employed to settle on the controller bandwidth: $k_p = \omega_c = (1/5 - -1/2)\omega_0$ [13].

LESO affording the estimate of the system total disturbance is the key part of LADRC, whose property will impact on the efficacy of LADRC. Therefore, the goodness of LESO estimation disturbances becomes vital. The transfer function (24) between the estimated value z_2 and the actual value of the total disturbance f can be acquired from Fig. 4.

$$\frac{z_2(s)}{f(s)} = \frac{\omega_0^2}{(s + \omega_0)^2} \tag{24}$$

5   Experimental Results

In order to verify the effectiveness of proposed control strategy, the simulation model of DAB is built by the simulation experiment platform, and Table 2 shows parameters used for converter simulation.

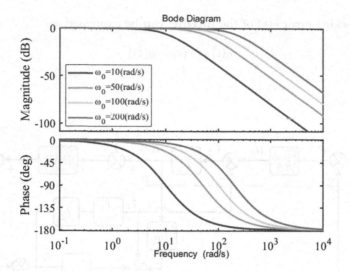

Fig. 5. Total disturbance estimation capability of z_2 by LESO in frequency domain bode diagram

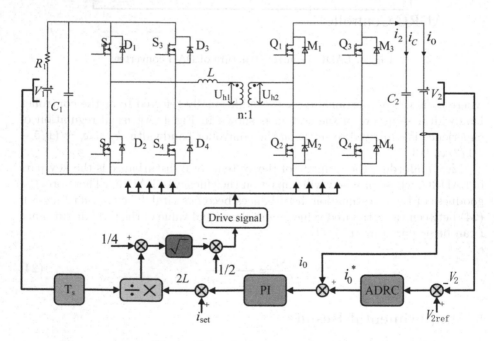

Fig. 6. Control schematic diagram

Table 2. Parameters of DABC converter prototype

Parameters	Value
Input voltage V_1	60 V
Input capacitance C_1	2000 μF
Output capacitance C_2	2000 μF
Resistance R	20 Ω
Leakage inductance L	0.1 mH
Switching frequency f_s	5 kHz
Turn ratio	1:1

The comparison of low bus voltage under the same experimental conditions and simulation experiment platform is shown in Fig. 7, where the capacitor voltage achieves the given value in 0.09 s without overshoot when LADRC control is implemented and in 0.18s when PI control is applied. It can be observed that when LADRC control is applied, the response time of system to reach steady state is faster than that of PI control. When the load rapidly changes, the load resistance changes from 20Ω to 5.7Ω, and the overshoot amount and regulation time corresponding to the low bus voltage are monitored under the operation of two controllers. For a low bus voltage of 16.5 V, the overshoot under PI control is 0.9V, while the overshoot under LADRC control is 0.68V. During transient response of the system, the overshoot of LADRC control is reduced. The low bus voltage regulation time under PI control is 0.135s. The low bus voltage regulation time under LADRC control is 0.046s. As a result, the regulation time of LADRC control in transient response the DAB system is reduced. LADRC control not only lowers bus voltage overshoot, but it also reduces low bus voltage regulation time.

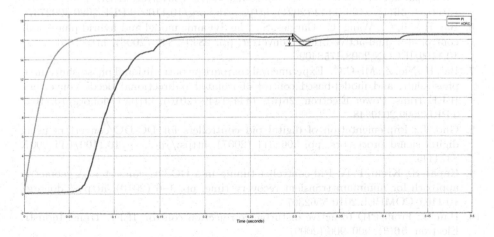

Fig. 7. Control schematic diagram

6   Conclusion

In order to improve the dynamic performance of DAB converter, this paper proposes a feed-forward control method based on input voltage and output current on the basis of LADRC control, gives a detailed derivation process, and conducts a comparative experimental study in a simulation-based platform. In this paper, LADRC controller is designed for closed-loop control strategy in DAB converter. The output current control method based on input voltage feed-forward proposed in this paper can significantly improve the dynamic performance of the converter during sudden changes in input voltage, sudden changes in load and given sudden changes. The output current control method based on input voltage feedforward proposed in this paper introduces inductance parameters for calculation, but the dependence on inductance parameters is small, which improves the compatibility of the control method.

References

1. Lu, J., Wang, Y., Wang, H., Wei, M., Zhou, Y., Zhang, Y.: Modulation strategy for improving the voltage gain of the dual-active-bridge converter. IET Power Electron. **13**(8), 1630–1638 (2020). https://doi.org/10.1049/iet-pel.2019.1250
2. Zhan, H., et al.: Model predictive control of input-series output-parallel dual active bridge converters based dc transformer. IET Power Electron. **13**(6), 1144–1152 (2020). https://doi.org/10.1049/iet-pel.2019.1061
3. Hou, N., Song, W., Wu, M.: Minimum-current-stress scheme of dual active bridge dcdc converter with unified phase-shift control. IEEE Trans. Power Electron. **31**(12), 8552–8561 (2016). https://doi.org/10.1109/TPEL.2016.2521410
4. Zhao, B., Yu, Q., Sun, W.: Extended-phase-shift control of isolated bidirectional DC-DC converter for power distribution in microgrid. IEEE Trans. Power Electron. **27**(11), 4667–4680 (2012). https://doi.org/10.1109/TPEL.2011.2180928
5. Hou, N., Li, Y.: A direct current control scheme with compensation operation and circuit-parameter estimation for full-bridge dcdc converter. IEEE Trans. Power Electron. **36**(1), 1130–1142 (2021). https://doi.org/10.1109/TPEL.2020.3002737
6. Bai, H., Mi, C., Wang, C., Gargies, S.: The dynamic model and hybrid phase-shift control of a dual-active-bridge converter, pp. 2840–2845 (2008). https://doi.org/10.1109/IECON.2008.4758409
7. Bai, H., Nie, Z., Mi, C.C.: Experimental comparison of traditional phase-shift, dual-phase-shift, and model-based control of isolated bidirectional dcdc converters. IEEE Trans. Power Electron. **25**(6), 1444–1449 (2010). https://doi.org/10.1109/TPEL.2009.2039648
8. Guo, L.: Implementation of digital pid controllers for DC-DC converters using digital signal processors, pp. 306–311 (2007). https://doi.org/10.1109/EIT.2007.4374445
9. Kapat, S., Krein, P.T.: Pid controller tuning in a DC-DC converter: A geometric approach for minimum transient recovery time, pp. 1–6 (2010). https://doi.org/10.1109/COMPEL.2010.5562367
10. Han, J.: From PID to active disturbance rejection control. IEEE Trans. Industr. Electron. **56**(3), 900–906 (2009)

11. Gao, Z.: Active disturbance rejection control: a paradigm shift in feedback control system design, p. 7 (2006). https://doi.org/10.1109/ACC.2006.1656579
12. Song, W., Hou, N., Wu, M.: Virtual direct power control scheme of dual active bridge DC-DC converters for fast dynamic response. IEEE Trans. Power Electron. **33**(2), 1750–1759 (2018). https://doi.org/10.1109/TPEL.2017.2682982
13. Gao, Z.: Scaling and bandwidth-parameterization based controller tuning **6**, 4989–4996 (2003). https://doi.org/10.1109/ACC.2003.1242516
14. Cao, Y., Zhao, Q., Ye, Y., Xiong, Y.: Adrc-based current control for grid-tied inverters: design, analysis, and verification. IEEE Trans. Industr. Electron. **67**(10), 8428–8437 (2020). https://doi.org/10.1109/TIE.2019.2949513

Adaptive Fuzzy Distributed Formation Tracking for Second-order Nonlinear Multi-agent Systems with Prescribed Performance

Binghe An[1] , Zongzhun Zheng[2] , Bo Wang[1(✉)], Huijin Fan[1], Lei Liu[1], and Yongji Wang[1]

[1] School of Artificial Intelligence and Automation, Huazhong University of Science and Technology, Wuhan, China
{wb8517,ehjfan,liulei,wangyjch}@hust.edu.cn
[2] Beijing Aerospace Automatic Control Institute, Beijing 100854, China

Abstract. The paper investigates the distributed prescribed performance output formation tracking problem of second-order nonlinear multi-agent systems subject to uncertain disturbances. The formation is realized in a leader-follower structure, which means all followers can form a desired formation pattern while tracking the leader. For accomplishing the formation with prescribed performance, firstly, a time-varying barrier Lyapunov function(TV-BLF) consisting of formation error and performance function is introduced. Then, an adaptive formation protocol is proposed based on the TV-BLF considering both matched and mismatched disturbances. Besides, unknown nonlinear terms in the dynamic models of agents are approximated by adaptive fuzzy logic systems. Further, it is proved rigorously that under the proposed method, the formation errors can satisfy the preset performance and converge to a predefined small region around the origin. At last, a simulation example is performed to validate the performance and the superiority of the developed scheme.

Keywords: Nonlinear multi-agent systems · Formation tracking · Prescribed performance · Adaptive fuzzy approximation

1 Introduction

Formation control has been a hot research topic owing to its applications in different fields such as unmanned aerial vehicles [1], spacecrafts [2] and unmanned surface vehicles [3], etc. With the development of distributed control methods, distributed formation controllers, which only need the neighbors' information were widely investigated [4–6]. Compared with the centralized methods, the distributed formation algorithms have improved the robustness and the scalability of the controlled system.

As is known to us, the transient performance (overshoot, convergence rate, undershot) is important for controlled systems. To prevent the system states from running into unstable areas, a small overshoot is desired. Fast convergence rate is necessary for some applications which need to be finished in finite time such as missile guidance [7]. Thus, it is critical to develop an effective control scheme to satisfy the requirement of the transient performance. However, it is not easy to manage the transient response because the dynamic differential equations of systems can't be solved in general. To deal with this issue, prescribed performance control was proposed in [8] by a predefined performance function and some error transform techniques. Further, in [9], prescribed performance control was extended to multi-agent systems for achieving the consensus of agents. For formation control, distributed formation strategy for multi-quadrotor with guaranteed performance was explored in [10]. Besides, distributed formation control with prescribed error constraint was investigated in [11] based on the sliding mode control, nevertheless, no mismatched disturbances are considered in the designing of the controller.

It is worth pointing out that in practical formation applications, agents are affected by uncertain external disturbances. Thus, one challenge to accomplish the formation control is how to deal with external disturbances. Some robust control methods [12,13] and observer-based methods [14,15] have been used to address the disturbance rejection problem. However, only the matched disturbances were considered in [11–15]. In addition to matched disturbances, practical systems may also be affected by mismatched disturbances, which can't be compensated directly by the control input. Furthermore, there usually exist unknown nonlinearities such as unmodeled dynamics in the system models. Hence, it's of great significance to design a robust control strategy to realize the formation with matched/mismatched disturbances and unknown nonlinearities.

Motivated by the discussions above, the output formation tracking for second-order multi-agent systems is studied in this paper. To improve the robustness of the formation system and obtain satisfactory transient performance, both mismatched/matched disturbances and prescribed performance are considered in the designing process of the controller. As we know, the fuzzy logic system based on human experience is an effective way to approximate nonlinear functions [16], therefore, we exploit adaptive fuzzy logic systems to approximate uncertain nonlinearities. The formation is accomplished in a leader-follower structure, as a result, the outputs of all followers can form a desired formation pattern while tracking the leader under the proposed method. The contributions and distinct features are as follows:

Firstly, a novel distributed prescribed performance formation controller is developed with the help of the backstepping technology and TV-BLF [17]. Under the proposed method, the transient performance in the response process can be predefined by some adjustable parameters. Secondly, both matched and mismatched disturbances are considered, resulting in better robust properties of the formation system. In fact, the agents in multi-agent systems need to resist disturbances coordinately, therefore, a novel adaptive estimation using neighbor's information is designed in the controller for estimating the mismatched disturbances. Thirdly, the adaptive fuzzy logic system is exploited to approximate the unknown nonlinear dynamics, and all the signals in the close-loop control system are proven to be uniformly bounded rigorously.

The contents of this paper are organized as follows: in Sect. 2, some basic graph theory and useful lemmas are introduced and the control objective of this paper is formulated. In Sect. 3, the proposed prescribed performance formation controller is proposed in detail and the stability of the controlled formation system is proved. A simulation example is given in Sect. 4 to validate the performance of the provided performance-guaranteed formation algorithm. Section 5 is reserved for conclusions.

Notions: In the following contents, $diag\{a_1, ..., a_N\} \in R^{N \times N}$ is used to described a diagonal matrix with $a_i(i = 1, 2...N)$ as its diagonal elements. $\lambda_{max}(M)$ and $\lambda_{min}(M)$ denote the maximum and minimum eigenvalues of the matrix M, respectively. 1_N is used to denote a column vector with size N composed of 1. $\otimes$ means the Kronecker product.

2 Problem Formulation and Preliminaries

In this section, basic topology description and necessary lemmas are given first, which will be exploited later to design the formation control scheme. Then the dynamic models of agents and the control objective are formulated.

2.1 Algebraic Graph Theory

The formation system considered in this paper consists of N followers and a leader. The information exchange among N followers is described by an undirected graph $G(\mathcal{V}, \mathcal{E})$. $\mathcal{V} = \{1, 2...N\}$ is the set of nodes and $\mathcal{E}$ means the set of edges in the graph G. The edge pair $(i, j) \in \mathcal{E}$ if agent i can communicate with agent j. In an undirected graph, $(j, i) \in \mathcal{E}$ equals to $(i, j) \in \mathcal{E}$. The adjacency matrix of the graph G is defined as $A = [a_{ij}] \in R^{N \times N}$ with $a_{ij} = 1$ if $(i, j) \in \mathcal{E}$ and $a_{ij} = 0 (i \neq j)$ if not. We assume there is no self-loop in the graph G, thus, $a_{ii} = 0$. Another important matrix to describe the interaction network is the Laplace matrix $L = [l_{ij}] \in R^{N \times N}$ of the graph G, which is defined as $l_{ii} = \sum\limits_{j=1}^{N} a_{ij}$ and $l_{ij} = -a_{ij}(i \neq j)$. Under the distributed leader-following framework, only part of followers can receive leader's states, therefore, we set $a_{i,0} = 1$ if agent can get leader's information and $a_{i,0} = 0$ otherwise. The matrix $H = [h_{ij}] \in R^{N \times N} = L + B$ is utilized to characterize the interaction relationship among the leader and followers with $B = diag\{a_{10}, ..., a_{N0}\}$.

Assumption 1: The graph G connected and there exists a path from the leader to each follower.

2.2 Useful Lemmas

Lemma 1. *[17]: For $x \in R$ and $k > 0$, if $x \in (-k, k)$, then the following relationship is true:*

$$\log \frac{k^2}{k^2 - z^2} \leq \frac{z^2}{k^2 - z^2}. \tag{1}$$

Lemma 2. *[18]: For $z \in R$ and $\sigma > 0$, it holds:*

$$0 \leq |z| - \frac{z^2}{\sqrt{z^2 + \sigma}} \leq \sigma. \tag{2}$$

Lemma 3. *[18]: For any given vectors m, n, it holds that:*

$$\pm 2m^T n \leq m^T M m + n^T M^{-1} n. \tag{3}$$

where M is any positive definite matrix with compatible dimension.

Lemma 4. *[20]: Under Assumption 1, H is a positive definite matrix.*

2.3 The Dynamic Models of Multi-agent Systems

Considering a group of second order multi-agent systems, the dynamics of follower $i(i = 1, 2...n)$ is modeled as:

$$\dot{p}_i = q_i + d_i; \quad \dot{q}_i = f_i(p_i, q_i) + u_i + \omega_i; \quad y_i = p_i \tag{4}$$

where $p_i = [x_{i,1}, ..., x_{i,n}]^T \in R^n$ and $q_i = [v_{i,1}, ..., v_{i,n}]^T \in R^n$ represent the states of agent i. $d_i = [d_{i,1}, ..., d_{i,n}]^T \in R^n$ is the mismatched disturbance in the formation system and $\omega_i = [\omega_{i,1}, ..., \omega_{i,n}]^T$ is the matched disturbance. $f_i(p_i, q_i) = [f_{i,1}(p_i, q_i), ..., f_{i,n}(p_i, q_i)]^T$ denotes the unknown nonlinear function vector. $u_i = [u_{i,1}, ..., u_{i,n}]^T$ gives the control input of agent i.

The dynamic model of the leader is:

$$\dot{p}_0 = q_0; \quad \dot{q}_0 = u_0; \quad y_0 = p_0 \tag{5}$$

where $p_0 = [x_{0,1}, ..., x_{0,n}]^T \in R^n$ and $q_0 = [v_{0,1}, ..., v_{0,n}]^T \in R^n$ represent the states of the leader. $u_0 = [u_{0,1}, ..., u_{0,n}]^T \in R^n$ is the input , which is unknown to all followers. y_0 denotes the leader's output.

Assumption 2: The mismatched disturbance $d_{i.m}$ and matched disturbance $\omega_{i.m}$ are bounded, which means there are unknown positive constants $\bar{d}$ and $\bar{\omega}$ satisfying $|d_{i.m}| \leq \bar{d}$ and $|\omega_{i.m}| \leq \bar{\omega}(i = 1, 2...N; m = 1, 2...n)$.

Assumption 3: The unknown input of the leader is bounded, that is there existing an unknown positive constant $\bar{u}$ satisfying $|u_{0,m}| \leq \bar{u}(m = 1, 2...n)$.

The formation output tracking is said to be accomplished if all outputs of followers form a desired formation pattern while tracking the output of the leader. In this paper, the desired formation pattern is described by a time-varying vector $\Theta = [\Theta_1^T, ..., \Theta_N^T]^T \in R^{nN}$. $\Theta_i = [\Theta_{i,1}, ..., \Theta_{i,n}]^T$ represents the desired output bias between the follower $i(i = 1, 2...N)$ and the leader. We assume Θ_i is twice differentiable. It can be seen that the formation pattern considered in this paper is time-varying, which is more general than the fixed formation pattern. By choosing $n = 2$, a time-varying square formation pattern Θ consisting of five agents is shown in Fig. 1.

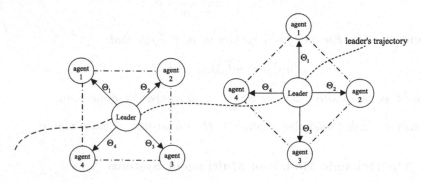

Fig. 1. Square formation with time-varying formation pattern

According to [13], the formation tracking errors of follower agent i can be defined as:

$$e_{i.m} = \sum_{j=1}^{N} a_{ij}(x_{i,m} - \Theta_{i,m} - x_{j,m} + \Theta_{j,m}) + a_{i0}(x_{i,m} - \Theta_{i,m} - x_{0,m}). \qquad m = 1, 2...n$$
$$(6)$$

Further, to guarantee a satisfactory transient response with allowable steady-state errors, the following performance function is introduced:

$$\alpha(t) = (\alpha_0 - \alpha_\infty)\exp(-ct) + \alpha_\infty \qquad (7)$$

where α_0, c, and α_∞ are positive constants. c is used to adjust the convergence rate and overshoot. α_∞ is utilized to specify the upper bound of steady-state error. The prescribed performance is said to be satisfied if for all $i = 1, 2...N; m = 1, 2...n$, the following relationship is satisfied:

$$|e_{i.m}| \leq \alpha(t). \qquad (8)$$

2.4 Control Objective

Based on the above discussions, the control objective of this paper is summarized as to designed a distributed formation controller u_i, such that the relationship (8) can be satisfied always.

Assumption 4: The initial formation error satisfy the prescribed performance, which means $|e_{i.m}(0)| \leq \alpha(t)$ holds.

In the following contents, $\alpha(t)$ is abbreviated as α for brevity.

3   Main Results

In this section, an adaptive prescribed performance controller is put forward in detail for accomplishing the time-varying formation.

Firstly, we rewrite (6) in the following compact form:

$$e_m = H \left(x_m - 1_N \otimes x_{0,m} - \widehat{\Theta}_m \right) \tag{9}$$

where $x_m = [x_{1,m}, ..., x_{N,m}]^T, v_m = [v_{1,m}, ..., v_{N,m}]^T, e_m = [e_{1,m}, ..., e_{N,m}]^T$ and $\widehat{\Theta}_m = [\Theta_{1,m}, ..., \Theta_{N,m}]^T$.

According to (4) and (9) , we have:

$$\dot{e}_m = H \left(v_m - 1_N \otimes v_{0,m} - \dot{\widehat{\Theta}}_m + d_m \right) \tag{10}$$

where $v_m = [v_{1,m}, ..., v_{N,m}]^T$ and $d_m = [d_{1,m}, ..., d_{N,m}]^T$.

Now, we consider the following TV-BLF:

$$V_1 = \sum_{m=1}^{n} \sum_{i=1}^{N} \log \frac{\alpha^2}{\alpha^2 - e_{i,m}^2} + \frac{1}{2\tau_1} \tilde{d}_{i,m}^2 \tag{11}$$

where $\tau_1 > 0$ is a constant. $\hat{d}_{i,m}$ denotes the estimation of $\bar{d}$ and $\tilde{d}_{i,.m} = \hat{d}_{i,m} - \bar{d}$. The estimation law for $\hat{d}_{i,m}$ will be given later. As indicated in [17], if V_1 is ensured to be bounded, then the prescribed performance can be satisfied.

The time derivative of V_1 can be derived as:

$$\dot{V}_1 = \sum_{m=1}^{n} \sum_{i=1}^{N} \frac{2e_{i,m}\dot{e}_{i,m}}{\alpha^2 - e_{i,m}^2} - \frac{2\dot{\alpha}e_{i,m}^2}{(\alpha^2 - e_{i,m}^2)\,\alpha} + \frac{1}{\tau_1}\tilde{d}_{i,m}\dot{\hat{d}}_{i,m}. \tag{12}$$

Define:

$$\eta_m = [\frac{2e_{1,m}}{\alpha^2 - e_{1,m}^2}, ..., \frac{2e_{N,m}}{\alpha^2 - e_{N,m}^2}]^T. \tag{13}$$

Substituting (10) and (13) into (12), one has that:

$$\dot{V}_1 = \sum_{m=1}^{n} [\eta_m^T[\dot{e}_m - \frac{\dot{\alpha}}{\alpha} \otimes e_m] + \sum_{i=1}^{N} \frac{1}{\tau_1}\tilde{d}_{i,m}\dot{\hat{d}}_{i,m}]$$

$$= \sum_{m=1}^{n} [\eta_m^T[H(v_m - 1_N \otimes v_{0.m} - \dot{\widehat{\Theta}}_m + d_m) - \frac{\dot{\alpha}}{\alpha} \otimes e_m] + \sum_{i=1}^{N} \frac{1}{\tau_1}\tilde{d}_{i,m}\dot{\hat{d}}_{i,m}]. \tag{14}$$

For the designing of the formation controller, firstly, a virtual control input $\xi_m \in R^n$ is defined as:

$$\xi_m = H\left(v_m - \mathbf{1}_N \otimes v_{0,m}\right) \tag{15}$$

and the desired virtual control input is designed as:

$$\xi_m^* = H\dot{\hat{\Theta}}_m + \frac{\dot{\alpha}}{\alpha} \otimes e_m - k_1 \otimes e_m - H\Lambda_m \tag{16}$$

where $\Lambda_m = [\Lambda_{1,m}, ..., \Lambda_{N,m}]^T$ with $\Lambda_{i,m} = \dfrac{\hat{d}_{i,m}\left(\sum\limits_{j=1}^{N} h_{ij}\eta_{j,m}\right)}{\sqrt{\left(\sum\limits_{j=1}^{N} h_{ij}\eta_{j,m}\right)^2 + \sigma_1}}$. $\eta_{j,m}$ is the jth

element of the vector η_m. σ_1 and k_1 are positive constants. $\hat{d}_{i,m}$ is governed by the adaptive law (17) with $\tau_1 > 0, k_2 > 0$.

$$\dot{\hat{d}}_{i,m} = \tau_1\left(\frac{\left(\sum\limits_{j=1}^{N} h_{ij}\eta_{j,m}\right)^2}{\sqrt{\left(\sum\limits_{j=1}^{N} h_{ij}\eta_{j,m}\right)^2 + \sigma_1}} - k_2\hat{d}_{i,m}\right) \tag{17}$$

Substituting (16) into (14), it yields:

$$\dot{V}_1 = \sum_{m=1}^{n}\left[\eta_m{}^T\left(\xi_m^* + \tilde{\xi}_m - H\dot{\hat{\Theta}}_m + Hd_m - \frac{\dot{\alpha}}{\alpha} \otimes e_m\right) + \sum_{i=1}^{N}\frac{1}{\tau_1}\tilde{d}_{i,m}\dot{\hat{d}}_{i,m}\right]$$

$$= \sum_{m=1}^{n}\sum_{i=1}^{N}\left[\eta_{i,m}\tilde{\xi}_{i,m} - \frac{2k_1 e_{i,m}^2}{\alpha^2 - e_{i,m}^2} + d_{i,m}\left(\sum_{j=1}^{N} h_{ij}\eta_{j,m}\right)\right.$$

$$\left. - \left(\sum_{j=1}^{N} h_{ij}\eta_{j,m}\right)\frac{\hat{d}_{i,m}\left(\sum\limits_{j=1}^{N} h_{ij}\eta_{j,m}\right)}{\sqrt{\left(\sum\limits_{j=1}^{N} h_{ij}\eta_{j,m}\right)^2 + \sigma_1}} + \frac{1}{\tau_1}\tilde{d}_{i,m}\dot{\hat{d}}_{i,m}\right] \tag{18}$$

where $\tilde{\xi}_m = \xi_m^* - \xi_m = [\tilde{\xi}_{1,m}, ..., \tilde{\xi}_{N,m}]^T \in R^N$.

Invoking the adaptive law (17), it follows that:

$$d_{i,m}\left(\sum_{j=1}^{N} h_{ij}\eta_{j,m}\right) - \left(\sum_{j=1}^{N} h_{ij}\eta_{j,m}\right)\frac{\hat{d}_{i,m}\left(\sum\limits_{j=1}^{N} h_{ij}\eta_{j,m}\right)}{\sqrt{\left(\sum\limits_{j=1}^{N} h_{ij}\eta_{j,m}\right)^2 + \sigma_1}} + \frac{1}{\tau_1}\tilde{d}_{i,m}\dot{\hat{d}}_{i,m}$$

$$= d_{i,m}\left(\sum_{j=1}^{N} h_{ij}\eta_{j,m}\right) - \left(\sum_{j=1}^{N} h_{ij}\eta_{j,m}\right)\frac{\hat{d}_{i,m}\left(\sum\limits_{j=1}^{N} h_{ij}\eta_{j,m}\right)}{\sqrt{\left(\sum\limits_{j=1}^{N} h_{ij}\eta_{j,m}\right)^2 + \sigma_1}}$$

$$+ (\hat{d}_{i,m} - \bar{d})(\frac{(\sum\limits_{j=1}^{N} h_{ij}\eta_{j,m})^2}{\sqrt{(\sum\limits_{j=1}^{N} h_{ij}\eta_{j,m})^2 + \sigma_1}} - k_2\hat{d}_{i,m})$$

$$= d_{i,m}(\sum\limits_{j=1}^{N} h_{ij}\eta_{j,m}) - \frac{\bar{d}(\sum\limits_{j=1}^{N} h_{ij}\eta_{j,m})^2}{\sqrt{(\sum\limits_{j=1}^{N} h_{ij}\eta_{j,m})^2 + \sigma_1}} - k_2\tilde{d}_{i,m}\hat{d}_{i,m}$$

$$\leq \bar{d}\left(\left|\sum\limits_{j=1}^{N} h_{ij}\eta_{j,m}\right| - \frac{(\sum\limits_{j=1}^{N} h_{ij}\eta_{i,m})^2}{\sqrt{(\sum\limits_{j=1}^{N} h_{ij}\eta_{j,m})^2 + \sigma_1}}\right) - k_2 d_{i,m} d_{i,m}. \tag{19}$$

Further, by Lemmas 2–3 and the last inequality in (19), one has that:

$$d_{i,m}(\sum\limits_{j=1}^{N} h_{ij}\eta_{j,m}) - (\sum\limits_{j=1}^{N} h_{ij}\eta_{j,m})\frac{\hat{d}_{i,m}(\sum\limits_{j=1}^{N} h_{ij}\eta_{j,m})}{\sqrt{(\sum\limits_{j=1}^{N} h_{ij}\eta_{j,m})^2 + \sigma_1}} + \tilde{d}_{i,m}\dot{\hat{d}}_{i,m}$$

$$\leq \bar{d}\sigma_1 - k_2\tilde{d}_{i,m}\hat{d}_{i,m} = \bar{d}\sigma_1 - k_2\tilde{d}_{i,m}(\tilde{d}_{i,m} + \bar{d}) \tag{20}$$

$$= \bar{d}\sigma_1 - k_2\tilde{d}_{i,m}^2 - k_2\tilde{d}_{i,m}\bar{d} \leq \bar{d}\sigma_1 - k_2\tilde{d}_{i,m}^2 + \frac{1}{2}k_2\tilde{d}_{i,m}^2 + \frac{1}{2}k_2\bar{d}^2$$

$$= -\frac{1}{2}k_2\tilde{d}_{i,m}^2 + \bar{d}\sigma_1 + \frac{1}{2}k_2\bar{d}^2.$$

Substituing (20) into (18), it yields that:

$$\dot{V}_1 \leq \sum\limits_{m=1}^{n}\sum\limits_{i=1}^{N}[\eta_{i,m}\tilde{\xi}_{i,m} - \frac{2k_1e_{i,m}^2}{\alpha^2 - e_{i,m}^2} - \frac{k_2}{2}\tilde{d}_{i,m}^2] + \gamma_1 \tag{21}$$

where $\gamma_1 = nN(\bar{d}\sigma_1 + \frac{1}{2}k_2\bar{d}^2)$.

According to the Lemma 1 in [19], it can be obtained that the unknown nonlinear term $f_{i,m}(p_i, q_i)$ can be approximated by the following fuzzy logic system:

$$f_{i,m}(p_i, q_i) = (\theta_{i,m}^*)^T \varphi_{i,m}(p_i, q_i) + \varepsilon_{i,m} \tag{22}$$

where $\theta_{i,m}^*$ denotes the optimal parameter vector and $\varphi_{i,m}(p_i, q_i)$ represents a fuzzy basis function vector. The minimum approximation error is given by $\varepsilon_{i,m}$. Moreover, there exists a constant $\bar{\varepsilon}$ satisfying $\bar{\varepsilon} \geq \varepsilon_{i,m}$ [19]. Thus, we can use $\theta_{i,m}^T \varphi_{i,m}(p_i, q_i)$ to approximate the nonlinear function $f_{i,m}(p_i, q_i)$. $\theta_{i,m}$ is an adaptive parameter vector, which is the estimation of $\theta_{i,m}^*$.

Then, the prescribed performance formation controller $u_{i,m}$ is proposed as follows:

$$u_{i,m} = \eta_{i,m} - k_3\xi_{i,m} - \frac{\hat{\omega}_{i,m}\tilde{\xi}_{i,m}}{\sqrt{\tilde{\xi}_{i,m}^2 + \sigma_2}} - \theta_{i,m}^T\varphi_{i,m}(p_i, q_i) \tag{23}$$

where $\sigma_2 > 0$, $\omega_{i,m}$ is the estimation of $\bar{\omega} + \bar{u} + \bar{\varepsilon}$. k_3 is a positive feedback gain. The adaptive laws of $\dot{\omega}_{i,m}$ and $\theta_{i,m}$ are put forward as:

$$\dot{\hat{\omega}}_{i,m} = \tau_2\left(\frac{\tilde{\xi}_{i,m}^2}{\sqrt{\tilde{\xi}_{i,m}^2 + \sigma_2}} - k_4\hat{\omega}_{i,m}\right) \tag{24}$$

$$\dot{\theta}_{i,m} = \tau_3(\tilde{\xi}_{i,m}\varphi_{i.m}(p_i, q_i) - k_5\theta_{i,m}) \tag{25}$$

where τ_2, τ_3 and $k_l(l = 4, 5)$ are positive constants.

Theorem 1. *Considering the leader-follower system (4)–(5) under Assumptions 1–4, if each follower is steered by the controller (23) with adaptive laws (24)–(25), then the desired formation pattern can be formed and the prescribed performance (8) can be guaranteed.*

Proof. Firstly, the following function V_2 is chosen as Lyapunov candidate:

$$V_2 = V_1 + \sum_{m=1}^{n}(\tilde{\xi}_m^T H^{-1}\tilde{\xi}_m + \sum_{i=1}^{N}[\frac{1}{2\tau_2}\tilde{\omega}_{i,m}^2 + \frac{1}{2\tau_3}\tilde{\theta}_{i,m}^T\tilde{\theta}_{i,m}]) \tag{26}$$

with $\tilde{\omega}_{i,m} = \hat{\omega}_{i,m} - (\bar{u} + \bar{\omega} + \bar{\varepsilon})$ and $\tilde{\theta}_{i,m} = \theta_{i,m} - \theta_{i,m}^*$.
Then, the derivative of V_2 with respect to time is calculated as:

$$\dot{V}_2 = \dot{V}_1 + \sum_{m=1}^{n}(\tilde{\xi}_m^T H^{-1}\dot{\tilde{\xi}}_m + \sum_{i=1}^{N}[\frac{1}{\tau_2}\tilde{\omega}_{i,m}\dot{\hat{\omega}}_{i,m} + \frac{1}{\tau_3}\tilde{\theta}_{i,m}^T\dot{\theta}_{i,m}]). \tag{27}$$

From (4) and (15), one has that:

$$\dot{\tilde{\xi}}_m = \dot{\xi}_m - \dot{\xi}_m^* = H(\hat{u}_m - 1_N \otimes u_{0,m} + f_m + \omega_m) - \dot{\xi}_m^* \tag{28}$$

where $f_m = [f_{1,m}(p_1, q_1), ..., f_{N,m}(p_N, q_N)]^T$, $\hat{u}_m = [u_{1,m}, ..., u_{N,m}]^T$ and $\omega_m = [\omega_{1,m}, ..., \omega_{N,m}]^T$.
Substituting (28) into (27), it follows that:

$$\dot{V}_2 = \dot{V}_1 + \sum_{m=1}^{n}(\tilde{\xi}^T(\hat{u}_m - 1_N \otimes u_{0,m} + f_m + \omega_m) + \tilde{\xi}_m^T H^{-1}\dot{\xi}_m^*$$
$$+ \sum_{i=1}^{N}[\frac{1}{\tau_2}\tilde{\omega}_{i,m}\dot{\hat{\omega}}_{i,m} + \frac{1}{\tau_3}\tilde{\theta}_{i,m}^T\dot{\theta}_{i,m}]). \tag{29}$$

Further, in according with (21)–(23) and (29), it yields:

$$\dot{V}_2 \leq \sum_{m=1}^{n} \sum_{i=1}^{N} [-\frac{2k_1 e_{i,m}^2}{\alpha^2 - e_{i,m}^2} + \tilde{\xi}_{i,m}[(\theta_{i,m}^*)^T \varphi_{i,m}(p_i, q_i) + \varepsilon_{i,m} + u_{0,m}$$

$$+\omega_{i,m} - \theta_{i,m}^T \varphi_{i,m}(p_i, q_i) - \frac{\hat{\omega}_{i,m} \tilde{\xi}_{i,m}}{\sqrt{\tilde{\xi}_{i,m}^2 + \sigma_2}} - k_3 \tilde{\xi}_{i,m}] \tag{30}$$

$$+\frac{1}{\tau_2} \tilde{\omega}_{i,m} \dot{\hat{\omega}}_{i,m} + \frac{1}{\tau_3} \tilde{\theta}_{i,m}^T \dot{\theta}_{i,m}] + \gamma_1 + \sum_{m=1}^{n} \tilde{\xi}_m^T H^{-1} \dot{\xi}_m^*$$

Considering the adaptive law (25), the (30) can be transformed to:

$$\dot{V}_2 \leq \sum_{m=1}^{n} \sum_{i=1}^{N} [-\frac{2k_1 e_{i,m}^2}{\alpha^2 - e_{i,m}^2} + \tilde{\xi}_{i,m}(-k_3 \tilde{\xi}_{i,m} + \varepsilon_{i,m} + u_{0,m} + \omega_{i,m} - \frac{\hat{\omega}_{i,m} \tilde{\xi}_{i,m}}{\sqrt{\tilde{\xi}_{i,m}^2 + \sigma_2}})$$

$$+\frac{1}{\tau_2} \tilde{\omega}_{i,m} \dot{\hat{\omega}}_{i,m} - k_5 \tilde{\theta}_{i,m}^T \theta_{i,m}] + \gamma_1 + \sum_{m=1}^{n} \tilde{\xi}_m^T H^{-1} \dot{\xi}_m^*. \tag{31}$$

It follows from Lemma 3 that:

$$-k_5 \tilde{\theta}_{i,m}^T \theta_{i,m} = -k_5 \tilde{\theta}_{i,m}^T (\tilde{\theta}_{i,m} + \theta_{i,m}^*)$$

$$\leq -k_5 \tilde{\theta}_{i,m}^T \tilde{\theta}_{i,m} + \frac{k_5}{2} \tilde{\theta}_{i,m}^T \tilde{\theta}_{i,m} + \frac{k_5}{2} (\theta_{i,m}^*)^T \theta_{i,m}^* \tag{32}$$

$$= -\frac{k_5}{2} \tilde{\theta}_{i,m}^T \tilde{\theta}_{i,m} + \frac{k_5}{2} (\theta_{i,m}^*)^T \theta_{i,m}^*.$$

Suppose $\dot{\xi}_{i,m}^*$ is bounded, then, by Lemma 3, we have:

$$\tilde{\xi}_m^T H^{-1} \dot{\xi}_m^* \leq \frac{1}{2} \tilde{\xi}_m^T \tilde{\xi}_m + \frac{1}{2} (\dot{\xi}_m^*)^T H^{-1} H^{-1} \dot{\xi}_m^*$$

$$\leq \frac{1}{2} \tilde{\xi}_m^T \tilde{\xi}_m + \frac{1}{2} [\lambda_{\max}(H^{-1})]^2 (\dot{\xi}_m^*)^T \dot{\xi}_m^* \tag{33}$$

$$\leq \sum_{i=1}^{N} [\frac{1}{2} \tilde{\xi}_{i,m}^2] + \frac{1}{2} [\lambda_{\max}(H^{-1})]^2 N \bar{\xi}^2.$$

where $\bar{\xi} = \max\{\dot{\xi}_{i,m}^*\}$.

Then, from (31)–(33), one has that:

$$\dot{V}_2 \leq \sum_{m=1}^{n} \sum_{i=1}^{N} [-\frac{2k_1 e_{i,m}^2}{\alpha^2 - e_{i,m}^2} - (k_3 - \frac{1}{2}) \tilde{\xi}_{i,m}^2 - \frac{k_5}{2} \tilde{\theta}_{i,m}^T \tilde{\theta}_{i,m} +$$

$$\tilde{\xi}_{i,m}[\varepsilon_{i,m} + u_{0,m} + \omega_{i,m} - \frac{\hat{\omega}_{i,m} \tilde{\xi}_{i,m}}{\sqrt{\tilde{\xi}_{i,m}^2 + \sigma_2}}) + \frac{1}{\tau_2} \tilde{\omega}_{i,m} \dot{\hat{\omega}}_{i,m}] + \gamma_2 \tag{34}$$

where $\gamma_2 = \gamma_1 + \frac{1}{2} [\lambda_{\max}(H^{-1})]^2 n N \bar{\xi}^2 + \sum_{m=1}^{n} \sum_{i=1}^{N} \frac{k_5}{2} (\theta_{i,m}^*)^T \theta_{i,m}^*.$

Now, invoking the adaptive law (24) and based on the similar analysis as shown in (19)–(20), the following relationship can be obtained:

$$
\tilde{\xi}_{i,m}[\varepsilon_{i,m} + u_{0,m} + \omega_{i,m} - \frac{\hat{\omega}_{i,m}\tilde{\xi}_{i,m}}{\sqrt{\tilde{\xi}_{i,m}^2 + \sigma_2}}) + \frac{1}{\tau_2}\tilde{\omega}_{i,m}\dot{\hat{\omega}}_{i,m}
$$
$$
\leq -\frac{1}{2}k_4\tilde{\omega}_{i,m}^2 + \delta\sigma_2 + \frac{1}{2}k_4\delta^2
\tag{35}
$$

where $\delta = \bar{\omega} + \bar{u} + \bar{\varepsilon}$.

Substituting (35) into (34), it follows that:

$$
\dot{V}_2 \leq \sum_{m=1}^{n}\sum_{i=1}^{N}[-\frac{2k_1 e_{i,m}^2}{\alpha^2 - e_{i,m}^2} - (k_3 - \frac{1}{2})\tilde{\xi}_{i,m}^2 - \frac{k_4}{2}\tilde{\omega}_{i,m}^2 - \frac{k_5}{2}\tilde{\theta}_{i,m}^T\tilde{\theta}_{i,m}]+\gamma_3
\tag{36}
$$

where $\gamma_3 = \gamma_2 + nN(\delta\sigma_2 + \frac{1}{2}k_4\delta^2)$.

By Lemma 1 and choosing $k_3 - \frac{1}{2}>0$, we have:

$$
\dot{V}_2 \leq -\beta V_2 + \gamma_3.
\tag{37}
$$

From (37), there is:

$$
V_2(t) \leq V_2(0)\exp(-\beta(t - t_0)) + \frac{\gamma_3}{\beta}[1 - \exp(-\beta(t - t_0))].
\tag{38}
$$

That is, V_2 is bounded, therefore, the prescribed performance is guaranteed.

4   Numerical Simulation

A numerical simulation is performed in this section to demonstrate the performance and feasibility of the proposed formation controller (23)–(25). Considering there are five agents in the formation system, the communication network among agents is shown in Fig. 2.

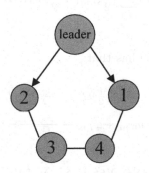

Fig. 2. Communication network among the leader and followers

We choose that $n = 2$ for simulation and the parameters in the proposed controller are chosen as: $k_1 = 2.5, k_2 = 0.1, k_3 = 2, k_4 = k_5 = 0.001, \sigma_1 = 1, \sigma_2 = 0.1, \tau_1 = 5, \tau_2 = 200, \alpha_0 = 12.2, \alpha_\infty = 0.2, c = 3.5$. The initial states of four followers are set as $p_1(0) = [6.5, 5]^T, p_2(0) = [4.5, -1]^T, p_3(0) = [4, -1]^T, p_4(0) = [5, 3]^T$ and the initial state of the leader is $p_0(0) = [3, 3]^T$. The input of the leader is $[0.5\sin(t/3), 0.5\cos(t/3)]^T$. Take $d_{i,1} = \cos(t), d_{i,2} = \sin(t)$ and $\omega_{i,1} = \sin(t/2) + \cos(t) + 1, \omega_{i,2} = \cos(t/2) + \sin(t) + 1$. The nonlinear functions are set as $f_{i,1}(p_i, q_i) = 0.8v_{i,1}\sin(v_{i,1}) + 2\sin(x_{i,1}) + 0.5\sin(x_{i,2})$ and $f_{i,2}(p_i, q_i) = 0.5x_{i,2}\sin(v_{i,2}) + \cos(x_{i,2}) + 0.3v_{i,1}\cos(v_{i,1})$. The desired formation pattern is chosen as $\Theta_{i,1} = [4\sin(0.5t + 0.5(i-1)\pi)], \Theta_{i,2} = [4\cos(0.5t + 0.5(i-1)\pi)](i = 1, 2, 3, 4)$. The simulation results are depicted in Figs. 3–6.

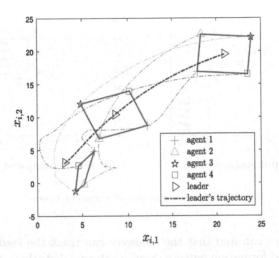

Fig. 3. The trajectories of five agents

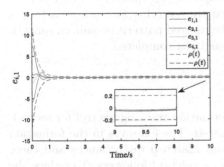

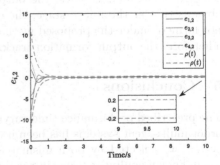

Fig. 4. Formation errors $e_{i,1}$ **Fig. 5.** Formation errors $e_{i,2}$

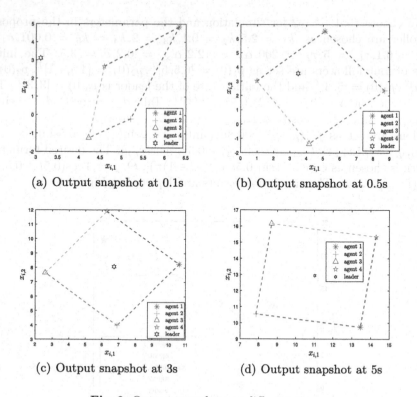

(a) Output snapshot at 0.1s

(b) Output snapshot at 0.5s

(c) Output snapshot at 3s

(d) Output snapshot at 5s

Fig. 6. Output snapshots at different times

From Fig. 3, we can find that the followers can track the leader while maintaining the desired formation pattern despite external disturbances. As indicated in Figs. 4 and 5, by the developed prescribed performance formation method, both the stability of the formation system and the desired transient performance are guaranteed, moreover, the steady formation errors are smaller than the preset error boundary. Besides, the output snapshots are provided in Fig. 6, it's obvious that at the initial stage, the desired time-varying formation pattern is not formed, under the proposed method, the desired pattern is realized quickly. Therefore, the output formation tracking has been completed.

5   Conclusions

The prescribed performance time-varying formation tracking control for second-order multi-agent systems has been investigated. The followers in the formation system are affected by both mismatched and matched disturbances and there are unknown nonlinear terms in the dynamic models of followers. To reduce the influence of disturbances on the controlled system, adaptive estimation technology is exploited to estimate the upper bound of the disturbances. Further, by combining the backstepping method with TV-BLF, a distributed adaptive prescribed performance formation controller is proposed with the aid of fuzzy logic approximations. In addition, the stability of the formation system has been analyzed rigorously. Last, a numerical simulation is carried out, which show the formation tracking can be achieved with prescribed performance under the proposed scheme.

References

1. Cong, Y.Z., Du, H.B., Jin, Q.C., Zhu, W.W., Lin, X.Z.: Formation control for multiquadrotor aircraft: connectivity preserving and collision avoidance. Int. J. Robust Nonlin. 30(6), 2352–2366 (2020)
2. Hu, Q.L., Dong, H.Y., Zhang, Y.M., Ma, G.F.: Tracking control of spacecraft formation flying with collision avoidance. Aerosp. Sci. Technol. 42, 353–364 (2015)
3. He, S.D., Wang, M., Dai, S.L., Luo, F.: Leader-follower formation control of USVs with prescribed performance and collision avoidance. IEEE Trans Ind Inform. 15(1), 572–581 (2019)
4. Du, H.B., Zhu, W.W., Wen, G.H., Duan, Z.S., Lu, J.H.: Distributed formation control of multiple quadrotor aircraft based on nonsmooth consensus algorithms. IEEE T Cybern. 49(1), 342–353 (2019)
5. Oh, K.K., Park, M.C., Ahn, H.S.: A survey of multi-agent formation control. Automatica 53, 424–440 (2015)
6. Kartal, Y., Subbarao, K., Gans, N.R., Dogan, A., Lewis, F.: Distributed backstepping based control of multiple UAV formation flight subject to time delays. IET Contr. Theory Appl. 14(12), 1628–1638 (2020)
7. Cheng, Z.T., Wu, H., Wang, B., Liu, L., Wang, Y.J.: Fixed-time convergent guidance law with impact angle control. Complexity 2020, 1–9 (2020)
8. Bechlioulis, C.P., Rovithakis, G.A.: Robust adaptive control of feedback linearizable MIMO nonlinear systems with prescribed performance. IEEE Trans. Autom. Control 53(9), 2090–2099 (2008)
9. Bechlioulis, C.P., Rovithakis, G.A.: Decentralized robust synchronization of unknown high order nonlinear multi-agent systems with prescribed transient and steady state performance. IEEE Trans. Autom. Control 62(1), 123–134 (2017)
10. Wei, C.S., Luo, J.J., Yin, Z.Y., Yuan, J.P.: Leader-following consensus of second-order multi-agent systems with arbitrarily appointed-time prescribed performance. IET Contr. Theory Appl. 12(16), 2276–2286 (2018)
11. Han, S.I.: Prescribed consensus and formation error constrained finite-time sliding mode control for multi-agent mobile robot systems. IET Contr. Theory Appl. 12(2), 282–290 (2018)
12. Abdoli, H.M.H., Najafi, M., Izadi, I., Sheikholeslam, F.: Sliding mode approach for formation control of multi-agent systems with unknown nonlinear interactions. ISA Trans. 80, 65–72 (2018)
13. Khoo, S.Y., Xie, L.H., Man, Z.H.: Robust finite-time consensus tracking algorithm for multirobot systems. IEEE-ASME Trans. Mechatron. 14(2), 219–228 (2009)
14. Liu, Y.A., Wang, Q., Dong, C.Y., Ran, M.P.: Time-varying formation control for unmanned aerial vehicles with external disturbances. Trans. Inst. Meas. Control. 41(13), 3777–3786 (2019)
15. Zhang, X.H., Gao, J.L., Zhang, W.F., Zeng, T., Ye, L.P.: Distributed formation control for multiple quadrotor based on multi-agent theory and disturbance observer. Math. Probl. Eng. 2019, 1–11 (2019)

16. Cui, Y., Liu, X., Deng, X.: Composite adaptive fuzzy decentralized tracking control for pure-feedback interconnected large-scale nonlinear systems. Neural Comput. Appl. **33**(14), 8735–8751 (2021). https://doi.org/10.1007/s00521-020-05622-y
17. Tee, K.P., Ren, B.B., Ge, S.S.: Control of nonlinear systems with time-varying output constraints. Automatica **47**(11), 2511–2516 (2011)
18. Yu, J.L., Dong, X.W., Li, Q.D., Ren, Z.: Practical time-varying formation tracking for second-order nonlinear multiagent systems with multiple leaders using adaptive neural networks. IEEE Trans. Neural Netw. Learn. Syst. **29**(12), 6015–6025 (2018)
19. Wang, W., Li, Y.: Observer-based event-triggered adaptive fuzzy control for leader-following consensus of nonlinear strict-feedback systems. IEEE T Cybern. **51**(4), 2131–2141 (2021)
20. Li, S.H., Wang, X.Y.: Finite-time consensus and collision avoidance control algorithms for multiple AUVs. Automatica **49**, 3359–3367 (2013)

Path Planning for Mobile Robots Based on Improved A* Algorithm

Yuxiang Hou[1,2], Huanbing Gao[1,2(✉)], Zijian Wang[1,2], and Chuansheng Du[1,2]

[1] School of Information and Electrical Engineering, Shandong Jianzhu University, Jinan 250101, China
gaohuanbing2004@sdjzu.edu.cn
[2] Shandong Key Laboratory of Intelligent Building Technology, Jinan 250101, China

Abstract. Aiming at the problem of low efficiency of mobile robot path planning in complex environments, based on the traditional A* algorithm and combined with the divide and conquer strategy algorithm, A four-way A* algorithm for a two-dimensional raster map is proposed in this paper. First, use random sorting and preprocessing to optimize the traditional A* algorithm and change the termination condition of the two-way A* algorithm expansion. Finally, use the start and end points to calculate the third node. The original problem is decomposed into a subproblem that simultaneously extends the four search trees from the starting point, intermediate point, and target point. After the pathfinding is successful, the paths of the subproblems are merged to get the optimal path. In addition, termination conditions have been added. While planning the four-way A* algorithm, it will also judge the one-way and two-way path planning so that the algorithm can find a path faster in a complex space. In order to verify the effectiveness of the improved algorithm, the improved algorithm and other algorithms are simulated in Matlab. The simulation results show that the path planning efficiency of the algorithm has been significantly improved, and as the scale of the environment increases, the advantages of the improved algorithm are more prominent.

Keywords: Mobile robot · Path planning · A* algorithm · Third Node · Four-way A* algorithm

1 Introduction

With the popularity of mobile robots, the path planning efficiency of mobile robots is becoming more and more important, and how to quickly plan an optimal path is of great research significance and practical value. Path planning plays an essential role in the navigation of mobile robots. Accurate path planning enables mobile robots to reach target points quickly without colliding with obstacles [1, 2].

The core of path planning is path planning algorithms, mainly the raster method, topological method, free space planning method, potential field method, etc. [3–5]. Among them, the raster method is commonly used in the navigation of mobile robots, such as path planning and real-time obstacle avoidance. The existing planning algorithms

© The Author(s), under exclusive license to Springer Nature Singapore Pte Ltd. 2022
H. Zhang et al. (Eds.): NCAA 2022, CCIS 1637, pp. 169–183, 2022.
https://doi.org/10.1007/978-981-19-6142-7_13

based on the raster method mainly include the A* algorithm, D* algorithm, Dijkstra algorithm, etc. [6–9]. A* algorithm is the classic heuristic search algorithm in the path planning algorithm, which can plan the best path faster and more efficiently [10].

However, the A* algorithm has some shortcomings because A* is a one-way recursive search. If the starting point and the target point are far away and the environment is complex, it will generate a huge amount of computation and need to store many nodes to be expanded, leading to a long path planning time. For such issues, a variant of the A* algorithm is proposed, which does not calculate the heuristic function value of each node, but only calculates the function value before the collision point, reducing the computation time of the algorithm. However, in the case of a complex environment, the algorithm runs a cumbersome process, resulting in inefficient pathfinding [11]. An algorithm combining A* and jump point search algorithm is proposed to reduce the computation of many unnecessary extension nodes and thus improve the pathfinding efficiency. However, the reduction of the computation of some nodes leads to an increase in the magnitude of the planned path turns, which will lead to the appearance of a path smoothing problem [12]. A*-connect algorithm is proposed [13]. An improved A*-connect algorithm is proposed, extending the traditional A* algorithm with a two-way search. The A*-connect algorithm can greatly reduce the pathfinding efficiency while ensuring the optimal path [14–16]. However, during the operation of the bidirectional A* algorithm, there is a certain chance that the forward and reverse search extension nodes cannot meet in the middle due to the complex environment, resulting in the algorithm needing to perform twice the computational A* and reducing the pathfinding efficiency.

In summary, this paper improves the traditional A* algorithm and the bidirectional A* algorithm to address the shortcomings of the A* algorithm and the A*-connect algorithm. First, the traditional sorting method in the algorithm is changed to random fast sorting to reduce the algorithm's complexity. Second, the A*-connect termination condition is added to improve the success probability of the algorithm. Finally, the A*-connect is extended to a four-way A* by combining the partitioning strategy. The intermediate nodes are calculated, and four search paths are extended from the starting node, intermediate node, and target node, respectively, to increase the algorithm's efficiency. The experimental results show that the improved A* algorithm can plan the optimal path quickly, and the advantage becomes more evident as the complexity of the environment increases.

2   Related Work

2.1   A* Algorithm

In robotics, the A* algorithm, which is mainly used in grid environments, is an algorithm that combines heuristics and conventional methods to ensure optimal paths as well as path planning efficiency [17, 18]. A* algorithm based on a valuation function defined in advance [19], planning optimal paths in a static environment. The valuation function is:

$$f(n) = g(n) + h(n) \tag{1}$$

where n represents the current node, $f(n)$ is the cost estimation function from the starting point through the current node to the target node, $g(n)$ represents the actual cost from the starting node to the current node, and $h(n)$ represents the estimated cost from the current node to the target node. Where $h(n)$ is the essential part of the A* heuristic function. When $h(n) = 0$, then the A* algorithm is transformed into the Dijkstra algorithm, so the correct choice of $h(n)$ can increase the accuracy and success rate of the A* algorithm. In this paper, we mainly use the Euclidean distance as the valuation function of $h(n)$, which is given by:

$$h(n) = \sqrt{(x_n - x_g)^2 + (y_n - y_g)^2} \tag{2}$$

where (x_n, y_n) represents the raster center coordinates of the current node and (x_g, y_g) represents the raster center coordinates of the target node.

The A* algorithm needs to calculate the valuation function $h(n)$ of 8 neighboring nodes for each node expansion during the operation until the end of the algorithm, and these computations primarily affect the pathfinding efficiency of the algorithm. In addition, each extension needs to select a minimum estimated value node from openist as the next extension node. The values in the openlist are first sorted, and then the smallest value is selected as the next extended knot. The time consumed for sorting in a small-scale environment is negligible, but when the environment is complex and large, the amount of data in the openlist will increase significantly and lead to a decrease in the efficiency of the A* algorithm path planning.

2.2 A*-Connect

The A* algorithm is an extended one-way search, which results in a long pathfinding time due to the large number of nodes traversed and low computational efficiency. Therefore, the A*-connect algorithm is generated, which reduces the redundant nodes in the computation process to enhance the search efficiency by alternating the search in both positive and negative directions. The algorithm runs by the valuation function of the A* algorithm.

$$f_1(n) = g_1(n) + h_1(n) \tag{3}$$

$$f_2(n) = g_2(n) + h_2(n) \tag{4}$$

The forward search T1 in the A*-connect algorithm is to use the starting point as the starting point S1 of T1 and the target point as the target point E1 of T1; the reverse search T2 is to use the target point as the starting point S2 of T2 and the starting point as the target point E2 of T2. Openlist1, closelist1, openlist2, and closelist2 were created for T1 and T2. The algorithm starts to run with T1 expansion, adding S1 to openlist1, then adding the reachable nodes around S1 to openlist1 as well, moving S1 from openlist1 to closelist1, and finally selecting the smallest estimated generation value in openlist1 as the next expanded node; then starting T2 expansion, adding S2 to openlist2, then add the nodes around S2 to openlist2, move S2 from openlist2 to closelist2, and finally select the smallest estimated generation value in openlist2 as the next expansion node. This process is repeated until T1 and T2 meet on the way to the expansion. This path is the optimal path, forward and reverse, which has always reached its target point, and the algorithm stops.

The traditional A*-connect algorithm sets two termination conditions: an identical extension node is searched for in both the forward search of T1 and the reverse search of the T2 extension path; this extension node is identified as the respective minimum cost node in both T1 and T2. The A*-connect algorithm can stop only when these two conditions are satisfied simultaneously. However, in the case of large scale and complex environment, T1 and T2, in the process of expansion, will appear that the expansion nodes cannot meet in the middle and finally end up reaching the target point, resulting in the algorithm needing to calculate double the amount of computation compared to A*, thus reducing the pathfinding efficiency.

3 Four-Way A* Algorithm

3.1 Improvement I

In order to solve the problem of low efficiency in the pathfinding process of the traditional A* algorithm, this paper selects the random fast sorting algorithm to sort the values in openlist and optimize the traditional A* algorithm by calculating each raster valuation function $h(n)$ in advance.

Optimize A*. For the A* algorithm, how to choose the appropriate sorting algorithm is often overlooked. Because quick sort is currently recognized as a better sorting algorithm, most people use the library function sort () when programming in C++ and M.A.T.L.A.B. languages. However, considering the actual situation of the A* algorithm, the fast algorithm is not the most suitable sorting algorithm. Fast sort is an improvement from bubble sort, but it differs in that the quick sort algorithm eliminates multiple inverse orders in a single swap, thereby significantly speeding up the sort [20].

The time complexity of the best-case fast sort is:

$$T(n) \leq nlog_2n + nT(1) \approx O(nlog_2n) \tag{5}$$

The time complexity of the worst case is:

$$KCN = \sum_{i=1}^{n}(n-i) = \frac{(n(n-1))}{2} \approx \frac{n^2}{2} = O(n^2) \tag{6}$$

since the fast sort is a recursive algorithm, a stack is needed to store the data during execution, so the space complexity of the fast sort is O(log n) in the best case and O(n) in the worst case. In addition, since the running time of the fast sort is determined by order of the input sequence, we can only assume that the input data are arranged with equal probability. However, before storing the neighbor nodes of the current node, the openlist stores the last ordered values, which increases the time complexity of the fast sort by a factor of n.

In this paper, we choose a randomized algorithm to add randomness to the fast sort so that the algorithm changes each time the main element is selected, making the algorithm have the same sorting complexity each time. T_n Denotes the time required to use random fast sort for an array of size n. x_n Denotes the number of comparisons that occurred in this sort. $x_{i,j}$ Indicates an indicator of whether a comparison has occurred. The value is one if a comparison has occurred and 0 if not:

$$E(T_n) = \theta\left(E\left(\sum_{i=1}^{n-1}\sum_{j=i+1}^{n} x_{i,j}\right)\right) = \theta\left(\left(\sum_{i=1}^{n-1}\sum_{j=i+1}^{n} E(x_{i,j})\right)\right) \quad (7)$$

Among them $E(x_{i,j}) = p(x_{i,j} = 1) = \frac{2}{j-i+1}$.

$$E(T_n) = \theta\left(E\left(\sum_{i=1}^{n-1}\sum_{k=2}^{n-i+1}\frac{1}{k}\right)\right) \quad (8)$$

also $\ln(n - i + 2) - ln2 = \int_2^{n-i-1}\frac{1}{x}dx < \sum_{k=2}^{n-i+1}\frac{1}{k} < \int_1^{n-i-1}\frac{1}{x}dx = \ln(n - i + 1)$ therefore:

$$\sum_{i=1}^{n-1}\sum_{k=2}^{n-i+1}\frac{1}{k} < \sum_{i=1}^{n-1}\ln n = n\ln n \quad (9)$$

The other direction:

$$\sum_{i=1}^{n-1}\sum_{k=2}^{n-i+1}\frac{1}{k} > \sum_{i=1}^{n-1}\ln(n - i + 2) - (n - 1)ln2 > \int_2^{n+1} lnxdx$$
$$- (n - 1)ln2 = n\ln n \quad (10)$$

Thus we end up with the expected time complexity of O(n ln(n)) [21].

Optimize A*. The calculation of h(n) during the algorithm's operation is also an essential factor affecting the algorithm's efficiency. When the A* algorithm runs, h(n) must be calculated for eight neighboring nodes for each node expansion. Such an improved path of the partitioning algorithm [22] is also an effective way to improve the algorithm's efficiency if particular work is preprocessed during the algorithm design to reduce the workload within the algorithm. Therefore, this paper uses the h(n) of each node of the raster map to be calculated in advance and called directly when the A* algorithm is run. Compared with the traditional A* algorithm, the algorithm can save part of the computational workload and improve the algorithm's efficiency at each node expansion. Significantly, the effect brought by preprocessing is more significant when the environment is large in scale.

3.2 Improvement II

In response to the problem that the forward and reverse directions cannot meet during the operation of the traditional two-way A* algorithm resulting in serious memory occupation and low pathfinding efficiency. A termination condition is added to the traditional A*-connect algorithm to improve the efficiency of the algorithm pathfinding.

When the A*-connect algorithm is running, check whether the forward and reverse expansion nodes have the same nodes in closelist1 and closelist2 and check openlist1 and openlist2. For example, when expanding forward to point α1, determine whether the coordinates of point α1 exist in closelist2 or openlist2, and if its condition holds, the algorithm terminates immediately, and the pathfinding succeeds; or when expanding backward to point $\alpha 2$, determine whether the coordinates of point $\alpha 2$ exist in closelist1 or openlist1 and if the condition holds the algorithm terminates immediately and the pathfinding succeeds. This enables the A*-connect algorithm to improve the chance of successful T1 and T2 encounters when expanding nodes in forward and reverse directions.

As shown in Fig. 1, the grid size is 50×50. Figure 1(a) shows the path planning result of the traditional A*-connect algorithm, and Fig. 1(b) shows the optimal path found by the optimized A*-connect algorithm. The blue and green in the figure represent the extended paths of T1 and T2, respectively; the red nodes at both ends are the starting point, and the target point and the red dots in the middle are the extended nodes where the forward and reverse search meet. As can be seen from Fig. 1, the expansion nodes of the A*-connect algorithm forward and reverse search fail to meet the ideal intermediate nodes; on the contrary, the optimized A*-connect algorithm runs as the current node searched to compare with the values in the opposing direction openlist and closelist, the optimized A*-connect algorithm forward and reverse expansion nodes almost meet in the middle, avoiding too many nodes of forward and reverse search, reducing the amount of computation and memory consumption.

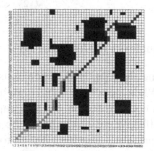 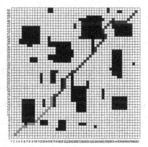

Fig. 1. Algorithm path planning results.

3.3 Optimize A*-Connect Algorithm

In this paper, based on the optimization of the above algorithm, the four-way A* algorithm is proposed in combination with the partitioning strategy. In addition, the four-way A* algorithm runs while the A* and A*-connect algorithm termination conditions are also being judged, significantly improving the efficiency of path planning.

Figure 2 shows the planning knot path of the A* algorithm. A and b diagrams start at opposite ends, and the final planned optimal paths are not the same. In addition, the A*-connect algorithm characteristics leading to the forward and reverse directions are difficult to meet or meet at the end, as shown in Fig. 3. In the absence of obstacles, the forward and reverse extensions eventually meet at the absence of obstacles target point neighboring nodes. Although the optimized A*-connect algorithm eventually finds the optimal path, it takes up twice the amount of storage, which can lead to a decrease in pathfinding efficiency.

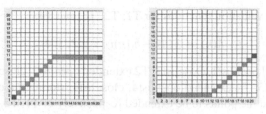

Fig. 2. A* path planning results.

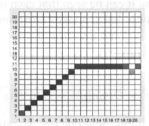

Fig. 3. A*-connect path planning result.

For the above problems, this paper proposes the four-way A* algorithm, whose main ideas are:

(1) Dividing or reducing the original problem into smaller sub-problems.
(2) Solving each subproblem recursively or iteratively.
(3) The solution of the original problem is obtained by combining the solutions of the subproblems.

The sub-problems are of the exact nature as the original problem, and the sub-problems solve the problem independently of each other; finally, the solutions of the sub-problems are combined into the solution of the original problem. It can be applied to the algorithm to reduce the path finding time of the algorithm effectively, avoid traversing too many invalid nodes, and improve the efficiency of the algorithm pathfinding.

With the start and end points known, a third node coordinate (x_m, y_m) is found using the following equation.

$$x_m = (x_s + x_e)/2 \tag{11}$$

$$y_m = (y_s + y_e)/2 \tag{12}$$

where $x_m, y_m, x_s, y_s x_e, y_e$ Represents the third node, the starting point, and the endpoint, respectively. The extended search direction is divided into T1, T2, T3, and T4, where T1 takes the starting point as the starting point 1 and the third node as the target point 1; T2 takes the third node as the starting point 2, and the starting point as the target point 2; T3 takes the third node as the starting point 3 and the endpoint as the target point 3; T4 takes the endpoint as the starting point 4 and the third node as the target point 4.

Establish heuristic valuation functions for T1, T2, T3, and T4:

$$f_i(n) = g_i(n) + h_i(n) i = 1, 2, 3, 4 \tag{13}$$

Create T1 extended openlist1, closelist1; T2 extended openlist2, closelist2; T3 extended openlist3, closelist3; T4 extended openlist4, closelist4, storing the extended points of T1, T2, T3, T4, and the points to be extended respectively of T1, T2, T3, and T4.

Set an obstacle avoidance principle. When the calculated third node is within the obstacle, it is replaced with the third node's unobstructed neighbor as the third node.

From the A* algorithm idea, it can be seen that each time an element is taken from the openlist, assuming a total of n times, it takes $O(n)$ times to expand the subsequent nodes of a node each time while doing a sort on the openlist is $O(n)$, in addition, this algorithm is based on the optimization A* algorithm, so the algorithm's randomized fast sorting algorithm $O(n\ln(n))$, then the total complexity of the four-way A* algorithm is:

$$O(n) \times O(n\ln(n)) = O(n^2 \ln(n)) \tag{14}$$

For a problem of size n, divide it into subproblems. The size of each subproblem is n/b. At this point, the complexity of the algorithm is:

$$\begin{cases} T(n) = aT\left(\frac{n}{b}\right) + f(n) \\ \qquad T(1) = 0 \end{cases} \tag{15}$$

$T(1)$ as the algorithm's complexity is 0 when there is only one node. $f(n)$ represents the cost of the algorithm decomposition and merging. In recursion, each time the number of subproblems is multiplied back by a when partitioning down, the size of the subproblem will be divided by b. Then when the size of the number of subproblems reaches 1, the subproblem is $a^{\log_b n}$:

$$T(n) = aT\left(\frac{n}{b}\right) + f(n) = a^2 T\left(\frac{n^2}{b}\right) + af\left(\frac{n}{b}\right) + f(n) = \cdots = \theta\left(n^{\log_b a}\right)$$

$$+ \sum_{j=0}^{\log_b n - 1} a^j f(\frac{n^j}{b}) \tag{16}$$

The pseudo-code is as follows:

Algorithm 1: Pseudo Code of Four-way A*
1   Initialization;
2   Calculate the third node;
3   If the third node in the barrier
4   Replacement of the third node;
5    $openlist_i \leftarrow T_i$ i =1,2,3,4;
6   While $openlist_1 \neq$null & $openlist_2 \neq$null & $openlist_3 \neq$null & $openlist_4 \neq$null
7   Computer T_i child node x_i; i=1,2,3,4;
8   Update $openlist_i$; i=1,2,3,4;
9   While $openlist_1 \neq$null & $openlist_2 \neq$null
10   if x_1 is in the $closelist_2$ or $openlist_2$
11   Break;
12   endif
13   find $\min f_1(x)$ in $openlist_1$;
14   T_1=x;
15           $closelist_1 \leftarrow T_1$
16   if x_2 is in the $closelist_1$ or $openlist_1$
17   Break;
18   endif
19   find $\min f_2(x)$ in $openlist_2$;
20   T_2=x;
21               $openlist_2 \leftarrow T_2$
22   While $openlist_3 \neq$null & $openlist_4 \neq$null
23   if x_3 is in the $closelist_4$ or $openlist_4$
24   Break;
25   endif
26   find $\min f_3(x)$ in $openlist_3$;
27   T_3=x;
28               $closelist_3 \leftarrow T_3$
29   if x_4 is in the $closelist_3$ or $openlist_3$
30   Break;
31   endif
32   find $\min f_4(x)$ in $openlist_4$;
33   T_4=x;
34               $openlist_4 \leftarrow T_4$
35   Get path/Fail to get a path

From the above equation, we can see that when we decompose the n-dimensional problem into multiple subproblems, the algorithm's complexity can be significantly reduced. The A* algorithm is divided into four subproblems considering the problem of storage capacity of the A* algorithm.

$$\begin{cases} T(n) = 4T\left(\frac{n}{4}\right) + f(n) \\ \quad\quad T(1) = 0 \end{cases} \tag{17}$$

From the A* algorithm idea, it can be seen that each time an element is taken from the openlist, assuming a total of n times, it takes O(n) times to expand the subsequent nodes of a node each time while doing a sort on the openlist is O(n), in addition, this algorithm is based on the optimization A* algorithm, so the algorithm's randomized fast sorting algorithm O(nln(n)), then the total complexity of the four-way A* algorithm is:

$$O(n) \times O(nln(n)) = O(n^2 ln(n)) \tag{18}$$

For a problem of size n, divide it into subproblems. The size of each subproblem is n/b. At this point, the complexity of the algorithm is:

$$\begin{cases} T(n) = aT\left(\frac{n}{b}\right) + f(n) \\ \quad T(1) = 0 \end{cases} \tag{19}$$

T(1) as the algorithm's complexity is 0 when there is only one node. f(n) represents the cost of the algorithm decomposition and merging. In recursion, each time the number of subproblems is multiplied back by a when partitioning down, the size of the subproblem will be divided by b. Then when the size of the number of subproblems reaches 1, the subproblem is $a^{log_b n}$:

$$T(n) = aT\left(\frac{n}{b}\right) + f(n) = a^2 T\left(\frac{n^2}{b}\right) + af\left(\frac{n}{b}\right) + f(n) = \cdots = \theta\left(n^{log_b a}\right)$$
$$+ \sum_{j=0}^{log_b n - 1} a^j f\left(\frac{n^j}{b}\right) \tag{20}$$

From the above equation, we can see that when we decompose the n-dimensional problem into multiple subproblems, the algorithm's complexity can be significantly reduced. The A* algorithm is divided into four subproblems considering the problem of storage capacity of the A* algorithm.

$$\begin{cases} T(n) = 4T\left(\frac{n}{4}\right) + f(n) \\ \quad T(1) = 0 \end{cases} \tag{21}$$

4   Experimental Verification

4.1   Simulation Analysis

In order to verify the effectiveness of the four-way search A* algorithm in this paper, simulations are conducted with the improved A* algorithm of literature [13] and the A*-connect algorithm of literature [16] in different sizes of raster maps, and all experiments in this paper are simulated and verified in Matlab 2018a environment.

This paper mainly constructs 50 × 50 and 30 × 30 raster maps containing known obstacles. The black raster represents the obstacles in the raster map, and the white raster represents the feasible interval without obstacles. The left side of Fig. 4 shows the optimized A* algorithm path planning result. The green grid represents the feasible paths searched; the right side shows the optimized A*-connect algorithm path planning result, the blue grid represents the feasible paths searched in the forward direction, and the green grid represents the feasible paths searched in the reverse direction except for the starting point and the ending point, the red grid is the forward and reverse encounter nodes; Fig. 5 shows the four-way A* algorithm path planning result. The green paths in Fig. 5 are the feasible paths searched by T1 and T3, and the blue paths are the feasible paths searched by T2 and T4. From left to right, the first and the last points are the starting and ending points, the second redpoint is the point where T1 and T2 meet by extension, the third red point is the calculated midpoint, and the fourth red point is the point where T3 and T4 meet by extension. Table 2 shows the search time, the number of extended nodes, and path nodes in different raster maps for the A*, A*-connect, and four-way A* algorithms.

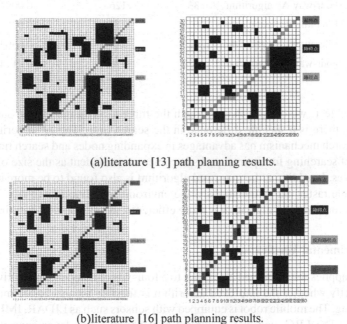

(a)literature [13] path planning results.

(b)literature [16] path planning results.

Fig. 4. Algorithm path planning results.

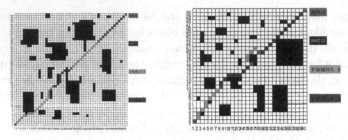

Fig. 5. Four-way A* algorithm path planning result.

Table 1. Comparison of experimental simulation results.

Map size	Algorithm	Path nodes	Extension node	Search time (s)
30 × 30	Literature [13] algorithm	33	152	0.04334
	Literature [16] algorithm	33	129	0.0228
	Four-way A* algorithm	33	126	0.0091
50 × 50	Literature [13] algorithm	61	267	0.0783
	Literature [16] algorithm	61	219	0.0305
	Four-way A* algorithm	61	170	0.0114

From Table 1, we can see that although the three algorithms find the same optical path length, there is a specific difference in the search time. The A* algorithm with a four-way search mechanism has advantages in expanding nodes and search time, and the efficiency of searching for optimal paths becomes more evident as the size of the raster map increases. In addition, the improved algorithm is also found to be more effective in other obstacle raster map simulations if the environment is more complex. Therefore, this paper's four-way A* algorithm is more efficient and has a shorter pathfinding time.

4.2 Experimental Verification

This paper applies the improved algorithm to a four-wheel independently driven mobile robot to verify whether the improved algorithm is feasible for inaccurate mobile robot path planning. The mobile robot is equipped with sensors such as LIDAR, IMU, encoder, camera, etc. The I.P.C. and N.X.P. control board form a two-layer control structure, as shown in Fig. 6(a). ROS. (Robot Operating System) is running on the IPC. And is responsible for the calculation of path planning; The NPX. Control board communicates with the I.P.C. through the serial port, receives the path information from the IPC., and uses the PWM. to control the motor and the servo to realize the robot movement; At the same time, the data on the laser radar is collected and sent to the IPC., and the IMU and encoder data are combined to obtain the odometer information of the undercarriage using the track extrapolation algorithm. The mobile robot uses LIDAR, IMU, encoder, and camera to obtain information about the outside world. The amcl and gmapping modules are used for localization and 2-dimensional map construction, and then the global planner and local planner in the move_base module are used for global and local path planning, respectively. In this paper, the default path planning algorithm in the global planner module is changed to an improved algorithm for global path planning.

The experimental scenario in this paper is a two-dimensional map constructed by the mobile robot in the laboratory site, as shown in Figure (b). The effectiveness of the improved algorithm is demonstrated by comparing the computational efficiency of the algorithms under the same path. Figure 6(c) and (d) show the pathfinding results of the improved algorithm under different paths, and the two figures show the 2-dimensional map and path planning results in RVIZ., the visualization tool of ROS.

The experimental results demonstrate that the four-way A* algorithm can effectively complete the path planning of mobile robots with faster computation speed and higher pathfinding efficiency than the traditional A* algorithm.

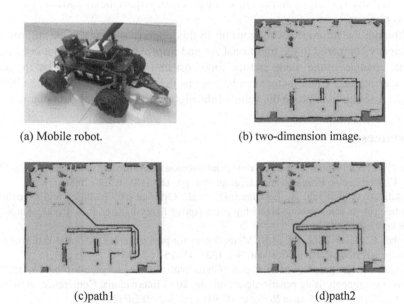

(a) Mobile robot.   (b) two-dimension image.

(c)path1   (d)path2

Fig. 6. Path planning results of improved A* algorithm.

Table 2. Path planning comparison.

Path	Algorithm	Path length(m)	Search time (s)
Path1	A*	13.8	178.35
	Four-way A*	13.6	97.60
Path2	A*	21.6	231.58
	Four-way A*	21.5	101.36

5 Conclusion

Since the A* algorithm needs to traverse too many nodes in the path planning process, which leads to excessive computation and memory consumption in the search process, in addition to the large scale of the mobile robot and the complex environment, in order to improve the efficiency of the path planning, this paper proposes an improved A* algorithm based on the optimized A* algorithm and the optimized A*-connect algorithm and uses a partitioning strategy. An improved A* algorithm for four-way search is proposed. The traditional A* algorithm simulation and the improved algorithm on the Matlab. The platform using raster maps of different scales and different obstacle spread has demonstrated that the proposed algorithm can effectively improve the path planning speed with the same length of optimal paths, and the effect is more evident in more significant scenarios. The application of the improved algorithm to mobile robots shows that the improved A* algorithm can meet the practical requirements and the optimization effect is obvious.

Although the improved A* algorithm in this paper has a significant improvement in efficiency compared to the traditional A* and improved A*, it does not consider the dynamic changes in the scene and the time cost and power cost spent by the mobile robot steering, etc. These factors can be used as the basis for improving the algorithm in future research to improve the value of the algorithm in practical path planning.

References

1. Sariff, N., Buniyamin, N.: An overview of autonomous mobile robot path planning algorithms. In: Conference on Research and Development, pp. 183–185. IEEE (2006)
2. Bakdi, A., Abdelfetah, H., Boutamai, H., et al.: Optimal path planning and execution for mobile robots using genetic algorithm and adaptive fuzzy-logic control. Robot. Auton. Syst. 89(1), 95–109 (2017)
3. Mohd, A., Nayab Zafar, J.C.M.B.: Methodology for path planning and optimization of mobile robots: a review. Procedia Comput. Sci. 133, 141–152 (2018)
4. Alajlan, M., Koubaa, A., Chaari, I., et al.: Global path planning for mobile robots in large-scale grid environments using genetic algorithms. In: 2013 International Conference on Individual and Collective Behaviors in Robotics (ICBR), pp. 3–5. IEEE (2013)
5. Mei, L.: Research on robot path optimization based on rasterized vision. Comput. Digit. Eng. 46(008), 1548–1552 (2018)
6. Stentz, A.: Optimal and efficient path planning for partially-known environments. In: Proceedings of 1994 IEEE International Conference on Robotics and Automation, pp. 3–8. IEEE (1994):
7. Dijkstra, E.W.: A note on two problems in connexion with graphs. Numer. Math. 1(1), 270–275 (1959)
8. Wang, H., Yuan, Y., Yuan, Q.: Application of Dijkstra algorithm in robot path-planning, pp. 1067–1069. IEEE (2011)
9. Hart, P.E., Nilsson, N.J., Raphael, B.: A formal basis for the heuristic determination of minimum cost paths. IEEE Trans. Syst. Sci. Cybern. 4(2), 28–31 (1972)
10. Aine, S., Swaminathan, S., Narayanan, V., et al.: Multi-heuristic A*. Int. J. Robot. Res. 35(1–3), 224–243 (2014)
11. Guruji, A.K., Agarwal, H., Parsediya, D.K.: Time-efficient A* algorithm for robot path planning. Procedia Technol. 23, 144–149 (2016)

12. Zhao, X., Wang, Z., Huang, C.K., et al.: Mobile robot path planning based on an improved A* algorithm. Robot **40**(6), 137–144 (2018)
13. Islam, F., Narayanan, V., Likhachev, M.: A*-Connect: Bounded suboptimal bidirectional heuristic search. In: 2016 IEEE International Conference on Robotics and Automation (ICRA), pp. 2753–2755. IEEE (2016)
14. Wu, P., Sang, C., Lu, Z., Yu, S., Fang, L., Zhang, Y.: Research on mobile robot path planning based on improved A* algorithm. Comput. Eng. Appl. **55**(21), 227–233 (2019)
15. Gao, M., Zhang, Y., Zhu, L.: Bidirectional time-efficient A* algorithm for robot path planning, Appl. Res. Comput. **36**(329(03)), 159–162+167 (2019)
16. Kong, J., Zhang, P., Liu, X.: Research on improved A* algorithm of bidirectional search mechanism. Comput. Eng. Appl. **57**(08), 231–237 (2021)
17. Sturtevant, N.R., Felner, A., Barrer, M., et al.: Memory-based heuristics for explicit state spaces. In: IJCAI 2009, Proceedings of the 21st International Joint Conference on Artificial Intelligence, Pasadena, California, USA, 11 17 July 2009, p. 612. Morgan Kaufmann Publishers Inc. (2009)
18. Ferguson, D., Likhachev, M.: A guide to heuristic-based path planning, pp. 9–13 (2005)
19. Xin, Y., Liang, H., Du, M., Mei, T., Wang, Z., Jiang, R.: An improved A* algorithm for searching infinite neighbourhoods. Robot **36**(05), 627–633 (2014)
20. Yan, W., Li, D., Wu, W.: Data Structure, vol. 281, pp. 234–269. Posts and Telecommunications Press, Beijing (2015)
21. Cormen, T.H., Leiserson, C.E., Rivest, R.L., Stein, C.: Introduction to Algorithms, pp. 100–102. MIT Press, Cambridge (2005)
22. Qu, W.: The Design and Analysis of Algorithm, pp. 26–34. Tsinghua University Press, Beijing (2016)

ML-TFN: Multi Layers Tensor Fusion Network for Affective Video Content Analysis

Qi Wang, Xiaohong Xiang$^{(\boxtimes)}$, and Jun Zhao

Department of Computer Science and Technology, Chongqing University of Posts and Telecommunications, Chongqing, China
{zhaojun,xiangxh}@cqupt.edu.cn, S200201065@stu.cqupt.edu.cn

Abstract. Affective video content analysis is a task of automatically recognizing the emotions induced by the video, which is important in video content analysis and other applications. Video is an inherently multimodal media and contains audio and visual features, which can depict the elicited emotions. Many previous works have been proposed for this task, but when it comes to multimodal fusion, they only use early fusion (concatenate, etc.) or late fusion (vote, etc.) which can not well study the interactions between multiple modalities. To address this, we propose a framework named ML-TFN (Multi Layers Tensor Fusion Network) to model the inter-modality dynamics through Tensor Fusion Network. Specifically, Tensor Fusion approach explicitly aggregates unimodal and bimodal interactions through audio and visual features extracted from Convolutional Neural Network (CNN), including RseNet-101 and VGGish. It takes advantage of tensor outer product and low-rank decomposition to capture the interactions between multiple modalities. In order to better model the inter-modality dynamics, a Multi Layers Fusion scheme is designed to fusion features at different layers in unimodal embedding subnetwork. Our proposed framework is evaluated on the MediaEval 2015 Affective Impact of Movies Task (AIMT15) dataset, where better results is achieved compared with other methods.

Keywords: Modality interaction · Tensor Fusion Network · Mutli layers fusion · Affective video content analysis

1 Introduction

Nowadays, the number of videos is increasing rapidly, due to the development of science and technology. In this case, automatic analysis of video content becomes particularly important. A large number of videos not only convey information, but also bring entertainment to users, which will inevitably affect the emotional state of users. Therefore, affective video content analysis has become an important research topic and attracted increasing attention in the research field of video content analysis. Meanwhile, it has been widely used in video content retrieval [7], personalized video recommendation and video summarization [35].

H. Zhang et al. (Eds.): NCAA 2022, CCIS 1637, pp. 184–196, 2022.
https://doi.org/10.1007/978-981-19-6142-7_14

The affective content of video is defined as the intensity and type of emotion expected to arise when people watch it [4]. Affective video content analysis is a task of automatically recognizing the emotions induced by video, whose process is to map the information contained in the video to the emotions generated by the user after watching the video. There are two methods to measure emotions, the discrete approach and the dimensional approach [4,28]. The discrete approach commonly classifies emotions into different categories, such as happiness, sadness, surprise, disgust, anger, fear, etc. The dimensional approach [10] maps emotions into continuous spaces and the arousal-valence model is the commonly used dimensional approach, where arousal measures the intensity of the emotions, valence represents the type of emotions [4,28].

Generally speaking, most existing studies focus on direct method in affective content analysis which has two parts: extracting features from the video, and then using the features for classification or regression. Thanks to the competition [24] held by MediaEval in recent years about affective video content analysis, many studies [1,8,31] have achieved promising progress. Rather than only using single modality information for affective video content analysis, recent researches tend to combine multiple modalities information to improve performance [22]. They often extract audio and visual features, simply fusing mulitple modalities information with early fusion or late fusion. However, this kind of method ignores the interactions between multiple modalities where our proposed method sloves this by capturing the interactions between multiple modlities.

In order to address the issue, a Multi Layers Tensor Fusion Network is proposed, which can capture the interactions between multiple modalities in an end to end manner. To achieve this, we propose a new multimodal fusion approach named Tensor Fusion [32], which explicitly captures the interactions. We suppose that using features from different layers in unimodal embedding subnetwork will potentially increase performance. Thus, we utilize the multi layers fusion scheme to model inter-modality dynamics at different layers in unimodal embedding subnetwork. In order to depict emotions, two features, which consider the visual and audio features of video content are calculated from Convolutional Neural Network (CNN), including ResNet-101 [11] and VGGish [12]. Extensive experiments are conducted on the AIMT15 [24] dataset and the results revealed the superiority of our method. In short, the contributions of our proposed method can be summarized as follows:

- In order to make better use of features from multiple modalities to predict emotion, a framework composed of Tensor Fusion Network and Multi Layers Fusion scheme is proposed, which has better performances.
- Tensor Fusion Network is utilized to model the inter-modality dynamics and Multi Layers Fusion scheme is designed to enhance the fused features by exploiting the information embedded at different layers in unimodal embedding subnetwork.

The remainder of this article is structured as follows. Section 2 gives an overview of related work. Section 3 presents our proposed method. Section 4 provides experimental results on the public AIMT15 dataset, followed by the conclusions in Sect. 5.

2   Related Work

Hanjalic and Xu in [10] first proposed a framework for emotional video content analysis by directly mapping video features into the valence-arousal space. They manually selected low-level features such as motion intensity, shot lengths, and audio features (loudness, speech rate, rhythm, etc.) for modeling arousal and valence, but it is not easy and time consuming to select features manually. Acar et al. [1] used CNNs to learn mid-level representations from low-level feature, including MFCC and color. And then, multi-class SVM was used to classify the video clip with the mid-level representations as input. They are the first to use deep learning based method but their method also need to extract low-level features from video. MediaEval has organized the challenges AIMT2015, where eight teams [6,8,16,18,19,23,25,31] participated in it. And they utilized multiple features with early fusion or late fusion for affective video content analysis.

Since video is an inherently multimodal media and contains audio and visual features, affective video content analysis can be regarded as a special mulit-modal learning task. Early fusion is a technique that concatenate features from different modalities. Yi et al. [30] extracted appearance and motion features from video frames with the two-stream temporal segment network (TSN) as ConvNets feature, manually designed MKT feature to depict motion information of video frames and selected the EmoBase10 feature to depict audio cues; then, the early fusion strategy was adopted and the fused feature was used as input to classify video clips with SVM. Ou et al. [20] used four kinds of representation from video clips, including visual appearance, motion, audio and tone; and adopted the local and global attention mechanism to take early fusion on four modalities representation. However, the early fusion still loses much useful information. Late fusion builds separate models for each modality and then integrates the outputs together using a method such as majority voting or weighted averaging. Baecchi et al. [3] extracted deep sentiment features with deep SentiBank network, deep face features with fine-tuned AlexNet, and deep object features with VGGNet; subsequently, each type of feature was used as input and SVM was used to classify video clips, finally the late fusion strategy was used in the decision-making stage.

Unlike the aforementioned methods [29,33,34], we consider and emphasize the interactions between the modalities. Specifically, Tensor Fusion Network and Multi Layers Fusion scheme are introduced to effectively model the interactions between multiple modalities.

3   Approach

3.1   Modality Embedding Subnetworks

Visual Embedding Subnetwork: Visual appearance in the image contains basic information which the videos bring to the viewer. In order to capture the visual appearance of a video, given the visual stream of a video, we first slice it into T segments called local frames based on the sparse sampling method

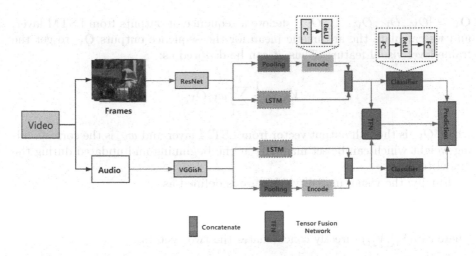

Fig. 1. Overview of proposed framework. Two kinds of cnns including ResNet-101 [11] and VGGish [12] are as the modality embedding subnetwork that is used to get the visual feature and audio feature. Tensor Fusion Network models the inter-modality dynamics through visual feature and audio feature. In the end, the classification results from unimodal classifier of visual and audio modality are combined with the output of tensor fusion network to get the final prediction.

described in TSN [27]. Then, we extract the local features from those local frames by using ResNet-101 [11] network which is pre-trained on the large-scale ImageNet [14] dataset.

Let $\mathbf{V} = [V_1, V_2, V_3, ..., V_T]$ denote a sequence of vectors which are extracted from local frames, where V_i is the i-th feature vector and T is the length of sequence. Then, we separately calculate the arithmetic mean and the standard deviation (denoted as pooling in Fig. 1) for the sequence features $\mathbf{V}$, which are denoted as $E(\mathbf{V})$ and $S(\mathbf{V})$, respectively. Finally, $E(\mathbf{V})$ and $S(\mathbf{V})$ are connected to get a fixed-length video-level visual feature. Therefore, the sequence $\mathbf{V}$ is indicated by

$$V_{vl} = En(con(E(\mathbf{V}), S(\mathbf{V}))) \tag{1}$$

where the function $con(E(\mathbf{V}), S(\mathbf{V}))$ directly concatenates the two vectors, function $En()$ reduces the dimension of concatenated vectors which is composed of FC layer and ReLU activation layer (denoted as Encode in Fig. 1).

The video-level visual features can efficiently capture visual concepts. However, the temporal or motion information is also very important for emotional description. An effective approach to learn temporal dynamics is using the LSTM [13]. Moreover, when people watch a video, the video content affects the viewer's emotion at all time. Therefore, the outputs of the LSTM layer are averaged. Let

$\mathbf{O}_V = [O_{V_1}, O_{V_2}, O_{V_3}, \ldots, O_{V_T}]$ denote a sequence of outputs from LSTM layer, and we calculate the arithmetic mean for the sequence outputs $\mathbf{O}_V$ to get the frame-level visual feature. Thus, it can be denoted as:

$$V_{fl} = \frac{1}{T} \sum_{i=1}^{T} w_{V_i} O_{V_i} \tag{2}$$

where O_{V_i} is the i-th output vector from LSTM layer and w_{V_i} is the corresponding weight which can be set manually at the beginning and updated during the model training.

Finally, the visual modality's feature is defined as

$$X_V = con(V_{vl}, V_{fl}) \tag{3}$$

where $con(V_{vl}, V_{fl})$ directly concatenates the two vectors.

Audio Embedding Subnetwork: As for a raw audio clip, firstly, we also slice it into T segments called local frames based on the sparse sampling method described in TSN. For each local frame, we extract its features using the state-of-the-art VGGish [12] pre-trained on AudioSet [9] which provides a 128 dimensional feature. The VGGish pre-trained on AudioSet can provide abundant audio category information, such as machines, animals, humans and so on.

Let $\mathbf{A} = [A_1, A_2, A_3, \ldots, A_T]$ denote a sequence of vectors which are extracted from local frames of audio modality, then we can get the video-level audio feature by calculating the arithmetic mean and the standard deviation and feed it into Encode network, which is defined as

$$A_{vl} = En(con(E(\mathbf{A}), S(\mathbf{A}))) \tag{4}$$

The frame-level audio feature is also calculated through the sequence of outputs $\mathbf{O}_A = [O_{A_1}, O_{A_2}, O_{A_3}, \ldots, O_{A_T}]$ from LSTM layer. Then the arithmetic mean is computed for the sequence outputs $\mathbf{O}_A$ using the Eq. 6.

$$A_{fl} = \frac{1}{T} \sum_{i=1}^{T} w_{A_i} O_{A_i} \tag{5}$$

In the end, the audio modality's feature is defined as

$$X_A = con(A_{vl}, A_{fl}) \tag{6}$$

where $con(A_{vl}, A_{fl})$ directly concatenates the two vectors.

3.2   Tensor Fusion Network

Tensor representation is one successful approach for multimodal fusion. It first requires a transformation of the input representations into a high dimensional tensor and then map it back to a lower-dimensional output vector space. Tensors are usually created by taking the outer product over the input modalities.

Given a set of vector representations, $\{X_m\}_{m=1}^{M}$ which are encoding unimodal information of the M different modalities, and the goal of multimodal fusion is to integrate the unimodal representation into one compact multimodal representation for downstream tasks. In addition, in order to be able to model the interactions between any subset of modalities using one tensor, [32] proposed a simple extension to append 1 s to the unimodal representations before taking the outer product. The input tensor formed by the unimodal representation is computed by:

$$\mathbf{X} = \bigotimes_{m=1}^{M} X_m \tag{7}$$

where $\bigotimes_{m=1}^{M}$ denotes the tensor outer product over a set of vectors indexed by m, and X_m is the input representation with appended 1 s. The input tensor $\mathbf{X}$ is then passed through a linear layer $g(\cdot)$ to produce a vector representation:

$$h = g(\mathbf{X}; \mathcal{W}, b) = \mathcal{W} \cdot \mathbf{X} + b \tag{8}$$

where $\mathcal{W}$ is the weight of this layer and b is the bias. An example of tensor fusion for the bi-modal case is illustrated in Fig. 2.

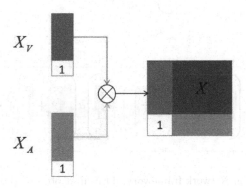

Fig. 2. Tensor fusion via tensor outer product for bimodal case. X_V and X_A are the visual and audio feature respectively.

To efficiently calculate the tensor outer product, we utilize low-rank decomposition proposed in [17] for approximating the high-order tensor $\mathcal{W}$, which defined as:

$$\mathcal{W} = \sum_{i=1}^{r} \bigotimes_{m=1}^{M} \mathbf{w}_m^{(i)} \tag{9}$$

where the minimal r that makes the decomposition valid is called the rank of the tensor and $\left\{w_m^{(i)}\right\}_{i=1}^{r}$ is the corresponding low-rank factors for modality m.

Therefore, considering the Eqs. 8, 9 and 10, we can get the final representation denoted as:

$$h = \Lambda_{m=1}^{M}\left[\sum_{i=1}^{r} \mathbf{w}_m^{(i)} \cdot X_m\right] \tag{10}$$

where $\Lambda_{m=1}^{M}$ denotes the element-wise product over a sequence of tensors: $\Lambda_{t=1}^{2} x_t = x_1 \circ x_2$. An illustration of the bimodal case of Eq. 11 is shown in Fig. 3. We can derive Eq. 11 for a bimodal case to clarify what it does:

$$h = \left(\sum_{i=1}^{r} w_A^{(i)} \cdot X_A\right) \circ \left(\sum_{i=1}^{r} w_V^{(i)} \cdot X_V\right) \tag{11}$$

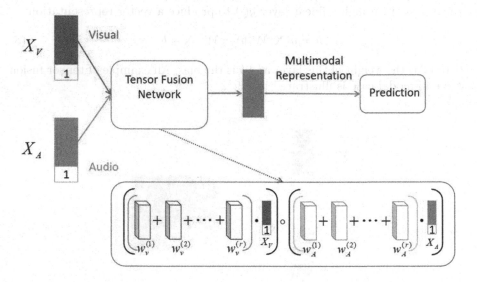

Fig. 3. Tensor Fusion Network framework. TFN first obtains the unimodal representation and through modality embedding subnetwork respectively. TFN produces the multimodal output representation by performing tensor outer production and utilizing low-rank decomposition method. The multimodal representation can be then used for generating prediction tasks.

3.3   Multi-layer Fusion

In this work, we start from the assumption of having a multi layers feature extractor for each one of the involved modalities. In practice, this means that we start from a multi layers neural network (as Fig. 4 shows) for each modality, which we assume to be already pre-trained. We argue in this paper that considering features extracted from all the hidden layers of unimodal embedding subnetwork could potentially increase performance and features from different layers at different modalities can give different insights from the input data.

Given this, we combine the proposed tensor fusion network and multi layers fusion scheme. In Fig. 4, each fusion layer l combines three inputs: the rank of the tensor R and one output from each modality respectively. This is done according to the following equation:

$$h_l = F(x_{\gamma_l^a}, y_{\gamma_l^b}, R_l) \tag{12}$$

where x and y are the general two inputs, $\gamma_l = (\gamma_l^a, \gamma_l^b)$ is a double of variable indices, which indicates the feature from the first modality and the second modality, R_l indicates the rank of the tensor and function $F()$ is the tensor fusion network. Furthermore, in Fig. 4, functions f and g are composed of a and b layers respectively, and $\gamma_l^a \in \{1, \cdots, a\}$, $\gamma_l^b \in \{1, \cdots, b\}$. The final results are the average of l layers' outputs.

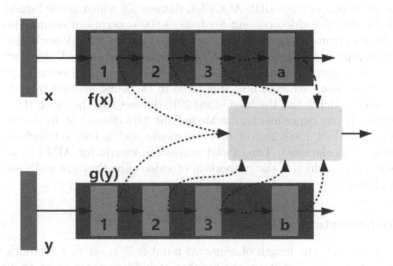

Fig. 4. General structure of a multi layers fusion network. Top: A neural network with several hidden layers (grey boxes). Bottom: A second network with several hidden layers (grey boxes). The yellow box and dotted lines form the multi layers fusion scheme. (Color figure online)

4   Experiments

The datasets and evaluation metrics are introduced in Sect. 4.1. Implementation details and three sets of experiments on AIMT15 dataset are elaborated in Sect. 4.2. Finally, the experiments results are detailed analyzed and discussed in Sect. 4.3.

Table 1. Performance on AIMT15 dataset for Valence and Arousal domain. The upper part shows the results with a single input modality (modality embedding subnetwork) and the bottom part lists the results with multiple modalities(ML-TFN fusion scheme).

Modality	Valence (Acc %)	Arousal (Acc %)
Unimodal (Visual)	**44.61**	56.42
Unimodal (Audio)	43.46	**56.54**
ML-TFN (Visual+Audio)	47.34	58.24

4.1 Datasets

The AIMT15 dataset [24] is used for the classification task in our experiments. It is the extension of the LIRIS-ACCEDE dataset [5] which is the largest video dataset for affective video content analysis. 9,800 excerpts of around 10 s have been extracted from 160 movies, and discretely classified for valence and arousal. The video clips from the LIRIS-ACCEDE dataset are ranked along with the arousal and valence axis, initially ranking from 0 to 9,799. Besides, 1100 new excerpts have been extracted and annotated in the same manner from 39 additional videos, which forms the MediaEval 2015 dataset together with the original 9800 videos. In our experiments, the MediaEval 2015 dataset is divided into two sets, a training set including of 6,144 elements and a test set including the remaining 4,756 elements. The official evaluation metric for AIMT15 is global accuracy (ACC), which is the proportion of video clips that are assigned to the correct class [24].

4.2 Implementation Details

In the experiment, the length of segments number T is set to 8 for both visual and audio feature. In addition, the number of hidden states is set to 1024 for visual modality embedding subnetwork, and 512 for audio modality embedding subnetwork and the number of LSTM layers is set to 3 for both visual and audio modality. Here we set rank to be 4 and 32 for valence domain and arousal domain, respectively. For the AIMT15 dataset, the batch size for network training is set to 16 and the learning rate is configured to 3e-4. The Adam [15] optimizer is used to update model's parameters during training and the whole model trains 20 epochs.

4.3 Experimental Results

On the AIMT-15 dataset, the performances of the unimodal of visual and audio modality is firstly evaluated separately. Then, the comparisons of the proposed framework with other fusion methods and other previous and recently proposed framework of affective video content analysis are shown respectively.

Table 2. Evaluation of different fusion methods on AIMT15 dataset for Valence and Arousal domain. GMU [2], CentralNet [26], MFAS [21] are three kinds of advanced multimodal fusion methods proposed recently.

Method	Valence (Acc %)	Arousal (Acc %)
Early fusion	47.04	57.80
Late fusion	46.88	57.66
GMU [2]	47.08	57.80
CentralNet [26]	47.02	57.82
MFAS [21]	**47.08**	**57.93**
TFN	**47.27**	**57.95**
ML-TFN	**47.34**	**58.24**

Table 3. Comparison with other methods on AIMT15 dataset for Valence ana Arousal domain. Previous studies tend to combine multiple features with early fusion and late fusion method.

Method	Fusion	Valence (Acc %)	Arousal (Acc %)
Chakraborty et al. [6]	Late fusion	35.66	48.95
Mironica et al. [19]	Late fusion	36.12	45.04
Seddati et al. [23]	Late fusion	37.28	52.44
Fudan-Huawei [8]	Early fusion	41.78	48.84
MIC-TJU [31]	Late fusion	41.95	55.93
Baecchi et al. [3]	Late fusion	45.31	55.98
Yi et al. [30]	Early fusion	46.22	57.40
Ou et al. [20]	Early fusion	46.57	57.51
ML-TFN	TFN & Multi layers	**47.34**	**58.24**

In order to evaluate the effect of the proposed two kinds of CNN based features of visual and audio modality, experiments are performed on AIMT15 with the results shown in Table 1. As shown in Table 1, the audio modality achieves better result than visual modality in the arousal domain, but it performs worse than the visual modality in the valence domain. By fusing the two modalities, the improvements of accuracy in the arousal and valence domains are 1.7% and 2.73%, respectively, these two input modalities complement each other and the performance is improved through modality fusion achieved by ML-TFN.

Early fusion and late fusion are simple and efficient techniques to combine multiple modalities, which are widely employed by a number of state-of-the-art methods [3,20]. GMU [2], CentralNet [26], MFAS [21] are the advanced multimodal fusion methods proposed recently. Therefore, we compare our ML-TFN framework with them. We simply concatenate the visual and audio feature, and use concatenated feature to make prediction for early fusion, and simply fuse the classification results from unimodal classifier for late fusion. The comparative experimental result are shown in Table 2. In order to make a fair comparison, all the evaluations utilize the same experimental setup except the fusion scheme. As shown in Table 2, GMU, CentralNet and MFAS all improve the performance of ACC on the arousal and valence domains with respect to early fusion method and late fusion method. But the improvements achieved by ML-TFN is higher, it's improvements of ACC on the arousal and valence domains are 0.3% and 0.44%, respectively. In conclusion, the proposed ML-TFN obtains the best performance in both two affective domain: Valence and Arousal, which demonstrates the effectiveness of the proposed Tensor Fusion Network and Multi Layers Fusion scheme.

A comparison of the proposed ML-TFN with other existing methods is shown in Table 3. The experimental results are separately evaluated in the two affective domains with the standard evaluation metrics (i.e., ACC for AIMT15). In summary, the approaches in Table 3 utilize multiple features and early fusion(late fusion) to combine multiple features. It can be seen from the Table 3 that the performance of the late fusion method [3] is not as well as the early fusion method [20,30], because the late fusion method prevents the model from learning inter-modality dynamics in an efficient way by assuming that simple weighted averaging is a proper fusion approach. This is in line with the experiment results shown in Table 2, in which the performance of late fusion method is lower than early fusion method. However, early fusion consists in simply concatenating multimodal features mostly at input level which does not allow the inter-modality dynamics to be efficiently modeled. Thus, the proposed ML-TFN framework which models the inter-modality dynamics through tensor fusion approach and multi layers fusion scheme obtains the best performance as shown in Table 3.

5   Conclusions

We proposed a framework named ML-TFN which can learn the interactions between multiple modalities in an end to end manner for affective video content analysis. The Tensor Fusion Network is designed to model the inter-modality dynamics between visual and audio modality. Two kinds of CNN based features are selected as the inputs to the Tensor Fusion Network. By using the Multi Layers Fusion scheme, we can better use features from different layers in unimodal embedding subnetwork to improve performance. Our method is evaluated on the AIMT15 dataset, where better results are obtained compared with other methods. Our future work will concentrate on exploiting more advanced networks to extract audio and visual features.

Acknowledgment. This work was supported in part by the Natural Science Foundation of Chongqing (No. cstc2020jcyj-msxmX0284, cstc2019jcyj-msxmX0021), the Scientific and Technological Research Program of Chongqing Municipal Education Commission (No. KJQN202000625, KJCXZD2020027, KJQN202100637), National Natural Science Foundation of China (61806033).

References

1. Acar, E., Hopfgartner, F., Albayrak, S.: Understanding affective content of music videos through learned representations. In: Gurrin, C., Hopfgartner, F., Hurst, W., Johansen, H., Lee, H., O'Connor, N. (eds.) MMM 2014. LNCS, vol. 8325, pp. 303–314. Springer, Cham (2014). https://doi.org/10.1007/978-3-319-04114-8_26
2. Arevalo, J., Solorio, T., Montes-y Gómez, M., González, F.A.: Gated multimodal units for information fusion. arXiv preprint arXiv:1702.01992 (2017)
3. Baecchi, C., Uricchio, T., Bertini, M., Del Bimbo, A.: Deep sentiment features of context and faces for affective video analysis. In: Proceedings of the 2017 ACM on International Conference on Multimedia Retrieval, pp. 72–77 (2017)
4. Baveye, Y., Chamaret, C., Dellandréa, E., Chen, L.: Affective video content analysis: a multidisciplinary insight. IEEE Trans. Affect. Comput. 9(4), 396–409 (2017)
5. Baveye, Y., Dellandréa, E., Chamaret, C., Chen, L.: LIRIS-ACCEDE: a video database for affective content analysis. IEEE Trans. Affect. Comput. 6(1), 43–55 (2015)
6. Chakraborty, R., Maurya, A.K., Pandharipande, M., Hassan, E., Ghosh, H., Kopparapu, S.K.: TCS-ILAB-mediaeval 2015: affective impact of movies and violent scene detection. In: MediaEval. Citeseer (2015)
7. Chan, C.H., Jones, G.J.: Affect-based indexing and retrieval of films. In: Proceedings of the 13th Annual ACM International Conference on Multimedia, pp. 427–430 (2005)
8. Dai, Q., et al.: Fudan-Huawei at mediaeval 2015: detecting violent scenes and affective impact in movies with deep learning. In: MediaEval (2015)
9. Gemmeke, J.F., et al.: Audio set: an ontology and human-labeled dataset for audio events. In: 2017 IEEE International Conference on Acoustics, Speech and Signal Processing (ICASSP), pp. 776–780. IEEE (2017)
10. Hanjalic, A., Xu, L.Q.: Affective video content representation and modeling. IEEE Trans. Multimed. 7(1), 143–154 (2005)
11. He, K., Zhang, X., Ren, S., Sun, J.: Deep residual learning for image recognition. In: Proceedings of the IEEE Conference on Computer Vision and Pattern Recognition, pp. 770–778 (2016)
12. Hershey, S., et al.: CNN architectures for large-scale audio classification. In: 2017 IEEE International Conference on Acoustics, Speech and Signal Processing (ICASSP), pp. 131–135. IEEE (2017)
13. Hochreiter, S., Schmidhuber, J.: Long short-term memory. Neural Comput. 9(8), 1735–1780 (1997)
14. Karpathy, A., Toderici, G., Shetty, S., Leung, T., Sukthankar, R., Fei-Fei, L.: Large-scale video classification with convolutional neural networks. In: Proceedings of the IEEE conference on Computer Vision and Pattern Recognition, pp. 1725–1732 (2014)
15. Kingma, D.P., Ba, J.: Adam: a method for stochastic optimization. arXiv preprint arXiv:1412.6980 (2014)
16. Lam, V., Le, S.P., Le, D.D., Satoh, S., Duong, D.A.: NII-UIT at mediaeval 2015 affective impact of movies task. In: MediaEval (2015)
17. Liu, Z., Shen, Y., Lakshminarasimhan, V.B., Liang, P.P., Zadeh, A., Morency, L.P.: Efficient low-rank multimodal fusion with modality-specific factors. arXiv preprint arXiv:1806.00064 (2018)
18. Marin Vlastelica, P., Hayrapetyan, S., Tapaswi, M., Stiefelhagen, R.: Kit at mediaeval 2015-evaluating visual cues for affective impact of movies task. In: MediaEval (2015)

19. Mironica, I., Ionescu, B., Sjöberg, M., Schedl, M., Skowron, M.: RFA at mediaeval 2015 affective impact of movies task: a multimodal approach. In: MediaEval (2015)
20. Ou, Y., Chen, Z., Wu, F.: Multimodal local-global attention network for affective video content analysis. IEEE Trans. Circuits Syst. Video Technol. **31**(5), 1901–1914 (2020)
21. Pérez-Rúa, J.M., Vielzeuf, V., Pateux, S., Baccouche, M., Jurie, F.: MFAS: multimodal fusion architecture search. In: Proceedings of the IEEE/CVF Conference on Computer Vision and Pattern Recognition, pp. 6966–6975 (2019)
22. Poria, S., Cambria, E., Bajpai, R., Hussain, A.: A review of affective computing: From unimodal analysis to multimodal fusion. Inf. Fusion **37**, 98–125 (2017)
23. Seddati, O., Kulah, E., Pironkov, G., Dupont, S., Mahmoudi, S., Dutoit, T.: UMONS at mediaeval 2015 Affective Impact of Movies Task Including Violent Scenes Detection. In: MediaEval (2015)
24. Sjöberg, M., et al.: The mediaeval 2015 affective impact of movies task. In: MediaEval (2015)
25. Trigeorgis, G., Coutinho, E., Ringeval, F., Marchi, E., Zafeiriou, S., Schuller, B.: The ICL-TUM-PASSAU approach for the mediaeval 2015 "Affective Impact of Movies" task. In: CEUR Workshop Proceedings, vol. 1436 (2015)
26. Vielzeuf, V., Lechervy, A., Pateux, S., Jurie, F.: CentralNet: a multilayer approach for multimodal fusion. In: Leal-Taixé, L., Roth, S. (eds.) ECCV 2018. LNCS, vol. 11134, pp. 575–589. Springer, Cham (2019). https://doi.org/10.1007/978-3-030-11024-6_44
27. Wang, L., et al.: Temporal segment networks for action recognition in videos. IEEE Trans. Pattern Anal. Mach. Intell. **41**(11), 2740–2755 (2018)
28. Wang, S., Ji, Q.: Video affective content analysis: a survey of state-of-the-art methods. IEEE Trans. Affect. Comput. **6**(4), 410–430 (2015)
29. Wei, Y., et al.: DeraincycleGAN: rain attentive cycleGAN for single image deraining and rainmaking. IEEE Trans. Image Process. **30**, 4788–4801 (2021)
30. Yi, Y., Wang, H.: Multi-modal learning for affective content analysis in movies. Multimed. Tools App. **78**(10), 13331–13350 (2018). https://doi.org/10.1007/s11042-018-5662-9
31. Yi, Y., Wang, H., Zhang, B., Yu, J.: MIC-TJU in mediaeval 2015 affective impact of movies task. In: CEUR Workshop Proceedings, vol. 1436 (2015)
32. Zadeh, A., Chen, M., Poria, S., Cambria, E., Morency, L.P.: Tensor fusion network for multimodal sentiment analysis. In: Proceedings of the 2017 Conference on Empirical Methods in Natural Language Processing, pp. 1103–1114 (2017)
33. Zhang, Y., et al.: Dual-constrained deep semi-supervised coupled factorization network with enriched prior. Int. J. Comput. Vision **129**(12), 3233–3254 (2021)
34. Zhang, Z., Ren, J., Zhang, Z., Liu, G.: Deep latent low-rank fusion network for progressive subspace discovery. In: Proceedings of the Twenty-Ninth International Conference on International Joint Conferences on Artificial Intelligence, pp. 2762–2768 (2021)
35. Zhao, S., Yao, H., Sun, X., Xu, P., Liu, X., Ji, R.: Video indexing and recommendation based on affective analysis of viewers. In: Proceedings of the 19th ACM International Conference on Multimedia, pp. 1473–1476 (2011)

A Dominance-Based Many-Objective Artificial Bee Colony Algorithm

Tingyu Ye, Hui Wang$^{(\boxtimes)}$, Tao Zeng, Zichen Wei, Shuai Wang, Hai Zhang, Jia Zhao, and Min Hu

School of Information Engineering, Nanchang Institute of Technology, Nanchang 330099, China
huiwang@whu.edu.cn

Abstract. Artificial bee colony (ABC) is a popular swarm intelligence algorithm, because it has been widely applied to many optimization problems. Nevertheless, solving many-objective optimization problems (MaOPs) using ABC is challenging. In this paper, a dominance-based many-objective ABC (called DMaOABC) algorithm is proposed to deal with MaOPs. Firstly, a secondary dominance criterion is used to redefine the fitness function. This is helpful to strengthen the selection pressure of solutions. Then, an improved search strategy is proposed to balance exploration and exploitation. The probability selection is modified based on the new fitness function. Moreover, a new mating selection is utilized to replace the greedy selection in the original ABC. The reference point strategy is introduced to improve the distribution. To investigate the performance of DMaOABC, a set of WFG benchmark problems with 3, 8, and 10 objectives are tested. The performance of DMaOABC is compared with four other many-objective evolutionary algorithms (MaOEAs). Experimental results show that DMaOABC outperforms the compared algorithms according to two performance indicators.

Keywords: Swarm intelligence optimization algorithm · Artificial bee colony algorithm · Many-objective optimization · Secondary dominance

1 Introduction

Multi-objective optimization problems (MOPs) have been applied in many scientific fields including engineering, economy, logistics and so on. It is difficult to achieve the best of all objectives at the same time for MOPs. In this case, the objective function is considered to be conflicting. The detailed descriptions of a general MOP are described as bellow.

$$\min F(X) = [f_1(X), f_2(X), \ldots, f_M(X)]$$
$$\text{S.t. } X = (x_1, x_2, \ldots, x_D) \in S \tag{1}$$

where X is the decision vector, $f_u(X)$ is the u-th optimization objective, S is the decision space, D is the number of decision variables, and M is number of objectives ($M \geq 2$). When $M > 3$, the MOPs are called as many-objective optimization problems (MaOPs). Though many multi-objective evolutionary algorithms show good performance for MOPs with two and three objectives, they encounter difficulties for MaOPs in high-dimensional objective spaces. With the increase of the number of objective, the difficulty of solving the problem gradually increase. So, it is challenging to effectively solve MaOPs.

For MOPs, Pareto dominance is usually utilized to compare the quality of solutions. The vectors a and b, a is deemed to dominate b $(a \prec b)$ if and only if the following equation is satisfied.

$$\{\forall i \in K : f_i(a) \le f_i(b)\} \wedge \{\exists j \in K : f_j(a) < f_j(b)\} \tag{2}$$

where $K \in \{1, 2, \ldots, M\}$. If there is no decision vector dominates a decision vector $X*$, $X*$ is called Pareto optimal solution. In the MOPs, there is more than one Pareto optimal solution generally. All Pareto optimal solutions corresponding to the objective functions constitute the Pareto Front (PF).

However, the efficiency of such Pareto-based MOEAs will seriously degrade when solving MaOPs. Early multi-objective evolutionary algorithms (MOEAs) for MOPs with two or three objectives fail to solve MaOPs. The main reason is that solutions are difficult to be distinguished according to the original dominance relationship in the early stage of the search and the majority of solutions in the population are non-dominated. To tackle this issue, three main methods were proposed: 1) using secondary criterion in dominance comparison [1]; 2) introduction of reference weights, vectors or points [2]; and 3) preference information [3]. With the aid of those strategies, the performance of MOEAs is improved on MaOPs.

Artificial bee colony (ABC) is a population intelligent optimization algorithm (IOA) inspired by bee honey collection mechanism [4]. Compared with other IOAs, such as evolutionary algorithm [5], differential evolution algorithm [6], particle swarm optimization [7,8], grey wolf optimizer [9], and pigeon-inspired optimization [10], artificial bee colony has the strong points of simple operation, few control parameters, and powerful search capability. So, it is usually utilize to solve single objective optimization problems by scholars [11–16]. Recently, it was developed to dispose of optimization problems with multiple objectives [17–20]. However, previous studies mainly focused on ordinary MOPs with two or three objectives, and it is rarely used to MaOPs.

In this paper, a dominance-based many-objective ABC algorithm (namely DMaOABC) is proposed for solving MaOPs. The framework of DMaOABC is similar to NSGA-III [21]. First, a novel secondary dominance criterion is used to redefine the fitness function. Then, a modified search strategy on the basis of elite set is proposed to enhance the search ability. Moreover, the original probability selection in the onlooker bee phase is not applicable, and it is modified based on the new fitness function. Finally, a new mating selection is used to displace the greedy selection in the standard ABC. The diversity of solutions is maintained by the reference point strategy [21]. In the experiment, the propped DMaOABC is compared with four popular MaOEAs on the WFG test problems. Simulation results demonstrate that DMaOABC is superior to the compared algorithms in terms of two performance indicators.

The rest parts of this paper are arranged as below. Section 2 introduces the original ABC. Our approach DMaOABC is described in Sect. 3. The experimental part and conclusion are given in Sects. 4 and 5, respectively.

2   Artificial Bee Colony Algorithm

Artificial bee colony algorithm (ABC) is a fashionable swarm intelligence optimization algorithm at present. In ABC, there are three types of bees: employed bee, onlooker bee and scout bee. Different bees undertake different tasks to assure the orderly implementation of the algorithm. The search process of ABC consists of four phases. The details of these search stages are gave as bellow.

(1) Initialization stage: A certain number of solutions are randomly generated in the population as below.

$$x_{i,j} = Min_j + rand \cdot (Max_j - Min_j) \tag{3}$$

where Min_j and Max_j are the bounds of the search range, and $rand$ is a random number between 0 and 1.

(2) Employed bee stage: At this stage, employed bees aim to search for different honey (food) sources, and judge whether there is a better one around the current honey source. If there is a better honey source, the current one will be replaced. The search strategy can be represented as follows.

$$v_{i,j} = x_{i,j} + \phi_{i,j} \cdot (x_{i,j} - x_{k,j}) \tag{4}$$

where X_i is the current solution, X_k is chosen randomly in the swarm ($i \neq k$), and $\phi_{i,j} \in [-1, 1]$ is a random number.

(3) Onlooker bee stage: The onlooker bee is responsible for further searching honey sources on the basis of employed bees. The onlooker bees choose select good honey sources by the probability p_i, the defined of p_i is gave as follows.

$$p_i = \frac{fit(X_i)}{\sum_{i=1}^{SN} fit(X_i)} \tag{5}$$

where $fit(X_i)$ represents the fitness value of the current honey source. Fitness value is the standard to evaluate the honey source. The larger the fitness value of honey source is, the better it is. The formula of fitness value is as follows.

$$fit(X_i) = \begin{cases} \frac{1}{1+f(X_i)}, & \text{if } f(X_i) \geq 0 \\ 1 + |f(X_i)|, & \text{if } f(X_i) < 0 \end{cases} \tag{6}$$

where $f(X_i)$ is the value of the objective function. Through this selection method, the higher the fitness value is, the higher the probability of honey source being selected is. It is often followed by bees to search many times, and the search formula is Eq. (4).

(4) Scout bee stage: When a honey source has not been updated in a preset number of iterations, the scout bee initialize it by Eq. (3).

Algorithm 1: Framework of DMaOABC

---

1 **Input:** Objectives M, population size N, problem dimension D, maximum number of fitness evaluations $maxFEs$;
2 **Output:** The final population P;
3 **while** $FEs <= maxFEs$ **do**
4     $E \leftarrow Employed\_Bee\_Search(P, D, M, N)$;
5     $E' \leftarrow Mating\_Selection(P, E, M, N)$;
6     $P' \leftarrow Environmental\_Selection(P, E', N, M, Z, Z_{min})$;
7     $O \leftarrow Onlooker\_Bee\_Search(P', D, M, N)$;
8     $O' \leftarrow Mating\_Selection(P', O, M, N)$;
9     $S \leftarrow Scout\_Bee\_Search(O', D, M, N)$;
10    $P \leftarrow Environmental\_Selection(P', S, M, N, Z, Z_{min})$;
11 **end**

---

3 Dominance-Based Many-Objective ABC (DMaOABC)

3.1 Framework of DMaOABC

As most existing dominance-based MaOEAs, such as NSGA-III [21], DMaOABC has a similar framework. It is noted that DMaOABC adds a secondary dominance and uses an improved ABC algorithm as the search engine. Algorithm 1 shows the framework of the proposed DMaOABC. As shown, reference points and idea points are generated at the initial stage [22]. Then, the improved ABC is used for population iteration, which is composed of key steps as follow. First, the employed bees aim to search for new solutions (line 4). A new mating selection is utilized to choose better solutions instead of the greedy choice of the original ABC (line 5), and environment selection based on reference point is used (line 6). Then, the onlooker bees search for new solutions based on a modified roulette selection mechanism (line 7), and the role of the scout bee is the equal to the standard ABC (line 9).

In the subsequent sections, the new mating selection, improved search strategy and selection mechanism used in DMaOABC are described in details.

3.2 New Fitness Function and Mating Selection Based on Secondary Dominance Criterion

In MaOPs, solutions are difficult to be distinguished according to the selection criterion on the basis of the original dominance relationship in the early search phase, because the majority of solutions in the population are non-dominated. In other words, almost all solutions are independent of each other in the population. In order to strengthen the selection pressure of solutions in the population, a secondary dominance criterion is used to redefine the fitness function. Then, the detailed definitions of the novel fitness function $S(\cdot)$ is described as bellow.

$$S(X_i) = \sum_{k=1}^{M} f_k(X_i) \tag{7}$$

Algorithm 2: *Mating_Selection*

---

1 **for** *each* $E(X_i)$ *in the swarm* **do**
2   **if** $E(X_i) \prec P(X_i)$ **then**
3 | $P(X_i) \leftarrow E(X_i)$;
4   **end**
5   **else if** $P(X_i) \prec E(X_i)$ **then**
6 | $P(X_i) \leftarrow P(X_i)$;
7 | $trial_i \leftarrow trial_i + 1$;
8   **end**
9   **else**
10   **if** $S(E(X_i)) < S(P(X_i))$ **then**
11 | $P(X_i) \leftarrow E(X_i)$;
12   **end**
13   **else if** $S(P(X_i)) < S(E(X_i))$ **then**
14 | $P(X_i) \leftarrow P(X_i)$;
15 | $trial_i \leftarrow trial_i + 1$;
16   **end**
17   **else**
18   **if** $rand(0,1) < 0.5$ **then**
19 | $P(X_i) \leftarrow E(X_i)$;
20   **end**
21   **else**
22 | $P(X_i) \leftarrow P(X_i)$;
23 | $trial_i \leftarrow trial_i + 1$;
24   **end**
25   **end**
26   **end**
27 **end**

---

where $S(X_i)$ the fitness value of X_i, $f_k(X_i)$ is the normalized value corresponding to each objective function. For the minimization optimization problem, the smaller $S(X_i)$, the closer it is to the true PF. For example, $F(X_1) = (5,3,4)$ and $F(X_2) = (1,3,5)$ are non-dominated solutions. But $S(X_2) < S(X_1)$, $S(X_2)$ is more potential to be retained. Through this method, the selection pressure is enhanced and better non-dominated solutions are selected. Noteworthy, all objective values should be normalized before this operation. In addition, maximization optimization problems can be converted into minimization problems.

In ABC, a greedy selection is employed to select excellent solutions. Nevertheless, the previous greedy selection cannot be directly applied to MaOPs. Therefore, a modified greedy selection called mating selection is designed in DMaOABC. In the mating selection, there are three main operations including dominance comparison, secondary selection criteria, and random selection. The concrete content of the mating selection method in DMaOABC is described in Algorithm 2. Firstly, parent population P and offspring population E are obtained. $P(X_i)$ and $E(X_i)$ are the i-th parent and offspring solution, respectively. If the offspring solution $E(X_i)$ dominates the parent solution $P(X_i)$, then $E(X_i)$ is chosen; otherwise $P(X_i)$ is retained (lines 4–10). When $E(X_i)$ and $P(X_i)$ are two non-dominated solutions, their fitness values are calculated by Eq. (7). If $S(E(X_i)) < S(P(X_i))$, then $E(X_i)$ is chosen (lines 12–14); If $S(P(X_i)) < S(E(X_i))$, then $P(X_i)$ is retained (lines 15–18). For the third case $S(P(X_i)) == S(E(X_i))$, a simple random method is used to choose a solution (lines 20–26).

Algorithm 3: *Employed_Bee_Search*

---

1 **for** *each P_i in the swarm* **do**
2 | Produce E_i by Eq. (8);
3 | Calculate $F(V_i)$ and set $FEs{+}{+}$;
4 **end**

Algorithm 4: *Onlooker_Bee_Search*

---

1 Calculate the probability p_i by Eq. (9);
2 Set $num = 0$ and $i = 1$;
3 **while** $num \leq SN$ **do**
4 | **if** $rand < p_i$ **then**
5 | | Produce E_i by Eq. (8);
6 | | Calculate $F(V_i)$ and set $FEs{+}{+}$;
7 | **end**
8 **end**

If the parent solution $P(X_i)$ cannot be renewed, the $trial_i$ is added by 1. When the maximum $trial_i$ is greater than the preset threshold $limit$, the scout bee search is activated. Then, the corresponding $P(X_i)$ is re-initialized by Eq. (3).

3.3 Improved ABC

The search strategy determines the performance of ABC. It is very important to use an appropriate search strategy to gain good optimization ability. Some references reported that ABC showed good exploration and weak exploitation [23–25]. So, it is difficult to use the original ABC to obtain the Pareto optimal solution. In our previous work [26], an efficient search strategy with strong exploitation ability was proposed. Then, a new search strategy is proposed as bellow.

$$v_{i,j} = x_{best,j} + \phi_{i,j}(x_{best,j} - x_{k,j}) \tag{8}$$

where X_k is selected from the population randomly ($k \neq i$), and X_{best} is randomly chosen from the elite set ES. For all non-dominated solutions, their fitness values are calculated by Eq. (7). The $m \cdot n$ non-dominated solutions with the smallest fitness values are selected into the elite set ES, where m is the number of non-dominated solutions and n is set to 0.2. The selected excellent solutions based on our secondary dominance criterion are used to guide the search. Through this strategy, it is helpful to find the excellent solutions.

Equation (5) defines the selection probability of apiece solution in the onlooker bee stage. It is obvious that Eq. (5) is not applicable for MaOPs, because there are more than two objective values for each solution. Consequently, a novel mode to calculate the probability p_i is designed as bellow.

$$p_i = 1 - \frac{S(X_i)}{MaxS(X_i)} \tag{9}$$

where $S(X_i)$ is the new fitness value of X_i, and $MaxS(X_i)$ is the maximum fitness value among all solutions. As seen, a good solution has a small $S(X_i)$ and large p_i. After calculating the probabilities of all solutions, the roulette method is utilized to select outstanding solutions for further search. The search strategy is the equal with the employed bee. For the details of the above operations, please refer to Algorithms 3 and 4.

4   Experimental Study

To verify the performance of DMaOABC, four popular MaOEAs, including MOEA/D [27], NSGA-III [21], RVEA [28], and SPEAR [29], are used for comparisons. Experiments are conducted on nine popular WFG benchmark problems with three, eighth and ten objectives [30]. Regarding the above nine problems, D is set to $K + L$, where K is equal to $M - 1$ and L is set to 10. The population size is set to 100 for 3-objective test instances, 230 for 8-objective test instances, and 240 for 10-objective test instances. Two popular indicators including GD and IGD are employed to measure the validity of DMaOABC.

Table 1 gives the average GD value of DMaOABC and four compared algorithms over 20 independent runs. At the bottom of the table, $+/-/=$ summarizes the overall performance comparison. On + instances, the compared MaOEAs outperforms our approach DMaOABC. On − instances, the proposed DMaOABC is better than the compared MaOEAs. On = instances, DMaOABC and the compared MaOEAs obtain similar performance. As seen, MOEA/D performs better than DMaOABC on 8 instances, but DMaOABC is better than MOEA/D on 16 instances. Compared with NSGA-III, DMaOABC achieves better results on 14 instances and they obtain the same performance on 7 instances. RVEA and DMaOABC gain the equal results on 2 instances, while DMaOABC is superior to RVEA on the other 16 instances. DMaOABC and SPEAR gain the equal results on 5 instances. For the rest of 22 instances, SPEAR is worse than DMaOABC on 14 instances. The above results demonstrate that DMaOABC can obtain competitive performance based on the GD values.

For further performance verification on DMaOABC, Table 2 shows the average IGD values of four compared algorithms and DMaOABC over 20 independent runs. IGD is an indicator employed for appraising the convergence and diversity. From the results of Table 2, DMaOABC still is superior to other algorithms on the majority of test instances, even though the performance indicator becomes more stringent. MOEA/D is better than DMaOABC on only 2 instances, but DMaOABC is superior to MOEA/D on the rest of 25 instances. NSGA-III wins on 4 instances, while DMaOABC is better on 14 instances. RVEA and DMaOABC gain the equal results on 4 instances. For the remaining 23 instances, DMaOABC outperforms RVEA on 16 instances. Compared with SPEAR, DMaOABC can obtain better results on 14 instances, while SPEAR wins on 8 instances. The above results show that DMaOABC performs better than four other MaOEAs based on the IGD values.

Table 1. The GD values of different algorithms.

Problems	M	MOEA/D	NSGA-III	RVEA	SPEAR	DMaOABC
WFG1	3	**1.1975e-3+**	5.9378e-2−	2.7966e-2−	1.5058e-2=	1.5133e-2
		(3.55e-4)	(1.15e-2)	(1.10e-2)	(4.24e-3)	(1.11e-2)
	8	3.5638e-2+	4.3218e-2=	**3.2036e-2+**	5.3318e-2−	4.2780e-2
		(2.71e-3)	(2.85e-3)	**(4.12e-3)**	(5.24e-3)	(3.09e-3)
	10	5.2006e-2−	5.1227e-2−	**2.7264e-2+**	5.1309e-2	4.0201e-2
		(7.36e-3)	(4.82e-3)	**(2.78e-3)**	(6.14e-3)	(7.45e-3)
WFG2	3	6.5275e-3−	7.6187e-3−	8.2278e-3−	6.4726e-3−	**2.2816e-3**
		(3.39e-3)	(1.09e-3)	(9.97e-4)	(4.03e-3)	**(2.67e-3)**
	8	5.9769e-2=	4.2594e-2=	**2.8927e-2+**	4.6546e-2=	4.1766e-2
		(4.79e-2)	(6.52e-3)	**(1.52e-3)**	(5.41e-2)	(9.91e-2)
	10	5.4508e-2=	4.5829e-2=	**2.8987e-2+**	4.7583e-2=	4.4749e-2
		(2.43e-2)	(7.86e-3)	**(2.85e-3)**	(8.09e-3)	(1.62e-2)
WFG3	3	1.0625e-1−	**8.3323-2+**	1.4643e-1−	9.2079e-2=	9.1491e-2
		(2.10e-3)	**(7.37e-3)**	(8.40e-3)	(6.95e-3)	(7.22e-3)
	8	**4.3025e-1+**	5.1215e-1=	5.1059e-1=	5.0795e-1−	4.9821e-1
		(9.20e-2)	(1.94e-2)	(7.72e-2)	(8.62e-3)	(1.85e-2)
	10	3.3419e-1+	**2.7887e-1+**	5.2131e-1+	5.5180e-1−	5.3592e-1
		(5.90e-2)	**(1.15e-1)**	(1.36e-2)	(7.13e-3)	(2.12e-2)
WFG4	3	**2.3529e-3+**	6.4184e-3−	5.7308e-3=	2.7383e-3+	5.5976e-3
		(1.87e-4)	(6.65e-4)	(7.42e-4)	(2.05e-4)	(3.75e-4)
	8	1.4348e-1−	1.1280e-1−	1.1564e-1−	1.1559e-1−	**1.0638e-1**
		(1.36e-2)	(5.11e-3)	(2.05e-3)	(1.20e-3)	**(2.08e-3)**
	10	1.8708e-1−	3.7147e-2+	5.4772e-2+	**2.5065e-2+**	3.9181e-2
		(2.22e-2)	(2.88e-2)	(1.23e-2)	**(1.50e-3)**	(2.79e-3)
WFG5	3	**7.3896e-3+**	9.2169e-3−	8.7303e-3−	7.6285e-3+	7.7396e-3
		(1.97e-4)	(2.86e-4)	(2.49e-4)	(5.00e-5)	(1.17e-4)
	8	1.2516e-1−	1.1946e-1−	1.1978e-1−	1.1798e-1−	**1.1306e-1**
		(6.52e-3)	(7.22e-4)	(1.34e-4)	(9.06e-4)	**(1.76e-3)**
	10	1.5167e-1−	2.7465e-2+	3.9546e-2+	**2.3415e-2+**	6.0729e-2
		(2.14e-2)	(2.49e-3)	(9.19e-3)	**(1.92e-3)**	(4.17e-3)
WFG6	3	1.1774e-2=	1.5345e-2−	1.2683e-2−	1.0890e-2=	**1.0772e-2**
		(8.16e-3)	(2.02e-3)	(2.31e-3)	(1.78e-3)	**(2.73e-3)**
	8	1.6194e-1−	1.1620e-1−	1.1679e-1−	1.1539e-1−	**1.0804e-1**
		(1.54e-2)	(1.05e-3)	(2.97e-3)	(1.99e-3)	**(4.41e-3)**
	10	2.1730e-1−	**2.8338e-2+**	8.1864e-2=	2.9114e-2+	4.9176e-2
		(3.80e-2)	**(3.60e-3)**	(1.03e-2)	(1.99e-3)	(6.86e-3)
WFG7	3	4.5831e-3−	6.5689e-3−	4.9863e-3−	4.5328e-3−	**2.6517e-3**
		(4.57e-4)	(6.14e-4)	(5.57e-4)	(7.87e-4)	**(2.17e-4)**
	8	1.3250e-1−	1.1333e-1−	1.1022e-1−	1.1308e-1−	**1.0650e-1**
		(1.07e-2)	(1.40e-3)	(2.51e-3)	(1.53e-3)	**(1.92e-3)**
	10	1.5905e-1−	4.7487e-2−	6.4144e-2−	3.8499e-2+	**3.1775e-2**
		(2.38e-2)	(6.21e-3)	(5.80e-3)	(1.57e-3)	**(1.75e-2)**
WFG8	3	2.9885e-2−	2.4067e-2−	2.3818e-2−	2.1912e-2−	**1.7457e-2**
		(4.11e-3)	(9.06e-4)	(9.09e-4)	(8.69e-4)	**(3.84e-4)**
	8	1.5539e-1−	1.2169e-1−	1.3754e-1−	1.2453e-1−	**1.0880e-1**
		(4.06e-2)	(8.94e-3)	(1.11e-2)	(8.24e-3)	**(8.46e-3)**
	10	1.4133e-1−	9.9853e-2−	1.0101e-1−	4.5430e-2+	**7.8229e-2**
		(4.20e-2)	(3.01e-2)	(1.18e-2)	(1.00e-2)	(1.90e-2)
WFG9	3	2.7828e-2−	1.0295e-2=	**6.7377e-3+**	6.9349e-3+	1.2240e-2
		(7.87e-3)	(3.16e-3)	**(1.21e-3)**	(9.84e-4)	(8.07e-3)
	8	**5.3885e-2+**	1.2564e-1=	1.1973e-1+	1.2640e-1−	**1.2165e-1**
		(1.48e-2)	(4.03e-3)	(2.76e-3)	(5.66e-3)	**(3.62e-3)**
	10	7.1851e-2+	8.1983e-2+	9.4604e-2+	**6.2864e-2+**	1.1374e-1
		(3.63e-2)	(1.30e-2)	(7.16e-3)	**(7.37e-3)**	(1.24e-2)
+/-/=		8/16/3	6/14/7	9/16/2	8/14/5	

Table 2. The IGD values of different algorithms.

Problems	M	MOEA/D	NSGA-III	RVEA	SPEAR	DMaOABC
WFG1	3	**3.1332e−1**+	6.9401e−1−	3.1765e−1+	3.1544e−1+	3.9490e−1
		(1.13e−2)	(9.02e−2)	(6.22e−2)	(5.70e−2)	(9.42e−2)
	8	1.4821e+0−	**1.5978e+0**=	1.0054e+0=	1.2811e+0−	1.0014e+0
		(4.47e−2)	**(8.41e−2)**	4.45e−2	(8.99e−2)	(6.86e−2)
	10	1.7086e+0−	1.1010e+0−	1.2247e+0−	1.4866e+0−	**1.1837e+0**
		(9.80e−2)	(6.82e−2)	(5.51e−2)	(7.34e−2)	**(7.11e−2)**
WFG2	3	3.0256e−1−	1.7020e−1−	1.8744e−1−	1.7048e−1−	**1.6401e−1**
		(3.87e−3)	4.85e−3	(7.41e−3)	(2.73e−3)	**(2.61e−3)**
	8	1.8427e+0−	1.1043e+0=	1.0234e+0=	**9.5443e−1**=	1.1629e+0
		(1.29e−1)	(2.26e−1)	3.09e−2	**(1.72e−2)**	(3.44e−1)
	10	1.8826e+0−	1.3225e+0=	**1.1348e+0**=	1.2268e+0=	1.3045e+0
		(1.35e−1)	(1.39e−1)	**(3.39e−2)**	(3.31e−2)	(2.40e−1)
WFG3	3	**1.1115e−1**+	1.5767e−1+	2.3919e−1−	1.5385e−1+	1.8320e−1
		(3.97e−3)	1.29e−2	(2.41e−2)	(1.46e−2)	(2.39e−2)
	8	2.4918e+0−	1.6928e+0=	2.3398e+0−	**1.6787e+0**+	1.8490e+0
		(1.09e−1)	(2.17e−1)	(5.22e−1)	**(1.50e−1)**	((3.32e−1)
	10	3.3116e+0−	**9.6206e−1**+	3.7361e+0−	1.8537e+0+	2.2085e+0
		(4.67e−1)	**(4.19e−1)**	(6.80e−1)	(5.21e−2)	(2.60e−1)
WFG4	3	3.5427e−1−	2.3161e−1=	2.4191e−1−	**2.2866e−1**+	2.3202e−1
		6.13e−3	(3.02e−3)	(5.42e−3)	**(4.09e−3)**	(3.28e−3)
	8	5.2956e+0−	2.9823e+0=	3.0063e+0−	2.9788e+0−	**2.9680e+0**
		(1.80e−1)	(7.36e−3)	(2.08e−2)	(9.09e−3)	**(1.48e−2)**
	10	7.5105e+0−	4.7623e+0+	4.6217e+0+	4.8021e+0−	4.7921e+0
		(4.30e−1)	(5.49e−2)	(6.65e−2)	8.84e−3	(2.09e−2)
WFG5	3	3.6013e−1−	2.3674e−1−	2.3638e−1−	2.3891e−1−	**2.3333e−1**
		8.26e−3	(1.56e−3)	(1.83e−3)	(4.75e−3)	**(1.90e−3)**
	8	4.5116e+0−	2.9420e+0=	2.9849e+0−	2.9501e+0−	2.9456e+0
		(2.65e−1)	(3.59e−3)	(1.70e−2)	(5.58e−3)	1.06e−2
	10	6.4998e+0−	4.7214e+0−	4.6754e+0−	4.7577e+0−	**4.6037e+0**
		(3.13e−1)	(1.35e−2)	(5.84e−2)	1.34e−2	**(2.10e−2)**
WFG6	3	3.7154e−1−	2.6914e−1−	2.6175e−1−	**2.5091e−1**+	2.5733e−1
		(8.20e−3)	(1.15e−2)	(1.05e−2)	**(8.26e−3)**	4.75e−2
	8	5.2909e+0−	2.9950e+0−	3.0376e+0−	3.0018e+0−	**2.9628e+0**
		(2.57e−1)	(3.86e−2)	(2.77e−2)	(1.80e−2)	**(7.24e−3)**
	10	7.5366e+0−	4.7799e+0=	**4.4615e+0**+	4.8024e+0−	4.7838e+0
		(6.23e−1)	1.05e−2	**(5.30e−2)**	(9.66e−3)	(1.66e−2)
WFG7	3	3.5528e−1−	2.3250e−1−	2.3640e−1−	2.3023e−1−	**2.2540e−1**
		(4.51e−3)	(2.99e−3)	(5.87e−3)	(3.17e−3)	**(1.20e−3)**
	8	5.3961e+0−	2.9133e+0−	3.0391e+0−	2.9856e+0=	**2.9791e+0**
		(3.21e−1)	(7.94e−3)	(2.05e−2)	(1.01e−2)	**(7.63e−3)**
	10	7.7565e+0−	4.7880e+0−	**4.5939e+0**+	4.8000e+0−	4.7722e+0
		(6.71e−1)	(2.03e−2)	**(1.57e−3)**	1.02e−2	3.56e−2
WFG8	3	3.9331e−1−	3.2018e−1−	3.1800e−1−	**2.7800e−1**+	3.0348e−1
		(3.99e−3)	(8.20e−3)	(6.70e−3)	**(2.34e−3)**	8.52e−3
	8	5.3449e+0−	3.2423e+0=	3.1076e+0+	**3.0962e+0**+	3.1995e+0
		(1.73e−1)	(1.63e−1)	(3.48e−2)	**(3.41e−2)**	4.26e−2
	10	8.8413e+0−	4.6996e+0−	**4.3975e+0**+	4.7940e+0−	4.5565e+0
		(3.82e−1)	(2.18e−1)	**(5.25e−2)**	(1.50e−2)	1.15e−1
WFG9	3	3.4756e−1−	2.4253e−1−	2.3214e−1−	2.2942e−1=	**2.2864e−1**
		(2.29e−2)	(1.61e−2)	(5.77e−3)	(3.40e−3)	**(3.02e−3)**
	8	5.4019e+0−	**2.9359e+0**+	2.9721e+0+	2.9420e+0=	2.9721e+0
		((2.46e−1)	**(9.50e−3)**	(2.52e−2)	(1.28e−2)	(8.28e−2)
	10	7.5046e+0−	4.5248e+0−	4.4625e+0=	4.6701e+0−	**4.4546e+0**
		(4.57e−1)	((3.81e−2)	(7.41e−2)	(1.08e−2)	**(1.06e−1)**
+/-/=		2/25/0	4/14/9	7/16/4	8/14/5	

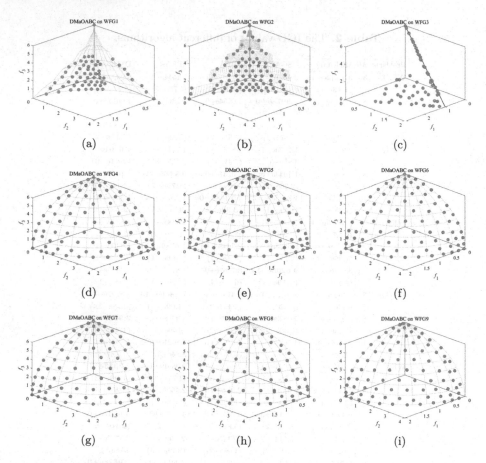

Fig. 1. The Parato fronts of DMaOABC on WFG1-WFG9 with 3 objectives.

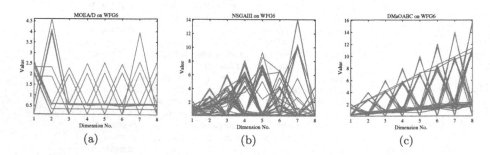

Fig. 2. Parallel coordinates plot of MOEA/D, NSGA-III and DMaOABC on WFG6 with 8 objectives.

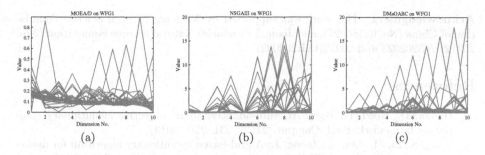

Fig. 3. Parallel coordinates plot of MOEA/D, NSGA-III and DMaOABC on WFG1 with 10 objectives.

Figure 1 shows intuitively the Pareto fronts obtained by DMaOABC on WFG1-WFG9 with 3 objectives. WFG1, WFG2, and WFG3 are three test problems with completely different characteristics, so the Pareto fronts of DMaOABC on WFG1-WFG3 have different shapes. WFG4-WFG9 are six test problems with similar characteristics, and Fig. 1(d)–Fig. 1(i) are similar. Figure 2 presents the parallel coordinates plot of MOEA/D, NSGA-III and DMaOABC on WFG6 with 8 objectives. Figure 3 gives the parallel coordinates plot of MOEA/D, NSGA-III and DMaOABC on WFG1 with 10 objectives. As seen, the solution distributions of DMaOABC are better than MOEA/D and NSGA-III.

5 Conclusion

In this paper, a new many-objective ABC algorithm (called DMaOABC) is proposed for solving MaOPs. There are main modifications in DMaOABC including: 1) the fitness function is redefined by a secondary dominance criterion. Based on the new fitness function, the probability selection used in the onlooker bees is modified to apply to MaOPs; 2) an improved search strategy based on the elite set is proposed to balance exploration and exploitation; 3) a new mating selection is used to displace the greedy selection in the standard ABC; and 4) the reference point strategy is introduced to maintain the solution distribution. To test and verify the validity of DMaOABC, nine popular WFG benchmark problems with 3, 8 and 10 objectives are tested.

From the comparison results, the proposed DMaOABC is better than RVEA, MOEA/D, NSGA-III and SPEAR on the majority of test instances in terms of GD and IGD values. DMaOABC can approximate the true Pareto fronts well and obtain good distributions for different number of objectives. So far, few studies on ABC focus on MaOPs. So, this paper is still an exploratory work to extend ABC to solve MaOPs. The main contribution of DMaOABC is attributed to the secondary domination. With aid of this method, excellent solutions can be easily chosen to guide the search process and improve the convergence performance. In the future, we will try to apply the proposed DMaOABC to more complex MaOPs with larger number of objectives.

Acknowledgment. This work was supported by National Natural Science Foundation of China (No. 62166027), and Jiangxi Provincial Natural Science Foundation (Nos. 20212BAB202023 and 20212BAB202022).

References

1. Hadka, D., Reed, P.: Borg: An auto-adaptive many-objective evolutionary computing framework. Evol. Comput. **21**(2), 231–259 (2013)
2. Yang, S., Li, M., Liu, X., Zheng, J.: A grid-based evolutionary algorithm for many-objective optimization. IEEE Trans. Evol. Comput. **17**(5), 721–736 (2013)
3. Bader, J., Zitzler, E.: HypE: an algorithm for fast hypervolume based many-objective optimization. Evol. Comput. **19**(1), 45–76 (2011)
4. Karaboga, D.: An idea based on honey bee swarm for numerical optimization, Technical report-TR06. Erciyes University, Engineering Faculty, Computer engineering Department (2005)
5. Liu, N.S., Pan, J.S., Sun, C.L., Chu, S.C.: An efficient surrogate-assisted quasi-affine transformation evolutionary algorithm for expensive optimization problems. Knowl. Based Syst. **209**, 106418 (2020)
6. Pan, J.S., Liu, N.S., Chu, S.C.: A hybrid differential evolution algorithm and its application in unmanned combat aerial vehicle path planning. IEEE Access **8**, 17691–17712 (2020)
7. Wang, F., Zhang, H., Li, K.S., Lin, Z.Y., Yang, J., Shen, X.L.: A hybrid particle swarm optimization algorithm using adaptive learning strategy. Inf. Sci. **436–437**, 162–177 (2018)
8. Wang, H., Wu, Z.J., Rahnamayan, S., Liu, Y., Ventresca, M.: Enhancing particle swarm optimization using generalized opposition-based learning. Inf. Sci. **181**(20), 4699–4714 (2011)
9. Hu, P., Pan, J.S., Chu, S.C.: Improved binary grey wolf optimizer and its application for feature selection. Knowl. Based Syst. **195**(11), 105746 (2020)
10. Tian, A.Q., Chu, S.C., Pan, J.S., Cui, H., Zheng, W.M.: A compact pigeon-inspired optimization for maximum shortterm generation mode in cascade hydroelectric power station. Sustainability **12**(3), 767 (2020)
11. Akay, B., Karaboga, D.: A modified artificial bee colony algorithm for real parameter optimization. Inf. Sci. **192**, 120–142 (2012)
12. Karaboga, D., Gorkemli, B.: A quick artificial bee colony (qABC) algorithm and its performance on optimization problems. Appl. Soft Comput. **23**(1), 227–238 (2014)
13. Wang, H., Wang, W.J., Xiao, S.Y., Cui, Z.H., Xu, M.Y., Zhou, X.Y.: Improving artificial Bee colony algorithm using a new neighborhood selection mechanism. Inf. Sci. **527**, 227–240 (2020)
14. Xiao, S., Wang, H., Wang, W., Huang, Z., Zhou, X., Xu, M.: Artificial bee colony algorithm based on adaptive neighborhood search and Gaussian perturbation. Appl. Soft Comput. **100**, 106955 (2021)
15. Ye, T., Zeng, T., Zhang, L., Xu, M., Wang, H., Hu, M.: Artificial bee colony algorithm with an adaptive search manner. In: Zhang, H., Yang, Z., Zhang, Z., Wu, Z., Hao, T. (eds.) NCAA 2021. CCIS, vol. 1449, pp. 486–497. Springer, Singapore (2021). https://doi.org/10.1007/978-981-16-5188-5_35
16. Zeng, T., Ye, T., Zhang, L., Xu, M., Wang, H., Hu, M.: Population diversity guided dimension perturbation for artificial bee colony algorithm. In: Zhang, H., Yang, Z., Zhang, Z., Wu, Z., Hao, T. (eds.) NCAA 2021. CCIS, vol. 1449, pp. 473–485. Springer, Singapore (2021). https://doi.org/10.1007/978-981-16-5188-5_34

17. Huo, Y., Zhuang, Y., Gu, J.J., Ni, S.R.: Elite-guided multi-objective artificial bee colony algorithm. Appl. Soft Comput. **32**, 199–210 (2015)
18. Xiang, Y., Zhou, Y.R.: A dynamic multi-colony artificial bee colony algorithm for multi-objective optimization. Appl. Soft Comput. **35**, 766–785 (2015)
19. Xiang, Y., Zhou, Y.R., Liu, H.L.: An elitism based multi-objective artificial bee colony algorithm. Eur. J. Oper. Res. **245**(1), 168–193 (2015)
20. Ye, T., Wang, H., Wang, W., Zeng, T., Zhang, L.: An improved bare-bones multi-objective artificial bee colony algorithm. In: Pan, L., Cui, Z., Cai, J., Li, L. (eds.) BIC-TA 2021. CCIS, vol. 1565, pp. 272–280. Springer, Singapore (2022). https://doi.org/10.1007/978-981-19-1256-6_20
21. Deb, K., Jain, H.: An evolutionary many-objective optimization algorithm using reference-point-based nondominated sorting approach, part I: solving problems with box constraints. IEEE Trans. Evol. Comput. **18**(4), 577–601 (2014)
22. Das, I., Dennis, J.E.: Normal boundary intersection: a new method for generating the Pareto surface in nonlinear multicriteria optimization problems. SIAM J. Optim. **8**(3), 631–657 (1998)
23. Zeng, T., et al.: Artificial bee colony based on adaptive search strategy and random grouping mechanism. Expert Syst. Appl. **192**, 116332 (2022)
24. Ye, T.Y., et al.: Artificial bee colony algorithm with efficient search strategy based on random neighborhood structure. Knowl. Based Syst. **241**, 108306 (2022)
25. Ye, T.Y., Wang, H., Wang, W.J., Zeng, T., Zhang, L.Q., Huang, Z.K.: Artificial bee colony algorithm with an adaptive search manner and dimension perturbation. Neural Comput. App. **34**, 16239–16253 (2022). https://doi.org/10.1007/s00521-022-06981-4
26. Wang, H., Wu, Z.J., Rahnamayan, S., Sun, H., Liu, Y., Pan, J.: Multi-strategy ensemble artificial bee colony algorithm. Inf. Sci. **27**, 587–603 (2014)
27. Zhang, Q., Li, H.: MOEA/D: a multi-objective evolutionary algorithm based on decomposition. IEEE Trans. Evol. Comput. **11**(6), 712–731 (2007)
28. Cheng, R., Jin, Y., Olhofer, M., Sendhoff, B.: A reference vector guided evolutionary algorithm for many-objective optimization. IEEE Trans. Evol. Comput. **20**(5), 773–791 (2016)
29. Jiang, S., Yang, S.: A strength Pareto evolutionary algorithm based on reference direction for multi-objective and many-objective optimization. IEEE Trans. Evol. Comput. **21**(3), 329–346 (2017)
30. Huband, S., Hingston, P., Barone, L., While, L.: A review of multi-objective test problems and a scalable test problem toolkit. IEEE Trans. Evol. Comput. **10**(5), 477–506 (2006)

Container Lead Seal Detection Based on Nano-CenterNet

Gang Zhang[✉], Jianming Guo, Qing Liu, and Haotian Wang

School of Automation, Wuhan University of Technology, Wuhan 430070, China
1480640365@qq.com

Abstract. In the process of container loading and unloading, manual inspection is still used in the process of container lead seal inspection, which has the problems of low efficiency, high labor cost, and high safety risk. Using visual object detection technology to replace manual lead seal automatic detection technology is an effective way to improve the efficiency of container operation. To address the problem of the tiny area of the seal in the image, the significant variation in scale, and the random location of its appearance, this paper proposes a Nano-CenterNet model. Based on the CenterNet, the lightweight feature extraction network is introduced, and the lightweight feature fusion network is added; the enhancement module was used to enhance the small object feature. The loss function of the algorithm is optimized to improve the imbalance between positive and negative samples. The Nano-CenterNet model was applied to the detection of container lead seals. The 3200 samples collected at the port entrance were used as the training set, and 400 samples were used as the test set. The measured precision rate was 96.5%, the recall rate was 95.4%, and the detection speed reached 18FPS, which met industrial application requirements.

Keywords: Nano-CenterNet · Container lead seal · Small object detect

1 Introduction

With the development of economic globalization, port transportation has become an essential link in global trade transportation. China's port container transport industry is developing rapidly and gradually moving toward automation and intelligence. In the port operation process, to verify the integrity of the container goods, The lead sealing and inspection of containers at the entrance bayonet is a link that cannot be ignored. At present, it still needs truck parking, and then the inspection work is carried out by field workers, which affects the efficiency of port automation operations. There are problems of high labor costs and high safety risks.

With the rapid development of deep learning and computer vision technology in recent years, visual algorithms based on deep learning can gradually replace human eyes to perform various detection tasks, with the advantages of high efficiency and broad application scenarios. Limited by the small scale and unobvious features of lead seal, there is no visual-based algorithm to replace manual inspection of the container lead

H. Zhang et al. (Eds.): NCAA 2022, CCIS 1637, pp. 210–221, 2022.
https://doi.org/10.1007/978-981-19-6142-7_16

seal. However, the small object detection algorithm based on deep learning is applied and studied in many other fields. For example, Ashraf [1] proposed a ground-to-air UAV detection algorithm based on the two-stage algorithm. Deshmukh [2] realized a real-time traffic sign detection algorithm based on the image segmentation method. Kampffmeyer [3] used deep convolutional neural networks for urban remote sensing image segmentation. BC Gonalves [4] proposed SealNet to implement seal detection from satellite images.

Due to the complex working conditions of the container terminal operation site, the angle and position of the camera are subject to various restrictions of the site situation. Usually, the camera has a high erection angle and large field of vision, resulting in smaller lead seal scale and large image resolution, which significantly increases the difficulty of object detection and puts forward higher requirements for the reasoning speed of the algorithm. Industry requires a minimum of 95% precision and recall for lead seal testing. Given many problems in container lead seal detection, this paper proposes the lightweight feature extraction network, which is used to accelerate the network reasoning speed, and the feature fusion network and enhancement module are introduced to enhance the small object feature. The network training process is optimized, and the data augment method is designed to improve the robustness of the network. Finally, the implementation of container lead seal inspection instead of field workers is realized, and the level of field automation is improved. Since the network proposed in this paper is very lightweight and optimized based on the CenterNet network, we refer to the improved network as Nano-CenterNet.

2 Container Lead Seal Detection Based on CenterNet

2.1 Image Analysis of Detection Target

The container lead seals that need to be detected are generally conventional high-protection seals, consisting of lock bodies and lock rods, typically colored cylindrical, locked in iron bolts behind the container box. With the help of a harbor staff, 3,600 sets of conventional seal images were collected. The number of images of different resolutions is shown in Table 1. The size distribution of the lead seal is shown in Fig. 1. The abscissa represents the size of the lead seal area, and the ordinate represents the number of lead seals. Analysis of the dataset, the check image collected by the pick-up is the following features:

- The resolution of the detected image is generally large, with a wide range, including 4096 × 2160, 1920 × 1080, 1280 × 960, 1280 × 720;
- The scale of lead sealing varies greatly, ranging from 247 to 3699;
- Due to the random location of the truck entering the entrance and the different camera erection angles at other entrances, the visual angle of the inspected image is variable, and the position of the lead seal in the image is random;
- When the color of the lead seal is similar to that of the box door, the characteristics of the lead seal are less noticeable. There are ropes, chains, and other foreign matters in the iron bolt behind the box door, which interfere with detecting the lead seal.

Table 1. Number of container lead seal samples of each resolution

Serial number	Resolution	Number of samples
1	4096 × 2160	1364
2	1920 × 1080	1157
3	1280 × 960	762
4	1280 × 720	317
Total	3600	

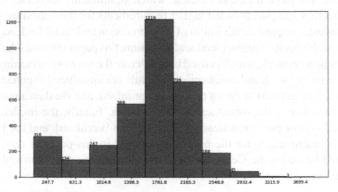

Fig. 1. Statistics of lead seal size

2.2 Container Lead Seal Detection Based on CenterNet

CenterNet [5] is a typical anchor-free method, which determines the bounding box by predicting the object's center point and the object width and height, and regards each element of the feature map as the center of the object. Compared with the anchor-based method, CenterNet is not affected by object scale, and does not need time-consuming post-processing operations such as NMS. It is a highly efficient object detector. The basic framework of CenterNet is shown in Fig. 2.

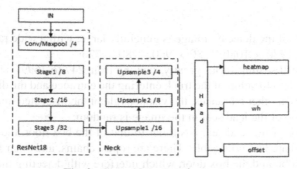

Fig. 2. CenterNet network structure

As shown in Fig. 2, CenterNet uses resnet18 as the feature extraction network to extract the 32 times down-sampled feature map, and then obtains the four times down-sampled feature map through the three times upsampling layer of the neck part, so as to increase the prediction accuracy of the object center point. Input the four times down-sampling feature map into the head part, and finally output the object center point coordinates, object width and height, and quantization error.

Due to the large sample resolution and tiny scale of the seal, if the original image is scaled directly into the network training, it will lead to severe loss or even disappearance of the seal features. Therefore, in the network training process, we first scaled the original image to 1024 × 540, and then randomly cropped a 512 × 512 size image block containing the seal as the network input; in the network inference process, we scaled the image to 1024 × 540, and then directly input the network to predict the detection results. Using CenterNet to train and test the dataset, the final precision rate was 87.8%, the recall rate was 89.6%, and the inference speed was 13 FPS, which could not meet the industrial demand.

In order to improve the detection accuracy of lead seals which is very small, this paper proposes Nano-CenterNet, which makes the following improvements on CenterNet:

1. Using ShuffleNetV2 as a feature extraction network to further improve the model inference speed.
2. Introducing a feature fusion network, using PANet as the feature fusion network and using the Ghost module instead of the basic convolution module of PANdet, which can enhance the prediction accuracy of the network for multi-scale objects and also reduce the large number of operations of PANet as much as possible.
3. Context detection module and CBAM module are added to the network. The context detection module can extract the contextual information around the small object and enhance the small object features; the CBAM module combines the channel attention mechanism and spatial attention mechanism to eliminate the background noise introduced to the small object features in the process of multi-scale feature fusion.
4. Introducing GHM loss function instead of Focal loss function to further balance the attention level of positive and negative samples as well as difficult and easy samples.
5. During the training process, the training samples are randomly panned. The boundary blank areas are filled with the training samples themselves to improve the generalization of the model and increase the number of positive samples.

3 Nano-CenterNet

3.1 ShuffleNetV2

ShuffleNetV1 uses point-wise group convolution to further reduce the number of model operations based on MobileNet, and uses the channel shuffle operation to improve the problem of lack of information flow between channels caused by point-wise group convolution, which significantly reduces the amount of network computation while ensuring computational accuracy. Based on ShuffleNetV1, Ma N [6] summarized four basic principles of lightweight network model design through theory and experiments:

1. The memory access cost is minimized when the input and output channel sizes are the same.
2. The use of group convolution increases the memory access cost.
3. Multi-branching of the network reduces parallelism and thus the speed of operations.
4. Element-level operations cannot be ignored. Element-level operations such as activation functions, summation, etc., tend to increase the memory access cost significantly.

According to the above four criteria, ShuffleNetV2 uses channel partitioning to split the input feature map into two parts by channel dimension, where one branch is directly mapped by constant mapping. In contrast, the other branch is operated by a depth-separable convolution module, thus ensuring that the input and output channels are identical. The point-wise grouping convolution operation in ShuffleNetV1 is re-moved. Feature fusion uses a concatenation operation instead of an element-level addition. The basic module of ShuffleNetV2 is shown in Fig. 3.

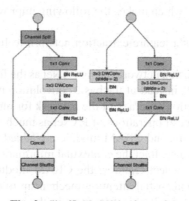

Fig. 3. ShuffleNetV2 basic module

The overall number of channels of ShuffleNetV2 is small, and the last layer of convolution is used for dimensionality enhancement. In the Nano-CenterNet model design, in order to enrich the shallow feature information while maintaining the lightweight of the model, this paper chooses to remove the last convolution layer and increase the number of channels in the lower layer accordingly. The network structure is shown in Table 2.

3.2 GhostPAN

The high-level feature map is rich in semantic features but low in positional information accuracy; the low-level feature map is simple in semantic features but rich in positional information. The first FPN proposed a feature pyramid network structure based on this characteristic for fusing multi-scale features. The feature pyramid network structure has a top-down network path, which can pass the high-level image features through the up-sampling method layer by layer and then fuse them with the low-level features. PANet

Table 2. ShuffleNetV2 network structure

Layer	Output size	Kernel size	Stride	Repeat	Output channels
ImageConv1	512×512				3
	256×256	3×3	2	1	24
MaxPool	128×128	3×3	2		
Stage2	64×64	5×5	2	1	132
	64×64	5×5	1	3	
Stage3	32×32	5×5	2	1	264
	32×32	5×5	1	7	
Stage4	16×16	5×5	2	1	528
	16×16	5×5	1	3	

[7] adds a bottom-up path to the FPN and adds a shortcut to each of the top-down and bottom-up paths to shorten the information transfer of the network and enhance feature fusion.

PANet has better performance but also adds a large number of operations, so we use the Ghost module instead of the basic convolution module of PANet. GhostNet [8] point out that redundancy of feature maps in the object detector is common. Such similar but not identical feature maps can ensure a more comprehensive understanding of the input data by the detector. Still, redundancy of feature maps can significantly increase the number of parameters and the number of operations in the network model. As shown in Fig. 4, the Ghost module divides the input feature map into two parts by channel, linearly transforms one part to obtain the redundant feature map, and then splices it with the other part, thus significantly reducing the computational effort of the redundant feature map.

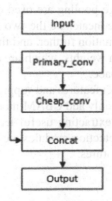

Fig. 4. GhostModule

3.3 Context Detection Module and CBAM Module

The context detection module [9] is divided into three branches, and the structure is shown in Fig. 5. The first branch contains a 3×3 convolution module, the second branch includes two 3×3 convolution modules, and the third branch contains three 3×3 convolution modules, feature maps are extracted from different perceptual fields by the three branches respectively and then stitched together as the final output, so the context detection module extracts and merges different ranges of contextual information of the object area and enriches the features of small objects.

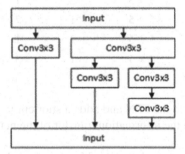

Fig. 5. Context detection module

The CBAM module [10] consists of two parts sequentially connected in series, the channel attention module and the spatial attention module, as shown in Fig. 6(a). Where the channel attention module is shown in Fig. 6(b), is to extract the global channel information by maximum global pooling and global average pooling of the input feature map, respectively, and then perform a nonlinear transformation through the fully connected network. The output of the transformation is summed and reactivated to obtain the channel attention weights. For the spatial attention module, as shown in Fig. 6(c), the maximum pooling and average pooling are used on the channels of the input feature map, respectively, and then concatenate the two feature maps together and use 7 $\times$ 7 convolution to fuse the information further, and finally use the sigmoid function to activate to get the spatial attention weights, which are used to strengthen or weaken the features at specific locations in space.

In order to enhance small object features and suppress background noise, while minimizing computational effort, we only add CBAM modules at the end of stage2, stage3 and stage4 of ShuffleNetV2 for extracting useful lead seal features and suppressing background noise. The context detection modules are added to the input of GhostPAN for extracting lead seal context features.

3.4 GHM Loss Function

GHM [11] proposed that the training process of the network has a high number of tough samples and elementary samples with high gradient densities, and that the sharing of losses by hard and easy samples should be reduced so as to reduce the problem of sample

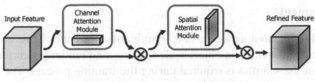

(a) Overall structure of CBAM

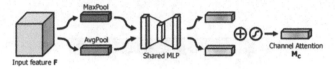

(b) Channel attention module

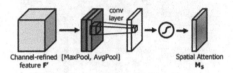

(c) Spatial attention module

Fig. 6. The structure of the CBAM module

imbalance, which is achieved by introducing a gradient density coordination parameter for the ith sample as follows.

$$\beta_{\varepsilon i} = \frac{N}{GD_{\varepsilon i}} \tag{1}$$

where gradient density:

$$GD_{\varepsilon i} = \frac{N_{\varepsilon i}}{l_{\varepsilon i}} \tag{2}$$

Denotes the sample format in the unit gradient modal length interval for the ith sample, where $N_{\varepsilon i} = \sum_{j=1}^{N} \delta_{\varepsilon}(g_j - g_i)$ is the number of samples whose gradient modulus length does not differ from g_i by more than $\frac{\varepsilon}{2}$, δ_{ε} is Instruction functions. The value is 1 in the interval $[-\frac{\varepsilon}{2}, \frac{\varepsilon}{2})$ and 0 for the rest. $g_i = |p_i - p_i^*|$ denotes the gradient modulus length of the ith sample; $l_{\varepsilon i}$ is the length of $[g_i - \frac{\varepsilon}{2}, g_i + \frac{\varepsilon}{2}] \cap [0, 1]$.

Ultimately, the GHM loss function used for classification is

$$L_{GHM-c} = \sum_{i=1}^{N} \frac{L_{CE}(p_i, p_i^*)}{GD(g_i)} \tag{3}$$

where N is the total number of samples.

GHM can adaptively adjust the sample loss weights according to the gradient density to further optimize the problem of positive and negative samples, hard and easy samples imbalance, and is insensitive to hyperparameters.

3.5 Data Augment

Due to the different shooting angles and significant differences in shooting resolution resulting in large differences in object scale features and random occurrence locations, data augment of the samples is required during the training process. The data augment mainly random translation and random scaling, and to increase the number of small objects, we use the image itself for boundary filling. The effect of the data augment is shown in Fig. 7.

Fig. 7. Data augment effect

4 Experiments and Analysis

Keeping the same training conditions such as input image size, learning strategy, optimizer, epoch size, etc., the effect of different improvement schemes on container lead seal detection was tested separately. The input image size was 512×512, the optimizer was AdamW, the learning rate was set to 0.01, the weight decay was set to 0.05, the test platform was Ubuntu 20.04, the graphics card was RTX 2080 8G, and a total of 50 epochs were trained. The experimental results are shown in Table 3.

From Table 3, it can be seen that:

- When the original CenterNet was used for container lead seal detection, the precision and recall rates were below 90%, which could not satisfy the inspection work of replacing manual workers.
- After replacing the feature extraction network of CenterNet with ShuffleNetV2, the precision rate decreased by 0.9%, and the recall rate decreased by 0.6%. Still, the network inference speed increased by 9 FPS, indicating that ShuffleNetV2 significantly improved the inference speed of the network at the expense of a small amount of precision.

Table 3. Comparison of network performance under different improvements

Group	Model	Precision	Recall	FPS
1	CenterNet	87.8%	89.6%	13
2	CenterNet+ShuffleNetV2	86.9%	89.0%	22
3	CenterNet+ShuffleNetV2+GhostPAN	91.3%	92.6%	19
4	CenterNet+ShuffleNetV2+GhostPAN+Context module	92.4%	93.5%	19
5	CenterNet+ShuffleNetV2+GhostPAN+Context module+CBAM	93.2%	94.0%	18
6	CenterNet+ShuffleNetV2+GhostPAN+Context module+CBAM+GHM	95.8%	93.6%	18
7	CenterNet+ShuffleNetV2+GhostPAN+Context module+CBAM+GHM+Data Augment	96.5%	95.4%	18

- The addition of GhostPAN significantly improved the network precision, with a 4.4% increase in precision and a 3.6% increase in recall, thanks to the enhanced multi-scale feature fusion, while the feature pyramid partitioning strategy im-proved the network's detection capability for objects at different scales. The lightweight GhostModule, with the addition of two branches from top-down and bottom-up, still kept the network inference speed at 19 FPS, which guaranteed the rate of inference.
- The addition of the context detection module to each of the input sections of GhostPAN increased the precision rate by 1.1% and the recall rate by 0.9%, indicating that the context detection module was able to extract valid contextual features and provide some enhancement to the lead seal features. Due to the need to control the inference speed of the network, a small number of context detection modules can only provide a limited enhancement.
- CBAM module was added after each stage of ShuffleNetV2 to adaptively focus on feature regions of small objects, suppressing background noise to some extent and increasing the feature weights of small objects regions, with a precision improvement of 0.8% and a recall improvement of 0.5%
- Using the GHM loss function instead of Focal loss, the network reduces the focus on easy and difficult samples during training, compensating for the lack of focus on outliers in Focal loss, resulting in a 2.6% increase in precision and a slight drop in the recall, down 0.4%.
- The addition of random panning as well as boundary padding effectively reduces the variability between different samples, greatly increases the number of lead-seal targets, and improves the robustness of the network to lead-seal locations and angles. The precision rate increased by 0.7%, and the recall rate increased by 1.8%.

The Nano-CenterNet module designed in this paper is applied to detecting lead seals during actual loading and unloading operations at container terminals. The actual results are shown in Fig. 8, where the detection frame is still relatively accurate even though the target is tiny.

Fig. 8. Nano-CenterNet container lead seal inspection results

5   Conclusion

This paper proposes a Nano-CenterNet model, based on the CenterNet, introducing ShuffleNetV2 as a feature extraction network and GhostPAN as a feature fusion network, adding a context detection module and CBAM module to enhance small object features and suppress background noise, using the GHM loss function to improve the sample imbalance problem. A random translation and boundary filling data augment method is designed to improve the algorithm's robustness. In the end, the Nano-CenterNet improves the precision rate by 8.7%, the recall rate by 5.8%, the inference speed by 5 FPS, the detection frame positioning precision, and the average IOU by 0.8, which can meet the demand of port container lead seal inspection.

References

1. Ashraf, M.W., Sultani, W., Shah, M.: Dogfight: detecting drones from drones videos (2021)
2. Deshmukh, V.R., Patnaik, G.K., Patil, M.E.: Real-time traffic sign recognition system based on colour image segmentation. Int. J. Comput. Appl. **83**(3), 30–35 (2013)
3. Kampffmeyer, M., Salberg, A.B., Jenssen, R.: Semantic segmentation of small objects and modeling of uncertainty in urban remote sensing images using deep convolutional neural networks. In: Proceedings of the IEEE Conference on Computer Vision and Pattern Recognition Workshops, pp. 1–9 (2016)
4. Gonalves, B.C., Spitzbart, B., Lynch, H.J.: SealNet: a fully-automated pack-ice seal detection pipeline for sub-meter satellite imagery. Remote Sens. Environ. **239**, 111617 (2020)
5. Zhou, X., Wang, D., Krhenbühl, P.: Objects as points (2019)
6. Ma, N., Zhang, X., Zheng, H.-T., Sun, J.: ShuffleNet V2: practical guidelines for efficient CNN architecture design. In: Ferrari, V., Hebert, M., Sminchisescu, C., Weiss, Y. (eds.) Computer Vision – ECCV 2018. LNCS, vol. 11218, pp. 122–138. Springer, Cham (2018). https://doi.org/10.1007/978-3-030-01264-9_8
7. Liu, S., Qi, L., Qin, H., et al.: Path aggregation network for instance segmentation. In: Proceedings of the IEEE Conference on Computer Vision and Pattern Recognition, pp. 8759–8768 (2018)
8. Han, K., Wang, Y., Tian, Q., et al.: GhostNet: more features from cheap operations. In: Proceedings of the IEEE/CVF Conference on Computer Vision and Pattern Recognition, pp. 1580–1589 (2020)
9. Najibi, M., Samangouei, P., Chellappa, R., et al.: SSH: Single-stage headless face detector. In: Proceedings of the IEEE International Conference on Computer Vision, pp. 4875–4884 (2017)
10. Woo, S., Park, J., Lee, J.-Y., Kweon, I.S.: CBAM: convolutional block attention module. In: Ferrari, V., Hebert, M., Sminchisescu, C., Weiss, Y. (eds.) ECCV 2018. LNCS, vol. 11211, pp. 3–19. Springer, Cham (2018). https://doi.org/10.1007/978-3-030-01234-2_1
11. Li, B., Liu, Y., Wang, X.: Gradient harmonized single-stage detector. In: Proceedings of the AAAI Conference on Artificial Intelligence, vol. 33, no. 01, pp. 8577–8584 (2019)

Backstepping Control of Air-Handling Unit for Indoor Temperature Regulation

Fang Shang[1](✉), Yongshuai Ji[1], Jingdong Duan[2], Chengdong Li[1](✉), and Wei Peng[1]

[1] Shandong Key Laboratory of Intelligent Buildings Technology, School of Information and Electrical Engineering, Shandong Jianzhu University, Jinan 250101, China
{shangfang,lichengdong,pengwei19}@sdjzu.edu.cn, yongshuai.ji@foxmail.com
[2] QingDao iESLab Electronic Co., LTD, Jinan, China
duanjingdong@ieslab.cn

Abstract. The control of indoor temperature has significant importance to maintain excellent thermal comfort and energy saving. At present, indoor temperature is mainly controlled by Air-Handling Unit (AHU), so a proper indoor temperature control strategy plays an important role in temperature regulation. In this paper, we consider the temperature regulation of a room effected by solar radiation, floor temperature and walls temperature. The nonlinear dynamic model of the room is analyzed, and the state space representation is given. Then, by introducing suitable state transformation, a new system is obtained which is convenient for controller design. Subsequently, the controller is designed for the new system using backstepping technique, and the control objective is realized for indoor temperature regulation. Finally, numerical simulations are provided to illustrate the effectiveness of the backstepping controller, and comparisons are also given.

Keywords: Indoor temperature · AHU · Controller design · Backstepping

1 Introduction

Energy has become the basis for the development of the society, thus energy consumption is a key concern for every country. At present, the building energy consumption accounts for about 40% of global energy consumption [1], and about 46% of the total energy consumption in China [2]. To maintain indoor temperature is one of the main reasons for building energy consumption. Therefore, it is important to design a suitable indoor temperature control strategy to maintain thermal comfort and save energy.

Nowadays, indoor temperature is mainly controlled by the Heating, Ventilation and Air Conditioning (HVAC) system, and researchers have kept working on investigating control strategies of HVAC system [3–7]. Due to the classicality and simplicity of PID control and on/off control, they are still the important control methods used in HVAC system control design. Cetin et al. used EnergyPlus to compare situations with and without on/off control in operation, and collected data in a room for verification, which showed that the HVAC system with on/off control was 19% of energy saving [3]. Xu et al. investigated the effect of on/off control on indoor temperature as well as airflow, and gave two solutions to deal with the problem of uneven indoor temperature [4]. PID control has been widely used in industry, which is also used in temperature control. Parameters of PID control have a great impact on control effect of the whole system. Most of the parameters rely on manual adjustments in practice, but they may not be the optimal ones. Many researchers explore the way of parameter optimization, such as using Bayesian to optimize PID parameters without human intervention [5]. In [6], a fast PID parameter tuning method was proposed by Almabrok et al., in combination with the Big-Bang-Big Crunch algorithm. Then it was implemented in FPGA devices for verification. G. Ulpiani et al. tested energy consumption on an experimental building with PID control, switching control, and fuzzy control [7]. Model Predictive Control (MPC) has also made much research progress in HVAC system [8–12], In [11,12], a detailed review of MPC applications in HVAC system was presented to discuss the impact of different MPC algorithms, and to highlight the advantages of MPC for building energy efficiency.

AHU is one of the key components of the HVAC system, which contacts the room directly. It is mainly used to regulate the indoor air quality and improve the indoor thermal comfort. A typical AHU system is a complex nonlinear system. The researchers usually use physics knowledge to develop a suitable mathematical model of the AHU system and design a controller for regulating indoor temperature. In [13], a simple controller was constructed based on pole configuration, and a robust controller was also designed to deal with model uncertainty. By analyzing, it is shown that the robust controller had better control performance and lower energy consumption than the simple one. In [14], an adaptive fuzzy logic controller was proposed based on genetic algorithm, to overcome the trade-off between stability and rise time in conventional PID controller. Setayesh et al. considered not only the control of temperature and humidity in the room, but the control of indoor CO_2 concentration [15]. In [16], Shah et al. designed a sliding mode controller to ensure the robustness of AHU with uncertainty in comparison with PID controller to highlight the effectiveness of the designed controller. In reality, other room temperatures may have impact on the temperature of the considered room. In [17], this case was considered, and a controller was studied using feedback linearization to regulate the room temperature.

On the other hand, more and more attention is paid to data-based control methods, especially machine learning methods, which do not require precise mathematical models. Biemann et al. evaluated the simulation of different reinforcement learning algorithms using Energyplus, and the results showed that all the reinforcement learning algorithms could achieve the room temperature control with good robustness. The algorithms reduce energy consumption by more than 13% compared to other algorithms [18]. And in [19], Demirezen et al. introduced a new Recurrent Neural Network (RNN) model for temperature control and designed an Integral-Proportional Differential (I-PD) controller to improve the stability of control. Meanwhile, the combination of machine learning methods with other control approaches has also been studied by many researchers [20–22].

According to the above discussion, the machine learning approach relies heavily on data, and it requires large amounts of data for learning training. Of course, it also takes much time for training. However, the model-based method avoids these problems. Currently, many researchers have developed various mathematical models for air conditioning systems. Some of them are relatively accurate, and we can design suitable controllers for these models to realize indoor temperature control. However, these mathematical models only consider the relevant parameters of the air conditioning system and do not analyze the whole room.

In this paper, indoor temperature regulation of a test room is investigated, considering the influence of the indoor temperature. For this test room, an appropriate nonlinear dynamic model is chosen and analyzed. And the flowchart of the controller design is shown in Fig. 1. First, the state transformation is introduced, then a new system is obtained which is convenient for controller design. Then, backstepping controller is constructed in a step-by-step manner. It is shown that,

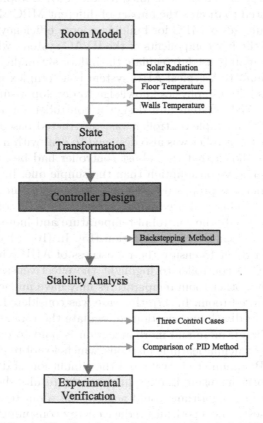

Fig. 1. The flowchart of the controller designed

by suitable choice of the design parameters, the indoor temperature is regulated to the desired temperature. Finally, simulation results are provided to illustrate the effectiveness of the proposed controller. Meanwhile, comparisons are made with PID controller to show the performance of the designed controller.

2   State Space Representation of the Test Room

The temperature of a building is affected by many factors, mainly including solar radiation, floor temperature, walls temperature, and changes of the indoor load. The temperature adjustment of the building is mainly carried out by the AHU system, which is an air treatment equipment widely used in public buildings and residential buildings. A schematic diagram of a typical test room is shown in Fig. 2. When the room temperature is low, hot air is sent into the room by AHU through the fan to increase the temperature. When the room temperature is high, cold wind is sent into the room by AHU for cooling. Finally, a comfortable indoor temperature can be obtained by the AHU system.

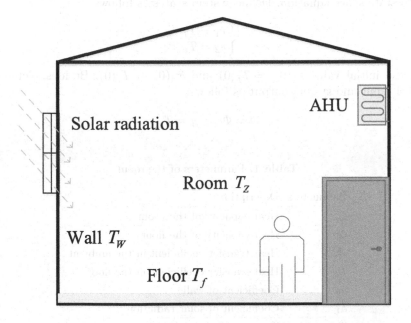

Fig. 2. Schematic diagram of the test room

The indoor environment is very complex, and in practice, indoor temperature is influenced by a variety of factors. So it is difficult to build an accurate model of the room considering all influencing factors. By considering the influence of solar radiation, walls temperature, floor temperature and some other factors on the room temperature, a dynamic model of the room is established based on the laws of thermodynamics and heat transfer: [23,24]:

$$\begin{cases} \dot{T}_f = \dfrac{1}{C_a} B_a(T_z - T_f) + \dfrac{1}{C_a} a_1 \Phi_s \\ \dot{T}_z = \dfrac{1}{C_i} a_0 B_w(T_w - T_z) + \dfrac{1}{C_i} B_a(T_f - T_z) + \dfrac{1}{C_i}(1 - a_1)\Phi_s + \dfrac{1}{C_i}\Phi_{ec} + \dfrac{1}{C_i}\Phi_r \end{cases}$$

(1)

where T_f and T_z represent the temperature of floor and room, respectively; Φ_r represents the power of AHU, Φ_{ec} is the power of others (for instance, the influence of the neighboring rooms) in the room. Other parameters of the room are listed in Table 1.

In reality, the room temperature is regulated by the AHU system. The objective is to design an appropriate controller, such that the floor and room temperature reaches the desired temperature. Suppose the desired temperature is T_t, and it has bounded second-order derivative.

System (1) is presented in the form of differential equation. We first convert it into state space equation. Define system states as follows:

$$\begin{cases} x_1 = T_f \\ x_2 = T_z \end{cases}$$

(2)

with the initial value $x_1(0) = T_f(0)$ and $x_2(0) = T_z(0)$. Besides, define the control input and system output as follows:

$$u = \Phi_r, \quad y = T_z$$

(3)

Table 1. Parameters of the room

Parameters	Description
C_i	Heat capacity of the room
C_a	Heat capacity of the floor
B_w	Heat transfer coefficient in the ambient
B_a	Heat transfer coefficient in the floor
a_0	Coefficient of wall
a_1	Coefficient of solar radiation
T_w	Temperature of the wall
Φ_s	Power of the solar radiation

Then, system (1) can be redescribed as:

$$\begin{cases} \dot{x}_1 = \alpha_1 - \beta_1 x_1 + \beta_1 x_2 \\ \dot{x}_2 = \alpha_2 + a_0\beta_2(T_w - x_2) + \beta_3(x_1 - x_2) + \beta_4 u \end{cases}$$

(4)

where $\alpha_1 = \frac{1}{C_a} a_1 \Phi_s$, $\alpha_2 = \frac{1}{C_i}(1 - a_1)\Phi_s + \frac{1}{C_i}\Phi_{ec}$, $\beta_1 = \frac{B_a}{C_a}$, $\beta_2 = \frac{B_w}{C_i}$, $\beta_3 = \frac{B_a}{C_i}$, and $\beta_4 = \frac{1}{C_i}$. As we know, solar radiation constantly exists, and solar radiation rate a_1 usually cannot reach full 100%. In addition, C_a, C_i, B_a and B_w are basic physical parameters that are non-zero. Therefore, α_1, α_2, and $\beta_i, i = 1, 2, \cdots, 4$ are nonzero constants.

For design convenience, the following state transformation is introduced:

$$\begin{cases} z_1 = x_1 - T_t \\ z_2 = x_2 - T_t \end{cases} \tag{5}$$

Then, a new system is obtained,

$$\begin{cases} \dot{z}_1 = \alpha_1 - \beta_1 z_1 + \beta_1 z_2 - \dot{T}_t \\ \dot{z}_2 = \alpha_2 + a_0 \beta_2 (T_w - T_t - z_2) + \beta_3(z_1 - z_2) + \beta_4 u - \dot{T}_t \end{cases} \tag{6}$$

This paper aims to design a controller for system (1), such that the floor and room temperature can be regulated to the desired temperature. According to the above transformation, it is clear that the problem of temperature regulation for system (1) can be solved by stabilization of system (6). Therefore, it suffices to design a stabilizing controller for system (6).

3   Control Design of the AHU System

Backstepping method is widely used in nonlinear system control [25–28], and the control idea is flexibly combining the design procedure of the controller with the selection of appropriate Lyapunov function. In this section, backstepping controller is constructed in a step-by-step manner to stabilize system (6).

3.1   Backstepping Controller Design

Step 1. Let $V_1 = \frac{1}{2} z_1^2$ be the Lyapunov function candidate for this step. Then the time derivative of V_1 along system (6) satisfies

$$\dot{V}_1 = z_1 \cdot \dot{z}_1 = z_1 \cdot (\alpha_1 - \beta_1 z_1 + \beta_1 z_2 - \dot{T}_t) \tag{7}$$

Define $\xi = z_2 - z_2^*$, where z_2^* is a virtual controller to be chosen as:

$$z_2^* = -\frac{1}{\beta_1}(\alpha_1 - \beta_1 z_1 + k_1 z_1 - \dot{T}_t) \tag{8}$$

where k_1 is a positive design parameter to be determined later.

Substituting (8) into (7) results in

$$\dot{V}_1 = -k_1 z_1^2 + \beta_1 z_1 \xi \tag{9}$$

Step 2. Let $V_2 = V_1 + \frac{1}{2}\xi^2$ be the Lyapunov function candidate for this step. Then the time derivative of V_2 along system (6) satisfies

$$\dot{V}_2 = \dot{V}_1 + \xi \cdot \dot{\xi} \tag{10}$$

From the definition of ξ, we know that

$$\dot{\xi} = \dot{z}_2 - \dot{z}_2^* \tag{11}$$

On the one hand, it is easily deduced from (6) that

$$\dot{z}_2 = \alpha_2 + a_0\beta_2(T_w - T_t - \xi - z_2^*) + \beta_3(z_1 - \xi - z_2^*) + \beta_4 u - \dot{T}_t$$

After carefully calculation, we have

$$\dot{z}_2 = \alpha_2 + a_0\beta_2 T_w + \frac{\alpha_1}{\beta_1}(a_0\beta_2 + \beta_3) - (a_0\beta_2 + \beta_3)\xi + (\frac{k_1}{\beta_1}a_0\beta_2 + \frac{k_1}{\beta_1}\beta_3 - a_0\beta_2)z_1$$
$$- a_0\beta_2 T_t + (-1 - \frac{1}{\beta_1}a_0\beta_2 - \frac{1}{\beta_1}\beta_3)\dot{T}_t + \beta_4 u \tag{12}$$

On the other hand, from (8), we have

$$\dot{z}_2^* = -\frac{1}{\beta_1}[(k_1 - \beta_1)\dot{z}_1 - \ddot{T}_t]$$
$$= -\frac{1}{\beta_1}[(k_1 - \beta_1)(\beta_1\xi - k_1 z_1) - \ddot{T}_t] \tag{13}$$

By (11), (12) and (13), we obtain

$$\dot{\xi} = \alpha_2 + a_0\beta_2 T_w + \frac{1}{\beta_1}a_0\alpha_1\beta_2 + \frac{1}{\beta_1}\alpha_1\beta_3 + (k_1 - \beta_1 - a_0\beta_2 - \beta_3)\xi$$
$$+ (k_1 - a_0\beta_2 - \frac{1}{\beta_1}k_1^2 + \frac{1}{\beta_1}a_0\beta_2 k_1 + \frac{1}{\beta_1}\beta_3 k_1)z_1$$
$$- a_0\beta_2 T_t + (-1 - \frac{1}{\beta_1}a_0\beta_2 - \frac{1}{\beta_1}\beta_3)\dot{T}_t - \frac{1}{\beta_1}\ddot{T}_t + \beta_4 u$$
$$= c_0 + c_1\xi + c_2 z_1 + c_3 T_t + c_4\dot{T}_t - \frac{1}{\beta_1}\ddot{T}_t + \beta_4 u \tag{14}$$

where $c_0 = \alpha_2 + a_0\beta_2 T_w + \frac{1}{\beta_1}a_0\alpha_1\beta_2 + \frac{1}{\beta_1}\alpha_1\beta_3$, $c_1 = k_1 - \beta_1 - a_0\beta_2 - \beta_3$, $c_2 = k_1 - a_0\beta_2 - \frac{1}{\beta_1}k_1^2 + \frac{1}{\beta_1}a_0\beta_2 k_1 + \frac{1}{\beta_1}\beta_3 k_1$, $c_3 = -a_0\beta_2$, $c_4 = -1 - \frac{1}{\beta_1}a_0\beta_2 - \frac{1}{\beta_1}\beta_3$. Obviously, $c_i, i = 0, 1, \cdots, 4$ are nonzero constants.

Thus, the following equation can be deduced from (10) and (14):

$$\dot{V}_2 = -k_1 z_1^2 + \beta_1 z_1\xi + \xi\left(c_0 + c_1\xi + c_2 z_1 + c_3 T_t + c_4\dot{T}_t - \frac{1}{\beta_1}\ddot{T}_t + \beta_4 u\right) \tag{15}$$

Now, we choose the actual controller as:

$$u = -\frac{1}{\beta_4}\left(c_0 + c_1\xi + c_2 z_1 + c_3 T_t + c_4\dot{T}_t - \frac{1}{\beta_1}\ddot{T}_t + k_2\xi + \beta_1 z_1\right) \tag{16}$$

where k_2 is a positive design parameter to be determined later.

Ultimately, controller (16) together with (15) results in

$$\dot{V}_2 = -k_1 z_1^2 - k_2\xi^2 \tag{17}$$

This complete the design procedure of the backstepping controller.

3.2   Main Results

In this subsection, we will give the main results of the paper and the corresponding proof. It can be clearly observed that $V_2 = \frac{1}{2}z_1^2 + \frac{1}{2}\xi^2$ is smooth, positive definite and radially unbounded. First of all, to guarantee the negative definiteness of $\dot{V}_2$, we have to choose appropriate design parameters k_1 and k_2.

At this point, we give the following theorem which is the summarization of the main contributions of the paper.

Theorem 1. *Consider system* (1). *Choose appropriate design parameters k_1 and k_2, which are positive constants. Then, the backstepping controller* (16) *guarantees that both the floor temperature T_f and the room temperature T_z can be regulated to the desired temperature T_t, exponentially.*

Proof. In order to guarantee the stability of the closed-loop system (6) and (16), we first choose the design parameters $k_1 > 0$ and $k_2 > 0$. By the system description, it is clear that the closed-loop system states are well defined on $[0, +\infty)$.

It is easily deduced from (17) that

$$\dot{V}_2 \leq -\min\{k_1, k_2\}(z_1^2 + \xi^2)$$

According to this and $V_2 = \frac{1}{2}(z_1^2 + \xi^2)$, we have

$$\dot{V}_2 \leq -2\min\{k_1, k_2\}V_2$$

Therefore, we have

$$V_2(z_1(t), \xi(t)) \leq V_2(z_1(0), \xi(0))e^{-2\min\{k_1, k_2\}t} \tag{18}$$

where the value of $V_2(z_1(0), \xi(0))$ depends on $z_1(0)$ and $\xi(0)$. According to (18), the states z_1 and ξ of the closed loop system exponentially approaches zero. This means that states z_1 and ξ are globally exponentially stable [29].

By (2) and (5), $z_1 = x_1 - T_t = T_f - T_t$, it is clearly known that $\lim_{t \to +\infty} T_f(t) = T_t$. Moreover, by (5), (6), and (8), the definition of $\xi = z_2 - z_2^*$, we know that $\lim_{t \to +\infty} z_2(t) = 0$. That is, $\lim_{t \to +\infty} T_z(t) = T_t$. Therefore, we can conclude that the temperature of the room and the floor finally reaches the desired temperature, exponentially.   □

4   Simulation Results and Discussions

In this section, in order to verify that the designed controller has a good control effect on indoor temperature, the control targets are set separately for different seasons. Besides, the results are compared with those using PID control. The corresponding parameters are listed in Table 2.

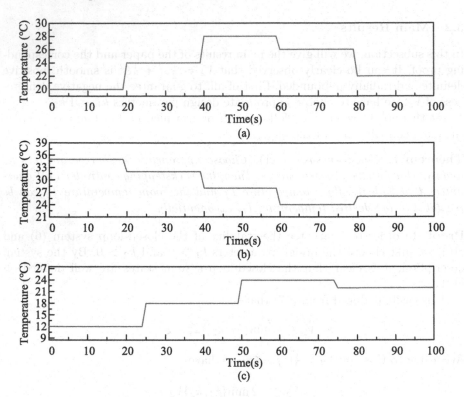

Fig. 3. Desired temperature in different seasons, (a) in spring and autumn, (b) in summer, and (c) in winter.

We take the case in Jinan, China, as an example, and we consider Fig. 3 to represent the expected temperatures of different seasons in Jinan. Figure 3 (a) shows the desired indoor temperature change in spring and autumn, and it is variable. The initial desired temperature is supposed to be 20 °C. As seen from this figure, the desired temperature rises twice and then it falls twice, which is conforming to the weather changes in Jinan. Figure 3 (b) represents the desired indoor temperature change in summer. Due to the hot weather and high temperature in summer, indoor cooling is required, which is executed by AHU system. Therefore, the desired indoor temperature gradually decreases from about 35 °C to eventually a suitable temperature. Figure 3 (c) represents the desired indoor temperature change in winter. In winter, the weather is cold and the temperature is low, the room needs to be heated by AHU system. Hence, the desired indoor temperature gradually increases from about 12 °C and finally stays around 23 °C.

Without loss of generality, we make the following assumptions: in spring and autumn, the average solar radiation is 253W and the initial room temperature is 22 °C. Figure 4 (a) represents the regulation effectiveness of indoor temperature in spring and autumn, and Fig. 4 (b) shows the control error of indoor temperature. From Fig. 4, it can be seen that the indoor temperature first expects to rise

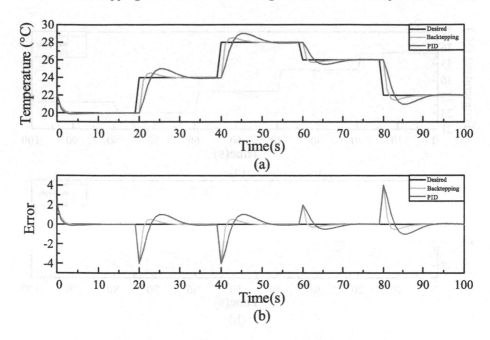

Fig. 4. Simulation results of spring and autumn, (a) represents the regulation effectiveness and (b) shows the control error of indoor temperature.

Table 2. Parameters value of the room

Symbol	Values	Unit
C_i	421.28	kJ/K
C_a	1531.85	kJ/K
B_w	12.65	$W/m^2 \cdot K$
B_a	542.18	$W/m^2 \cdot K$

and then fall, the designed controller is able to quickly take actions to achieve the target tracking. The control effect has faster response speed and control accuracy compared to PID control.

During summer, solar radiation is more powerful. Suppose it takes the value of 334W, and the indoor temperature is high, reaching 37 °C. Figure 5 (a) and (b) show the control tracking and control error in summer, respectively. From Fig. 5, the designed controller and PID control can both achieve rapid reduction of room temperature. However, the designed controller cools down much faster.

In fact, the controller can realize the rise in temperature, too. Figure 6 (a) and (b) indicate the control tracking and control error in winter, respectively, where the solar radiation value is 168W and the indoor temperature is 10 °C. From Fig. 6, we see that the designed controller realizes indoor temperature regulation faster with less overshoot, comparing with PID controller.

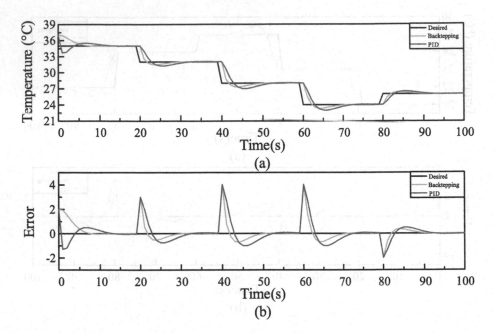

Fig. 5. Simulation results of summer, (a) represents the regulation effectiveness and (b) shows the control error of indoor temperature.

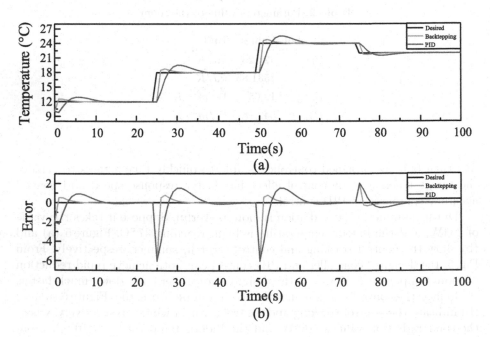

Fig. 6. Simulation results of winter, (a) represents the regulation effectiveness and (b) shows the control error of indoor temperature.

5 Conclusions

In this paper, we first consider the factors affecting the indoor temperature and choose the nonlinear dynamic model of the room. Next, the state space representation of this model is given. By introducing an appropriate state transformation, a new system is obtained which is convenience for controller design. Subsequently, the backstepping controller is constructed for the system, and the regulation of the indoor temperature is realized by choosing proper design parameters. Finally, the simulation is provided, which is carried out by MATLAB/SIMULINK software, to show the effectiveness of the designed controller. At the same time, the control targets are set separately for different seasons in Jinan. From simulation results, it is seen that the designed controller can achieve the control of indoor temperature in any season. Also, the proposed controller has higher accuracy, faster speed response and lower overshoot than PID control.

Acknowledgment. This study is partly supported by the National Natural Science Foundation of China (61903226, 61104069), the Key Research and Development Program of Shandong Province (No. 2021CXGC011205, 2021TSGC1053).

References

1. Brozovsky, J., Gustavsen, A., Gaitani, N.: Zero emission neighbourhoods and positive energy districts-a state-of-the-art review. Sustain. Cities Soc. **72**, 103013 (2021)
2. Xu, G., Wang, W.: China's energy consumption in construction and building sectors: an outlook to 2100. Energy **195**, 117045 (2020)
3. Cetin, K.S., Fathollahzadeh, M.H., Kunwar, N., et al.: Development and validation of an HVAC on/off controller in EnergyPlus for energy simulation of residential and small commercial buildings. Energy Build. **183**, 467–483 (2019)
4. Xu, Y.Q., Peet, Y.T.: Effect of an on/off HVAC control on indoor temperature distribution and cycle variability in a single-floor residential building. Energy Build. **251**, 111289 (2021)
5. Fiducioso, M., et al.: Safe contextual Bayesian optimization for sustainable room temperature PID control tuning. arXiv preprint arXiv:1906.12086 (2019)
6. Almabrok, A., Psarakis, M., Dounis, A.: Fast tuning of the PID controller in an HVAC system using the big bang-big crunch algorithm and FPGA technology. Algorithms **11**(10), 146 (2018)
7. Ulpiani, G., Borgognoni, M., Romagnoli, A., di Perna, C.: Comparing the performance of on/off, PID and fuzzy controllers applied to the heating system of an energy-efficient building. Energy Build. **116**, 1–17 (2016)
8. Raman, N.S., Devaprasad, K., Chen, B., et al.: Model predictive control for energy-efficient HVAC operation with humidity and latent heat considerations. Appl. Energy **279**, 115765 (2020)
9. Merema, B., Breesch, H., Saelens, D.: Comparison of model identification techniques for MPC in all-air HVAC systems in an educational building. In: E3S Web of Conferences, vol. 111, p. 01053. EDP Sciences (2019)
10. Wang, Z., Hu, G.: Economic MPC of nonlinear systems with nonmonotonic Lyapunov functions and its application to HVAC control. Int. J. Robust Nonlinear Control **28**(6), 2513–2527 (2018)

11. Serale, G., Fiorentini, M., Capozzoli, A., et al.: Model predictive control (MPC) for enhancing building and HVAC system energy efficiency: Problem formulation, applications and opportunities[J]. Energies **11**(3), 631 (2018)
12. Yao, Y., Shekhar, D.K.: State of the art review on model predictive control (MPC) in heating ventilation and air-conditioning (HVAC) field. Build. Environ. **200**, 107952 (2021)
13. Moradi, H., Bakhtiari-Nejad, F., Saffar-Avval, M.: Multivariable robust control of an air-handling unit: a comparison between pole-placement and H_∞ controllers. Energy Convers. Manage. **55**, 136–148 (2012)
14. Khan, M.W., Choudhry, M.A., Zeeshan, M., et al.: Adaptive fuzzy multivariable controller design based on genetic algorithm for an air handling unit. Energy **81**, 477–488 (2015)
15. Setayesh, H., Moradi, H., Alasty, A.: Nonlinear robust control of air handling units to improve the indoor air quality & co_2 concentration: a comparison between H_∞ & decoupled sliding mode controls. Appl. Thermal Eng. **160**, 113958 (2019)
16. Shah, A., Huang, D., Chen, Y., et al.: Robust sliding mode control of air handling unit for energy efficiency enhancement. Energies **10**(11), 1815 (2017)
17. Elnour, M., Meskin, N.: Multi-zone HVAC control system design using feedback linearization. In: 2017 5th International Conference on Control, Instrumentation, and Automation (ICCIA), pp. 249–254. IEEE (2017)
18. Biemann, M., Scheller, F., Liu, X., et al.: Experimental evaluation of model-free reinforcement learning algorithms for continuous HVAC control. Appl. Energy **298**, 117164 (2021)
19. Demirezen, G., Fung, A.S., Deprez, M.: Development and optimization of artificial neural network algorithms for the prediction of building specific local temperature for HVAC control. Int. J. Energy Res. **44**(11), 8513–8531 (2020)
20. Yang, S., Wan, M.P., Chen, W., et al.: Model predictive control with adaptive machine-learning-based model for building energy efficiency and comfort optimization. Appl. Energy **271**, 115147 (2020)
21. Yang, S., Wan, M.P., Chen, W., et al.: Experiment study of machine-learning-based approximate model predictive control for energy-efficient building control. Appl. Energy **288**, 116648 (2021)
22. Zou, Z., Yu, X., Ergan, S.: Towards optimal control of air handling units using deep reinforcement learning and recurrent neural network. Build. Environ. **168**, 106535 (2020)
23. Andersen, K.K., Madsen, H., Hansen, L.H.: Modelling the heat dynamics of a building using stochastic differential equations. Energy Build. **31**(1), 13–24 (2000)
24. Tashtoush, B., Molhim, M., Al-Rousan, M.: Dynamic model of an HVAC system for control analysis. Energy **30**(10), 1729–1745 (2005)
25. Fossen, T.I., Strand, J.P.: Tutorial on nonlinear backstepping: applications to ship control. (1999)
26. Kwan, C., Lewis, F.L.: Robust backstepping control of nonlinear systems using neural networks. IEEE Trans. Syst. Man Cybern. Part A Syst. Humans **30**(6), 753–766 (2000)
27. Fang, Y., Fei, J., Yang, Y.: Adaptive backstepping design of a microgyroscope. Micromachines **9**(7), 338 (2018)
28. Zhao, X., Wang, X., Zhang, S., et al.: Adaptive neural backstepping control design for a class of nonsmooth nonlinear systems. IEEE Trans. Syst. Man Cybern. Part A Syst. Humans **49**(9), 1820–1831 (2018)
29. Khalil, H.K.: Nonlinear Systems, 3rd edn., Prentice Hall, Upper Saddle River (2002)

Laplacain Pair-Weight Vector Projection with Adaptive Neighbor Graph for Semi-supervised Learning

Yangtao Xue and Li Zhang[✉]

The School of Computer Science and Technology and Joint International Research
Laboratory of Machine Learning and Neuromorphic Computing, Soochow University,
Suzhou 215006, China
zhangliml@suda.edu.cn

Abstract. Recently, Laplacian pair-weight vector projection (LapPVP) algorithm was proposed for semi-supervised classification. Although Lap-PVP achieves a good classification performance for semi-supervised learning, it may be sensitive to noise and outliers for using the neighbor graph with a fixed similarity matrix. To remedy it, this paper proposes a novel method named Laplacain pair-weight vector projection with adaptive neighbor graph (ANG-LapPVP), in which the graph induced by the Laplacian manifold regularization is adaptively constructed by solving an optimization problem. For binary classification problems, ANG-LapPVP learns a pair of projection vectors by solving the pair-wise optimal formulations in which we maximize the between-class scatter and minimize both the within-class scatter and the adaptive neighbor graph (ANG) regularization. The ANG regularization is to learn the ANG whose similarity matrix varies with iterations, which may solve the issue of LapPVP. Thus, ANG-LapPVP simultaneously learns adaptive similarity matrices and a pair of projection vectors with an iterative process. Experimental results on an artificial and real-world benchmark datasets show the superiority of ANG-LapPVP compared to the related methods. Thus, ANG-LapPVP is promising in semi-supervised learning.

Keywords: Semi-supervised learning · Binary classification · Manifold regularization · Adaptive neighbor graph

1 Introduction

Recently, more and more research attention acts in for semi-supervised classification tasks. On the basis of graph theory, the manifold regularization is becoming a popular technology to extend supervised learners to semi-supervised ones [2,4,6]. Considering the structure information provided by unlabeled data, semi-supervised learners outperform related supervised ones and have more applications in reality [1,3,12,15].

Because the performance of semi-supervised learners partly depends on the corresponding supervised ones, we need to choose a good supervised learner to

© The Author(s), under exclusive license to Springer Nature Singapore Pte Ltd. 2022
H. Zhang et al. (Eds.): NCAA 2022, CCIS 1637, pp. 235–246, 2022.
https://doi.org/10.1007/978-981-19-6142-7_18

construct a semi-supervised one. Presently, inspired by the idea of non-parallel planes, many supervised algorithms have been designed for dealing with binary classification problems, such as generalized proximal support vector machine (GEPSVM) [9,16], twin support vector machine (TSVM) [7], least squares twin support vector machine (LSTSVM) [8], multi-weight vector support vector machine (MVSVM) [18], and enhanced multi-weight vector projection support vector machine (EMVSVM) [17]. In these learners with non-parallel planes, each plane is as close as possible to samples from its own class and meanwhile as far as possible from samples belonging to the other class. Owing to the outstanding generalization performance of TSVM and LSTSVM, they have been extended to semi-supervised learning by using the manifold regularization framework [3,12]. In [12], Laplacian twin support vector machine (LapTSVM) constructs a more reasonable classifier from labeled and unlabeled data by integrating the manifold regularization. Chen et al. [3] proposed Laplacian least squares twin support vector machine (LapLSTSVM) based on LapTSVM and LSTSVM. Different from LapTSVM, LapLSTVM needs to solve only two systems of linear equations with remarkably less computational time. These semi-supervised learners have proved that the manifold regularization is a reasonable and effective technology. On the basis of EMVSVM and manifold regularization, Laplacian pair-weight vector projection (LapPVP) was extended to semi-supervised learning for binary classification problems [14]. LapPVP achieves a pair of projection vectors by maximizing the between-class scatter and minimizing the within-class scatter and manifold regularization.

These performance of semi-supervised learners is partly related to the neighbor graph induced by the manifold regularization. Generally, the neighbor graph is predefined and may be sensitive to noise and outliers [10]. To improve the robustness of LapPVP, we propose a novel semi-supervised learner, named Laplacain pair-weight vector projection with adaptive neighbor graph (ANG-LapPVP). ANG-LapPVP learns a pair of projection vectors by solving the pair-wise optimal formulations, where we maximize the between-class scatter and minimize both the within-class scatter and the adaptive neighbor graph (ANG) regularization. In ANG, the similarity matrix is not fixed but adaptively learned on both labeled and unlabeled data by solving an optimization problem [11,19,20]. Moreover, the between- and within-class scatter matrices are computed separably for each class, which can strengthen the discriminant capability of ANG-LapPVP. Therefore, it is easy for ANG-LapPVP to handle binary classification tasks and achieve a good performance.

2   Proposed Method

The propose method ANG-LapPVP is an enhanced version of LapPVP. In ANG-LapPVP, we learn an ANG based on the assumption that the smaller the distance between data points is, the greater the probability of being neighbors is. Likewise, ANG-LapPVP is to find a pair of the projection vectors by maximizing between-class scatter and minimizing both the within-class scatter and the ANG regularization.

Let $\mathbf{X} = [\mathbf{X}_\ell; \mathbf{X}_u] \in \mathbb{R}^{n \times m}$ be the training sample matrix, where n and m are the number of total samples and features, respectively; $\mathbf{X}_\ell \in \mathbb{R}^{\ell \times m}$ and $\mathbf{X}_u \in \mathbb{R}^{u \times m}$ are the labeled and unlabeled sample matrices, respectively; ℓ and u are the number of labeled and unlabeled samples, respectively, and $n = \ell + u$. For convenience, we use y_i to describe the label situation of sample $\mathbf{x}_i$. If $y_i = 1$, $\mathbf{x}_i$ is a labeled and positive sample; if $y_i = -1$, $\mathbf{x}_i$ is a labeled and negative sample; if $y_i = 0$, $\mathbf{x}_i$ is unlabeled. Furthermore, the labeled sample matrix $\mathbf{X}_\ell$ can be represented as $\mathbf{X}_\ell = [\mathbf{X}_1; \mathbf{X}_2]$, where $\mathbf{X}_1 = [\mathbf{x}_{11}, \mathbf{x}_{12}, \ldots, \mathbf{x}_{1\ell_1}]^T \in \mathbb{R}^{\ell_1 \times m}$ is the positive sample matrix with a label of 1, $\mathbf{X}_2 = [\mathbf{x}_{21}, \mathbf{x}_{22}, \ldots, \mathbf{x}_{2\ell_2}]^T \in \mathbb{R}^{\ell_2 \times m}$ is the negative sample matrix with a label of -1, $\ell = \ell_1 + \ell_2$, ℓ_1 and ℓ_2 are the number of positive and negative samples, respectively.

2.1   Formulations of ANG-LapPVP

For binary classification tasks, the goal of ANG-LapPVP is to find a pair of projection vectors similar to LapPVP. As mentioned above, the proposed ANG-LapPVP is an enhanced version of LapPVP. To better describe our method, we first briefly introduce LapPVP [14]. For the positive class, LapPVP is to solve the following optimization problem:

$$\max_{\mathbf{v}_1} \quad \mathbf{v}_1^T \mathbf{B}_1 \mathbf{v}_1 - \alpha_1 \mathbf{v}_1^T \mathbf{W}_1 \mathbf{v}_1 - \beta_1 \mathbf{v}_1^T \mathbf{X}^T \mathbf{L} \mathbf{X} \mathbf{v}_1$$
$$s.t. \quad \mathbf{v}_1^T \mathbf{v}_1 = 1 \tag{1}$$

where $\alpha_1 > 0$ and $\beta_1 > 0$ are regularization parameters, $\mathbf{L}$ is the Laplacian matrix of all training data, and $\mathbf{B}_1$ is the between-class scatter matrix and $\mathbf{W}_1$ is the within-class scatter matrix of the positive class, which can be calculated by

$$\mathbf{B}_1 = \left(\mathbf{X} - \mathbf{e}\mathbf{u}_1^T\right)^T \left(\mathbf{X} - \mathbf{e}\mathbf{u}_1^T\right) \tag{2}$$

and

$$\mathbf{W}_1 = \left(\mathbf{X}_1 - \mathbf{e}_1 \mathbf{u}_1^T\right)^T \left(\mathbf{X}_1 - \mathbf{e}_1 \mathbf{u}_1^T\right) \tag{3}$$

where $\mathbf{v}_1 \in \mathbb{R}^m$ is the projection vector of the positive class, $\mathbf{u}_1 = \frac{1}{\ell_1}\sum_{i=1}^{\ell_1} \mathbf{x}_{1i}$ is the mean vector of the positive samples, $\mathbf{e}_1 \in \mathbb{R}^{\ell_1}$ and $\mathbf{e} \in \mathbb{R}^\ell$ are the vectors of all ones with different length.

In the optimization problem (1), the laplacian matrix $\mathbf{L}$ is computed in advance and is independent of the objective function. The concept of ANG was proposed in [11], which has been applied to feature selection for unsupervised multi-view learning [19] and semi-supervised learning [20]. We incorporate this concept into LapPVP and form ANG-LapPVA.

Similarly, the pair of projection vectors of ANG-LapPVP is achieved by a pair-wise optimal formulations. On the basis of (1), the optimal formulation of ANG-LapPVP for the positive class is defined as:

$$\max_{\mathbf{v}_1, \mathbf{S}_1} \quad \mathbf{v}_1^T \mathbf{B}_1 \mathbf{v}_1 - \alpha_1 \mathbf{v}_1^T \mathbf{W}_1 \mathbf{v}_1 - \beta_1 (\mathbf{v}_1^T \mathbf{X}^T \mathbf{L}_{s_1} \mathbf{X} \mathbf{v}_1 + \gamma_1 \mathbf{S}_1^T \mathbf{S}_1)$$
$$s.t. \quad \mathbf{v}_1^T \mathbf{v}_1 = 1, \quad \mathbf{S}_1 \mathbf{e} = \mathbf{e}, \quad \mathbf{S}_1 > 0 \tag{4}$$

where $\mathbf{S}_1$ is the similarity matrix for the positive class, $\mathbf{L}_{s_1}$ is the Laplacian matrix related to $\mathbf{S}_1$ for the positive class, and $\gamma_1 > 0$ is a regularization parameter.

Compared with (1), (4) has a different term, or the third term, which is called the ANG regularization here. $\mathbf{S}_1$ varies with iterations, and then the Laplacian matrix $\mathbf{L}_{s_1} = \mathbf{D}_{s_1} - \mathbf{S}_1$ is changed, where $\mathbf{D}_{s_1}$ is a diagonal matrix with diagonal elements of $(D_{s_1})_{ii} = \sum_j (S_1)_{ij}$. The first and second terms represent the between- and within-class scatters of the positive class, and the regularization parameter α_1 is to make a balance between these two scatters. ANG-LapPVP can keep data points as near as possible in the same class while as far as possible from the other class by maximizing the between-class scatter and minimizing the within-class scatter.

For the negative class, ANG-LapPVP has the following similar problem:

$$\max_{\mathbf{v}_2, \mathbf{S}_2} \mathbf{v}_2^T \mathbf{B}_2 \mathbf{v}_2 - \alpha_2 \mathbf{v}_2^T \mathbf{W}_2 \mathbf{v}_2 - \beta_2 (\mathbf{v}_2^T \mathbf{X}^T \mathbf{L}_{s_2} \mathbf{X} \mathbf{v}_2 + \gamma_2 \mathbf{S}_2^T \mathbf{S}_2)$$

$$s.t. \quad \mathbf{v}_2^T \mathbf{v}_2 = 1, \quad \mathbf{S}_2 \mathbf{e} = \mathbf{e}, \quad \mathbf{S}_2 > 0 \tag{5}$$

where $\mathbf{v}_2$ is the projection vector for the negative class, α_2, β_2, and γ_2 are positive regularization parameters, $\mathbf{S}_2$ is the similarity matrix for the negative class, $\mathbf{L}_{s_2}$ is the Laplacian matrix related to $\mathbf{S}_2$, $\mathbf{B}_2$ and $\mathbf{W}_2$ are the between- and within-class scatter matrices for the negative class, respectively, which can be written as:

$$\mathbf{B}_2 = \left(\mathbf{X} - \mathbf{e}\mathbf{u}_2^T\right)^T \left(\mathbf{X} - \mathbf{e}\mathbf{u}_2^T\right) \tag{6}$$

and

$$\mathbf{W}_2 = \left(\mathbf{X}_2 - \mathbf{e}_2\mathbf{u}_2^T\right)^T \left(\mathbf{X}_2 - \mathbf{e}_2\mathbf{u}_2^T\right) \tag{7}$$

where $\mathbf{u}_2 = \frac{1}{\ell_2} \sum_{i=1}^{\ell_2} \mathbf{x}_{2i}$ is the mean vector of negative samples, $\mathbf{e}_2 \in \mathbb{R}^{\ell_c}$ is the vector of all ones.

2.2   Optimization of ANG-LapPVP

Problems (4) and (5) form the pair of optimization problems for ANG-LapPVP, where projection vectors $\mathbf{v}_1$ and $\mathbf{v}_2$ and similarity matrices $\mathbf{S}_1$ and $\mathbf{S}_2$ are unknown. It is difficult to find the optimal solution to them at the same time. Thus, we use an alternative optimization approach to solve (4) or (5). During the optimization procedure, we would fix a set of variables and solve the other set of ones.

When $\mathbf{S}_1$ and $\mathbf{S}_2$ are fixed, the optimization formulations of ANG-LapPVP can be reduced to

$$\max_{\mathbf{v}_1} \quad \mathbf{v}_1^T \mathbf{B}_1 \mathbf{v}_1 - \alpha_1 \mathbf{v}_1^T \mathbf{W}_1 \mathbf{v}_1 - \beta_1 \mathbf{v}_1^T \mathbf{X}^T \mathbf{L}_{s_1} \mathbf{X} \mathbf{v}_1$$

$$s.t. \quad \mathbf{v}_1^T \mathbf{v}_1 = 1 \tag{8}$$

and

$$\max_{\mathbf{v}_2} \quad \mathbf{v}_2^T \mathbf{B}_2 \mathbf{v}_2 - \alpha_2 \mathbf{v}_2^T \mathbf{W}_2 \mathbf{v}_2 - \beta_2 \mathbf{v}_2^T \mathbf{X}^T \mathbf{L}_{s_2} \mathbf{X} \mathbf{v}_2$$

$$s.t. \quad \mathbf{v}_2^T \mathbf{v}_2 = 1 \tag{9}$$

which are exactly LapPVP.

According to the way in [14], we can find the solutions $\mathbf{v}_1$ and $\mathbf{v}_2$ to (8) and (9), respectively. In [14], (8) and (9) can be respectively converted to the following eigenvalue decomposition problems:

$$\mathbf{B}_1\mathbf{v}_1 - \alpha_1\mathbf{W}_1\mathbf{v}_1 - \beta_1\mathbf{X}^T\mathbf{L}_{s_1}\mathbf{X}\mathbf{v}_1 = \lambda_1\mathbf{v}_1$$
$$\mathbf{B}_2\mathbf{v}_2 - \alpha_2\mathbf{W}_2\mathbf{v}_2 - \beta_2\mathbf{X}^T\mathbf{L}_{s_2}\mathbf{X}\mathbf{v}_2 = \lambda_2\mathbf{v}_2 \tag{10}$$

where λ_1 and λ_2 are eigenvalues for the positive and negative classes, respectively. Thus, the optimal solutions here are the eigenvectors corresponding to the largest eigenvalues.

Once we get the projection vectors $\mathbf{v}_1$ and $\mathbf{v}_2$, we take them as fixed variables, then solve $\mathbf{S}_1$ and $\mathbf{S}_2$. In this case, the optimization formulations of ANG-LapPVP are reduced to

$$\min_{\mathbf{S}_1} \quad \mathbf{v}_1^T\mathbf{X}^T\mathbf{L}_{s_1}\mathbf{X}\mathbf{v}_1 + \gamma_1\mathbf{S}_1^T\mathbf{S}_1$$
$$s.t. \quad \mathbf{S}_1\mathbf{e} = \mathbf{e}, \quad \mathbf{S}_1 > 0 \tag{11}$$

and

$$\min_{\mathbf{S}_2} \quad \mathbf{v}_2^T\mathbf{X}^T\mathbf{L}_{s_2}\mathbf{X}\mathbf{v}_2 + \gamma_2\mathbf{S}_2^T\mathbf{S}_2$$
$$s.t. \quad \mathbf{S}_2\mathbf{e} = \mathbf{e}, \quad \mathbf{S}_2 > 0 \tag{12}$$

For simplicity, let $(Z_1)_{ij} = ||\mathbf{v}_1^T\mathbf{x}_i - \mathbf{v}_1^T\mathbf{x}_j||^2$ and $(Z_2)_{ij} = ||\mathbf{v}_2^T\mathbf{x}_i - \mathbf{v}_2^T\mathbf{x}_j||^2$. Then matrices $\mathbf{Z}_1$ and $\mathbf{Z}_2$ are constant when $\mathbf{v}_1$ and $\mathbf{v}_2$ are fixed. Thus, (11) and (12) can be rewritten as:

$$\min_{\mathbf{S}_1} \quad \frac{1}{2}(\mathbf{S}_1 + \frac{1}{2\gamma_1}\mathbf{Z}_1)^T(\mathbf{S}_1 + \frac{1}{2\gamma_1}\mathbf{Z}_1)$$
$$s.t. \quad \mathbf{S}_1\mathbf{e} = \mathbf{e}, \quad \mathbf{S}_1 > 0 \tag{13}$$

and

$$\min_{\mathbf{S}_2} \quad \frac{1}{2}(\mathbf{S}_2 + \frac{1}{2\gamma_2}\mathbf{Z}_2)^T(\mathbf{S}_2 + \frac{1}{2\gamma_2}\mathbf{Z}_2)$$
$$s.t. \quad \mathbf{S}_2\mathbf{e} = \mathbf{e}, \quad \mathbf{S}_2 > 0 \tag{14}$$

Because (13) and (14) are similar, we describe the optimization procedure only for (13). First, we generate the Lagrangian function of (13) with multipliers δ_1 and ζ_1 as follows:

$$L(\mathbf{S}_1, \delta_1, \zeta_1) = \left(\mathbf{S}_1 + \frac{1}{2\gamma_1}\mathbf{Z}_1\right)^T\left(\mathbf{S}_1 + \frac{1}{2\gamma_1}\mathbf{Z}_1\right) - \delta_1(\mathbf{S}_1\mathbf{e} - \mathbf{e}) - \zeta_1\mathbf{S}_1 \tag{15}$$

According to the KKT condition [13], we derive the partial derivative of $L(\mathbf{S}_1, \delta_1, \zeta_1)$ with respect to the primal variables $\mathbf{S}_1$ and make it vanish, which results in

$$\mathbf{S}_1 = \delta_1 + \zeta_1 - \frac{1}{2\gamma_1}\mathbf{Z}_1 \tag{16}$$

Similarly, the similarity matrix $\mathbf{S}_2$ is achieved by

$$\mathbf{S}_2 = \delta_2 + \zeta_2 - \frac{1}{2\gamma_2}\mathbf{Z}_2 \tag{17}$$

where δ_2 and ζ_2 are positive Lagrange multipliers.

Since the constraint $\mathbf{Se} = \mathbf{e}$, we have

$$\delta_1 + \zeta_1 = \frac{1}{n} + \frac{1}{2n\gamma_1}\mathbf{Z}_1 \tag{18}$$

and

$$\delta_2 + \zeta_2 = \frac{1}{n} + \frac{1}{2n\gamma_2}\mathbf{Z}_2 \tag{19}$$

where the parameters γ_1 and γ_2 can be computed as follows [11]:

$$\gamma_1 = \frac{1}{n}\mathbf{e}^T\left(\frac{k}{2}\mathbf{Z}_1\mathbf{q}_k - \frac{1}{2}\mathbf{Z}_1\tilde{\mathbf{q}}_{k-1}\right) \tag{20}$$

and

$$\gamma_2 = \frac{1}{n}\mathbf{e}^T\left(\frac{k}{2}\mathbf{Z}_1\mathbf{q}_k - \frac{1}{2}\mathbf{Z}_1\tilde{\mathbf{q}}_{k-1}\right) \tag{21}$$

where k is the neighbor number in the graph, $\mathbf{q}_k$ and $\tilde{\mathbf{q}}_{k-1}$ are indicator vectors. We set $\mathbf{q}_k = [0,0,\cdots,0,1,0,\cdots,0,0]^T \in \mathbb{R}^n$ in which the k-th element is one and the others are zero and $\tilde{\mathbf{q}}_{k-1} = [1,1,\cdots 1,0,\cdots,0,0]^T \in \mathbb{R}^n$ in which the first $(k-1)$ elements are one and the others are zero.

2.3 Strategy of Classification

The pair of projection vectors $(\mathbf{v}_1,\mathbf{v}_2)$ can project data points into two different subspaces. The distance measurement is a reasonable way to estimate the class label of an unknown data point $\mathbf{x} \in \mathbb{R}^m$. Here, we define the strategy of classification using the minimum distance.

For an unknown point $\mathbf{x}$, we project it into two subspaces induced by $\mathbf{v}_1$ and $\mathbf{v}_2$. In the subspace induced by $\mathbf{v}_1$, the projection distance between $\mathbf{x}$ and positive samples is defined as:

$$d_1 = \min_{i=1,2,\cdots,\ell_1}\left(\mathbf{v}_1^T\mathbf{x} - \mathbf{v}_1^T\mathbf{x}_{1i}\right)^2 \tag{22}$$

In the subspace induced by $\mathbf{v}_2$, the projection distance between $\mathbf{x}$ and negative samples is computed as:

$$d_2 = \min_{i=1,2,\cdots,\ell_2}\left(\mathbf{v}_2^T\mathbf{x} - \mathbf{v}_2^T\mathbf{x}_{2i}\right)^2 \tag{23}$$

It is reasonable that $\mathbf{x}$ is taken as the positive point if $d_1 < d_2$, which is the minimum distance strategy. Thus, we assign a label to $\mathbf{x}$ by the following rule:

$$\hat{y} = \begin{cases} 1, & if \ \ d_1 \leq d_2 \\ -1, & Otherwise \end{cases} \tag{24}$$

2.4 Computational Complexity

Here, we analyze the computational complexity of ANG-LapPVP. Problems (4) and (5) are non-convex. The ultimate optimal pair-wise projection vectors are obtained by applying an iterative method. In iterations, the optimization problems of ANG-LapPVP can be decomposed into eigenvalue decomposition ones and quadratic programming ones with constraints.

The computational complexities of an eigenvalue decomposition problem and a quadratic programming one are $O\left(m^2\right)$ and $O\left(n^2\right)$, respectively, where m is the number of features and n is the number of samples. Let t be the iteration times. Then, the total computational complexity of ANG-LapPVP is $O\left(t\left(m^2 + n^2\right)\right)$. In the iteration process of ANG-LapPVP, the convergence condition is set as the difference between current and previous projection vectors, i.e., $||\mathbf{v}_1^t - \mathbf{v}_1^{t-1}|| \leq 0.001$ or $||\mathbf{v}_2^t - \mathbf{v}_2^{t-1}|| \leq 0.001$.

3 Experiments

We conduct experiments in this section. Firstly, we compare ANG-LapPVP with LapPVP on an artificial dataset to illustrate the improvement achieved by the ANG regularization. The comparison with other non-parallel planes algorithms on benchmark datasets is then implemented to analyze the performance of ANG-LapPVP.

3.1 Experiments on Artificial Dataset

An artificial dataset, called CrossPlane, is generated by perturbing points originally lying on two intersecting planes. CrossPlane contains 400 instances with only 2 labeled and 198 unlabeled ones for each class. The distribution of Cross-Plane is shown in Fig. 1. Obviously, some data points belonging to Class +1 are surrounded by the data points of Class −1 and vice versa.

Figure 2 plots the projection vectors learned by LapPVP and ANG-LapPVP. We can see that projection vectors learned by ANG-LapPVP are more suitable than LapPVP. The accuracy of LapPVP is 91.50%, and that of ANG-LapPVP is 97.00%. Clearly, ANG-LapPVP has a better classification performance on the CrossPlane dataset. In other words, ANG-LapPVP is robust to noise and outliers. In to all, the ANG regularization can improve the performance of LapPVP, which makes ANG-LapPVP better.

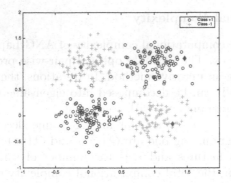

Fig. 1. Distribution of CrossPlane.

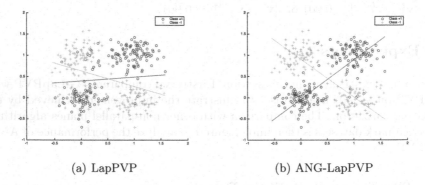

(a) LapPVP   (b) ANG-LapPVP

Fig. 2. Projection vectors obtained by LapPVP (a) and ANG-LapPVP (b).

3.2 Experiments on Benchmark Datasets

In the following experiments, we compare ANG-LapPVP with supervised algorithms, including GEPSVM, MVSVM, EMVSVM, TSVM and LSTSVM to evaluate the effectiveness of ANG-LapPVP, and compare it with semi-supervised algorithms (LapTSVM, LapLSTSVM and LapPVP) to verify the superiority of ANG-LapPVP on ten benchmark datasets. The benchmark datasets are collected from the UCI Machine Learning Repository [5]. We normalize the datasets so that all features range in the interval $[0, 1]$.

Each experiment is run 10 times with random 70% training data and the rest 30% test data. The average classification results are reported as the final ones. The grid search method is applied to finding the optimal hyper-parameters in each trial. Parameters β_1 and β_2 in both LapPVP and ANG-LapPVP are selected from $\{2^{-10}, 2^{-9}, \ldots, 2^0\}$, and other regularization parameters in all methods are selected from the set $\{2^{-5}, 2^{-4}, \ldots, 2^5\}$. In semi-supervised methods, the number of nearest neighbors is selected from the set $\{3, 5, 7, 9\}$.

Table 1. Mean accuracy and standard deviation (%) obtained by supervised algorithms with different scale of labeled data.

Dataset	Percent	GEPSVM	MVSVM	EMVSVM	TSVM	LSTSVM	ANG-LapPVP
Breast	10%	97.08 ± 1.15	95.92 ± 1.27	96.64 ± 0.91	96.02 ± 1.05	97.06 ± 0.66	**98.25 ± 1.18**
	30%	97.39 ± 0.98	96.73 ± 2.74	95.78 ± 2.08	**97.54 ± 1.16**	97.49 ± 0.87	**97.54 ± 0.80**
	50%	97.25 ± 0.66	96.58 ± 1.33	95.55 ± 1.82	97.87 ± 1.03	**97.91 ± 0.46**	97.44 ± 0.46
Check	10%	53.36 ± 1.82	54.68 ± 2.79	54.22 ± 2.43	61.13 ± 0.01	54.45 ± 3.30	**61.76 ± 3.15**
	30%	52.26 ± 1.50	51.46 ± 2.11	54.58 ± 3.76	59.80 ± 3.49	54.62 ± 3.58	**73.62 ± 3.82**
	50%	55.81 ± 2.07	50.53 ± 1.80	53.29 ± 2.70	55.61 ± 1.79	53.89 ± 1.79	**75.51 ± 2.34**
German	10%	71.13 ± 1.32	69.90 ± 1.21	60.00 ± 2.48	**73.80 ± 1.74**	73.60 ± 2.09	72.93 ± 1.72
	30%	74.63 ± 2.09	38.13 ± 8.83	61.87 ± 2.23	**76.10 ± 2.20**	75.97 ± 2.47	73.67 ± 1.04
	50%	75.00 ± 1.89	68.03 ± 5.16	61.13 ± 2.97	76.93 ± 1.00	**76.67 ± 1.72**	73.93 ± 0.90
Haberman	10%	66.88 ± 7.43	65.70 ± 7.12	58.82 ± 4.18	75.91 ± 3.17	76.02 ± 2.73	**77.67 ± 2.33**
	30%	71.72 ± 5.51	67.31 ± 6.04	55.70 ± 5.73	77.42 ± 2.27	78.17 ± 2.49	**78.71 ± 2.81**
	50%	75.81 ± 1.98	73.87 ± 3.51	55.91 ± 3.83	78.49 ± 2.03	77.96 ± 2.22	**79.14 ± 3.17**
Heart	10%	76.63 ± 7.69	73.04 ± 6.08	81.52 ± 3.90	67.61 ± 10.56	78.26 ± 4.38	**81.96 ± 3.77**
	30%	80.87 ± 2.36	60.33 ± 7.48	81.41 ± 3.18	80.89 ± 3.34	83.46 ± 3.40	**84.13 ± 2.77**
	50%	82.89 ± 2.08	82.39 ± 4.46	82.61 ± 1.85	82.65 ± 3.90	84.07 ± 3.60	**84.24 ± 2.47**
Hepatiti	10%	60.00 ± 8.79	67.23 ± 6.20	68.72 ± 3.62	60.43 ± 8.64	57.87 ± 8.26	**75.11 ± 2.00**
	30%	62.34 ± 7.31	64.47 ± 6.95	70.00 ± 5.16	62.77 ± 5.87	70.21 ± 4.91	**76.26 ± 3.81**
	50%	69.15 ± 1.81	67.23 ± 3.64	70.21 ± 4.60	68.09 ± 1.00	65.32 ± 3.02	**82.13 ± 2.69**
Liver	10%	58.56 ± 3.06	59.90 ± 5.07	57.60 ± 2.24	70.29 ± 1.15	68.65 ± 4.60	**71.35 ± 2.02**
	30%	57.50 ± 4.34	61.83 ± 2.87	55.10 ± 2.27	70.73 ± 2.87	69.52 ± 2.31	**71.06 ± 3.19**
	50%	61.83 ± 4.28	65.00 ± 2.80	52.02 ± 3.28	71.62 ± 1.82	70.29 ± 2.92	**72.40 ± 2.03**
Sonar	10%	62.03 ± 7.14	73.28 ± 7.81	61.41 ± 5.85	72.34 ± 3.04	67.81 ± 4.67	**74.84 ± 5.13**
	30%	73.13 ± 3.81	75.63 ± 2.11	62.34 ± 5.02	72.50 ± 3.70	63.28 ± 4.67	**79.22 ± 3.76**
	50%	80.47 ± 4.12	81.08 ± 5.37	59.22 ± 5.07	66.72 ± 6.08	72.50 ± 1.32	**81.41 ± 2.80**
Wdbc	10%	95.64 ± 2.16	94.77 ± 1.60	94.48 ± 1.00	84.48 ± 6.69	85.93 ± 4.49	**96.45 ± 0.89**
	30%	95.70 ± 1.26	79.07 ± 19.63	94.94 ± 0.99	96.51 ± 1.78	96.28 ± 1.51	**97.73 ± 0.70**
	50%	97.97 ± 0.88	96.86 ± 0.78	94.07 ± 2.14	97.38 ± 1.43	98.26 ± 0.27	**98.38 ± 0.92**
Wpbc	10%	76.23 ± 1.77	76.23 ± 4.84	66.07 ± 7.00	72.95 ± 4.59	68.20 ± 2.34	**78.69 ± 1.73**
	30%	75.08 ± 3.52	72.79 ± 3.39	63.93 ± 8.29	70.00 ± 8.07	71.31 ± 3.56	**80.49 ± 0.52**
	50%	77.70 ± 3.64	77.70 ± 2.07	63.93 ± 6.60	76.39 ± 1.15	79.67 ± 4.18	**80.67 ± 0.85**

Comparison with Supervised Algorithms. We first compare ANG-LapPVP with GEPSVM, MVSVM, EMVSVM, TSVM and LSTSVM to investigate the performance of adaptive neighbors graph. Specially, we discuss the impact of the different scale of labeled data on these algorithms. Additionally, ANG-LapPVP has 50% training samples as unlabeled ones.

Table 1 lists the results of supervised algorithms and ANG-LapPVP with 10%, 30% and 50% of training data as labeled samples, where the best results are highlighted. Experimental results in Table 1 show the effectiveness of ANG-LapPVP. With the increasing number of labeled data, the accuracy of ANG-LapPVP on most datasets goes up gradually, which indicates that the labeled data can provide more discriminant information. Moreover, we observe that ANG-LapPVP trained with unlabeled data has the best classification performance on all ten datasets except Breast with 50% labeled data and German with three situations, which fully demonstrates the significance of the adaptive

Table 2. Mean accuracy and standard deviation (%) obtained by semi-supervised algorithms on 30% unlabeled data.

Dataset	LapTSVM	LapLSTSVM	LapPVP	ANG-LapPVP
Breast	95.17 ± 1.58	97.30 ± 1.63	97.77 ± 0.95	**98.25 ± 1.18**
Check	50.90 ± 5.01	58.04 ± 3.36	60.00 ± 1.50	**75.78 ± 3.09**
German	70.17 ± 1.97	**73.87 ± 2.22**	70.93 ± 0.21	72.43 ± 1.16
Haberman	69.46 ± 3.59	74.52 ± 2.03	71.83 ± 7.07	**77.53 ± 2.51**
Heart	72.83 ± 6.87	81.41 ± 6.33	**84.02 ± 2.18**	81.96 ± 4.73
Hepatiti	57.87 ± 6.17	66.17 ± 6.22	73.40 ± 4.84	**75.11 ± 2.02**
Liver	59.23 ± 6.80	66.35 ± 3.42	60.19 ± 2.65	**70.67 ± 2.28**
Sonar	67.34 ± 4.51	74.22 ± 3.40	68.91 ± 5.28	**75.78 ± 4.12**
Wdbc	94.42 ± 1.82	96.51 ± 1.37	95.47 ± 0.77	**96.80 ± 0.63**
Wpbc	73.77 ± 5.01	79.02 ± 2.16	77.21 ± 0.52	**79.18 ± 1.74**

similarity matrices provided by labeled and unlabeled training data. Generally speaking, semi-supervised algorithms outperform the related supervised ones, and the proposed ANG-LapPVP gains the most promising classification performance.

Comparison with Semi-supervised Algorithms. To validate the superiority of ANG-LapPVP, we further analyze experimental results of LapTSVM, LapLSTSVM, LapPVP and ANG-LapPVP. Tables 2 and 3 list mean accuracy and standard deviation obtained by semi-supervised algorithms on 30% and 50% training samples as unlabeled ones, respectively, where the best results are in bold. Additionally, there are 20% training samples as labeled ones.

From the results in Tables 2 and 3, we can see that ANG-LapPVP has a higher accuracy than LapPVP on all ten datasets except Heart with 30% unlabeled data. The evidence further indicates that ANG-LapPVP with the ANG regularization well preserves the structure of training data and has a better classification performance than LapPVP. Moreover, compared with the other semi-supervised algorithms, ANG-LapPVP has the highest accuracy on eight datasets in Table 2 and on nine datasets in Table 3. That is to say, ANG-LapPVP has substantial advantages over LapTSVM and LapLSTSVM. On the whole, ANG-LapPVP has an excellent ability in binary classification tasks.

Table 3. Mean accuracy and standard deviation (%) obtained by semi-supervised algorithms on 50% unlabeled data.

Dataset	LapTSVM	LapLSTSVM	LapPVP	ANG-LapPVP
Breast	95.45 ± 2.00	97.77 ± 0.74	96.16 ± 1.54	**98.58 ± 0.55**
Check	50.86 ± 5.04	56.64 ± 2.64	59.63 ± 0.80	**61.76 ± 3.15**
German	71.17 ± 2.33	**74.10 ± 0.74**	72.67 ± 1.51	73.20 ± 1.00
Haberman	70.11 ± 4.35	66.45 ± 4.72	74.30 ± 2.12	**79.46 ± 2.89**
Heart	73.80 ± 7.02	77.93 ± 8.39	78.04 ± 5.22	**82.39 ± 2.28**
Hepatiti	59.79 ± 6.06	65.53 ± 5.92	67.66 ± 4.89	**71.06 ± 4.04**
Liver	59.62 ± 5.91	63.85 ± 5.48	62.21 ± 4.44	**71.35 ± 2.02**
Sonar	66.09 ± 3.76	72.03 ± 2.01	70.00 ± 1.77	**73.13 ± 5.25**
Wdbc	94.01 ± 1.40	94.88 ± 1.19	95.41 ± 0.75	**96.45 ± 0.89**
Wpbc	74.10 ± 3.26	90.68 ± 3.55	90.30 ± 1.28	**93.03 ± 3.27**

4 Conclusion

In this paper, we propose ANG-LapPVP for binary classification tasks. As the extension of LapPVP, ANG-LapPVP improves its classification performance by introducing the ANG regularization. The ANG regularization induces an adaptive neighbor graph where the similarity matrix is changed with iterations. Experimental results on the artificial and benchmark datasets validate that the effectiveness and superiority of the proposed algorithm. In a nutshell, ANG-LapPVP has a better classification performance than LapPVP and is a promising semi-supervised algorithm.

Although ANG-LapPVP achieves a good classification performance on datasets used here, the projection vectors obtained by ANG-LapPVP may be not enough when handling with a large scale dataset. In this case, we could consider projection matrices that may provide more discriminant information. Therefore, the dimensionality of projection matrices is a practical problem to be addressed in our following work. In addition, multi-class classification tasks in reality are also in consideration.

Acknowledgments. This work was supported in part by the Natural Science Foundation of the Jiangsu Higher Education Institutions of China under Grant Nos. 19KJA550002 and 19KJA610002, by the Priority Academic Program Development of Jiangsu Higher Education Institutions, and by the Collaborative Innovation Center of Novel Software Technology and Industrialization.

References

1. Belkin, M., Niyogi, P., Sindhwani, V.: Manifold regularization: a geometric framework for learning from labeled and unlabeled examples. J. Mach. Learn. Res. **7**, 2399–2434 (2006)

2. Chapelle, O., Schölkopf, B., Zien, A.: Introduction to semi-supervised learning. In: Chapelle, O., Schölkopf, B., Zien, A. (eds.) Semi-Supervised Learning, pp. 1–12. The MIT Press, Cambridge (2006)

3. Chen, W., Shao, Y., Deng, N., Feng, Z.: Laplacian least squares twin support vector machine for semi-supervised classification. Neurocomputing **145**, 465–476 (2014)

4. Culp, M.V., Michailidis, G.: Graph-based semisupervised learning. IEEE Trans. Pattern Anal. Mach. Intell. **30**(1), 174–179 (2008)

5. Dua, D., Graff, C.: UCI machine learning repository (2017). https://archive.ics.uci.edu/ml

6. Fan, M., Gu, N., Qiao, H., Zhang, B.: Sparse regularization for semi-supervised classification. Pattern Recogn. **44**(8), 1777–1784 (2011)

7. Jayadeva, Khemchandani, R., Chandra, S.: Twin support vector machines for pattern classification. IEEE Trans. Pattern Anal. Mach. Intell. **29**(5), 905–910 (2007)

8. Kumar, M.A., Gopal, M.: Least squares twin support vector machines for pattern classification. Expert Syst. Appl. **36**(4), 7535–7543 (2009)

9. Mangasarian, O.L., Wild, E.W.: Multisurface proximal support vector machine classification via generalized eigenvalues. IEEE Trans. Pattern Anal. Mach. Intell. **28**(1), 69–74 (2006)

10. Nie, F., Dong, X., Li, X.: Unsupervised and semisupervised projection with graph optimization. IEEE Trans. Neural Netw. Learn. Syst. **32**(4), 1547–1559 (2021)

11. Nie, F., Wang, X., Huang, H.: Clustering and projected clustering with adaptive neighbors. In: Macskassy, S.A., Perlich, C., Leskovec, J., Wang, W., Ghani, R. (eds.) The 20th ACM SIGKDD International Conference on Knowledge Discovery and Data Mining, KDD 2014, 24–27 August 2014, pp. 977–986. ACM, New York (2014)

12. Qi, Z., Tian, Y., Shi, Y.: Laplacian twin support vector machine for semi-supervised classification. Neural Netw. **35**, 46–53 (2012)

13. Vapnik, V.: Statistical Learning Theory. Wiley, Hoboken (1998)

14. Xue, Y., Zhang, L.: Laplacian pair-weight vector projection for semi-supervised learning. Inf. Sci. **573**, 1–19 (2021)

15. Yang, Z., Xu, Y.: Laplacian twin parametric-margin support vector machine for semi-supervised classification. Neurocomputing **171**, 325–334 (2016)

16. Yang, Z.: Nonparallel hyperplanes proximal classifiers based on manifold regularization for labeled and unlabeled examples. Int. J. Pattern Recognit. Artif. Intell. **27**(5), 1350015 (2013)

17. Ye, Q., Ye, N., Yin, T.: Enhanced multi-weight vector projection support vector machine. Pattern Recogn. Lett. **42**, 91–100 (2014)

18. Ye, Q., Zhao, C., Ye, N., Chen, Y.: Multi-weight vector projection support vector machines. Pattern Recogn. Lett. **31**(13), 2006–2011 (2010)

19. Zhang, H., Wu, D., Nie, F., Wang, R., Li, X.: Multilevel projections with adaptive neighbor graph for unsupervised multi-view feature selection. Inf. Fusion **70**, 129–140 (2021)

20. Zhong, W., Chen, X., Nie, F., Huang, J.Z.: Adaptive discriminant analysis for semi-supervised feature selection. Inf. Sci. **566**, 178–194 (2021)

A Collaborators Recommendation Method Based on Multi-feature Fusion

Qi Yuan[1], Lujiao Shao[1], Xinyu Zhang[1], Xinrui Yu[1], Huiyue Sun[1], Jianghong Ma[1], Weizhi Meng[2], Xiao-Zhi Gao[3], and Haijun Zhang[1](✉)

[1] Departmant of Computer Science, Harbin Institudte of Technology,
Shenzhen, China
hjzhang@hit.edu.cn

[2] Department of Applied Mathematics and Computer Science, Technical University
of Denmark, Kongens Lyngby, Denmark

[3] School of Computing, University of Eastern Finland, Kuopio, Finland

Abstract. This research introduces a new collaborators recommendation model based on multi-feature fusion. Specifically, we use a tree structure to integrate scholar information and extract content features from a scholar tree by using a Tree2vector-CLE model. Then, from the heterogeneous academic network, we extract the meta-path feature between scholars, which quantifies the similarity and co-operation potential between scholars from a multidimensional perspective. By combining content and meta-path features, we reconstruct a co-authorship network. Finally, we use the network representation learning method to represent the nodes in the reconstructed co-authorship network where the top-k collaborators are recommended for the target scholar with the random walk strategy controlled by the meta-path feature weighting. Experimental results on a real dataset demonstrate that our proposed method is effective in the task of collaborators recommendation.

Keywords: Collaborators recommendation · Scholars representation · Meta-path features · Network representation learning

1 Introduction

Information acquisition has become very convenient in the era of academic big data, and collaboration between researchers has become more convenient and frequent than ever. Scholars can inspire each other with ideas and help each other with resources by working together, thereby increasing the output of academic achievements. The rapid development of academic big data, on the other hand, brings the problem of information overload. Given the varying quality of network data, how to mine effective information for scholars to recommend collaborators has become a challenging problem.

Collaborators recommendations is a type of expert recommendations in the field of science that is used to recommend suitable collaborators for the target scholar. The recommendation task for collaborators can promote academic communication among scholars and improve research quality, making it a research

H. Zhang et al. (Eds.): NCAA 2022, CCIS 1637, pp. 247–261, 2022.
https://doi.org/10.1007/978-981-19-6142-7_19

task with high application value. Currently, collaborators recommendations primarily use two types of features: scholar content features and scholar network structure features [1]. For the extraction of features in the former kind, the existing investigations mainly focus on the information of scholars' published papers, personal information and keywords, but these approaches often overlook the inherent hierarchical structure and relationship of scholars information. For the extraction of features in the latter kind, the current studies mainly use network representation methods to represent scholars in the academic network where the larger similarity between the nodes suggests the higher relationship between the corresponding scholars. When extracting network structural features, the weight information of cooperative edges should be taken into account. If two authors have a cooperative relationship, it may be a weak cooperative edge. It should not be processed with a strong cooperative edge in this case. As a result, a worthwhile research direction is how to effectively integrate scholar information and explore valuable cooperative edges in academic networks.

In this paper, we propose a novel academic collaborators recommendation method based on multi-feature fusion to effectively integrate scholar information and explore high potential cooperative relationships in a unified model. First, a tree structure is used to integrate scholar information and extract content features from the scholar tree using a Tree2vector-CLE [2] model. Second, based on the academic heterogeneous network, we extract the meta-path features [3] between scholars. The meta-path features quantify the similarity and cooperation potential between scholars across multiple dimensions, and differentiate between strong and weak relationships in the cooperative network. Finally, the network representation learning method [4] is used to represent the features of the reconstructed network's nodes and mine high potential cooperative relationships. In particular, the contribution of this paper can be divided into two parts: 1) We construct a scholar tree, where the scholar information is connected through tree structure according to the hierarchical logical structure of scholar data; and 2) We propose a method to recommend collaborators for scholars based on the content and structure features of scholars. By using the proposed method, the scholar relationships in the co-authorship network through content features is enriched and the high-potential cooperative relationships through meta-path features is also mined.

The remainder of this article is organized as follows. Section 2 provides a brief overview of the related works. Next, we introduce the formulation of the proposed model for recommending collaborators in Sect. 3, followed by the comparative experiments conducted in Sect. 4. Section 5 concludes this article with suggestions for the future work.

2   Related Works

In this section, we review related state-of-the-arts for recommending collaborators. The existing collaborators recommendation methods are roughly categorized into three types: content-based methods, network based methods and hybrid methods.

1) Content-based methods: These methods recommend collaborators based on the similarity of scholar content features. If two scholars have similar content features, there is the high chance that they will be mutually recommended for academic collaboration. Gollapall et al. [5] extracted professional information from academic publications and homepages of scholars, and then proposed a model to calculate scholar similarity. Li et al. [6] hypothesized that when users utilize collaborative technologies, they will leave corresponding historical information which can be used to identify user situations and generate corresponding recommendations based on user context information. Zhang et al. [7] proposed a spatial-temporal restricted supervised learning model by combining publication time and academic impact information. Liu et al. [8] proposed the context-aware collaborator recommendation, which consists of a collaborative entity embedding network and a hierarchical decomposition model. The collaborative entity embedded network was used to capture the context-aware cooperation pattern of researchers as well as the potential semantics of topics. The hierarchical decomposition model extracted the activeness and conservatism of researchers, reflecting their level of academic cooperation and proclivity to collaborate with non-partners.

2) Network-based methods: Methods in this category make cooperator recommendation by measuring the similarity of two nodes in the network. Newman et al. [9] investigated many aspects of academic collaborative networks, such as the number of papers written by scholars, the number of authors per paper, and the number of collaborators that scholars have. Then, they built a co-authorship network among scholars, in which two scholars are connected if they have co-authored one or more papers. To represent a co-author network, Liu et al. [10] proposed a weighted directed network model and defined Author-Rank as the influence index of individual scholars in the network. Mohsen et al. [11] proposed a supervised random walk algorithm that assigns weights to network edges so that the random walk is more likely to visit nodes where new links will be created in the future. Chen et al. [12] used scholar text information to improve the content of cooperative networks in order to mine potential cooperative relationships.

3) Hybrid methods: This kind of methods considers both the content attributes and the structural information of scholars in the academic network. It combines the benefits of the two types of features to make recommendations. Rafiei et al. [13] proposed a combined model based on content analysis based on a concept map and social network analysis based on the PageRank algorithm. Pradhan et al. [14] proposed a multi-level fusion collaborator recommendation model that employs the topic model to extract scholar research interests and capture scholar dynamic research interests. They constructed an author network based on scholar similarities. The collaborators can be recommended through a random walk with restart model in this network. Kong et al. [15] proposed a model for recommending the most valuable collaborators, in which valuable collaborators who are excellent and active have relevant research interests and are capable of providing academic guidance to researchers. The model extracted scholar annual interest distribution as academic features and then recommended collaborators based on scholar influence.

3   Collaborators Recommendation Method Based on Multi-feature Fusion

This section introduces a method for recommending collaborators based on structural and content features of academic networks. First, the proposed collaborators recommendation framework will be illustrated. The extraction of content features, extraction of meta-path features, reconstruction of co-authorship networks, and representation of scholars based on reconstructed networks are then described sequentially.

3.1   Overview of Our Framework

The collaborators recommendation task can be defined as: given a target scholar, a group of scholars who are most likely to collaborate with the target scholar can be calculated using social network relationships and scholar attributes. The basic flow of the proposed method is illustrated in Fig. 1. First, we extract scholar papers from the database to create a scholar tree, and then we use the Tree2vector-CLE [2] model to extract scholar text features. The similarity between different scholars can be calculated by these text features. Second, we extract structural features from an academic heterogeneous network based on different meta-paths. Third, we reconstruct the original co-authorship networks by combining the similar edges and meta-path features among scholars, and obtain an academic collaborator network with enhanced multi-feature attributes. Finally, we use the Node2vec+ [16] to obtain the scholar feature representation through the edge weight controlled random walk strategy. The collaborative recommendations can then be made according to the similarity between scholar feature representations.

3.2   Content Features Extraction

In the face of an abundance of scholar information, how to effectively integrate the available scholar information is critical to the representation of scholar content features. We use a tree structure to integrate scholar information and Tree2Vector-CLE [2] to extract features of scholar tree in order to mine hierarchical information of scholars and effectively represent scholars.

We construct a three-layer scholar tree based on the relationship between scholar information. Figure 2 shows the tree structure of a scholar. The root node of the first layer represents a collection of the titles of papers published by the scholar. Each node in the second layer represents a paper published by the scholar. The feature of the node is extracted from the abstract of the paper. The number of nodes in this layer represents the total number of published papers. The notes in the third layer represent the titles of the reference cited in each paper in the second layer. As a result, the three kinds of scholar information form a scholar tree with the structure "paper collection -> paper abstract -> references." Each node in the tree contains information about the node index,

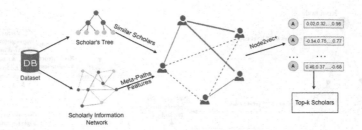

Fig. 1. The overview of the proposed collaborators recommendation framework (the solid lines indicate strong connections, i.e., direct cooperation, and the thickness of the lines indicates the frequency of the cooperation. The dash lines indicate weak connections, i.e., indirect cooperation).

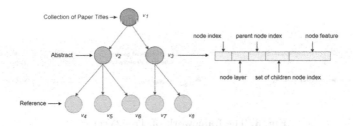

Fig. 2. Scholar representation by tree-structured feature.

node layer, parent node index, set of child node indexes, and node features. The original features of nodes in the scholar tree are text features. Here, we employ the Doc2vec [17] model to extract text features as a 100-dimensional low-dimensional vector.

After building the scholar tree, we used Tree2vector-CLE [2] to obtain vector representation of the scholar tree, with the model shown in Fig. 3. The local reconstruction model is built based on the parent node and children node. After obtaining the reconstruction coefficient, the parent node and children node are fused to obtain the new representation. The process starts from the bottom to up, and the root node is the final representation vector as scholar content features.

3.3 Meta-path Features Extraction

The traditional co-authorship network is a homogeneous network, with only one type of object (scholar) and one type of edge (cooperative relationship) in the network, whereas the real academic network has many types of objects (scholar, paper, venue, etc.) and many types of connections between these objects. We systematically define relationships in heterogeneous networks through meta-paths, and extract structural features in heterogeneous academic networks based on meta-paths. The meta-paths used in the model are shown in Table 1. Meta-path

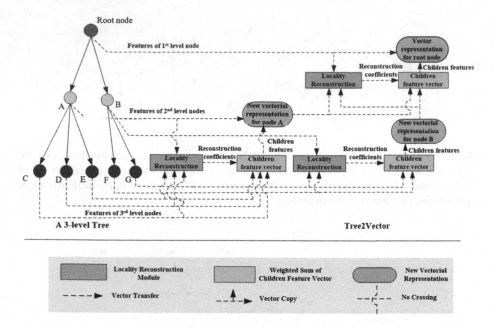

Fig. 3. The framework of Tree2Vector.

features can be classified into three types based on the type of meta-path in Table 1: direct reference similarity, co-reference similarity, and venue similarity.

Table 1. Meta-path description.

No.	Meta-path	Description
P_1	$A - P \rightarrow P - A$	a_i cites paper written by a_j
P_2	$A - P \leftarrow P - A$	Paper written by a_i is cited by a_j
P_3	$A - P - V - P - A$	a_i and a_j publish in the same venues
P_4	$A - P \rightarrow P \leftarrow P - A$	a_i and a_j cite the same paper
P_5	$A - P \leftarrow P \rightarrow P - A$	A paper cites both a_i and a_j

Direct Reference Similarity: Two scholars are highly likely to collaborate if they frequently cite each other, indicating that they have similar research fields and academic standards. The corresponding similarity is defined as follows:

$$Sim_{DR}(a_i, a_j) = PC_{P_1}(a_i, a_j) + PC_{P_1}(a_j, a_i) \qquad (1)$$

where $PC_{P_1}(a_i, a_j)$ is the number of paths following P_1 starting with a_i and ending with a_j, $PC_{P_1}(a_j, a_i)$ is the number of paths following P_1 starting with a_j and ending with a_i.

Co-Reference Similarity: If two scholars' papers are cited by the same paper, or if two scholars frequently cite the same paper, it indicates that they are likely to work in the same research field and may collaborate in the future. The corresponding similarity is defined as follows:

$$Sim_{CR}(a_i, a_j) = PC_{P_4}(a_i, a_j) + PC_{P_5}(a_i, a_j) \tag{2}$$

Venue Similarity: Venues play an important reference role in the cooperation between scholars. When scholars publish papers in the same venue, it reflects that their research fields and academic level are similar. Each scholar is represented by a vector $W_i \in R^{n \times 1}$, $n = \|V\|$. If scholar a_i has published a paper in venue l, then $W_{i,l} = 1$, otherwise $W_{i,l} = 0$. The corresponding similarity is defined as follows:

$$Sim_V(a_i, a_j) = \frac{\sum_{l=1}^{n}(W_{i,l} \times W_{j,l})}{\sqrt{\sum_{l=1}^{n} W_{i,l}^2} \times \sqrt{\sum_{l=1}^{n} W_{j,l}^2}} \tag{3}$$

It should be noted that the numerical range of the similarity of the three kinds of features is different. It is necessary to normalize the similarity. In reality, it is unlikely that the similarity between the two scholars is 1. Also, the random walk is sensitive to the edge weight. Therefore, we apply the normalization in the range of [0.1–0.9] as follows:

$$s_{norm} = \rho_1 + (\rho_2 - \rho_1)\frac{(s - s_{min})}{(s_{max} - s_{min})} \tag{4}$$

where $\rho_1 = 0.1$ and $\rho_2 = 0.9$.

The final meta-path features are shown as follows:

$$Sim_{MPF} = Sim'_{DR} + Sim'_{CR} + Sim'_{V} \tag{5}$$

3.4   Reconstruction of Co-authorship Network

A co-authorship network is represented as $G = (V, E)$. Scholars are represented as $V = \{v_1, v_2, ..v_m\}$, where m denotes the number of scholars. Relationships between scholars are represented as $E = \{(v_i, v_j)\}$, where (v_i, v_j) denotes v_i collaborated with v_j on one or more papers. Based on the original co-authorship network, we add similar edges based on text features and meta-path features of academic heterogeneous networks to construct an academic cooperative network with enhanced multi-feature attributes. The corresponding construction process is shown in Fig. 4: We first extract the Top-M scholars who are most similar to each other in research interests by calculating the similarity of content features among scholars. We add similarity edge E' into the original co-authorship network based on the sorting result of content features, and the construction formula is shown below:

$$e'_{ij} = \begin{cases} 1 & \text{if } sim(T_i, T_j) \text{ in Top-M} \\ 0 & \text{otherwise} \end{cases} \tag{6}$$

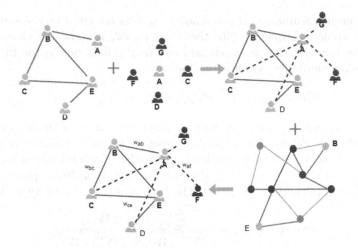

Fig. 4. The process of co-authorship network reconstruction.

If $e'_{ij} = 1$, an edge is added between scholar v_i and scholar v_j, which is a similar edge obtained based on scholar content features. Thus, the academic network is updated as $G' = (V, E \cup E')$ which combines contents features and structure information, enriching the information of co-authorship network, and lays the groundwork for the future feature fusion. Then, we calculate the meta-path features Sim_{MPF} of the two scholars as edge weight in the network to judge the strength attribute of cooperation between them.

The final multi-feature attribute enhanced academic network can be expressed as $G' = (V, E \cup E', W)$, where V is the node in the network, $E \cup E'$ is the cooperative edge integrating content features, W is the edge weight obtained based on heterogeneous academic information network and reflects the potential for collaboration among academics.

3.5   Representation of Scholar Based on Reconstructed Network

After obtaining the reconstructed network G', we use Node2vec+ [16] to obtain the low-dimensional vector representation of scholars. We hope that the representation of nodes with close distances or a similar structure in the network will be more similar in the academic cooperation network. Node2vec+ [16] performs biased random walk by adjusting the parameters q and p. When q is large, it prefers breadth first search and pays much attention to the structural equivalence of the network. When p is large, it prefers the depth first search and pays much attention to the homogeneity of the network. Node2vec+ [16] also considers the influence of edge weight on the probability of random walk, which can distinguish strong and weak cooperative edges in the reconstructed network.

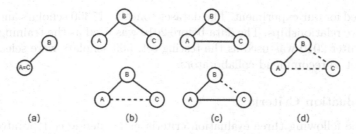

Fig. 5. Four connections in a network.

First, we define the strength of the cooperative edge. Suppose there is an edge (u, v). When $w(u, v) < \tilde{d}(u)$, (u, v) is a weak cooperative edge; when $w(u, v) \geq \tilde{d}(u)$, (u, v) is a strong cooperative edge, where $\tilde{d}(u) = \frac{\sum_{v' \in \mathcal{N}(u)} w(u, v')}{|\mathcal{N}(u)|}$. $\mathcal{N}(u)$ is the set of neighbors of u. According to the strength of cooperation among scholars, G' can be divided into four types as shown in figure Fig. 5, in which A is the last node of the random walk, B is the current node of the random walk, C is the next node visited by the random walk. The transfer probability of the final node B is shown in the following formula:

$$\alpha_{pq}(A, B, C) = \begin{cases} \frac{1}{p} & \text{if } A = C \\ \frac{1}{q} + \left(1 - \frac{1}{q}\right) \frac{w(C,A)}{\tilde{d}(C)} & \text{if } w(C,A) < \tilde{d}(C) \text{ and } w(B,C) \geq \tilde{d}(B) \\ 1 & \text{if } w(C,A) \geq \tilde{d}(C) \\ \min\left\{1, \frac{1}{q}\right\} & \text{if } w(C,A) < \tilde{d}(C) \text{ and } w(B,C) < \tilde{d}(B) \end{cases} \quad (7)$$

After the node sequence is obtained by the random walk, the final node representation is obtained by Skip-gram as:

$$\max_{f} \sum_{u \in V} \log \mathrm{P}\left(\mathcal{N}_S(u) \mid f(u)\right) \quad (8)$$

where $\mathcal{N}_S(u)$ is the set of neighbors of u, $f(u)$ final representation of u.

4 Experiment

In order to evaluate the performance of the proposed framework, we conducted intensive experiments on collaborators recommendation to exhibit the superiority of the proposed model compared to other related models.

4.1 Dataset

The experimental data is from DBLP-citation-NetworkV12 [18], which is extracted from DBLP, ACM and other sources. A total of 4,894,081 papers and 4,556,149 citations are included in this dataset. In this paper, a subset is constructed for our experiment. The dataset contains 15330 scholars and 123,201 cooperative relationships. The data before 2010 was used as the training set, and the data after 2010 was used as the testing set. 500 scholars were selected from the dataset to recommend collaborators.

4.2   Evaluation Criteria

We use the following three evaluation criteria as academic collaborator recommendation evaluation metrics.

Precision@k means the ratio of the number of scholars cooperating with target scholars to top-k recommended scholars. Precision@k is defined as follows:

$$Precision@k = \frac{1}{N} \sum_{i=1}^{N} \frac{|R| \cap |T|}{|R|} \quad (9)$$

where $|R|$ is the number of recommended samples and $|T|$ is the number of correct samples.

Recall@k means the ratio of the number of scholars cooperating with target scholars to all correct samples, which is defined as:

$$Recall@k = \frac{1}{N} \sum_{i=1}^{N} \frac{|R| \cap |T|}{|T|} \quad (10)$$

F1@k is the harmonic average value of Precision@k and Recall@k, which comprehensively reflects the results of Precision@k and Precision@k. The F1 value is computed as:

$$F1@k = \frac{2 \times Precision@k \times Recall@k}{Precision@k + Recall@k} \quad (11)$$

4.3   Results and Comparison

We compared MFRec with the following seven baselines:

1) DeepWalk [19]: DeepWalk is a network representation learning model that considers the relationship structure in academic cooperative networks. The more similar the structure of scholars in the academic network is, the more similar the node representation is.
2) GraphSAGE [20]: GraphSAGE is an unsupervised graph neural network model. In the unsupervised training, the vector representations of nodes with close distances are similar.
3) GATNE [21]: GATNE is a graph embedding model for heterogeneous networks. After training, we can obtain vector representations of nodes under different edge types.
4) RWR [22]: The RWR model starts from a certain starting point and returns to the starting point with a certain probability after each walk. After iteration, each node will get a correlation score, which is used to recommend collaborators.

5) CNEacR [12]: CNEacR is a cooperator recommendation model which considers semantic information of scholars and cooperative network relationship. The Content-enhanced cooperative network is constructed through the semantic relations of the scholars and the features of the scholars are obtained by network representation learning.
6) MRCR [14]: MRCR is an improved restart random walk model. First, the scholar network is constructed according to the similarity among scholars. Then, the edges are weighted with meta-path features to control the transition probability of the random walk.
7) MFRec-G: MFRec-G is a part of MFRec model, which only uses the structural information of cooperative network to recommend collaborators.

The experimental parameters of the MFRec model were set as follows: the number of semantic relations was set to 3, the random walk parameter p was set to 0.75, the parameter q was set to 1.5, and the dimension of feature representation was set to 128. Cosine similarity was used to calculate the similarity in the collaborators recommendation task.

Table 2. Comparative experiment results of collaborators recommendation (%).

Algorithm	Precision@k		Recall@k		F1@k	
	10	20	10	20	10	20
DeepWalk	14.98	10.11	13.77	18.58	14.35	13.10
GraphSAGE	16.84	11.84	15.51	21.76	16.14	15.33
GATNE	16.82	11.87	15.49	21.83	16.13	15.37
RMR	26.22	17.16	24.08	31.51	25.10	22.22
CNEacR	24.70	16.87	22.72	31.02	23.67	21.86
MRCR	**28.04**	17.93	25.75	32.93	26.84	23.21
MFRec-G (ours)	26.66	16.93	24.48	31.11	25.52	21.93
MFRec (ours)	28.01	**19.02**	**25.78**	**35.01**	**26.85**	**24.69**

The experimental results of the MFRec model and seven comparison algorithms are shown in Table 2. The proposed method outperforms the optimal baseline method by a substantial margin of 1.48% in F1@20. The table shows that the result of MFRec is better than that of MFREc-G, indicating that semantic relationships can be used to complement cooperative network. The results of MFRec are better than those of CNEacR, demonstrating that the meta-path feature can differentiate between strong and weak relationships and mine valuable cooperative relationships in cooperative networks. The performance of MFRec is also superior to RMR and MRCR, which prove that the random walk capable of distinguishing strong and weak relationships can effectively mine cooperative relationships in networks. The performance of DeepWalk is relatively poor indicating that cooperator recommendation is a relatively complex problem which is difficult to achieve favorable results through a network structure with only single relations. The results of MFRec are superior to GATNE and GraphSAGE, two graph neural network models. GATNE considers the influence of different edge types on node representation, but suitable training tasks are required to deliver better results. In the training process of GraphSAGE, the nodes with close distances have similar representation vectors. However, the similarity between the nodes with close distance and the similarity between nodes with similar structures should both be considered in the cooperative relationship mining. Figure 6 shows the results of a comparative experiment with different recommended lengths k. It can be seen that with the increase of k, the accuracy of the algorithms decreases continuously, while the recall rate increases continuously. The results of F1 increase gradually at first and then decrease slowly, which is consistent with the general situation of recommendation results. Under these three metrics, when the number of recommendations is greater than 10, MFRec performs better than all the compared algorithms, and when the recommendation length is 10, F1 of MFRec reaches the maximum value.

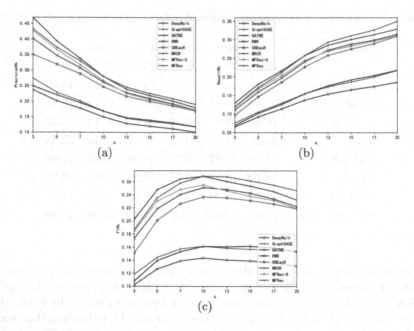

Fig. 6. Results of different methods on collaborators recommendation in terms of (a)Precision@k, (b)Recall@k and (c)F1@k

4.4 Parametric Study

The reconstruction of academic cooperative network involves the selection of the number of semantic relations, which are added to the academic network as a supplement of content features. This subsection conducted a sensitivity analysis on the influence of the number of semantic relations and recommendation results. The values of semantic relation m are set from 0 to 5 respectively.

Table 3. Recommendation results for different numbers of semantic relations (%).

Number of Semantic relationship	Precision@k		Recall@k		F1@k	
	10	20	10	20	10	20
0	26.66	16.93	24.48	31.11	25.52	21.93
1	27.46	18.05	25.25	33.16	26.30	23.37
2	27.86	18.63	25.65	34.26	26.71	24.13
3	**28.01**	**19.02**	**25.78**	**35.01**	**26.85**	**24.69**
4	27.78	18.89	25.54	34.75	26.61	24.47
5	27.45	18.89	25.30	34.78	26.34	24.48

The experimental results are shown in Table 3. When the number of semantic relationships is set at 3, the cooperator recommendation achieves the best result; when the number of semantic relationships exceeds 3, all indicators decline, but they are better than the recommendation results without semantic relationships. It implies that semantic relationships can enrich the information of cooperative network. However, when there are too many semantic relationships, these relationships will bring interference information to the reconstructed network due to the decrease of similarity of selected nodes.

Table 4. Recommendation results for combination of different meta-path features (%).

Method	Meta-path feature combination	Precision@k		Recall@k		F1@k	
		10	20	10	20	10	20
MFRec1	$None$	26.24	17.50	24.14	32.14	25.14	22.66
MFRec2	Sim_{DR}	26.48	18.58	24.37	34.18	25.38	24.08
MFRec3	Sim_V	26.59	18.53	24.46	34.09	25.48	24.01
MFRec4	Sim_{CR}	26.56	18.47	24.44	33.98	25.45	23.94
MFRec5	$Sim_{DR} + Sim_V$	26.94	18.54	24.81	34.10	25.83	24.03
MFRec6	$Sim_{DR} + Sim_{CR}$	26.76	18.62	24.61	34.22	25.64	24.12
MFRec7	$Sim_V + Sim_{CR}$	26.92	18.69	24.78	34.39	25.80	24.21
MFRec8	$Sim_{DR} + Sim_{CR} + Sim_V$	**28.01**	**19.02**	**25.78**	**35.01**	**26.85**	**24.69**

In addition, we combined three meta-path features to verify the influence of different meta-path features on the recommendation effect. Table 4 shows eight meta-path combinations. We carried out experiments on these 8 combination methods with experimental results shown in Table 4. It can be seen from the table that the best recommendation result is achieved when the three meta-path features are added together, while the recommendation performace of the method without meta-path features is poorer. The recommendation result using only one meta-path feature or combining two meta-path features is better than the recommendation result without meta-path feature, but is worse than the method combining three meta-path features.

5   Conclusion

In this paper, we introduced an academic cooperation recommendation method based on multi-feature fusion to explore high potential cooperative relationships in academic networks. The method can be divided into two steps: multi-feature attribute enhanced academic cooperative network constructing and cooperative network representation learning. The original cooperative network is reconstructed by using the semantic relations and meta-path features of scholars. The reconstructed network contains the content features of scholars and cooperation potential between scholars. Then, the vector representation of the scholar is extracted through the Node2Vec+, where the meta-path feature controls the transition probability of the random walk to excavate the network structure with high cooperative potential. We showed that the collaborator recommendation associated with MFRec is superior to other methods on DBLP data set. In the future work, we will focus on cross-disciplinary collaborator recommendations.

Acknowledgements. This work was supported in part by the National Natural Science Foundation of China under Grant no. 61972112 and no. 61832004, the Guangdong Basic and Applied Basic Research Foundation under Grant no. 2021B1515020088, the Shenzhen Science and Technology Program under Grant no. JCYJ20210324131203009, and the HITSZ-J&A Joint Laboratory of Digital Design and Intelligent Fabrication under Grant no. HITSZ-J&A-2021A01.

References

1. Roozbahani, Z., Rezaeenour, J., Emamgholizadeh, H., Bidgoly, A.: A systematic survey on collaborator finding systems in scientific social networks. Knowl. Inf. Syst. **62**(10), 3837–3879 (2020). https://doi.org/10.1007/s10115-020-01483-y
2. Zhang, H., Wang, S., Zhao, M., Xu, X., Ye, Y.: Locality reconstruction models for book representation. IEEE Trans. Knowl. Data Eng. **30**(10), 1873–1886 (2018)
3. Sun, Y., Barber, R., Gupta, M., Aggarwal, C., Han, J.: Co-author relationship prediction in heterogeneous bibliographic networks. In: Proceedings of the International Conference on Advances in Social Networks Analysis and Mining, pp. 121–128 (2011)
4. Cui, P., Wang, X., Pei, J., Zhu, W.: A survey on network embedding. IEEE Trans. Knowl. Data Eng. **31**(5), 833–852 (2018)

5. Gollapalli, S., Mitra, P., Giles, C.: Similar researcher search in academic environments. In: Proceedings of the IEEE-CS Joint Conference on Digital Libraries, pp. 167–170 (2012)
6. Li, S., Abel, M.-H., Negre, E.: Using user contextual profile for recommendation in collaborations. In: Visvizi, A., Lytras, M.D. (eds.) RIIFORUM 2019. SPC, pp. 199–209. Springer, Cham (2019). https://doi.org/10.1007/978-3-030-30809-4_19
7. Zhang, Q., Mao, R., Li, R.: Spatial-temporal restricted supervised learning for collaboration recommendation. Scientometrics 119(3), 1497–1517 (2016)
8. Liu, Z., Xie, X., Chen, L.: Context-aware academic collaborator recommendation. In: Proceedings of the ACM SIGKDD International Conference on Knowledge Discovery and Data Mining, pp. 1870–1879 (2018)
9. Newman, M.: Scientific collaboration networks. I. Network construction and fundamental results. Phys. Rev. E 64(1), 016131 (2001)
10. Liu, X., Bollen, J., Nelson, M., van de Sompel, H.: Co-authorship networks in the digital library research community. Inf. Process. Manag. 41(6), 1462–1480 (2005)
11. Backstrom, L., Leskovec, J.: Supervised random walks: predicting and recommending links in social networks. In: Proceedings of the ACM International Conference on Web Search and Data Mining, pp. 635–644 (2011)
12. Chen, J., Wang, X., Zhao, S., Zhang, Y.: Content-enhanced network embedding for academic collaborator recommendation. Complexity 2021, 1–12 (2021)
13. Rafiei, M., Kardan, A.: A novel method for expert finding in online communities based on concept map and PageRank. Hum.-Centric Comput. Inf. Sci. 5(1), 1–18 (2015)
14. Pradhan, T., Pal, S.: A multi-level fusion based decision support system for academic collaborator recommendation. Knowl.-Based Syst. 197, 105784 (2020)
15. Kong, X., Jiang, H., Wang, W., Bekele, T., Xu, Z., Wang, M.: Exploring dynamic research interest and academic influence for scientific collaborator recommendation. Scientometrics 113(1), 369–385 (2017)
16. Liu, R., Hirn, M., Krishnan, A.: Accurately modeling biased random walks on weighted graphs using Node2vec+. arXiv preprint arXiv 2109.08031 (2021)
17. Le, Q., Mikolov, T.: Distributed representations of sentences and documents. In: Proceedings of the International Conference on Machine Learning, pp. 1188–1196 (2014)
18. Sinha, A., et al.: An overview of microsoft academic service and applications. In: Proceedings of the International Conference on World Wide Web, pp. 243–246 (2015)
19. Perozzi, B., Al-Rfou, R., Skiena, S.: Deepwalk: online learning of social representations. In: Proceedings of the ACM SIGKDD International Conference on Knowledge Discovery and Data Mining, pp. 701–710 (2014)
20. Hamilton, W., Ying, R., Leskovec, J.: Inductive representation learning on large graphs. In: Proceedings of the International Conference on Neural Information Processing Systems, pp. 1025–1035 (2017)
21. Cen, Y., Zou, X., Zhang, J., Yang, H., Zhou, J., Tang, J.: Representation learning for attributed multiplex heterogeneous network. In: Proceedings of the ACM SIGKDD International Conference on Knowledge Discovery and Data Mining, pp. 1358–1368 (2019)
22. Konstas, I., Stathopoulos, V., Jose, J.: On social networks and collaborative recommendation. In: Proceedings of the International ACM SIGIR Conference on Research and Development in Information Retrieval, pp. 195–202 (2009)

Design of Online ESN-ADP for Dissolved Oxygen Control in WWTP

Yingxing Wan$^{(\boxtimes)}$, Cuili Yang, and Yilong Liang

Beijing University of Technology, Chaoyang District, Beijing, China
wanyx@emails.bjut.edu.cn, clyang5@bjut.edu.cn

Abstract. In this paper, the online optimal control is discussed to control the dissolved oxygen concentration (DO) in wastewater treatment process (WWTP). The echo state network (ESN) and the online adaptive dynamic programming (ADP) (ESN-ADP for short) are combine, which is adapted to the nonlinear and dynamic conditions of WWTP. Firstly, the ESN is proposed to replace the conventional BP network, which is not proper for the nonlinear, dynamic property of WWTP. Then, the online training algorithm is designed for ESN to ensure the adaptive adjustment ability of ADP controller. The simulation results illustrate that the ESN-ADP controller has obtained better control performance other control schemes.

Keywords: Online optimal control · Echo state networks · Online training · Wastewater treatment process

1 Introduction

Due to the serious shortage of fresh water resources, many wastewater treatment process (WWTP) plants has been built, which can be used to clean sewage and improve its recycling capacity. Because of the drastic changes of water discharge and pollutants, the WWTP can be regarded as a complex nonlinear industrial processing process, which shows strong nonlinear and dynamic characteristics [3]. Furthermore, the sludge bulking method has been widely used in WWTP, in which the dissolved oxygen (DO) concentration is important [1–4]. On one hand, the too high DO concentration will accelerate the consumption of organic matter in sewage, which reduces the flocculation performance and adsorption capacity of activated sludge. On the other hand, the lower DO concentration will inhibit the degradation of organic matter by organisms, which may produce sludge expansion [1]. Thus, how to realize the accurate control of DO concentration has become an important and difficult problem.

To control DO concentration, the Proportional-Integral-Derivative (PID) [5, 6] and many other strategies have been studied, such as the output feedback predictive control [7], feedforward control [7], and so on. For example, in [7], the feedforward-feedback and PID control scheme has been designed to control the DO value, in which the different control rules are made. These traditional control strategies always have simple design and are easily operated. However, the parameters of these method are always fixed, which cannot be adjusted according to the complex and changeable environment in WWTP.

Recently, the neural network based controller has been widely proposed, which has strong nonlinear approximation ability for nonlinear and dynamic system. For example, in [10,11], the fuzzy neural network based controller has been discussed and used to control DO. At the same time, the better control effect has been achieved than PID scheme. However, the fuzzy information processing is very simple, which may decrease control precision. In [8,9], the model predictive control has been proposed to solve DO control problems, which is based on neural network theory and modeling theory. However, the accurately model is required by MPC controller, which is difficult due to the nonlinear property of WWTP.

As an important branch of modern control theory, the adaptive dynamic programming (ADP) based control has been widely studied, which combines the adaptive evaluation algorithm, neural network and reinforcement learning. For example, the ADP controller is proposed in [12] to achieve the minimum cumulative tracking error of wastewater treatment plant. As compared with traditional PID controller, the control performance of ADP is greatly improved.

As an approximation of the ADP controller, the neural network can be considered as the most important part. Because when the neural network is used as an approximator, the learning process of the controller can be equivalently regarded as the process of network weight adjustment. Due to the nonlinear and dynamical property of WWTP, the training of neural network is a difficult problem. Fortunately, in recent years, the echo state network (ESN) has been proposed. In ESN, a randomly generated recursive pool is used to replace the recursive layer of the traditional recursive network. In the ESN training process, only the weights of the output layer need to be trained, and the weights in the recursive pool are randomly generated and fixed. In [13], the ESN and ADP (ESN-ADP) are combined to realize the online control of dissolved oxygen, the online recurrent least square (ORLS) algorithm is used to train the output weights of each ESN. However, the importance of system state is neglected. The accumulated error by historical date may reduce the control effectiveness. Thus, how to improve the control accuracy of ESN-ADP controller is still an unsolved problem.

To realize the online control of DO, the forgetting recursive least squares algorithm is proposed to train the output weights of each modles in ESN-ADP. The main contribution is as follows. Firstly, the forgetting technique is introduced to improve the performance of ESN. Secondly, the iterative and adaptive evaluation of online ESN-ADP controller is established. Finally, the application results in WWTP are given.

The rest of the paper is organized as follows. In Sect. 2, the preliminary knowledge about ADP and ESN has been presented. In Sect. 3, the online training algorithm of ESN-ADP is proposed. In Sect. 4, the ESN-ADP is operated in experiment. Some conclusions are given in Sect. 5.

2 Preliminary Knowledge

2.1 Echo State Network

The original ESN is shown in Fig. 1. The original ESN is constructed by three components, including the input layer, recursive reservoir and output layer. Let K, N and n are the dimension of input, recursive reserve pool and output. At time step k, the input-output formula can be described as below,

$$\mathbf{s}(t) = \sigma(\mathbf{W}_i \mathbf{z}(t) + \mathbf{W}_r \mathbf{s}(t-1)) \tag{1}$$

$$\mathbf{y}(t) = \mathbf{s}^T \mathbf{W}_o \tag{2}$$

where σ is the activation function, $\mathbf{s}(t) = [s_1(t), \ldots, s_N(t)] \in \mathbb{R}^N$ is the state of the reserve neuron, $\mathbf{z}(t) = [z_1(t), \ldots, z_K(t)]^T \in \mathbb{R}^K$ and $\mathbf{y}(t) = [y_t(t), \ldots, y_L(t)] \in \mathbb{R}^n$ are inputs and outputs, respectively. $\mathbf{W}_i$, $\mathbf{W}_r$, $\mathbf{W}_o$ stand for the input weight matrix, the internal weight matrix of recursive reserve and the output weight matrix. Only $\mathbf{W}_o$ should be calculated when train ESN, while $\mathbf{W}_i$ and $\mathbf{W}_r$ are randomly generated and keep constant during training process.

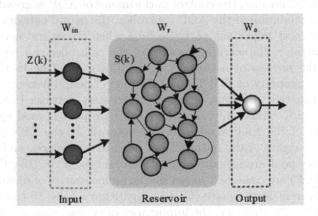

Fig. 1. The architecture of an original ESN.

The least square method is always used to calculated the output weight matrix $\mathbf{W}_o$ of ESN. Suppose there are L training samples, Let $\mathbf{D} = [\mathbf{d}(1), \cdots, \mathbf{d}(L)]^T$ is the ideal output value, set $\mathbf{H} = [\mathbf{s}(1), \mathbf{s}(2), \cdots, \mathbf{s}(L)]^T$, one obtains,

$$\mathbf{T} = \mathbf{H}\mathbf{W}_o \tag{3}$$

Then, the solution of Eq. (3) is represented as below

$$\mathbf{W}_o = (\mathbf{H}^T\mathbf{H})^{-1}\mathbf{H}^T\mathbf{D} \tag{4}$$

The Eq. (4) is always used to realize the offline training of ESN.

2.2 ESN-ADP

Generally speaking, the nonlinear discrete dynamics system can be described by,

$$\mathbf{x}(t+1) = F[\mathbf{x}(t), \mathbf{u}(t)] \tag{5}$$

where t is time step, $\mathbf{x}(t)$ is the system state at time step t, $\mathbf{u}(t)$ is the decision variable or control input, $F(\cdot)$ is nonlinear transformation function in which $\mathbf{x}(t)$ and $\mathbf{u}(t)$ are independent variables.

Let $J[\mathbf{x}(t), \mathbf{u}(t)]$ be the total discounted cost function as below,

$$J[\mathbf{x}(t), \mathbf{u}(t)] = \sum_{i=t}^{\infty} \gamma^{i-t} U[\mathbf{x}(i), \mathbf{u}(i)] \tag{6}$$

where $U[\mathbf{x}(i), \mathbf{u}(i)]$ is the cost function, $0 < \gamma \le 1$ is discount factor. Furthermore, Eq. (6) can also be rewritten as below,

$$J[\mathbf{x}(t), \mathbf{u}(t)] = U[\mathbf{x}(t), \mathbf{u}(t)] + \gamma J[\mathbf{x}(t+1), \mathbf{u}(t+1)] \tag{7}$$

To achieve the optimal control of system (5), a sequence of control quantities $\mathbf{u}^*(t) \in \mathbf{\Omega}(t = 0, 1, \cdots, \infty)$ should be chosen to minimize the total cost $J[\mathbf{x}(t), \mathbf{u}(t)]$,

$$J^*[\boldsymbol{x}(t), \boldsymbol{u}(t)] = \min_{\boldsymbol{u}(t) \in \boldsymbol{\Omega}} J[\boldsymbol{x}(t), \boldsymbol{u}(t)] \tag{8}$$

Based on Hamilton-Jacobi-Bellman (HJB) equation, the optimal control strategy should satisfies the following condition,

$$\mathbf{u}^*(t) = \arg \min_{\boldsymbol{u}(t) \in \boldsymbol{\Omega}} J[\boldsymbol{x}(t), \boldsymbol{u}(t)] \tag{9}$$

The working flow chart of ESN-ADP controller is shown in the Fig. 2. It contains three modules, namely actor-ESN, critical-ESN and model-ESN. The model-ESN is used to model the dynamic system as shown in the Eq. (5), the actor ESN and critic ESN are used to calculate the control law and cost function, respectively. The working process of ESN-ADP controller is described as below.

- Step 1. Actor-ESN generates the control law $u(t)$ at time step t.
- Step 2. The critic-ESN$_1$ evaluation the cost $J(\boldsymbol{x}(t), \boldsymbol{u}(t))$ at time step t.
- Step 3. UTILITY module calculates the utility function $U[\mathbf{x}(t), \mathbf{u}(t)]$ at time step t.
- Step 4. Model-ESN approximates the system state $\mathbf{x}(t+1)$
- Step 5. Actor-ESN inputs the dynamic state $\mathbf{x}(t+1)$ from dynamic system and outputs the control law $\mathbf{u}(t+1)$.
- Step 6. The Critic-ESN$_2$ evaluation the cost $J(\boldsymbol{x}(t+1), \boldsymbol{u}(t+1))$.
- Step 7. Update the output weight matrix of each ESN module by the online updating rule.

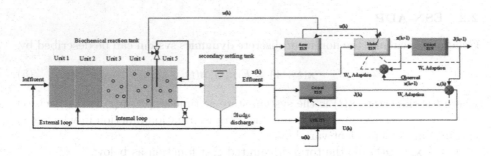

Fig. 2. The schematic diagram of ESN-ADP.

3 The Online Training Algorithm of ESN-ADP

3.1 The FORLS with Forgetting Parameter for ESN

Although the offline training algorithm is easily operated to calculate the output wight matrix of each ESN module, it is difficult for the obtained ESN-ADP controller to cope with the dynamical changes of system model. Furthermore, in some existing online training algorithms [13], the importance of current state or data is omit, which may reduce the control effectiveness.

To update the ESN parameters by the sequentially arrived value and reduce the effect of holistic system states, the online recurrent least square algorithm with forgetting parameter is designed. In the following, the detail calculation process is introduced.

At time step t, let $\mathbf{W}_o(t)$ represent the output weights matrix, the cost function for traditional FORLS is given as below,

$$\Theta(\mathbf{W}_o(t)) = \frac{1}{2} \sum_{i=1}^{t} \left(\mathbf{d}(i) - \mathbf{s}^T(i)\mathbf{W}_o(t) \right)^2 \tag{10}$$

To reduce the accumulated error by historical system state, the forgetting factor λ is introduced, which satisfies $0 < \lambda < 1$. Then the Eq. (10) is modified as below,

$$\Gamma(\mathbf{W}_o(t)) = \frac{1}{2} \sum_{i=1}^{t} \lambda^{t-i} \left(\mathbf{d}(i) - \mathbf{s}^T(i)\mathbf{W}_o(t) \right)^2 \tag{11}$$

To get the extremum, calculate the partial derivatives of $\Gamma(\mathbf{W}_o(t))$ with respect to $\mathbf{W}_o(t)$,

$$\frac{\partial \Gamma(\mathbf{W}_o(t))}{\partial \mathbf{W}_o(t)} = -\sum_{i=1}^{t} \lambda^{t-i} \mathbf{s}(i)(\mathbf{d}(i) - \mathbf{s}^T(i)\mathbf{W}_o(t)) \tag{12}$$

The optimal value of $\mathbf{W}_o(t)$, which minimize the cost function $\Gamma(\mathbf{W}_o(t))$, can be obtained by setting the partial derivatives in Eq. (12) as $\mathbf{0}$.

$$\mathbf{W}_o(t) = \left[\sum_{i=1}^{t} \lambda^{t-i} \mathbf{s}(i)\mathbf{s}^T(i) \right]^{-1} \sum_{i=1}^{t} \lambda^{t-i} \mathbf{s}(i)\mathbf{d}(i) \tag{13}$$

To further the computational analysis, the new variable $\mathbf{P}(t)$ is introduced,

$$\mathbf{P}(t) = \left(\sum_{i=1}^{t} \lambda^{t-i} \mathbf{s}(i) \mathbf{s}^{\mathrm{T}}(i) \right)^{-1} \tag{14}$$

Furthermore, the inverse of $\mathbf{P}(t)$ is represented as below,

$$\mathbf{P}^{-1}(t) = \sum_{i=1}^{t} \lambda^{t-i} \mathbf{s}(i) \mathbf{s}^{\mathrm{T}}(i) \tag{15}$$

Then, the $\mathbf{W}_o(t)$ in Eq. (13) is replaced as,

$$\mathbf{W}_o(t) = \mathbf{P}(t) \sum_{i=1}^{t} \lambda^{t-i} \mathbf{s}(i) \mathbf{d}(i) \tag{16}$$

Based on Eq. (15), one gets

$$\mathbf{P}^{-1}(t) = \lambda \mathbf{P}^{-1}(t-1) + \mathbf{s}(t) \mathbf{s}^{\mathrm{T}}(t) \tag{17}$$

Based on Eq. (16), one obtains

$$\mathbf{W}_o(t) = \mathbf{P}(t) \sum_{i=1}^{t-1} \lambda^{t-i} \mathbf{s}(i) \mathbf{d}(i) + \mathbf{s}(t) \mathbf{d}(t) \tag{18}$$

Thus, the iterative updating rule of $\mathbf{W}_o(t)$ is calculated as below,

$$\mathbf{W}_o(t) = \mathbf{W}_o(t-1) + \mathbf{P}(t) \mathbf{s}(t) (\mathbf{d}(t) - \mathbf{s}^{\mathrm{T}}(t) \mathbf{W}_o(t-1)) \tag{19}$$

Furthermore, the recursive calculation way of $\mathbf{P}(t)$ is given,

$$\begin{aligned} \mathbf{P}(t) &= \left[\mathbf{P}(t-1) + \mathbf{s}^{\mathrm{T}}(t) \mathbf{s}(t) \right]^{-1} \\ &= \left(\mathbf{I} - \frac{\mathbf{P}(t-1) \mathbf{s}(t) \mathbf{s}^{\mathrm{T}}(t)}{\lambda + \mathbf{s}^{\mathrm{T}}(t) \mathbf{P}(t-1) \mathbf{s}(t)} \right) \frac{\mathbf{P}(t-1)}{\lambda} \end{aligned} \tag{20}$$

According to Eqs. (19) and (20), it is easily found that $\mathbf{W}_o(t)$ can be recursively derived. Moreover, the $\mathbf{W}_o(t)$ is updated by the recurrent arrived value, thus the system model dynamics can be learned by the proposed online learning algorithm.

3.2   The Online Training Process for ADP Controller

As mentioned in the previous section, online training of ESN-ADP is actually for online adjustment of the output weights of each ESN in the control process. When $\mathbf{W}_o(0)$ and $\mathbf{P}(0)$ are initialized, $\mathbf{e}(t)$ becomes the only parameter required in the online training process of the network.

- For Model-ESN, the observed system states error is defined as $\mathbf{e}_m(t)$, which is given as

$$\mathbf{e}_m(t) = \mathbf{x}_o(t) - \mathbf{x}_m(t) \tag{21}$$

where $\mathbf{x}_o(t)$ and $\mathbf{x}_m(t)$ are the observed state and network output, respectively.

- The Critic-ESN is used to approximate the cost function in Eq. (7). The approximation error $\mathbf{e}_c(t)$ is given as,

$$\mathbf{e}_c(t) = U(t) + \gamma J(t+1) - J(t) \tag{22}$$

- For actor ESN, the network should approximate the optimal control scheme in Eq. (9). Thus, define the $\mathbf{e}(t)$ as $\mathbf{e}_a(t)$,

$$\mathbf{e}_a(t) = -\mu \frac{\partial J(t)}{\partial \mathbf{u}(t)} - \Delta \mathbf{u}(t+1) \tag{23}$$

where μ is the step length.

4   Experiment and Discussion

In this section, the BSM1 (Benchmark Simulation Model No. 1) is taken as system platform to study the application of ESN-ADP. As an experimental platform for wastewater treatment process, the BSM1 is consisted of a 5-layer integrated biochemical reaction tank and a 10-layer secondary sedimentation tank, as shown in Fig. 3. Particularly, the biochemical tank is constructed of two anaerobic and three aerobic zones. The K_{La_5} is used to realize the DO concentration control in the fifth reaction tank. The 14 days of storm weather is chosen to verify control performance.

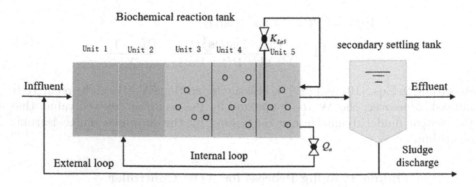

Fig. 3. The main layout of BSM1.

To evaluate the control performance, three performance indices are chosen, including the integral of absolute error (IAE), integral of square error (ISE), Max deviation from set point ($\text{Dev}^{\max}$), which are represented as below,

$$\text{IAE} = \frac{1}{t} \sum_{i=1}^{t} |y(i) - d(i)| \tag{24}$$

$$\text{ISE} = \sum_{i=1}^{t} (y(i) - d(i))^2 \tag{25}$$

$$\text{Dev}^{\max} = \max\{|y(i) - d(i)|\} \tag{26}$$

in which t is the size of dataset, $y(i)$ is the ESN output, $d(i)$ is the reference variable at time t. The smaller value of IAE, ISE or $\text{Dev}^{\max}$ imply better control performance.

Table 1. Performance comparison of different controllers

	PID	RLS-ESN
Set point	2.0	2.0
ISE	0.0019	0.000063
IAE	0.0150	0.0025
$\text{Dev}^{\max}$	0.1107	0.0091
Mean value of K_{La_5}	140.42	140.48

Firstly, the control performance of ESN-ADP controller and traditional PID controller are tested. The control results of ESN-ADP and PID for DO concentration is shown in Fig. 4(a), the related value of K_{La_5} is shown in Fig. 4(b), in which the sampling of dataset is 30 min. It is easily found that there exist obvious fluctuation with the large dynamic change for PID controller, the concentration range of DO is $[1.4, 2.4]\,\text{g(COD)}/\text{m}^3$. While for ESN-ADP controller, the DO concentration range is $[1.997, 2.006]\,\text{g(COD)}/\text{m}^3$. Compared with PID control, using ESN-ADP controller can obtain a smoother control curve and can quickly implement smooth control of DO.

Furthermore, the control results of ESN-ADP controller and PID controller are compared by evaluation indexes, which are recorded in Table 1. As compared with PID, the ESN-ADP controller has smaller values in terms of ISE, IAE, as well as $\text{Dev}^{\max}$. While the control accuracy of ESN-ADP controller is obtained at the cost of energy consumption, which is represented by larger mean value of K_{La_5}.

In order to study the adjustment process of the output weights of each module, the evolving process of $\mathbf{e}_m(t)$ for model ESN is shown in Fig. 5(a), the evolving process of $e_a(t)$ for actor ESN is illustrated in Fig. 5(b), the evolving

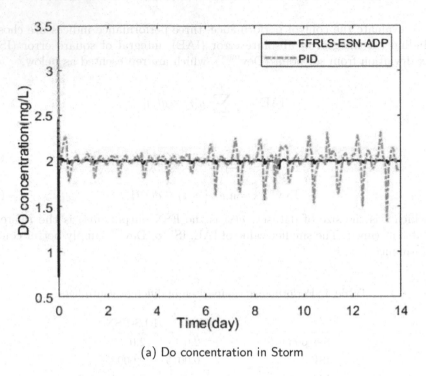

(a) Do concentration in Storm

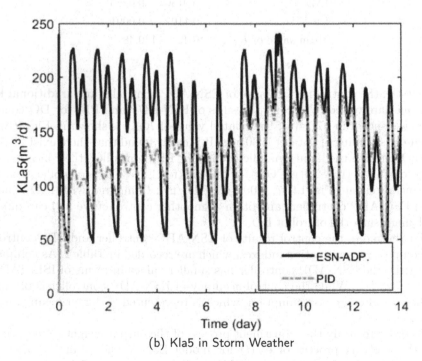

(b) Kla5 in Storm Weather

Fig. 4. Controller control performance comparison

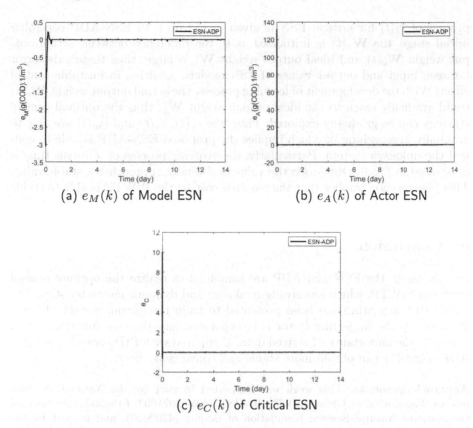

(a) $e_M(k)$ of Model ESN

(b) $e_A(k)$ of Actor ESN

(c) $e_C(k)$ of Critical ESN

Fig. 5. Learning process of ESN-ADP controller

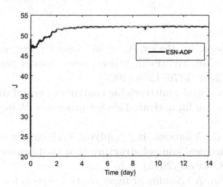

Fig. 6. Evolving process of J for ESN-ADP

process of $e_c(t)$ for critical ESN is given in Fig. 5(c). In ESN-ADP controller initial stage, the $\mathbf{W}_o(t)$ is initialized as $\mathbf{0}$, the difference between actual output weight $\mathbf{W}_o(t)$ and ideal output weight $\mathbf{W}_o^*$ is large, thus there exist error between input and output values of ESN models, resulting in unstable control effect. With the development of learning process, the actual output weight $\mathbf{W}_o(t)$ could gradually reach to the ideal output weight $\mathbf{W}_o^*$, thus the optimal control strategy can be gradually explored. Thus, the $e_c(t)$, $e_a(t)$ and $e_m(t)$ are able to gradually approaching 0, which implies the proposed ESN-ADP is able to control the unknown system. Particularly, the evolving process of J versus time is illustrated in Fig. 6. Obviously, the value of J finally approaches a stable value. This phenomenon implies that the required cost by the WWTP is able to stable at a constant value.

5   Conclusion

In this paper, the ESN and ADP are combined to realize the optimal control of DO in WWTP, which has strong nonlinear and dynamic characteristics. The online RLS algorithm has been presented to train the output weight of ESN. Particularly, the forgetting factor is incorporated into the cost function, which considers the importance of arrived data. Compared with PID control, the ESN-ADP controller can obtain more stable and convergent effect.

Acknowledgements. This work was supported in part by the National Natural Science Foundation of China (61973010, 618909305,62021003, 61533002), in part by the National Natural Science Foundation of Beijing (4202006), and in part by the National Key Research and Development Project (2021ZD0112302, 2019YFC1906002, 2018YFC1900802).

References

1. Holenda, B., Domokos, E., Redey, A., et al.: Dissolved oxygen control of the activated sludge wastewater treatment process using model predictive control. Comput. Chem. Eng. **32**(6), 1270–1278 (2008)
2. Piotrowski, R.: Two-Level multivariable control system of dissolved oxygen tracking and aeration system for activated sludge processes. Water Environ. Res. **87**(1), 3–13 (2015)
3. Santin, I., Pedret, C., Vilanova, R.: Applying variable dissolved oxygen set point in a two level hierarchical control structure to a wastewater treatment process. J. Process Control **28**, 40–55 (2015)
4. Piotrowski, R., Skiba, A.: Nonlinear fuzzy control system for dissolved oxygen with aeration system in sequencing batch reactor. Inf. Technol. Control **44**(2), 182–195 (2015)
5. Wahab, N.A., Katebi, R., Balderud, J.: Multivariable PID control design for activated sludge process with nitrification and denitrification. Biochem. Eng. J. **45**(3), 239–248 (2009)

6. Luo, F., Hoang, B.L., Tien, D.N., et al.: Hybrid PI controller design and hedge alge-bras for control problem of dissolved oxygen in the wastewater treatment system using activated sludge method. Int. Res. J. Eng. Technol. **2**(7), 733–738 (2015)
7. Tang, W., Feng, Q., Wang, M., Hou, Q., Wang, L.: Expert system based dissolved oxygen control in APMP wastewater aerobic treatment process. In: 2008 IEEE International Conference on Automation and Logistics, pp. 1308–1313 (2008)
8. Yang, T., Qiu, W., Ma, Y., et al.: Fuzzy model-based predictive control of dissolved oxygen in activated sludge processes. Neurocomputing **136**, 88–95 (2014)
9. Han, H., Qiao, J.: Nonlinear model-predictive control for industrial processes: an application to wastewater treatment process. IEEE Trans. Ind. Electron. **61**(4), 1970–1982 (2014)
10. Xu, J., Yang, C., Qiao, J.: A novel dissolve oxygen control method based on fuzzy neural network. In: 2017 36th Chinese Control Conference (CCC), pp. 4363–4368 (2017)
11 Fu, W. T., Qiao, J.-F., Han, G.-T., Meng: Dissolved oxygen control system based on the T-S fuzzy neural network. In: 2015 International Joint Conference on Neural Networks (IJCNN), pp. 1–7 (2015)
12. Qiao, J., Wang, Y., Chai, W.: Optimal control based on iterative ADP for wastew-ater treatment process. J. Beijing Univ. Technol. **44**(2), 200–206 (2018)
13. Bo, Y., Qiao, J.: Heuristic dynamic programming using echo state network for multivariable tracking control of wastewater treatment process. Asian. J. Control **17**(5), 1654–1666 (2015)

Deep Echo State Network Based Neuroadaptive Control for Uncertain Systems

Baolei Xu[1] and Qing Chen[2(✉)]

[1] Key Laboratory of Exploitation and Study of Distinctive Plants in Education Department of Sichuan Province, Sichuan University of Arts and Sciences, Dazhou 635000, China
20100012@sasu.edu.cn
[2] School of Electrical Engineering, Chongqing University of Science and Technology, Chongqing 401331, China
2021049@cqust.edu.cn

Abstract. This work presents a deep echo state network (DESN) based neuroadaptive control approach for a class of single-input single-output (SISO) uncertain system. In which, a DESN based on multiple reservoirs is applied for approximating the uncertain parts of the control system and the rigorous stability condition under the presented control strategy is analyzed. The availability of the approach is proved by comparison with the control technique using radial basis function neural network (RBFNN) and the control scheme using traditional echo state network (ESN) via numerical simulations, demonstrating that superior tracking performance is achieved by the proposed method.

Keywords: Uncertain control system · Neuroadaptive control · DESN

1 Introduction

The control design of uncertain system has aroused growing interests in control theory research community because of its practical importance. In particular, neural network (NN), which possesses universal approximation and learning capabilities, has been widely used to handle the coupling nonlinearities and uncertainties in nonlinear control system [1–3]. Different type of neural networks have been applied in the NN based control studies. For example, in [4], a tracking controller using back propagation neural network is designed for the uncertain systems subjected to input saturation constraint and external disturbances. In [5–7], RBFNN is applied for approximating the uncertain functions of nonlinear control system. In [8], a model predictive control using recurrent neural network is developed.

In particular, owing to its low complexity in computation and fast speed in convergence, ESN has gained increasing attention in recent years [9]. In [10,11],

© The Author(s), under exclusive license to Springer Nature Singapore Pte Ltd. 2022
H. Zhang et al. (Eds.): NCAA 2022, CCIS 1637, pp. 274–285, 2022.
https://doi.org/10.1007/978-981-19-6142-7_21

upon using fuzzy system together with echo state network, the authors propose a strict feedback controller for a class of multi-input multi-output nonlinear dynamic system. In [12], an ESN based adaptive control method is developed for a class of constrained pure-feedback system. In [13], an optimal control method using ESN is presented to solve the control problem of a class of continuous-time nonlinear systems. In [14], a backstepping adaptive iterative learning control based on ESN is established for nonlinear strict-feedback system. In [15], to identify the periodic discrete-time dynamic nonlinear systems with noise, the authors propose a sinusoidal echo state network-based identification method. In [16], an ESN based control method is designed to optimize the proportional-integral-derivative (PID) parameters.

However, these existing studies are commonly based on the shallow ESNs. As DESNs have been proved effective in addressing challenging real-world problems from several application fields [17], it is more reasonable consider echo state network with multiple reservoir layers as introduced in [18–21] and applied in time series prediction [22], fault diagnosis [23], destination prediction [24], and diagnosis of parkinson's disease [25]. Motivated by the above analysis, a neuradaptive control strategy using DESN is developed in this work. Compared with the widely used control scheme based on RBFNN and control technique based on traditional ESN, the proposed method based on DESN has a better control performance as showed in the numerical simulation.

2   Deep Echo State Network

The traditional echo state network is given in Fig. 1, including an input component, a reservoir, and a readout component, while the model of DESN consists of a input component, several layers of reservoirs, and a readout component. Specifically, the external input $z(k)$ is connected to the first layer of reservoir for each time step k, then the response signal of the first reservoir layer is treated as the input signal of the second reservoir layer, and the response signal of the second reservoir layer is treated as the input signal of the third reservoir layer, so on so forth. At last, all the response signals of the reservoir layers are connected to the readout component. Thus the output signal of the first reservoir layer is described by

$$h^1(k) = \tanh(W_{in}^T z(k) + W_1^T h^{(1)}(k-1)) \tag{1}$$

while for every layer $l > 1$ it is defined as:

$$h^l(k) = \tanh(W_{(l-1,l)}^T h^{l-1}(k) + W_l^T h^{(l)}(k-1)) \tag{2}$$

The output signal of the DESN is computed by:

$$y_{out}(k) = W_{out}^T H(k) \tag{3}$$

where $h^{(l)}(k) \in R^N$ represents the output state of layer l, $W_{in} \in R^{N_{in} \times N}$ represents the input weight matrix, $W_l \in R^{N \times N}$ represents the weight matrix of layer l. $W_{(l-1,l)}$ represents the weight matrix from $l-1$ layer to l layer of reservoir, $l = 2, ..., m$, m is the amount of reservoir layers, N represents the amount of neurons in each reservoir layer. tanh represents hyperbolic tangent function, $W_{out} \in R^{Nm \times N_{out}}$ is the output weights matrix, $H(k) = (h^1(k), ..., h^m(k)) \in R^{Nm}$ contains the states of each reservoir layer, N_{in} represents the amount of neuron in input component, N_{out} represents the amount of neuron in readout component. It is worthy noting that the output weight matrix W_{out} is trained by some learning algorithms, while the weights W_{in}, W_l, $W_{l-1,l}$ are usually chosen randomly.

In this work, the DESN will be used for approximating the unknown parts of the control system. As stated in [26], there exists an ideal DESN capable of approximating the unknown and continuous function $f(z) : R^n \to R$ defined on a compact set Ω_z, such that

$$f(z) = W_{out}^{*T} h(z) + \epsilon^*, \forall z \in \Omega \qquad (4)$$

where z is the external input, W_{out}^* represents the optimal output weight, h the output state of reservoir layers, ϵ^* represents the estimated error and $|\epsilon^*| \leq \epsilon$, $\epsilon > 0$ being a unknown constant.

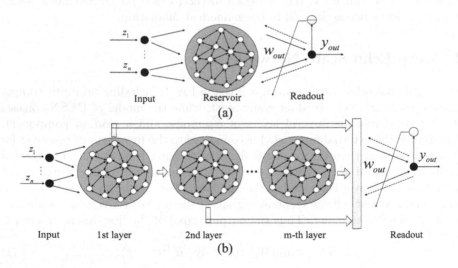

Fig. 1. The diagrammatic sketch of traditional ESN (a) and DESN (b).

3   System Description and Problem Statement

We consider the SISO uncertain systems modeled by:

$$\dot{x}_i = x_{i+1}, \quad i = 1, 2, ..., n-1$$
$$\dot{x}_n = g(x,t)u + f(x,t) \qquad (5)$$

where $x = [x_1, ..., x_n]^T \in R^n$ and $x_i \in R, i = 1, ..., n$ represent the system states, $u \in R$ represents the control input signal, $g(x,t) \in R$ and $f(x,t) \in R$ are the nonlinear functions with $g(x,t) \in R$ being the uncertain time-varying control gain and $f(x,t) \in R$ being the unknown uncertainties.

The objective is to develop a DESN-based adaptive control scheme for SISO system (5) such that all the closed-loop signals remain bounded and the tracking error $e = x_1 - x_{d1}$ converges to a small residual set containing the origin for any given desired trajectory x_{d1}. To this end, the following assumptions are needed.

Assumption 1. The desired trajectory signal x_{d1} and its derivatives $\dot{x}_{d1}, ..., x_{d1}^{(n-1)}$ are assumed to be known and bounded functions.

Assumption 2. The control gain $g(x,t)$ is unknown but bounded, i.e. there exits a unknown constant g_1 such that $|g(x,t)| > g_1 > 0$, and $g(x,t)$ is sign-definite. In particular, in this work, $sgn(g) = +1$ is assumed without loss of generality.

To proceed, we first define a filtered variable χ indicated as below:

$$\chi = \lambda_1 e + \lambda_2 \dot{e} + ... + \lambda_{n-1} e^{n-2} + e^{n-1} \tag{6}$$

thus

$$\dot{\chi} = \lambda_1 \dot{e} + \cdots + \lambda_{n-1} e^{n-1} + g(x,t)u + f(x,t) - x_{d1}^{(n)}$$
$$= g(x,t)u + l(x,t) \tag{7}$$

where $\lambda_i, i = 1, 2, ..., n-1$ are some constants selected by the designer such that the polynomial $s^{n-1} + \lambda_{n-1} s^{n-2} + \cdots + \lambda_1$ is Hurwitz. $l(x,t) = \lambda_1 \dot{e} + \cdots + \lambda_{n-1} e^{n-1} + f(x,t) - x_{d1}^{(n)}$, where the term $l(x,t)$ is always unknown, thus the corresponding control strategy cannot be built upon $l(x,t)$ directly. In this study, by applying the property that NN is able to estimate any unknown continuous function, $l(x,t)$ is approximated by the proposed DESN. Thus we can obtain that

$$l(x,t) = w_a^{*T} \phi(z_a) + \epsilon_a$$
$$\leq \|w_a^*\| \|\phi(z_a)\| + \epsilon_{am} \tag{8}$$
$$\leq \varpi_1 \Xi(z_a)$$

where w_a^* is the output weight of DESN, z_a is the input signal of DESN, $\phi(\cdot)$ is the hyperbolic tangent function, ϵ_a is the estimated error and $\epsilon_a \leq \epsilon_{am}$ with $\epsilon_{am} > 0$ being a unknown constant, $\varpi_1 = \max\{\|w_a^*\|, \epsilon_{am}\}$ and $\Xi(z_a) = 1 + \|\phi(z_a)\|$.

4   Controller Design and Stability Analysis

According to the above analysis, the following control scheme is proposed in this work,

$$u = -k_1 \chi - c_1 \hat{\varpi}_1 \Xi^2(z_a) \chi \tag{9}$$

with
$$\dot{\hat{\varpi}}_1 = -\beta_1 \hat{\varpi}_1 + c_1 |\chi|^2 \Xi^2(z_a) \tag{10}$$

where $k_1 > 0$, $\beta_1 > 0$ and $c_1 > 0$ are constants selected by the user. $\hat{\varpi}_1$ is the estimated value of ϖ_1. $\hat{\varpi}_1(0) \geq 0$ is the arbitrarily chosen initial value.

Theorem 1. *Consider the SISO system (5) with the error function described by (6). Suppose that the conditions as imposed in Assumptions 1–2 hold. If the neuroadaptive controller (9) and (10) is applied, then the closed-loop signals remain bounded and the tracking error converges to a small residual set containing the origin for any given desired trajectory.*

Proof. Firstly, define Lyapunov function as follows:
$$V_1 = \frac{1}{2}\chi^2 + \frac{1}{2g_1}\tilde{\varpi}_1^2 \tag{11}$$

where $\tilde{\varpi}_1 = \varpi_1 - g_1\hat{\varpi}_1$. Differentiating V_1 and substituting (7) and (8) into it yields

$$\dot{V}_1 = \chi\dot{\chi} - \tilde{\varpi}_1\dot{\hat{\varpi}}_1$$
$$\leq \chi g u + |\chi|\varpi_1 \Xi(z_a) - \tilde{\varpi}_1\dot{\hat{\varpi}}_1 \tag{12}$$

By Young's inequality, for any $c_1 > 0$, $|\chi|\Xi(z_a)$ becomes

$$|\chi|\Xi(z_a) \leq c_1|\chi|^2 \Xi^2(z_a) + \frac{1}{4c_1} \tag{13}$$

Applying (9) and (13) to (12) yields that

$$\dot{V}_1 \leq -k_1 g_1 \chi^2 - c_1 g_1 \hat{\varpi}_1 |\chi|^2 \Xi^2(z_a) + \frac{\varpi_1}{4c_1}$$
$$+ c_1 \varpi_1 |\chi|^2 \Xi^2(z_a) - \tilde{\varpi}_1\dot{\hat{\varpi}}_1 \tag{14}$$

Inserting (10) into (14), it can be obtained that

$$\dot{V}_1 \leq -k_1 g_1 \chi^2 + \frac{\varpi_1}{4c_1} + \beta_1 \tilde{\varpi}_1 \hat{\varpi}_1 \tag{15}$$

where

$$\tilde{\varpi}_1 \hat{\varpi}_1 = \frac{1}{g_1}\tilde{\varpi}_1(\varpi_1 - \tilde{\varpi}_1)$$
$$\leq \frac{1}{g_1}\left(\frac{1}{2}\varpi_1^2 + \frac{1}{2}\tilde{\varpi}_1^2 - \tilde{\varpi}_1^2\right) \tag{16}$$
$$\leq \frac{1}{2g_1}(\varpi_1^2 - \tilde{\varpi}_1^2)$$

Thus (15) can be further expressed as

$$\dot{V}_1 \leq -k_1 g_1 \chi^2 - \frac{\beta_1}{2g_1}\tilde{\varpi}_1^2 + \frac{\varpi_1}{4c_1} + \frac{\beta_1}{2g_1}\varpi_1^2$$
$$\leq \varrho_1 V_1 + \varrho_2 \tag{17}$$

where $\varrho_1 = \min\{2k_1g_1, \beta_1\}$, $\varrho_2 = \dfrac{\varpi_1}{4c_1} + \dfrac{\beta_1}{2g_1}\varpi_1^2$. From (17), it is ensured that $V_1 \in \ell_\infty$, which indicates that $\chi \in \ell_\infty$ and $\tilde{\varpi}_1(\hat{\varpi}_1) \in \ell_\infty$. Then it can be concluded that $x \in \ell_\infty$ and $\varXi \in \ell_\infty$. From (9), it is shown that $u \in \ell_\infty$. Further, it follows from (17) that $\dot{V}_1 < 0$ as long as $\chi^2 > \dfrac{\varpi_1}{4c_1k_1g_1} + \dfrac{\beta_1}{2k_1g_1^2}\varpi_1^2$ and $\tilde{\varpi}_1^2 > \dfrac{g_1\varpi_1}{2c_1\beta_1} + \varpi_1^2$. Thus χ and $\tilde{\varpi}_1$ are uniformly ultimately bounded as $|\chi| < (\dfrac{\varpi_1}{4c_1k_1g_1} + \dfrac{\beta_1}{2k_1g_1^2}\varpi_1^2)^{1/2}$ and $|\tilde{\varpi}_1| < (\dfrac{g_1\varpi_1}{2c_1\beta_1} + \varpi_1^2)^{1/2}$. Therefore, $e^{(i)}, i = 1, 2, ..., n-1$ are uniformly ultimately bounded.

5 Simulation

Two numerical simulation examples are conducted in this section to confirm the availability of the control approach using DESN.

Example 1

Given the second-order SISO uncertain system as follows:

$$\begin{cases} \dot{x}_1 = x_2 \\ \dot{x}_2 = g(x)u + f(x) \end{cases}$$

where x_1 and x_2 are position and speed, respectively. $f(x) = -25x_2$ and $g(x) = 133$. u is control input. The desired trajectory is $x_{d1} = \sin(t) + 0.1\cos(t)$. The initial values are $x_1(0) = 0.5$, $x_2(0) = 0.5$, $\varpi_1(0) = 0$. Control parameters are designed as $\lambda_1 = 2$, $k_1 = 1$, $c_1 = 1$, and $\beta_1 = 1.2$. The DESN consists of 10 layers of reservoir, and the amount of neuron in each layer is 10. Simulation results are shown in Figs. 2, 3, 4 and 5. It can be easily observed from Fig. 2 that good tracking result is obtained under the proposed DESN based control scheme. For comparison, both traditional ESN based control and the RBFNN based control are tested. In the simulation, the amount of neuron used in ESN and RBFNN is 100. Figure 3 depicts the tracking error, from which we can see that the control scheme based on DESN has a higher tacking precision than the one based on RBFNN and the one based on traditional ESN under the similar control input signal as shown in Fig. 4. The evolution process of estimate parameter $\hat{\varpi}_1$ is presented in Fig. 5.

Example 2

In this example, the well-known inverted pendulum described as follows is used.

$$\begin{cases} \dot{x}_1 = x_2 \\ \dot{x}_2 = g(x_1, x_2)u + f(x_1, x_2) \end{cases}$$

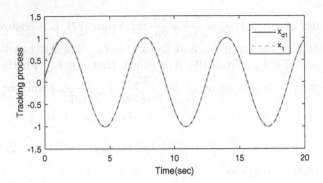

Fig. 2. Tracking processing under the proposed DESN based control scheme (Example 1).

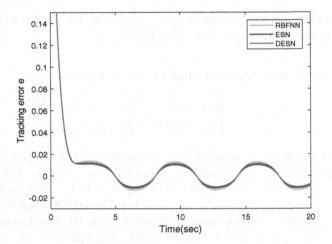

Fig. 3. Tracking error comparison (Example 1).

where the detailed information are:

$$g(x_1, x_2) = \frac{\frac{\cos(x_1)}{m_c + m}}{m_l\left(\frac{4}{3} - \frac{m\cos^2(x_1)}{m_c + m}\right)}$$

$$f(x_1, x_2) = \frac{9.8\sin(x_1) - \frac{mm_l x_2^2 \cos(x_1)\sin(x_1)}{m_c + m}}{m_l\left(\frac{4}{3} - \frac{m\cos^2(x_1)}{m_c + m}\right)}$$

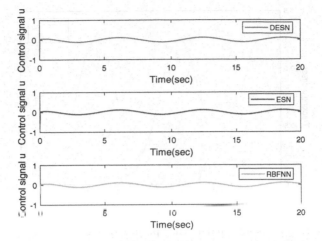

Fig. 4. Control input signal comparison (Example 1).

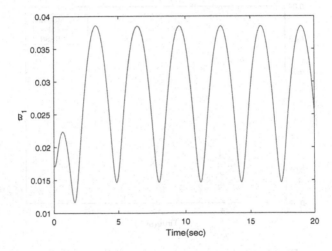

Fig. 5. The estimate parameter $\hat{\varpi}_1$ updating process (Example 1).

where $m_l = 0.5$ represents the half length of the pole, $m = 0.1$ and $m_c = 1$ represents the mass of pole and cart, respectively. The desired trajectory $x_{d1} = \sin(t)$. The initialization parameters are $x_1(0) = 0.1$, $x_2(0) = 0.1$, $\varpi_1(0) = 0$, $\lambda_1 = 20$, $k_1 = 10$, $c_1 = 1$, and $\beta_1 = 1.2$. The DESN consists of 10 layers of reservoir, and the amount of neuron of each reservoir is 10. The simulation results have been shown in Figs. 6, 7, 8 and 9. Figure 6 plots the tracking processing under the DESN based control strategy, from which it can be obtained that the actual

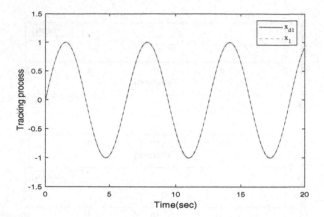

Fig. 6. Tracking processing under the proposed DESN based control scheme (Example 2).

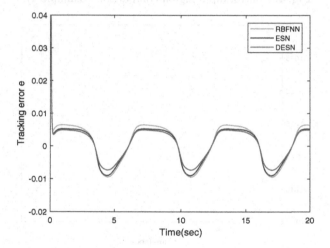

Fig. 7. Tracking error comparison (Example 2).

trajectory converges to the desired position successfully. Figure 7 depicts the tracking error compared with ESN based control and RBFNN based control, demonstrating the enhanced control performance of the DESN based method. Figure 8 plots the control input signal u, and Fig. 9 depicts the evolution trajectory of estimate parameter $\hat{\varpi}_1$.

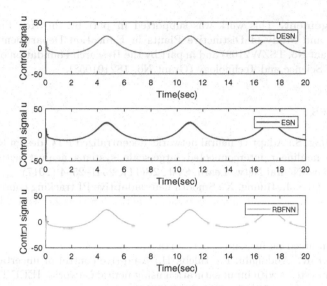

Fig. 8. Control input signal comparison (Example 2).

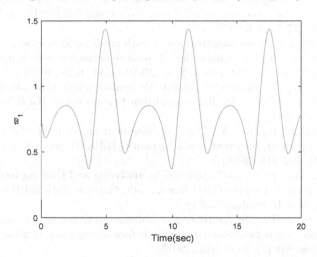

Fig. 9. The estimate parameter $\hat{\varpi}_1$ updating process (Example 2).

6   Conclusion

Neuroadaptive control schemes based on DESN is proposed in this study to handle the trajectory tracking control issue of a class of SISO uncertain control systems. Compared with the RBFNN based control and traditional echo state network based technique, the tracking precision of the DESN based approach is improved due to the deep reservoirs structure. Extension of DESN based approach to more general uncertain systems is a potential research in the future.

Acknowledgement. This work was supported in part by the Key Laboratory of Exploitation and Study of Distinctive Plants in Education Department of Sichuan Province (Grant No. TSZW2109) and in part by the Research Foundation of Chongqing University of Science and Technology (Grant No. 182101058).

References

1. Li, Y., Tong, S.: Adaptive neural networks decentralized FTC design for nonstrict-feedback nonlinear interconnected large-scale systems against actuator faults. IEEE Trans. Neural Netw. Learn. Syst. **28**(11), 2541–2554 (2017)
2. Song, Y., Guo, J., Huang, X.: Smooth neuroadaptive PI tracking control of nonlinear systems with unknown and nonsmooth actuation characteristics. IEEE Trans. Neural Netw. Learn. Syst. **28**(9), 2183–2195 (2017)
3. Yang, C., Wang, X., Cheng, L., Ma, H.: Neural-learning-based telerobot control with guaranteed performance. IEEE Trans. Cybern. **47**(10), 3148–3159 (2017)
4. Esfandiari, K., Abdollahi, F., Talebi, H.: Adaptive control of uncertain nonaffine nonlinear systems with input saturation using neural networks. IEEE Trans. Neural Netw. Learn. Syst. **26**(10), 2311–2322 (2015)
5. Liu, Y., Li, J., Tong, S., Chen, C.: Neural network control-based adaptive learning design for nonlinear systems with full-state constraints. IEEE Trans. Neural Netw. Learn. Syst. **27**(7), 1562–1571 (2016)
6. Song, Y., Zhou, S.: Neuroadaptive control with given performance specifications for MIMO strict-feedback systems under nonsmooth actuation and output constraints. IEEE Trans. Neural Netw. Learn. Syst. **29**(9), 4414–4425 (2018)
7. Zhao, K., Song, Y.: Neuroadaptive robotic control under time-varying asymmetric motion constraints: a feasibility-condition-free approach. IEEE Trans. Cybern. **50**(1), 15–24 (2020)
8. Han, H., Zhang, L., Hou, Y., Qiao, J.: Nonlinear model predictive control based on a self-organizing recurrent neural network. IEEE Trans. Neural Netw. Learn. Syst. **27**(2), 402–415 (2016)
9. Jaeger, H.: The "echo state" approach to analysing and training recurrent neural networks. Technical report GMD Report 148, German National Research Center for Information Technology (2001)
10. Han, S., Lee, J.: Precise positioning of nonsmooth dynamic systems using fuzzy wavelet echo state networks and dynamic surface sliding mode control. IEEE Trans. Ind. Electron. **60**(11), 5124–5136 (2013)
11. Han, S., Lee, J.: Fuzzy echo state neural networks and funnel dynamic surface control for prescribed performance of a nonlinear dynamic system. IEEE Trans. Ind. Electron. **61**(2), 1099–1112 (2014)
12. Chen, Q., Shi, L., Na, J., Ren, X., Nan, Y.: Adaptive echo state network control for a class of pure-feedback systems with input and output constraints. Neurocomputing **275**, 1370–1382 (2017)
13. Liu, C., Zhang, H., Luo, Y., Su, H.: Dual heuristic programming for optimal control of continuous-time nonlinear systems using single echo state network. IEEE Trans. Cybern. (2020) https://doi.org/10.1109/TCYB.2020.2984952
14. Chen, Q., Shi, H., Sun, M.: Echo state network-based backstepping adaptive iterative learning control for strict-feedback systems: an error-tracking approach. IEEE Trans. Cybern. **50**(7), 3009–3022 (2020)

15. Yao, X., Wang, Z., Zhang, H.: Identification method for a class of periodic discrete-time dynamic nonlinear systems based on Sinusoidal ESN. Neurocomputing **275**, 1511–1521 (2018)
16. Wang, Z., Yao, X., Li, T., Zhang, H.: Design of PID controller based on echo state network with time-varying reservoir parameter. IEEE Trans. Cybern. (2021). https://doi.org/10.1109/TCYB.2021.3090812
17. Hermans, M., Schrauwen, B.: Training and analyzing deep recurrent neural networks. In: Proceedings of the 27th Conference on Neural Information Processing Systems, pp. 190–198 (2013)
18. Gallicchio, C., Micheli, A.: Deep reservoir computing: a critical analysis. In: Proceedings of the 24th European Symposium on Artificial Neural Networks, pp. 497–502 (2016)
19. Gallicchio, C., Micheli, A.: Deep echo state network (DeepESN): a brief survey. arXiv preprint arXiv: 1712.04323 (2017)
20. Gallicchio, C., Micheli, A., Pedrelli, L.: Deep reservoir computing: a critical experimental analysis. Neurocomputing **268**, 87–99 (2017)
21. Claudio, G., Alessio, M., Luca, P.: Design of deep echo state networks. Neural Netw. **108**, 33–47 (2018)
22. Kim, T., King, B.R.: Time series prediction using deep echo state networks. Neural Comput. Appl. **32**(23), 17769–17787 (2020). https://doi.org/10.1007/s00521-020-04948-x
23. Long, J., Zhang, S., Li, C.: Evolving deep echo state networks for intelligent fault diagnosis. IEEE Trans. Industr. Inf. **16**(7), 4928–4937 (2020)
24. Song, Z., Wu, K., Shao, J.: Destination prediction using deep echo state network. Neurocomputing **406**, 343–353 (2020)
25. Gallicchio, C., Micheli, A., Pedrelli, L.: Deep echo state networks for diagnosis of Parkinson's disease. In: Proceedings of the 26th European Symposium on Artificial Neural Networks, pp. 397–402 (2018)
26. Funahashi, K., Nakamura, Y.: Approximation of dynamical systems by continuous time recurrent neural networks. Neural Netw. **6**, 801–806 (1993)

Bolt Loosening Detection Based on Principal Component Analysis and Support Vector Machine

Shiwei Wu, Sisi Xing, Fei Du[✉], and Chao Xu

School of Astronautics, Northwestern Polytechnical University, Xi'an, China
{wushiwei,xss}@mail.nwpu.edu.cn, {dufei,chao_xu}@nwpu.edu.cn

Abstract. Online monitoring of bolt preload is essential to ensure the proper functioning of bolted structures. Ultrasonic guided wave has the advantages of high sensitivity and wide monitoring range, so it is widely used in the study of bolt loosening monitoring. However, the propagation mechanism of ultrasonic guided waves in bolted connection structure is complicated, and it is difficult to establish a direct relationship between guided wave signal and bolt loosening state directly. In recent years, machine learning and other artificial intelligence technologies have flourished, and a more effective bolt loosening detection technique can be established by using machine learning combined with the principle of guided wave damage detection. In this paper, a bolt loosening identification method based on principal component analysis (PCA) and support vector machine (SVM) is proposed, aiming to achieve end-to-end bolt loosening monitoring with few samples. The ultrasonic wave-guided experimental results of the bolted joint lap plate show that the proposed PCA and SVM technique achieves a loosening recognition accuracy of 92.5%, which is higher than other machine learning methods, and the effects of signal length, number of principal components and the choice of kernel function on the classification performance are explored.

Keywords: Machine learning · Bolt loosening monitoring · Ultrasound-guided wave · SVM

1 Introduction

Bolted connections are widely used in aerospace and civil equipment, due to installation errors, external loads and temperature effects, fatigue cracks, material aging, bolt loosening, and other damage that will inevitably occur in the equipment structure. These damages will affect the serviceability, safety, and reliability of the equipment structure if not found and disposed of in time, therefore, it is important to carry out timely and accurate health monitoring and management of the aircraft structure.

Ultrasonic guided wave is an elastic wave propagating in bounded thin-walled structures, which has the advantages of high frequency, small wavelength, and long propagation distance, making it ideal for health monitoring of bolted structures [1, 2]. However, ultrasonic guided waves have the characteristics of dispersion and multimodality, while

H. Zhang et al. (Eds.): NCAA 2022, CCIS 1637, pp. 286–300, 2022.
https://doi.org/10.1007/978-981-19-6142-7_22

the bolt connection structure is complex, with many connection interfaces, and the reflection and mode conversion of the guided waves are complicated, making the collected guided wave signals quite complex, and simple signal waveform-based methods are difficult to effectively perform bolt preload detection [3]. Yang et al. [4] first proposed the use of ultrasonic guided waves to detect bolt preload, and they used the transmitted guided wave signal energy as a damage indicator and successfully identified the loosening of plate and frame bolts. After their work, the "guided wave signal energy" was widely used as a measure of bolt preload. Wang et al. [5] used a similar method for individual bolts for bolt preload monitoring, and the obtained guided wave transmission energy was essentially proportional to the torque level, and the experimental results showed that saturation occurs when the applied torque reaches a certain value. Similarly, Amerini et al. [6] calculated the frequency domain energy of the transmitted guided wave to assess the bolt loosening level and still saw a more pronounced saturation phenomenon. Haynes et al. [7] used laser vibrometry to perform full-field measurements of guided wave propagation in a bolted structure and still observed saturation. The ultrasonic guided wave detection method based on the energy of the guided wave signal is difficult to detect the early recession of the bolt preload [8, 9]. For this reason, Parvasi et al. [10] developed a time-reversal-based method to monitor the bolt preload force, and the experimental results showed that the tightening index increased with the increase of the bolt torque. Du et al. [8] from Northwestern Polytechnic University proposed an improved time-reversal method that is more sensitive to bolt loosening, especially early loosening.

In recent years, machine learning methods have been increasingly applied to the loosening detection of multi-bolt connection structures due to the complexity of the guided wave signals in bolted connections. Nazarko et al. [11] proposed a method for bolt axial force identification using PCA and artificial neural networks for bolted flange structures. Experiments were conducted on a single-bolt flange structure, and a total of 602 sets of data were collected for four working conditions, and the relationship between signal changes and force changes could be found. Liang et al. [12] developed a decision fusion based loosening detection method for multi-bolt connection structures, and the decision fusion method combined with the decision of the selected classifier to give the final evaluation, and they used a large aerospace aluminum plate structure as the detection object, and only one of the bolts was considered to be completely loosened in each working condition, which verified the Mita A et al. [13] proposed a SVM based ultrasonic guided wave detection method for bolt preload, taking a 10-bolt lap aluminum plate as the object of study and considering one of the bolts fully loosened, half loosened, and tightened in three working conditions to verify the effectiveness of the method, but the method requires the use of running Fourier. In 2020, Wang et al. [14] proposed the use of entropy-based tightening index and least squares SVM based on the genetic algorithm to monitor multi-bolt connection loosening, and the effectiveness of the method was verified by experiments. Later, Wang et al. [15] proposed a new entropy index and a stacking-based integrated learning classifier to detect multi-bolt loosening in underwater structures using guided waves.

Therefore, combined with machine learning methods, ultrasonic guided waves can accurately achieve bolt loosening monitoring, however, existing methods, usually require manual feature extraction. Deep convolutional neural networks, for example, do not require manual extraction of features, however, the amount of data required is large. To address the above problems, this paper proposes an end-to-end detection method for bolt loosening based on PCA and SVM techniques, and experimental validation is performed with a single bolt connector to compare the effects of different guide wave signal lengths, etc.

The contents of this paper are organized as follows. Section 2 introduces the basic principles of guided bolt loosening detection. Section 3 demonstrates the single bolt loosening detection method based on PCA and SVM. Section 4 performs experimental validation. The detection results are analyzed and discussed in Sect. 5. Finally, conclusions are drawn in Sect. 6.

2   The Basic Principle of Wave-Guided Bolt Loosening Detection

The following equation [16] is satisfied between the bolt preload force P and the torque T applied to the bolt.

$$P = \frac{T}{KD} \tag{1}$$

where
  P—bolt preload
  T—bolt torque
  K—coefficient of friction between the bolt and the nut
  D—bolt diameter

From a microscopic point of view, the contact surface of a bolted joint is not completely flat and can be seen as consisting of many tiny convex bodies (Fig. 1), so the true contact area of the joint interface is the sum of the contact areas of the micro-convex bodies.

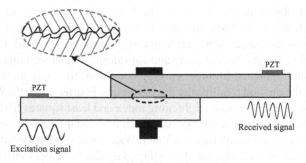

Fig. 1. Schematic diagram of guided wave propagation across a bolted joint

The transmitted energy of the ultrasound-guided Wave can be used to characterize the magnitude of the bolt preload in a lap joint structure, which is usually used widely as a tightening indicator for bolt loosening detection and is defined as follows:

$$I = \frac{\Omega_{leak}}{\Omega_{leak0}} = \frac{\int W(f) \cdot df}{\int W_0(f) \cdot df} \tag{2}$$

where

Ω_{leak}—energy of transmitted guided waves

$W(f)$—energy spectral density of the received signal

Subscript 0—working conditions at rated torque

The damage index I varies linearly with the bolt preload force T, i.e., a larger I indicates a larger bolt preload force.

However, according to the theory of rough contact mechanics, the true contact area at the connection interface will reach saturation when the contact pressure reaches a certain value [17], at which point the true contact area will no longer change with the change in contact pressure. Therefore, when using transmitted energy as a tightening indicator, the detection sensitivity is poor when the bolt loosening degree is small.

3   Bolt Loosening Monitoring Method Based on PCA and SVM

3.1   PCA

The PCA is one of the most commonly used methods for data dimensionality reduction. This method uses orthogonal transformations to convert observed data represented by linearly correlated variables into data represented by a few linearly uncorrelated variables, and the linearly uncorrelated variables are called principal components. For example, when projecting a set of two-dimensional data into a one-dimensional space, the only way to maximize the variance of the data and retain more information about the original data is to choose the direction with the largest variance of the data for projection.

Suppose there is a set of D-dimensional samples $x^{(n)} \in R^D$, $1 \leq n \leq N$, hoping to project it into one dimension, the projection vector is $\omega \in R^D$, limit the length of ω to 1, i.e. $\omega^T \omega = 1$. The representation of each sample point $x^{(n)}$ after projection is:

$$z^{(n)} = \omega^T x^{(n)} \tag{3}$$

The input samples are represented by the matrix $X = \left[x^{(1)}, x^{(2)}, \ldots, x^{(N)} \right]$, $\bar{x} = \frac{1}{N} \sum_{n=1}^N x^{(n)}$ is the center point of the original sample, the variance of all samples after projection is:

$$\sigma(X; \omega) = \omega^T \sum \omega \tag{4}$$

where

$\bar{X} = \bar{x} 1_D^T$ is the outer product of the vector $\bar{x}$ and the D-dimensional all-1 vector 1_D;

$\sum = \frac{1}{N}(X - \bar{X})(X - \bar{X})^T$ is the covariance matrix of the original sample;

Maximize the projection variance $\sigma(X; \omega)$ and satisfy $\omega^T \omega = 1$. At this time, ω is the eigenvector of the covariance matrix Σ, and λ is the eigenvalue and:

$$\sigma(X; \omega) = \omega^T \sum \omega = \omega^T \lambda \omega = \lambda \tag{5}$$

That is, λ is also the variance of the sample after projection. Therefore, PCA can be transformed into a matrix eigenvalue decomposition problem, and the projection vector ω is the eigenvector corresponding to the largest eigenvalue of the matrix Σ.

If the sample is to be projected into the D' dimensional space by the projection matrix $W \in R^{D \times D'}$, the projection matrix satisfies $W^T W = I$ as a unit array, and only the eigenvalues of $\sum$ need to be arranged from largest to smallest, the first D' eigenvectors are retained and their corresponding eigenvectors are the optimal projection matrix [18].

3.2 SVM

The purpose of SVM is to find a hyperplane to segment the samples, and the segmentation principle is interval maximization, which is eventually converted into a convex quadratic programming problem to be solved. The segmentation hyperplane found by this model has good robustness and does not require a large amount of data, so it is widely used for many tasks and has shown strong advantages.

Given a binary classification training sample set as follows, $D = \{(x_1, y_1), (x_2, y_2), \ldots, (x_N, y_N)\}, y_i \in \{-1, +1\}$. If the two classes of samples are linearly separable, i.e., there exists a hyperplane $\omega^T x + b = 0$ separating the two classes of samples, then for each sample there is $y_i(\omega^T x_i + b) > 0$.

The distance from each sample x_i in dataset D to the segmentation hyperplane is:

$$\gamma_i = \frac{|\omega^T x_i + b|}{\|\omega\|} = \frac{y_i(\omega^T x_i + b)}{\|\omega\|} \tag{6}$$

Define the interval γ as the shortest distance from all samples in the whole dataset D to the segmentation hyperplane, i.e., $\gamma = \min \gamma_i$, then the larger the interval γ is, the more stable its segmentation hyperplane is for dividing the two datasets, and the less susceptible to noise and other factors. The goal of a SVM is to find a hyperplane such that γ is maximum, i.e.

$$\max_{\omega, b} \frac{1}{\|\omega\|^2}$$
$$s.t. \quad y_i\left(\omega^T x_i + b\right) \geq \gamma, \forall i \in \{1, \ldots, N\} \tag{7}$$

All sample points in the dataset that satisfy $y_i(\omega^T x_i + b) = 1$ are called support vectors.

For a linearly divisible dataset, there are many segmentation hyperplanes, but the hyperplane with the largest interval is unique. To find the maximally spaced segmentation hyperplane, the objective function of Eq. 7 is written as a convex optimization problem.

According to the complementary relaxation condition in the KKT (Karush-Kuhn-Tucker) condition of the Lagrange multiplier method, the optimal solution satisfies $\lambda_i^* \left(1 - y_i \left(\omega^{*T} x_i + b^* \right)\right) = 0$. If the sample x_i is not on the constraint boundary, $\lambda_i^* = 0$ and its constraint fails; If the sample x_i is on the constraint boundary, $\lambda_i^* \geq 0$. These points on the constraint boundary are called support vectors, i.e., the points that are closest to the decision plane.

After computing λ^* and the optimal weight ω^*, the optimal bias b^* can be computed by choosing any of the support vectors $(\tilde{x}, \tilde{y})$:

$$b^* = \tilde{y} - \omega^{*T} \tilde{x} \tag{8}$$

The decision function of the SVM with optimal parameters is:

$$f(x) = sgn\left(\omega^{*T} x + b^*\right)$$
$$= sgn\left(\sum_{i=1}^{N} \lambda_i^* y_i x_i^T x + b^*\right) \tag{9}$$

The decision function of a SVM depends only on the sample points with $\lambda_i^* > 0$, i.e., the support vector. The objective function of the SVM can obtain the global optimal solution by optimization methods such as SMO, and thus is more efficient than other classifiers in learning. In addition, the decision function of the SVM depends only on the support vector, independent of the total number of training samples, and the classification speed is faster.

3.3 Identification Method of Bolt Loosening Detection Based on SVM and PCA

(1) Signal pre-processing

The preliminary data processing flow is shown in Fig. 2. It is divided into four steps as follows:

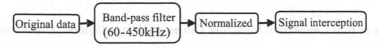

Fig. 2. Flow chart of preliminary data processing

FIR bandpass filtering of the experimentally acquired signal in the band range of 60 to 450 kHz. Normalization of the data based on the following two equations:

$$X_{i\_s} = \frac{X_i - min(X)}{max(X) - min(X)} \tag{10}$$

$$X_{i\_new} = \frac{X_{i\_s} - mean(X_s)}{X_{i\_s} - std(X_s)} \tag{11}$$

where

X_{i_s}—ith element of the new signal sequence after normalization i = 1, 2, ...n;

X_i—ith element of the original signal sequence

X—original signal sequence

X_{i_new}—ith element of the new signal sequence after standardization

$mean(X_s)$—mean value of the new signal sequence after normalization

$std(X_s)$—variance of the new signal sequence after normalization

(2) **Bolt loosen detection**

In order to identify the bolt loosening condition using the guide wave signal, the pre-processed guide wave data is used as the input, and the first n principal components of the input data are extracted using the PCA technique, and the data is downscaled into a (n × 1) feature vector, and then this vector is used as the input of the SVM classifier to classify it, where the RBF function is selected as the kernel function. The method is implemented as shown in Fig. 3.

Fig. 3. Schematic diagram of SVM+PCA method implementation

4 Experimental Verification

In the following, the effectiveness of the method is verified with a single-bolt connection structure, relying on the PyTorch platform.

4.1 Single-Bolt Connection Structure Ultrasonic Wave-Guiding Experiment

The experimental object in this section is an M6 bolt connection structure with piezoelectric sheets glued on both sides of the bolt for ultrasonic signal excitation and reception, and the experimental parts are shown in Fig. 4(a), and the experimental study is carried out at room temperature. Among them, the material of the plate is aluminum, the size of the two plates is identical, the strength grade of the bolt is 8.8, the model of the piezoelectric sheet is P5–1, the diameter is 8 mm, the size of the experimental parts and the position of the piezoelectric sheet adhesion are shown in Fig. 4(b).

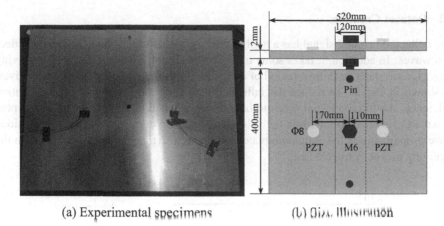

(a) Experimental specimens   (b) Size illustration

Fig. 4. M6 single bolt connector

The ultrasonic guided wave experimental process is shown in Fig. 5. Under a given bolt torque level, the signal is transmitted using NI 5413, the signal generator module of NI PXIe-1082, and the signal is applied to PZTA and received by PZTB after the bolt connection interface, and the received guided wave data is collected via the oscilloscope module NI 5105 with a sampling frequency of 20 MHz. Where the bolt torque controls the Stanley SD-030–22 digital display torque wrench is used to achieve this.

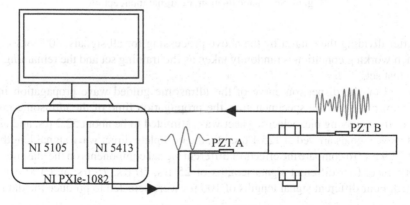

Fig. 5. Experimental flow

Four working conditions of 2 Nm, 4 Nm, 6 Nm and 8 Nm are considered in this experiment, where 8 Nm is the standard torque. At room temperature, each working condition is repeated 30 times, and a total of 120 sets of data can be obtained. The excitation signal in the experiment is a 150 kHz sinusoidal signal with a center frequency of Hanning window modulation and a peak signal of 5.5 V, as shown in Fig. 5. The total length of the acquired signal is 500 μs, and the sampling frequency is 20 MHz.

4.2 Dataset Creation

The collected original signal has a total of 500 μs and contains many boundary reflection waves. In addition, it takes some time for the signal to propagate to the receiving piezoelectric sheet, making a period of data with amplitude 0 before the signal arrives, as shown in Fig. 6. To reduce the influencing factors and improve the classification efficiency, the filtered signal needs to be intercepted, and the length of the intercepted signal can be selected based on the velocity dispersion curve of the ultrasonic guided wave propagating in the aluminum plate to estimate the time for the signal to reach the receiving piezoelectric element.

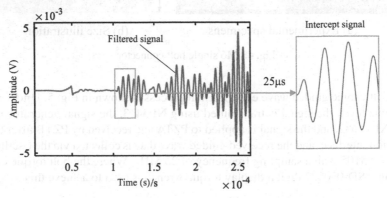

Fig. 6. Schematic diagram of signal interception

After dividing the dataset by the above processing for all signals, 80% of the data for each working condition is randomly taken as the training set and the remaining 20% as the test set.

Based on the dispersion curve of the ultrasonic guided wave propagation in the aluminum plate and the specimen size, the propagation time of the ultrasonic guided wave to the receiving piezoelectric sheet was estimated to be about 52.2 μs, and all the direct wave signals arrived at 125.1 μs, i.e., the complete direct wave signal length was about 72.9 μs. To compare the effects of different signal components on the classification performance, four different signal lengths of 25 μs, 50 μs, 75 μs, and 100 μs were selected. Four different signal lengths of 100 μs were selected to produce the dataset.

5 Experimental Results

The task of this section is to use the small amount of experimental data obtained in Sect. 4.1 to identify four different bolt loosening states, that is, to build a model to solve a four-classification problem, and evaluate the classification accuracy of the model. For the two proposed classification methods, the classification performance of the proposed two methods is compared, and the influence of the number of principal components and the kernel function on the classification performance is explored, which is introduced in detail below.

5.1 Results of PCA

Firstly, the time domain signals of ultrasonic guide waves collected at different torque levels were compared, as shown in Fig. 7. It is obvious that the ultrasonic guide wave signals at different torque levels do not show a more obvious pattern, and it is almost impossible to distinguish the signals of each working condition by waveforms only.

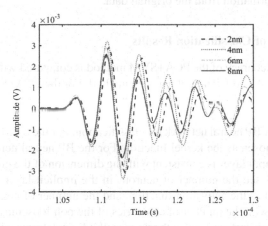

Fig. 7. Comparison of time-domain guided wave signals at different torques

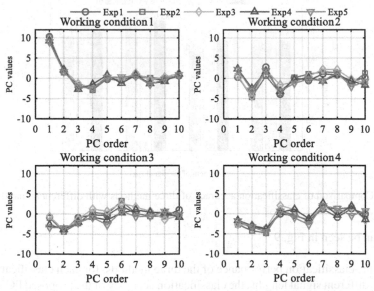

Fig. 8. Comparison of the first tenth-order principal components at different torque levels (75 μs for example)

Figure 8 shows the comparison of the first 10 orders of principal components of the guided wave signal at different torque levels. Here, a 75 μs-long signal is used as an example, and five sets of experimental data are selected for each working condition. It can be seen that after extracting the principal components, each working condition presents a significant difference. This is because the PCA technique selects the direction with the largest data variance for projection, maximizing the variance of the data while retaining more information from the original data.

5.2   Comparison of Classification Results

The classification accuracy of the PCA+SVM method is compared with that of the SVM method without PCA and BP neural network method. For the PCA+SVM method, here the first 10th order principal components of the input data are extracted as the input to the SVM; for the SVM and BP neural network methods, the original data are directly input to the SVM or BP neural network for classification. For the SVM method, the same RBF function is chosen as the kernel function. For the BP neural network, the number of neurons in the input layer is consistent with the dimension of the guided wave signal, an implicit layer is set, the number of neurons in the implicit layer is 128, the ReLU function is selected as the activation function, and the number of neurons in the output layer is consistent with the number of categories of the bolt loosening.

The classification performance of the three methods on datasets with different length signals is collated as shown in Fig. 9.

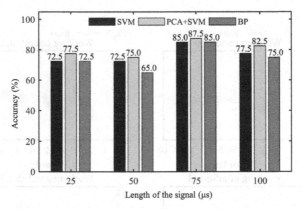

Fig. 9. Comparison of classification accuracy of three methods with different signal duration

As can be seen in Fig. 9

(1)   For the classification performance of the three methods, for each classification task with different signal lengths, the classification accuracy of the proposed PCA+SVM method is higher than that of the SVM and BP neural network methods only, which indicates that for this classification task, the use of PCA to extract the principal components of the original ultrasonic guide wave signal is beneficial for the SVM

classifier to better perform the classification task, reflecting the advantages of the proposed method.

(2) For different lengths of the guided wave signals, the classification accuracy reaches the highest when the selected signal length is 75 μs, reaching 85%, 87.5% and 85%, respectively, while the classification is poorer for the remaining three signal lengths. The reason for this analysis is that when the selected signal length is short (25 μs and 50 μs), the signal contains less structural information, which makes the classification accuracy lower, while when the selected signal is too long (100 μs), it contains part of the boundary reflection signal, which easily confuses the direct wave information and thus reduces the classification accuracy. This suggests that the optimal classification performance can be achieved by using the complete direct wave signal.

5.3 Effect of the Number of Principal Components on the Classification Performance of PCA+SVM Method

To investigate the effect of the number of principal components on the performance of the PCA+SVM method, a classification task with a signal length of 75 μs was selected for a comparative study, and the classification performance of the PCA+SVM method was compared when a series of different numbers of principal components were selected.

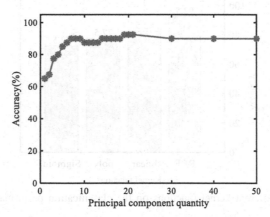

Fig. 10. Effect of the number of principal components on classification performance

It can be seen from Fig. 10 that the number of selected principal components has a great influence on the classification performance of the PCA+SVM method. When the number of selected principal components is small, the classification accuracy is low because the effective information contained in the principal component vector is small, and as the number of principal components increases, the effective information contained in the principal component vector gradually increases and the classification accuracy gradually increases.

After the number of principal components is greater than 7, the classification accuracy gradually tends to saturate with small fluctuations because the effective information has been fully extracted and the information contained no longer increases with the number of principal components. According to the results, for this classification task, the highest classification accuracy of 92.5% was achieved when the number of principal components was selected as 19/20/21.

5.4 Effect of Kernel Function on the Classification Performance of PCA+SVM Method

To investigate the effect of kernel function on the performance of the PCA+SVM method, using the above findings, the first 20 order principal components of the signal were selected and a classification task with a signal length of 75 μs was chosen for a comparative study to compare the classification performance of PCA+SVM method when linear function, polynomial, RBF function and Sigmoid function were chosen as kernel functions, respectively, and the comparison of classification accuracy is shown in Fig. 11.

From Fig. 11, it can be seen that different kernel functions have a large impact on the performance of the PCA+SVM method, and for this classification task, the best classification results can be obtained by choosing the RBF function.

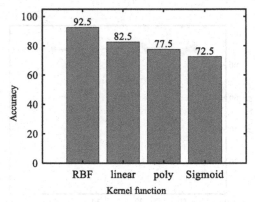

Fig. 11. Effect of different kernel functions on the classification performance of PCA+SVM method

6   Conclusion

In this paper, a single-bolt structural loosening detection method based on PCA and SVM techniques is proposed, and the method is validated with a single-bolt joint as the research object. Also, a comparison with the method using only SVM is made. In addition, the effects of signal length, the number of principal components, and the choice of kernel function on the classification performance are explored. End-to-end detection is achieved and the amount of data is reduced.

The experimental results show that the classification performance of the PCA and SVM technique proposed in this paper is significantly better than that of the method using only SVM, and the classification accuracy can be improved by 2%–5%. This method can automatically extract the characteristics of the guided wave signal and accurately identify the bolt loosening state. For different signal lengths, the best classification effect is achieved by the complete direct wave signal; the classification performance is poor when the number of principal components is small, which contains less information about the signal features, and the classification accuracy gradually increases and tends to saturate as the number of principal components increases; the highest accuracy is achieved by the kernel function RBF function.

Acknowledgement. This study is supported by the National Natural Science Foundation of China (Grant No. 52075445). This study is also supported by Key Research and Development Program of Shaanxi (Program No.2021ZDLGY11-10).

References

1. Wang, Y., Qin, X.: Progress in health monitoring technology of composite connected structures. J. Compos. Mater. **33**(1), 1–16 (2016)
2. Mitra, M., Gopalakrishnan, S.: Guided wave based structural health monitoring: a review. Smart Mater. Struct. **25**, 053001 (2016)
3. Zhu, Y., Li, F., Hu, Y.: The contact characteristics analysis for rod fastening rotors using ultrasonic guided waves. Measurement **151**, 107149 (2020)
4. Zhang, Z., Xiao, Y., Su, Z., et al.: Continuous monitoring of tightening condition of single-lap bolted composite joints using intrinsic mode functions of acoustic emission signals: a proof-of-concept study. Struct. Health Monit. **18**(4), 147592171879076 (2018)
5. Wang, T., Song, G., Wang, Z., et al.: Proof-of-concept study of monitoring bolt connection status using a piezoelectric based active sensing method. Smart Mater. Struct. **22**(8), 087001 (2013)
6. Amerini, F., Meo, M.: Structural health monitoring of bolted joints using linear and nonlinear acoustic/ultrasound methods. Struct. Health Monit. **10**(6), 659–672 (2011)
7. Kundu, T., Haynes, C., Yeager, M., et al.: Monitoring bolt torque levels through signal processing of full-field ultrasonic data. In: SPIE Smart Structures & Materials + Nondestructive Evaluation & Health Monitoring, p. 906428. International Society for Optics and Photonics (2014)
8. Du, F., Xu, C., Zhang, J.: A bolt preload monitoring method based on the refocusing capability of virtual time reversal. Struct. Control. Health Monit. **26**(8), e2370 (2019)
9. Wang, F., Ho, S.C.M., Song, G.: Monitoring of early looseness of multi-bolt connection: a new entropy-based active sensing method without saturation. Smart Mater. Struct. **28**(10), 10LT01 (2019)
10. Parvasi, S.M., Ho, S.C.M., Kong, Q., et al.: Real time bolt preload monitoring using piezoceramic transducers and time reversal technique—a numerical study with experimental verification. Smart Mater. Struct. **25**(8), 085015 (2016)
11. Nazarko, P., Ziemianski, L.: Force identification in bolts of flange connections for structural health monitoring and failure prevention. Procedia Struct. Integr. **5**, 460–467 (2017)
12. Liang, D., Yuan, S.F.: Decision fusion system for bolted joint monitoring. Shock Vib. **2015**(pt. 1), 1–11 (2015)

13. Mita, A., Fujimoto, A.: Active detection of loosened bolts using ultrasonic waves and support vector machines. In: Proceedings of the 5th International Workshop on Structural Health Monitoring, pp. 1017–1024. IWSHM (2005)
14. Wang, F., Chen, Z., Song, G.: Monitoring of multi-bolt connection looseness using entropy-based active sensing and genetic algorithm-based least square support vector machine. Mech. Syst. Signal Process. **136**, 106507 (2020)
15. Wang, F., Chen, Z., Song, G.: Smart crawfish: a concept of underwater multi-bolt looseness identification using entropy-enhanced active sensing and ensemble learning. Mech. Syst. Signal Process. **149**, 107186 (2021)
16. Mínguez, J.M., Vogwell, J.: Effect of torque tightening on the fatigue strength of bolted joints. Eng. Fail. Anal. **13**(8), 1410–1421 (2006)
17. Müser, M., Popov, V.L.: Contact mechanics and friction: physical principles and applications. Tribol. Lett. **40**(3), 395 (2010)
18. Zhou, Z.: Machine Learning. Tsinghua University Press, Beijing (2016)

Detection and Identification of Digital Display Meter of Distribution Cabinet Based on YOLOv5 Algorithm

Yanfei Zhou[1], Yunchu Zhang[1,2]([✉]), Chao Wang[1], Shaohan Sun[1], and Jimin Wang[1]

[1] Shandong Jianzhu University, Jinan 250101, China
yczhang@sdjzu.edu.cn
[2] Shandong Key Laboratory of Intelligent Buildings Technology, Jinan 250101, China

Abstract. Aiming at the problem of low recognition accuracy of digital display meter readings when the inspection robot performs inspection tasks, a YOLOv5-based digital display meter detection and recognition algorithm for distribution cabinets is proposed. For the digital display meter image captured by the inspection robot spherical camera, the YOLOv5 model is used to locate the target character area. After scale normalization, image correction, filtering and noise removal and other image pre-processing operations, the character recognition is completed in combination with the traditional machine vision algorithm, and the reading results are automatically output. For the characteristics of digital tube characters, the threading method, support vector machine algorithm and PaddleOCR algorithm are used for comparison, and a suitable algorithm model is selected to recognize numeric, alphabetic and decimal point characters. The experimental results show that the accuracy of detecting and identifying digital display meters using the YOLOv5 model and PaddleOCR algorithm is 95.3%.

Keywords: Digital display meter · YOLOv5 · Machine vision · Character recognition · Deep learning

1 Introduction

The power distribution cabinet is the final stage of electric power transmission and distribution system equipment, which is widely used in factories, building parks, urban infrastructure and other energy-using places. Digital display meter has the advantages of convenient meter reading, fast recording and accurate reading, which is widely used in power distribution cabinet. Due to technical limitations and economic cost constraints, manual observation of meter data has become the mainstream way of distribution cabinet inspection work. However, the manual observation method has the disadvantages of low efficiency, high leakage rate, and large workload to a certain extent. Therefore, accurate real-time detection of digital meter readings is important to ensure power supply safety, improve power supply efficiency, reduce line loss and save energy.

With the development of China's intelligent information era, manual inspection can no longer meet the needs of industrial sites. Real-time monitoring and accurate identification of digital meter data are the current research's main content. Traditional digital meter detection and identification algorithms include template matching, support vector machines (SVM), threading, etc. Ju Gao [1] et al. proposed a method to add character feature matching based on the template matching method, which solves the problem of misjudgment of similar characters to a certain extent, but the ability to recognize different tilted characters of the same font is very poor. Yanling Zhang [2] et al. proposed using SVM to recognize the instrument panel parameter symbols, which improved the accuracy level of the instrument panel and the recognition rate of special parameter symbols. However, for the instrument image data with different scenes and environmental noise, there are still problems such as misjudgment, which does not apply to the processing of large sample data. Wenliang Liu [3] proposed to use the improved multi-threshold localization segmentation method and threading method to recognize the digital characters of seven-segment digital tube type digital display meters, and the correct rate was greatly improved. However, under the influence of lighting conditions and shooting angles, the phenomenon of digital misjudgment and failure to judge the results may occur.

In recent years, the rapid development of deep learning algorithms has resulted in the emergence of many target detection algorithms based on deep learning that solve some bottlenecks that traditional machine vision methods would encounter [4]. Digital instrument detection and recognition algorithms based on deep learning mainly address two problems: digital area localization and character recognition. Xun Xiong [5] et al. proposed to use contour extraction algorithm to locate dial character regions and use the improved convolutional memory neural network model (CLSTM) for character recognition, which improved the accuracy by 4.2% compared with the traditional LSTM network. Longyu Zhang [6] proposed an improved Tiny-EAST algorithm based on scene text detection algorithm to detect and locate and identify target characters of digital meters with 99.7% character detection accuracy. Peng Tang [7] et al. proposed to use Mask-RCNN algorithm to detect and identify digital meters, which has worse detection rate than YOLOv3 algorithm, but has higher accuracy. Chaoran Qu [8] et al. proposed to use DB segmentation algorithm for character region detection, and by introducing the attention mechanism improved CRNN algorithm for character recognition of the detected data, and the results show that the character recognition rate can reach 96%, but the detection rate on the test set is low.

This paper proposes a digital display meter detection and recognition algorithm based on YOLOv5. Using the images of various devices in the distribution cabinet collected by the inspection robot as the data set, YOLOv5s deep learning framework is selected to determine the target character area of the digital display meter by training the model. Then the localized character image is segmented into single characters and input to the character recognition model in turn to complete the detection and recognition of the digital display meter. The experimental results show that digital display meter reading recognition accuracy reaches 95.3%.

2   Method

Industrial meter recognition is different from text recognition in general-purpose situations, which can be disturbed by complex environments and more factors, causing reading recognition errors. Using a traditional machine vision algorithm or OCR algorithm for instrument character detection and recognition does not solve the above problems, so this paper proposes a deep learning-based instrument character region detection and recognition algorithm for complex backgrounds. Digital display meter reading recognition mainly includes character region detection and positioning, character segmentation and character recognition. This section will introduce the algorithm model and method selection for each part in three parts.

2.1 Instrument Character Area Detection and Positioning

There are many different power devices on the distribution cabinet, such as pointer meters, digital display meters and indicators. If the reading recognition of the digital display meter is to be completed, the primary key is to detect the character area of the digital display meter from different devices.

The mainstream deep learning algorithms for target detection are the YOLO [9–12] series of algorithms. YOLO is a single-stage target detection algorithm that transforms the original target detection problem into a regression problem by directly performing classification probability regression and envelope coordinate regression on the input image to achieve target detection with fast detection capability. The YOLOv5 algorithm has comparable performance with YOLOv4 [12], but the YOLOv5 model uses the Pytorch framework, which makes the algorithm inference faster and more suitable for deployment in engineering.

The network structure of YOLOv5 consists of four parts: Input, Backbone, Neck and Head (see Fig. 1). After inputting the images, mosaic [13] data enhancement, adaptive anchor calculation, and image scaling are used sequentially to expand the dataset to improve the generalization of the target detection model. The backbone network uses Focus structure and C3Darknet-53 structure. Neck consists of FPN [14] and PANet [15] structures. Head uses CIoU_loss [16] as the loss function for bounding box regression, while redundant prediction boxes are filtered using DIoU_NMS [17].

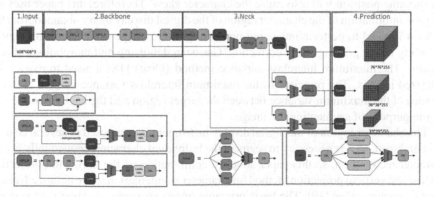

Fig. 1. YOLOv5s network diagram

According to the depth and width of the backbone network, YOLOv5 is divided into four magnitudes of models, which are noted as YOLOv5s, YOLOv5m, YOLOv5l and YOLOv5x. With the increase of model parameters and the model frame sequentially, the target detection capability gradually improves. To achieve the accurate localization of the digital display meter's character area and meet industrial sites' deployment requirements, this paper selects the YOLOv5s model to classify and locate the digital display meter on the distribution cabinet panel. YOLOv5s is the network with the smallest depth and width of the feature map in the YOLOv5 series, and its model parameter number is only 7.5M, which is suitable for deployment in the actual industrial sites. The results of the meter character area detection tested by the trained YOLOv5 model are shown in Fig. 2.

(a)Original image   (b)test result image

Fig. 2. Instrument character area detection results chart

2.2 Instrument Character Segmentation

In the actual inspection, the light source divergence of the digital tube will cause the character connection problem in the digital display meter image, and the incorrect camera shooting position will also cause the character skew. Therefore, this paper uses the position information of the character region of the digital display meter obtained from the YOLOv5 model to perform image pre-processing of the image of the character region, including image graying, skew correction, Gaussian denoising and morphological processing. The maximum interclass variance method (Otsu) [18] is used to extract the binarized images. The principle of the maximum interclass variance method is to use the idea of the maximum variance between the target region and the background region for the purpose of segmenting the image.

The character region of the digital display instrument is composed of several characters combined, so it is necessary to segment the individual characters sequentially before character recognition, so this paper adopts a combination of the horizontal projection method and vertical projection method for character segmentation of the binarized instrument character image [19]. The basic principle of the projection method is to perform the pixel statistics in horizontal and vertical directions on the binary image, and the peak and trough positions of the instrument character information in horizontal and vertical directions can be obtained. The continuous peaks are character regions, and the constant troughs are character intervals according to which individual characters can be segmented. The result of character segmentation by the projection method is shown in Fig. 3.

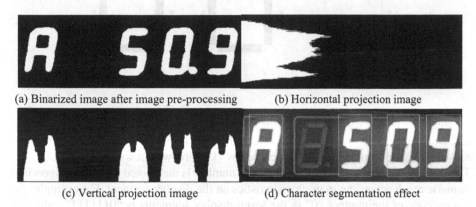

(a) Binarized image after image pre-processing   (b) Horizontal projection image

(c) Vertical projection image   (d) Character segmentation effect

Fig. 3. Projection method to split the character effect

2.3 Instrument Character Recognition

Traditional character recognition algorithms include template matching, SVM and threading method. Template matching requires many templates to be prepared in advance, and the character characteristics in different environments are different, so character recognition using a template matching algorithm is not very applicable. This section introduces the principles of the threading method and SVM to recognize characters. In addition, the PaddleOCR [20] algorithm introduced by Baidu Flying Pulp is also referred to as recognizing meter characters.

Threading Method. According to the characteristics of the digital tube-type characters, character recognition can be realized by the threading method. The principle of the threading method is to use the basic features of the 7 display segments in the 7-segment digital tube to complete the accurate judgment of the characters by extracting the feature information [21]. The seven display segments of the digital tube are labeled as ABCDEFG in the clockwise direction, as shown in Fig. 4., and the horizontal and vertical segmentation lines L1 to L7 are made in turn, corresponding to the seven display segments.

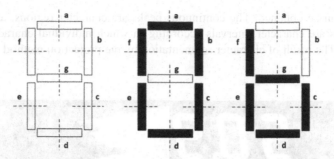

Fig. 4. Seven-segment digital tube

Scan the display segments, count the number of white points, and set the appropriate threshold value. If it is greater than or equal to the threshold value, that is, the field exists in the pen segment, it will be recorded as "1", otherwise it will be recorded as "0". This results in a series of binary codes. Each number is then noted as the corresponding numeric result according to its characteristics on the display segment. For example, the binary code of the number "0" in the seven display segments is "0111111", which is converted into decimal as "63" and recorded as "0". The binary code of the letter "b" in the seven display segments is "1111100", which is converted to "124" in decimal and is recorded as "B". The results of 0 to 9 and the corresponding codes of ABC are shown in Table 1.

Table 1. Digital tube character code correspondence table.

	a	b	c	d	e	f	g	Binary	Decimal
0	√	√	√	√	√	√		0111111	63
1		√	√					0000110	6
2	√	√		√	√		√	1011011	91
3	√	√	√	√			√	1001111	79
4		√	√			√	√	1100110	102
5	√		√	√		√	√	1101101	109
6	√		√	√	√	√	√	1111101	125
7	√	√	√					0000111	7
8	√	√	√	√	√	√	√	1111111	127
9	√	√	√	√		√	√	1101111	103
A	√	√	√		√	√	√	1110111	119
B			√	√	√	√	√	1111100	124
C	√			√	√	√		0111001	57

Due to the special position of the decimal point, it is easy to stick together with other characters after image pre-processing, which makes the threading method unable to realize the decimal point recognition. According to the location characteristics of the decimal point located in the lower right corner of the character, the process of lower right corner counting pixels is used for decimal point recognition. The specific method is to select a fixed region in the lower right corner of a single character picture, iterate the pixel value of the part, set a threshold value, and determine whether the decimal point exists by counting the number of white points.

SVM. SVM is a practical data classifier that is easier to apply than neural networks. The goal of SVM is to generate a model that can predict the target value [22]. Given a set of points $\{(x_1, y_1), (x_2, y_2), \ldots\ldots, (x_m, y_m)\}$, where $x_i \in R^n$ denotes the sample points, and $y_i \in \{-1, 1\}$ represents the class to which the corresponding sample point x_i belongs, the SVM requires solving this optimization problem as follows,

$$\min_{w,b,\xi} \frac{1}{2} w^T w + C \sum_{i=1}^{i} \xi_i \tag{1}$$

$$y_i(w^T \phi(x_i) + b) \geq 1 - \xi_i, \xi_i \geq 0 \tag{2}$$

The training vector x_i is projected to a high-dimensional space defined by ϕ. The SVM is to find a linearly differentiable hyperplane in this high-dimensional space. $C > 0$ is the penalty parameter of the training error. $K(x_i, x_j) = \phi(x_i)^T \phi(x_j)$ is the kernel function, and the radial basis kernel function (RBF) is chosen in this paper, where $K(x_i, x_j) = \exp(-\gamma \|x_i - x_j\|^2)$ and $\gamma = 0.01$ are the kernel parameters.

All images obtained by character segmentation, including numbers, letters and other characters, are uniformly normalized to a standard size of 28×28 and named with the value of each character as a classification folder as the sample images for SVM training in this paper and some of the instrumentation character samples are shown in Fig. 5.

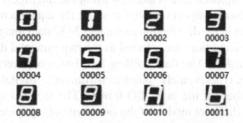

Fig. 5. Part of the instrument character samples

This paper combines the Principal Component Analysis (PCA) algorithm [23] for extracting character sample image features and then uses SVM for classification. The basic principle of PCA algorithm is to expand the n × n images into $1 \times n^2$ one-dimensional vectors. Suppose there are m samples, which are processed by the PCA algorithm to form an m × n2 array, with each row representing one sample. Each row represents one sample. In this paper, the image size is 28×28, the sample category is

14, and the training sample for each type is 240, so each image is expanded into a 1×784 one-dimensional vector, which eventually forms a 3360×784 array file as the data set for SVM training.

PaddleOCR. PaddleOCR is an ultra-lightweight OCR (Optical Character Recognition) system open-sourced by Baidu, which mainly consists of three parts: DB text detection [24], detection frame correction [25] and CRNN text recognition [26]. In this paper, we use PaddleOCR source code to recognize the meter character area detected by YOLOv5 directly, and the detection effect is shown in Fig. 6. To improve the recognition accuracy, this paper crops and saves the character region images based on the location information of the meter character regions detected by YOLOv5, construct the dataset, and retrains the model based on the PaddleOCR source code to achieve the character region recognition of meters.

3 Experimental Results and Analysis

3.1 Experimental Platform

In this paper, an intelligent inspection robot equipped with a visual inspection system is used as the experimental platform, and the camera equipment is a Hikvision thermal imaging dual-spectrum network intelligent dome camera. The experimental hardware is a PC, the operating system is Windows 10 64-bit system, the PC processor model is Intel(R) Core(TM) i7-7700, the graphics card model is NVIDIA GeForce GTX 1050Ti. The algorithm uses the deep learning Pythorch framework, and the programming language is Python.

3.2 Acquisition and Labeling of Experimental Data Sets

In this paper, the experimental data is taken by using the intelligent spherical camera of the hanging rail intelligent inspection robot to shoot the equipment operation status of the distribution cabinet in the distribution room, and 2432 data images of different types and different lighting conditions are selected as the experimental data set to ensure the reliability of the experiment. Use the LabelImg labelling tool to rectangular label boxes with a total of 3 device types, namely "pointer instrument", "digital meter", and "colour indicator", and save the text file in YOLO format. The labelled images are shown in Fig. 6. To distinguish different models of the same type of equipment, five other equipment categories are derived, namely "cos instrument", "pointer instrument A (pointer instrument with indicator)", "digital meter A (circuit breaker digital meter)", "white indicator", and "meter light (indicator on pointer meter)", a total of eight categories.

Fig. 6. Labelimg labeled images

3.3 Model Training

The YOLOv5 digital meter character region localization model uses the deep learning framework Pytorch, with SGD stochastic gradient descent and learning rate decay strategies selected as hyper-parameters to train the network. The initial learning rate is 0.01, the batch size is 8, the number of iterative rounds (epochs) is 150, and the input image resolution is 640 × 640. The learning rate momentum factor (SGD momentum) is 0.937, and the weights of the loss function are $\lambda_{box} = 0.05$, $\lambda_{cls} = 0.5$ and $\lambda_{obj} = 1.0$.

Considering the small sample size of characters, this paper uses the SVC (classification algorithm of support vector machine) of the SVM algorithm in the Scikit-learn machine learning library to complete the digital classification training. All single-character images obtained by projection segmentation are normalized to 28 × 28 pictures and then used as the data set for digital classification training. The kernel function (kernel) is selected as "RBF" (Gaussian kernel function), the kernel function coefficient gamma is 0.01, and the penalty coefficient C is 15.

The PaddleOCR character recognition model uses the deep learning framework Pytorch. The backbone network is selected as the CRNN recognition model of Resnet34_vd, and the hyperparameters are chosen as the stochastic gradient descent method SGD with the restart. The learning rate descent method is selected as the cosine annealing function Cosine. The initial learning rate is 0.001, and the number of iterations (epoch_num) is 200. The configuration file also has default data enhancement, including colour space transformation (cvtColor), blur, jitter, gauss noise, random crop, perspective, colour inversion (reverse), etc.

3.4 Results Analysis

Analysis of Instrument Character Area Detection Results. In this paper, a total of 2432 inspection images were selected as the dataset for YOLOv5 model training, of which 1946 were in the training set, accounting for 80%; 243 were in the test set, accounting for 10%; and 243 were in the validation set, accounting for 10%. The training was completed in 9 h using the YOLOv5s model on PC, and the training results are shown in Fig. 7.

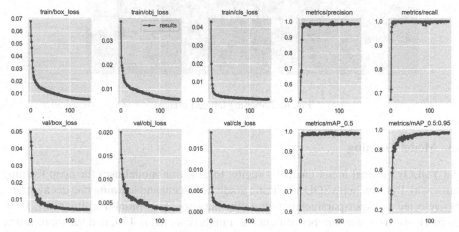

Fig. 7. YOLOv5s model training results image

The horizontal coordinates of YOLOv5s model training results represent the number of training rounds epoch, and the main indicators are the target and envelope loss curves loss, Precision, Recall, and mAP@0.5, mAP@0.5:0.95 for the training and validation sets in order. 0.5:0.95 value approaches 1 as the number of training rounds increases, and the training effect is good.

In this study, mAP (mean Average Precision) is used as the evaluation metric of the model. P(Precision) represents the proportion of correctly identified samples to all identified samples in the dataset, and R(Recall) represents the proportion of correctly identified samples to all models and is calculated as follows.

$$P = \frac{TP}{TP + FP} \times 100\% \tag{3}$$

$$R = \frac{TP}{TP + FN} \times 100\% \tag{4}$$

TP indicates the number of correctly identified distribution cabinet devices, FP shows the number of incorrectly identified distribution cabinet devices, and FN shows the number of missed identified distribution cabinet devices.

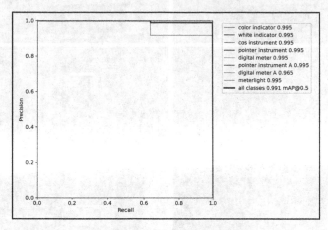

Fig. 8. P_R curve of YOLOv5 model training

The P-R (Precision-Recall) curve can be plotted according to Eq. (3), as shown in Fig. 8. The average precision AP (Average Precision) of a single category can be calculated by averaging the Precision values of the P-R curves. The mAP is the average value of each classification accuracy AP, and the relevant formula of mAP is as follows:

$$AP_i = \int_0^1 p(r)dr, i = 1, 2, ..., n \qquad (5)$$

$$mAP = \frac{\sum AP_i}{n}, i = 1, 2, ..., n \qquad (6)$$

where p is the precision rate, r is the recall rate, and n is the number of categories. The effect of using the YOLOv5s model to detect the character region of the distribution cabinet meter is shown in Fig. 9.

Fig. 9. Distribution cabinet instrumentation character area detection effect

Analysis of Instrument Character Recognition Results. In this paper, after completing the detection and positioning of the character area of the meter using YOLOv5, the character recognition is performed by character area cropping, image pre-processing and character segmentation operations, using the threading method, SVM and PaddleOCR, respectively. According to the character characteristics of the digital display meter in the data set of this paper, the decimal point is connected with the numeric characters in the segmented single character picture, which is difficult to recognize. The threading method and SVM algorithm determine whether the decimal point exists by scanning the lower right corner of the single character binarization picture to count the number of white points and output the character recognition results sequentially according to the coordinate order. The instrument character recognition results are shown in Fig. 10.

Fig. 10. Character recognition results

Character recognition was tested using the threading method, SVM and PaddleOCR for comparison, using a verification set of 240 digital display meter pictures, and the test results are shown in Table 2.

Table 2. Comparison of character recognition results of different algorithms

Algorithm	Accuracy (%)	Relative error (%)	FPS
Threading method	77.9	22.1	62
SVM	93.8	6.2	71
PaddleOCR	95.3	4.6	257

As shown in Table 2, the relative errors of the SVM algorithm model and PaddleOCR detection on the test set are significantly reduced, and the correct rates are improved by 15.9% and 17.4%, respectively, relative to the threading method. The recognition speed of the PaddleOCR algorithm is faster, and it only takes 4 milliseconds (ms) to process one frame of the image, which can ensure the real-time performance of the algorithm. The experimental results show that using the YOLOv5 algorithm to detect the meter character region, combined with the PaddleOCR algorithm to recognize characters, can improve the accuracy of digital display meter detection and recognition in complex scenes, and the final meter detection and recognition accuracy is 95.3%.

4 Conclusion and Prospect

For the actual needs of digital display meter detection and recognition in the process of robot inspection, this paper proposes a YOLOv5-based digital display meter detection and recognition algorithm. The method uses the YOLOv5s model to locate the character region of the digital display meter image, obtains the target character image, and uses the PaddleOCR algorithm model to complete the character recognition, improving the accuracy and speed of digital display meter recognition in complex scenes. Future work will consider improving the network structure to increase character recognition accuracy while maintaining speed.

References

1. Gao, J., Ye, H.: An effective algorithm for secondary recognition of digital images of water meters. J. Southeast Univ. (Nat. Sci. Ed.) **43**(S1), 153–157 (2016)
2. Zhang, Y.L.: Research on instrument dial parameter symbol recognition technology. Guangdong University of Technology (2008)
3. Liu, W.L.: Research and implementation of digital recognition in seven-segment digital display instrumentation. Dalian University of Technology (2013)
4. Hua, Z.X., Shi, H.B., Luo, Y., et al.: Identification of digital instrumentation detection in substations based on lightweight YOLO-v4 model. J. Southwest Jiaotong Univ., 1–11 (2021)
5. Xiong, X., Chen, X.D., Wu, L., Lin, X.L.: Digital dial reading recognition based on convolutional memory neural network. Comb. Mach. Tools Autom. Mach. Technol. **07**, 72–75 (2019)
6. Zhang, L.Y.: Research on detection and identification algorithm of substation instrumentation based on deep learning. Tianjin University of Technology (2021)
7. Tang, P., Liu, Y., Wei, H.G., Dong, X.Y., et al.: Automatic identification algorithm of digital meter readings based on mask-RCNN offshore booster station. Infrared Laser Eng. **50**(S2), 163–170 (2021)
8. Qu, C.R., Chen, L.W., Wang, J.S., Wang, S.G.: Research on industrial digital meter recognition algorithm based on deep learning. Appl. Sci. Technol., 1–7 (2022)
9. Redmon, J., Divvala, S., Girshick, R., et al.: You only look once: unified, real-time object detection. In: Computer Vision & Pattern Recognition (2016)
10. Redmon, J., Farhadi, A.: YOLO9000: better, faster, stronger. In: IEEE Conference on Computer Vision & Pattern Recognition, pp. 6517–6525 (2017)
11. Redmon, J., Farhadi, A.: Yolov3: an incremental improvement. arXiv preprint arXiv:1804.02767 (2018)
12. Bochkovskiy, A., Wang, C.Y., Liao, H.Y.M.: Yolov4: optimal speed and accuracy of object detection. arXiv preprint arXiv:2004.10934 (2020)
13. Yun, S., Han, D., Oh, S.J., et al.: CutMix: regularization strategy to train strong classifiers with localizable features. In: Proceedings of the IEEE/CVF International Conference on Computer Vision, pp. 6023–6032 (2019)
14. Lin, T.Y., Dollár, P., Girshick, R., et al.: Feature pyramid networks for object detection. In: Proceedings of the IEEE Conference on Computer Vision and Pattern Recognition, pp. 2117–2125 (2017)
15. Liu, S., Qi, L., Qin, H., et al.: Path aggregation network for instance segmentation. In: Proceedings of the IEEE Conference on Computer Vision and Pattern Recognition, pp. 8759–8768 (2018)

16. Rezatofighi, H., Tsoi, N., Gwak, J.Y., et al.: Generalized intersection over union: a metric and a loss for bounding box regression. In: Proceedings of the IEEE/CVF Conference on Computer Vision and Pattern Recognition, pp. 658–666 (2019)

17. Zheng, Z., Wang, P., Liu, W., et al.: Distance-IoU loss: faster and better learning for bounding box regression. In: Proceedings of the AAAI Conference on Artificial Intelligence, vol. 34, no. 07, pp. 12993–13000 (2020)

18. Otsu, N.: A threshold selection method from gray-level histograms. IEEE Trans. Syst. Man Cybern. **9**(1), 62–66 (1979)

19. Fang, X.W., Fu, X.W.: ID card segmentation algorithm based on projection method and Caffe framework. Comput. Eng. Appl. **53**(23), 113–117 (2017)

20. Qiu, D., Weng, M., Yang, H.-T.: A fast lane line detection method based on improved probabilistic Hough transform. Comput. Technol. Dev. **30**(5), 43–48 (2020)

21. Song, Y.Y., Tang, D.L., Wu, X.L., et al.: Study on digital tube image reading by combining improved threading method and HOG+ SVM method. Comput. Sci. **48**(S2), 396–399+440 (2021)

22. Wang, X.D., Wei, C.B., Feng, H.R., et al.: Digital identification method for digital display instruments in substation protection rooms based on SVM. Electron. Meas. Technol. **42**(2), 92–95 (2019)

23. Chen, X., Xu, S.Y., et al.: Identification of braided river reservoir inclusions based on support vector machine and principal component analysis. J. China Univ. Petr. (Nat. Sci. Edn) **45**(04), 22–31 (2021)

24. Liao, M., Wan, Z., Yao, C., et al.: Real-time scene text detection with differentiable binarization. In: Proceedings of the AAAI Conference on Artificial Intelligence, vol. 34, no. 07, pp. 11474–11481 (2020)

25. Yu, D., Li, X., Zhang, C., et al.: Towards accurate scene text recognition with semantic reasoning networks. In: Proceedings of the IEEE/CVF Conference on Computer Vision and Pattern Recognition, pp. 12113–12122 (2020)

26. Li, W., Cao, L., Zhao, D., et al.: CRNN: Integrating classification rules into neural network. In: The 2013 International Joint Conference on Neural Networks (IJCNN), pp. 1–8 IEEE (2013)

Analysis of Autoencoders
with Vapnik-Chervonenkis Dimension

Weiting Liu and Yimin Yang[✉]

Department of Computer Science, Lakehead University, Thunder Bay, Canada
{wliu27,yyang48}@lakeheadu.ca

Abstract. In statistical learning, the Vapnik-Chervonenkis(VC)-dimension has been widely used to analyze single-layer neural networks such as Perceptron and Support Vector Machine while utilizing it for multilayer networks has rarely been explored. This motivates us to introduce the VC-dimension method to autoencoder, one of important multilayer networks. The paper proposes several theoretical observations of analyzing the relationship among network architectures, activation functions, and the learning capacity and effectiveness of autoencoders. We also provide a theoretical VC-limitation result to quantify the boundary of hidden neurons in an Autoencoder.

Keywords: VC-dimension · Autoencoder · Space complexity

1 Introduction

The Vapnik-Chervonenkis (VC) theory is closely related to stability, aiming to explain the learning process for a neural network from a statistical learning view. The concept of VC-dimension was first proposed into netural network by Vapnik [14]. The VC dimension can predict the upper bound of the probability of the test error of the classification model. Furthermore, Vapnick [4] also investigated the mature computations for computing the VC dimension on the Perceptrons and the Support Vector Machine (SVM).

It is widely accepted that the VC-dimension could not only reflect the learning capacity of a neural network, but also can be used to evaluate the sample complexity, and to finalize the optimal model architecture. However, VC-dimension is not a one-fit-for-all solution, which needs to be further polished along with a flexible network structure. Although the results of VC-dimension for SVM or Preceptron are widely used to analysis the performance of the networks, we have been rarely found the theoretical analysis of Autoencoders with the VC-dimension.

In this paper, we exploit VC-dimension techniques to autoencoders for analyzing the optimal network architecture. We obtain a general solution of VC-dimension in Eq. 34 for any Autoencoders. Furthermore, by understanding the statistical principle of VC-dimension in Autoencoder, we have calculated a general solution of VC-dimension to quantify the optimal network structure of

H. Zhang et al. (Eds.): NCAA 2022, CCIS 1637, pp. 316–326, 2022.
https://doi.org/10.1007/978-981-19-6142-7_24

Autoencoders. In addition, in the process of calculating the VC-dimension, we also investigate that all the structural factors in Autoencoders significantly influence the VC-dimension limitations including neuron sizes, activation functions, etc. Therefore, the VC-dimension can be used as an evaluation metric to measure the performance of Autoencoders. There are many factors to consider when building a neural network system. Although It is still mainly used to test the same neural network with different datasets as evaluation, using VC-dimension for performance analysis is a more general and theoretical manner.

2 Related Work

In the past decades, many important results have been proposed in the field of VC-dimension to analysis several specific behaviors in neural networks. In 1989, Abu [1] indicated that VC-dimension could use as a theoretical measurement to quantify the learning capacity of neural networks. This years, using VC-dimension for evaluation is also a topic that has been studied. Bartlett [3] and Pinto [10] select sample complexity as assessed by the VC dimension. And transform the data by increasing the dimension of the input features based on the sample complexity evaluated by the VC dimension. Chen [7] provided theoretical insights that SVM is actually designed from both VC-dimension theory and principle of structural risk minimization, obtaining better generalization performance with small, non-linearity, high dimensionality samples.

2.1 Statistics Concepts on VC-Dimension

It is well known that two of the most important aspects of machine learning models are how well the model generalizes to unknown data, and how well the model scales with problem complexity [9]. For a neural network, the influence of variables may sometimes exceed the architecture of the entire neural network, which is the number of hidden layers, the number of neurons, the weights and the activation function. Moreover, overfitting is a fundamental issue in supervised machine learning which prevents us from perfectly generalizing the models to well fit observed data on training data, as well as unknown data on testing set. Overfitting occurred [13] because of presence of noise, limited size of training sets, and complexity of classifiers.

Here we firstly introduce several concepts, assuming that the hypothesis space as H, $\hat{Y}$ is the ideal output of the model, and Y is the actual output. The expected error is E_X, the empirical error is E_M. The goal is to make $\hat{Y}$ approximately equal to Y, and $E_M(\hat{Y}) = 0$. Which means that $E_M(Y) \approx 0$. The Hoeffding inequality that must be mentioned first [6]:

Definition 1. *For a group of independent random variables $X_1, ..., X_n \in \mathbb{R}$, assuming for all $a_i \leq i \leq b_i$, which is*

$$\mathbb{P}(X_i \in [a_i, b_i)) = 1 \tag{1}$$

The sum of random variables is:

$$S_n = X_1 + ... + X_n \qquad (2)$$

The expected value of S_n is $E(S_n)$ So for all $t \geq 0$:

$$\mathbb{P}(|S_n \quad E(S_n| \geq t) \leq 2exp(-\frac{2t^2}{\sum_{i=1}^{n})(b_i - a_i)^2}) \qquad (3)$$

For one hypothesis h in H, when the number of samples N is large enough, use the Hoeffding inequality to infer the overall expected error $E_X(h)$ through the empirical error $E_M(h)$ on the sample set:

$$\mathbb{P}[|E_X(h) - E_M(h)| > \epsilon] \leq 2exp(-2\epsilon^2 N) \qquad (4)$$

So that, when N is large enough, $E_X(h)$ will be close enough to $E_M(h)$. This situation only suit for only one hypothesis in H. Now let us assumed that there are M hypothesises in H, which is $h_1, h_2..., h_M$, $E_[h_i] = |E_X(h) - E_M(h)|$. The Hoeffding inequality will be:

$$\mathbb{P}[|[E_{h1}| > \epsilon \cup |E_{h2}| > \epsilon \cup ... \cup |E_{hM}| > \epsilon]$$
$$\leq \mathbb{P}[E_{h1} > \epsilon] + \mathbb{P}[E_{h2} > \epsilon] + ... + \mathbb{P}[E_{hM} > \epsilon] \qquad (5)$$
$$\leq 2Mexp(-2\epsilon^2 N)$$

It can be rewritten as:

$$\forall \, Y \in H, \; \mathbb{P}[|E_X(Y) - E_M(Y)| > \epsilon] \leq 2Mexp(-2\epsilon^2 N) \qquad (6)$$

The conclusion here is that the number of samples needs to be large enough under the assumption that the number M is finite. If the number of hypotheses M in the hypothesis space is infinite, then the limit $2Mexp(-2\epsilon^2 N)$ will also become infinite, which means that learning is meaningless. So equation (6) will be:

$$\forall \, Y \in H, \; \mathbb{P}[|E_X(g) - E_M(g)| > \epsilon] \leq 2eff(M)exp(-2\epsilon^2 N) \qquad (7)$$

In order to define a finite M, and get rid of dataset(No longer limited to any one particular dataset), a growth function need to be added [16] [11] [2],

$$m_H(N) = max_{X_1, X_2, ..., X_N \in X}|H(X_1, X_2, ..., X_N)| \qquad (8)$$

The growth function's superior bound is 2^N, So that M changed from limit to 2^n, to reduce the magnitude,we need to introduce break point:

Definition 2. *For the growth function $m_H(N)$ of the hypothesis space H, N is the sample size. When $N = k, m_H(N) < 2N$, k is the break point of H.*

Thus if the break point is available, growth function $m_H(N)$ will be a polynomial, the magnitude will reduce, which means the learning is meaningful. Using both a break point and the growth definition, we could change the Eq. 7 of VC bound as:

$$\forall\, Y \in H, \mathbb{P}[|E_X(Y) - E_M(Y)| > \epsilon] \leq 4M_H(2N)exp(-\frac{1}{8}\epsilon^2 N) \qquad (9)$$

It shows that, as N gradually increases, the exponential $exp(\cdot)$ decreases faster than the polynomial $M_H(2N)$ increases. According to this, we get the definition of the VC-dimension on the hypothesis space H [11]:

Definition 3. *Suppose the VC-dimension of space H is the size of the largest dataset that can be broken up by H, that is:*

$$VC(H) = max\{N : m_H(N) = 2^N\} \qquad (10)$$

So, the $VC(H) = k - 1$, k is the break point of H.

We know that the VC-dimension did not connect with the learning algorithm, the specific distribution of the dataset or the objective function. It only influenced by model itself and hypothesis space.

2.2   Single-layer Network with VC-dimension

The traditional definition for the VC-dimension is for an indicator function set. If there are H samples that can be separated by the functions in the function set in all possible forms of the H power of 2, then the function set is said to be able to break up the H samples.; The VC-dimension of the function set is the maximum number of samples H that it can break up. As for the connection between VC-dimension and neural network, I have to mention the article published by Sontag in 1998. First, he pointed out that the VC-dimension is oriented towards binary classification. The concept of VC-dimension can be generalized in a number of ways to deal with the problem of "learning" (approximating from data) real-valued functions. This also leads to pseudo-dimensions, fat-crushing dimensions, and several other concepts [12]. In his paper, he assumed that a set U, which has been called as the input space, U is also a subset of R^m. Definition of VC-dimension in Neural Network has been provided as:

Definition 4. *If F is a vector subspace of R^U, then VC-dimension of F $=dimF$*

This is directly applied to the perceptron [12], which is just a linear discriminator that exists in R^m, and its VC-dimension is defined as:

$$VCDP_m = m + 1 \qquad (11)$$

When this definitionis directly applied to Single Hidden Layer Nets with Fixed Input Weights, it becomes different.

Here is a defined single hidden layer netural network showing in Fig. 1, it has been defined as a n row, m vectors, input-layer weight $A_1,...,A_m$, input-layer bias $b_1,...,b_n$, output-layer weight $C_0,...,C_m$, $\sigma(A_iu+B_i)$, $i = 1,...,n$, so the dimension will be [12]:

$$VCDF_{n,\sigma,A,B} \leq n+1 \tag{12}$$

The conditions under which the above equations hold are related to the choice of activation function. This article exemplifies when tanh is selected as the activation function, and, $(A_i, b_i) \neq \pm(A_j, b_j)$ for all $i \neq j$ and that $A_i \neq 0$ for all i [12]. Here, the calculation of the VC-dimension of the neural network can be extended to more complex neural networks, such as Autoencoders, which can automatically generate weights and biases [8]. It is still important to note that the most important point in Sontag's theory is that in this proposition [8,12], different constraints need to be matched with different activation functions before the corresponding VC-dimension can be calculated.

3   VC-dimension in Autoencoders

For Autoencoders, it is more efficient to compute the VC-dimension separately for the encoder and decoder. When discussing the impact of VC-dimension on Autoencoders, the selected neural network architecture needs to be mentioned. Therefore, even the number of neurons, the number of hidden layers and other variables are all the same, the difference in activation functions will also lead to the difference in the VC-dimension and the VC bound. Therefore, reducing all the constant factors, and simply calculating the different representations of

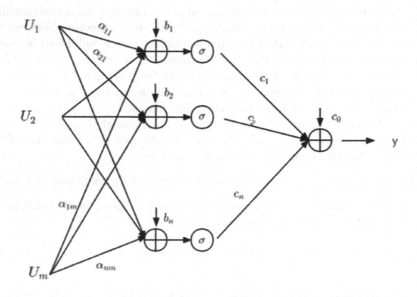

Fig. 1. Single-hidden layer netural net

VC bounds caused by different activation functions has become the first topic to be discussed in this section. We firstly simplify the structure of the Autoencoder structure as mentioned in Fig. 2, then will extend the solution to a general network architecture of Autoencoder at the end.

3.1   VC Dimension of Known Autoencoders

This Autoencoder has its input X, target Y and the neutral network output $\hat{Y}$. The dimension of $X, Y, \hat{Y}$ will be d_0. In Sect. 2.1, we learned that if the superior limit of the number of hypotheses in the hypothesis space can be calculated, it will be more possible to calculate the VC-dimension. The VC-dimension is a measure of the capacity (complexity, expressiveness, richness, or flexibility) of the space of functions that can be learned by statistical classification algorithms. Therefore, for the VC-dimension, the space complexity will affect the value to a certain extent, but this calculation is based on the same conditions as other variables. However, the main comparison in this article is in the encoder/decoder part, how will the VC-dimension change, and to what extent will it be affected when the selected activation function is different. The two activation function that will be used in is the sigmoid function [5]:

$$g(x) = \frac{1}{1 + e^{-x}} \tag{13}$$

And the hyperbolic tangent function [15]:

$$g(x) = \frac{e^x - e^{-x}}{e^x + e^{-x}} \tag{14}$$

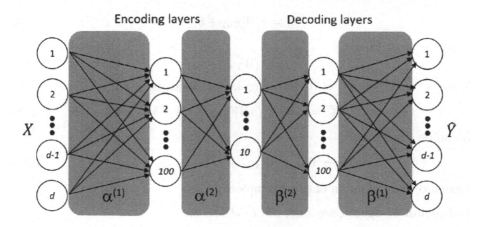

Fig. 2. The Autoencoder model

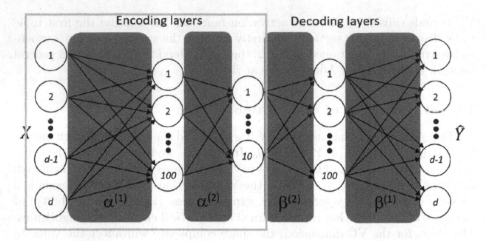

Fig. 3. Encoding layers

First divide this Autoencoder into two parts, encoder and decoder. Given a training dataset, the dimensionality of the original data (d) will be progressively reduced to d_1, and d_2 through the encoding layers, then will be increased to d_3, and d respectively. For the encoder layer, $X_{11}, X_{21},..., X_{n1}$ is the input data, the dimension will be d_0. $Y_{11}, Y_{21},..., Y_{n1}$, is on the first layer which will reduced to d_1 dimensions, the bias is b_1. Then $Y_{12}, Y_{22},..., Y_{n2}$ has the d_2 dimensions, the bias is b_2, σ as the activation function (Fig. 3).

Theorem 1. *For this Autoencoder encoder part, the VC-dimension will be:*

$$VCD_{f_{\sigma,b,X,Y_{n2}}} \leq d_1^2 + 2 \tag{15}$$

Proof. z_1 as the first layer, so:

$$z_1 = \sigma(W_{Xn1} + b_1) \tag{16}$$

z_2 as the second layer:

$$z_2 = \sigma(W_{Yn1} + b_2) \tag{17}$$

Make X' as the output of the encoding layers, which is:

$$X' = \sigma(W_{Yn1}(\sigma(W_{Xn1} + b1) + b_2) \tag{18}$$

It can be written as:

$$X' = \sigma^2 W_{Yn1} \cdot W_{Xn1} + \sigma^2 W_{Yn1} \cdot b_1 + \sigma b_2 \tag{19}$$

Introduce the Definition of space complexity

Definition 5. *Space complexity can be shown as:*

$$Space \sim O(\sum_{l=1}^{D} K_l^2 \cdot C_{(l-1)} \cdot C_l) \tag{20}$$

For the space complexity, the theory is usually used in convolutional neural networks, which also means that the space complexity affects the VC-dimension, while the sample size does not affect the space complexity. For this, the dimension for the first layer is d_1, so $O(Y_{n1}) = d_1^2$. Therefore, since we know that the essence of encoding is to reduce the dimension, the dimension of the final output X' is d_2, which means $O(X') = d_2$, much lower than $O(Y_{n1})$. So for the encoding part, the $O(X') = d_1^2$.

Based on (19), which is a multivariate quadratic equation, it let the VC-dimension becomes:

$$VCD_{f_{\sigma,b,X,X'}} \leq d_1^2 + 2 \tag{21}$$

When the equation holds, the VC-dimension reaches the maximum value, that 10 $d_1^2 + $ ₁ ?

Theorem 2.

$$VCD_{f_{\sigma,b_m,X_n,Y_{n3}}} = d_1^2 + 2, when\ b_2 = 0\ while\ \sigma = sigmoid \neq 0$$

Proof. When the activation function is $\sigma = sigmoid$, the equation (19) becomes:

$$\sigma^2 W_{Yn1} \cdot W_{Xn1} + \sigma^2 W_{Yn1} \cdot b_1 + \sigma \cdot b_2 = 0 \tag{22}$$

Assume $\sigma \neq 0$:

$$\sigma \cdot W_{Yn1} \cdot W_{Xn1} + \sigma \cdot W_{Yn1} \cdot b_1 + b_2 = 0 \tag{23}$$

Since $W_{Yn1} = (W_{Xn1} + b_1)$, so

$$\sigma(W_{Xn1} + b_1) \cdot W_{Xn1} + \sigma(W_{Xn1} + b_1) \cdot b_1 + b_2 = 0 \tag{24}$$

Take b_2 into another side:

$$W_{Xn1}^2 + 2 \cdot W_{Xn1} b_1 + b_1^2 = -b_2 \cdot (1 + e^{-x}) \tag{25}$$

So that it can be seen like:

$$(W_{Xn1} + b_1)^2 = -b_2 \cdot (1 + e^{-x}) \tag{26}$$

The left side of the equation becomes a quadratic polynomial, which means that the value of the left side of the equation is ≥ 0. On the right side of the equation, there are $(-b_2)$ and the sigmoid function. The value of the sigmoid function is $(0,1)$. It is assumed that the sigmoid function is not zero, so the only way to equal is that $b_2 = 0$.

What if the σ changed? The tanh function also is a popular option as an activation function. It cannot be ignore that the value of tanh function is $[-1, 1]$. It is worth noting that this is an important different from sigmoid function so:

Theorem 3. *When* $(W_{Xi1}, b_{i1}) \neq \pm(W_{Xj1}, b_{j1})$ *for all* $i \neq j$, *and* $b_{j2} \neq 0$ *for all* j *while* $\sigma = tanh \neq 0$ *in the encoder part, the VC-dimension will be:*

$$VCD_{f_{\sigma,b_m,X_n,Y_{n3}}} = d^2 + 2 \tag{27}$$

The proof process can be found in Sontag [12]. It can be seen from these two different activation functions, the constraints on reaching the upper limit of the VC-dimension are different (Fig. 4).

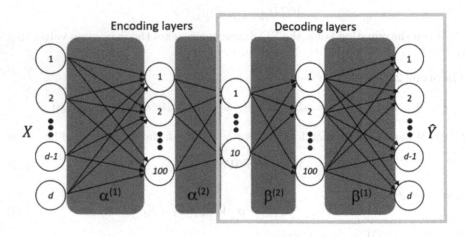

Fig. 4. Decoding layers

After the derivation of the encoder, for the VC-dimesion of the decoder, the factor that does not need to be considered is the activation function. In the usual sense, the decoder uses the inverse function of the encoder as the activation function, which is σ'. For decoder, $X'_1, X'_2, ..., X_n$ is the input data, the dimension will be d_2. $Y_{13}, Y_{23}, ..., Y_{n3}$, is the first decoding layer which will increase to $d_3 = d_1$ dimensions, the bias is b_3. Then $Y_{14}, Y_{24}, ..., Y_{n4}$ has the d_0 dimensions, the bias is b_4. The $\hat{Y}$, which is the output for the decoding part, will be:

$$\hat{Y} = \sigma'(W_{Y_{n4}}(\sigma'(W_{X'} + b3) + b_4) \tag{28}$$

The space complexity will change. Because the structure of decoding is different from encoding, 10 neurons in the first layer of decoding will be added to the space complexity, which is:

$$O'(\hat{Y}) = d_1^2 + d_2 \tag{29}$$

Theorem 4. *So the VC-dimension of decoding will be:*

$$VCD_{f_{\sigma',b_m,X',\hat{Y}}} \leq d_1^2 + d_2 + 2 \tag{30}$$

When $\sigma = sigmoid$, through Theorem 2,

$$VCD_{f_{\sigma,b_m,X',\hat{Y}}} = d_1^2 + d_2, when\ b_4 = 0\ while\ \sigma = sigmoid \neq 0 \qquad (31)$$

While the activation function is tanh, when $(W_{Yi2}, b_{i3}) \neq \pm(W_{Yj2}, b_{j3})$ for all $i \neq j$, and $b_{j4} \neq 0$ for all j while $\sigma = tahh \neq 0$, the VC-dimension will be:

$$VCD_{f_{\sigma',b_m,X',\hat{Y}}} \leq d_1^2 + d_2 \qquad (32)$$

So, for this whole Autoender, the VC-dimension will be:

$$VCD_{f_{\sigma',b_m,X',\hat{Y}}} \leq d_1^2 + d_2 + 2 \qquad (33)$$

3.2   VC Dimension of Autoencoders with Unfixed Structure

When the number of layers and the number of neurons in the hidden layer are unknown, how to calculate the VC-dimension becomes to the extended talks. We use an unfixed Autoencoder structure where $N_1, N_2, ..., N_n$ are the dimension of each hidden layer, M is the number of encoding layer, X denotes input data, Y denotes output data, σ is the activation function, and $b_1, b_2, ..., b_{2M}$ are the bias for each layer. Through Theorem 4, the VC-dimension will be:

$$VCD_{f_{\sigma',b_m,X,Y}} \leq N_1^M + N_2^{M-1} + ... + N_n^1 + M \qquad (34)$$

When the activaction function is sigmoid function, it will have a limitation, which is: $b_M, b_{2M} \neq 0$ while $\sigma \neq 0$. If σ changed as tanh, the limitation is: $(W_{YiM}, b_{i(M+1)}) \neq \pm(W_{YjM}, b_{j(M+1)})$ for all $i \neq j$, and $b_{j(2M)} \neq 0$ for all j.

4   Conclusion

In this paper, the VC-dimension on Autoencoders has been provided. In the case of choosing different activation functions, there will be other constraints. But in essence, for the Autoencoder, selecting the appropriate number of neurons for the hidden layer becomes the best way to optimize itself. When neuron number reaches a peak, the learning ability of the neural network declines, so selecting a suitable activation function and the number of neurons has become a way to optimize the VC-dimension. Therefore, these provided results could be used to solve the optimization problem. VC-dimension still has a massive impact on neural network systems. Through calculation, we know that when the space complexity of a nervous system is determined, the analysis of VC-dimension is obvious. However, this paper doesn't cover any VC-dimension results for deep networks. It is worth further investigating the VC-dimension for the deep network such as deep convolutional neural networks.

References

1. Abu-Mostafa, Y.S.: The Vapnik-Chervonenkis dimension: information versus complexity in learning. Neural Comput. **1**(3), 312–317 (1989)
2. Abumostafa, Y.S., Magdonismail M.L.H.T.: Learning from data: a short course. Amlbook (2012)
3. Bartlett, P.L., Harvey, N., Liaw, C., Mehrabian, Λ.: Nearly-tight vc-dimension and pseudodimension bounds for piecewise linear neural networks. J. Mach. Learn. Res. **20**(1), 2285–2301 (2019)
4. Blumer, A., Ehrenfeucht, A., Haussler, D., Warmuth, M.K.: Learnability and the Vapnik-Chervonenkis dimension. J. ACM (JACM) **36**(4), 929–965 (1989)
5. Han, J., Moraga, C.: The influence of the sigmoid function parameters on the speed of backpropagation learning. In: Mira, J., Sandoval, F. (eds.) IWANN 1995. LNCS, vol. 930, pp. 195–201. Springer, Heidelberg (1995). https://doi.org/10.1007/3-540-59497-3_175
6. Hoeffding, W.: Probability inequalities for sums of bounded random variables. In: The Collected Works of Wassily Hoeffding, pp. 409–426. Springer, Heidelberg (1994). https://doi.org/10.1007/978-1-4612-0865-5_26
7. Jinfeng, C.: Research and Application of Support Vector Machine Regression Algorithm. Ph.D. thesis, Master's Thesis. Jiangnan University, Wuxi (2008)
8. Kárnẏ, M., Warwick, K., Krková, V.: Recurrent neural networks: some systems-theoretic aspects. In: Dealing with Complexity, pp. 1–12. Springer, Heidelberg (1998).https://doi.org/10.1007/978-1-4471-1523-6_1
9. Lawrence, S., Giles, C.L., Tsoi, A.C.: Lessons in neural network training: Overfitting may be harder than expected. In: AAAI/IAAI, pp. 540–545. Citeseer (1997)
10. Pinto, L., Gopalan, S., Balasubramaniam, P.: On the stability and generalization of neural networks with VC dimension and fuzzy feature encoders. J. Franklin Inst. **358**(16), 8786–8810 (2021)
11. Shalef-Schwarz, S.B.D.S.: Deep Understanding of Machine Learning: From Principle to Algorithm. Machinery Industry Press, Beijing (2016)
12. Sontag, E.D., et al.: Vc dimension of neural networks. NATO ASI Series F Comput. Syst. Sci. **168**, 69–96 (1998)
13. Srivastava, N., Hinton, G., Krizhevsky, A., Sutskever, I., Salakhutdinov, R.: Dropout: a simple way to prevent neural networks from overfitting. J. Mach. Learn. Res. **15**(1), 1929–1958 (2014)
14. Vapnik, V.N., Chervonenkis, A.Y.: On the uniform convergence of relative frequencies of events to their probabilities. In: Vovk, V., Papadopoulos, H., Gammerman, A. (eds.) Measures of Complexity, pp. 11–30. Springer, Cham (2015). https://doi.org/10.1007/978-3-319-21852-6_3
15. Namin, A.H., Leboeuf, K., Muscedere, R.: Efficient hardware implementation of the hyperbolic tangent sigmoid function. In: 2009 IEEE International Symposium on Circuits and Systems, pp. 2117–2120. IEEE (2009)
16. Zhou, Z.: Machine Learning. Tsinghua University Press, Beijing (2016)

Broad Learning with Uniform Local Binary Pattern for Fingerprint Liveness Detection

Mingyu Chen[1,2], Chengsheng Yuan[1,2(✉)], Xinting Li[3], and Zhili Zhou[1,2]

[1] Engineering Research Center of Digital Forensics, Ministry of Education, Nanjing University of Information Science and Technology, Nanjing 210044, China
{201983290123,yuancs,002552}@nuist.edu.cn
[2] School of Computer Science, Nanjing University of Information Science and Technology, Nanjing 210044, China
[3] School of International Relations, National University of Defense Technology, Nanjing 210044, China

Abstract. Recently, with the widespread application of mobile communication devices, fingerprint identification is the most prevalent in all types of mobile computing. While they bring a huge convenience to our lives, the resulting security and privacy issues have caused widespread concern. Fraudulent attack using forged fingerprint is one of the typical attacks to realize illegal intrusion. Thus, fingerprint liveness detection (FLD) for True or Fake fingerprints is very essential. This paper proposes a novel fingerprint liveness detection method based on broad learning with uniform local binary pattern (ULBP). Compared to convolutional neural networks (CNN), training time is drastically reduced. Firstly, the region of interest of the fingerprint image is extracted to remove redundant information. Secondly, texture features in fingerprint images are extracted via ULBP descriptors as the input to the broad learning system (BLS). ULBP reduces the variety of binary patterns of fingerprint features without losing any key information. Finally, the extracted features are fed into the BLS for training. The BLS is a flat network, which transfers and places the original input as a mapped feature in feature nodes, generalizing the structure in augmentation nodes. Experiments show that in Livdet 2011 and Livdet 2013 datasets, the average training time is about 1 s and the performance of identifying real and fake fingerprints is effect. Compared to other advanced models, our method is faster and more miniature.

Keywords: Fingerprint liveness detection · Broad learning · ULBP · Biometrics · Real-time

---

C. Yuan and M. Chen—Contributed equally to this work and should be considered co-first authors.

1   Introduction

In recent years, with the rapid development of biotechnology, biometric technology has been widely used in the field of identity verification [1], such as fingerprint recognition, face recognition, voiceprint recognition and so on. The traditional authentication method generally adopts the user name and password authentication mode, but the problem is also obvious, that is, easy to be lost, stolen or deciphered. Therefore, it is urgent to design a more secure and convenient way of identity authentication. Human biometrics have the characteristics of accessibility, uniqueness, and universality, which can be used for user identity authentication. In the existing biometric identification technology, the application of fingerprint identification technology is more mature and common, such as mobile payment, attendance system, access control identification and other aspects of life [2–4] (Fig. 1).

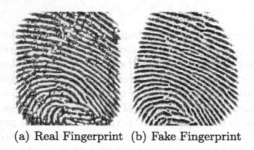

(a) Real Fingerprint   (b) Fake Fingerprint

Fig. 1. Real and fake fingerprint images. Forged fingerprint cannot be discriminated by the naked eye.

Despite its high security and reliability, fingerprint authentication still has certain security risks [5]. Especially in recent years, with the emergence of various emerging technologies such as high-resolution simulation, 3D printing, and generative adversarial networks, fingerprint recognition systems are extremely vulnerable to various attacks, such as the use of emerging technologies or special materials such as silica gel and gelatin to collect fingerprints [6,7]. Fingerprints are imitated to make fake fingerprints to deceive the fingerprint identification system, and successfully gain the trust of the fingerprint identification system. In order to solve the above problems, forged fingerprint detection technology came into being. In order to resist the forged fingerprint attack, FLD has aroused interest of people, considered as the main countermeasure to protect fingerprint identification system. FLD is an important auxiliary algorithm to protect user privacy, which has become a hotspot in academic research community. In recent years, the research on fingerprint liveness detection has attracted the attention of many scholars. Many competitions related to forged fingerprint detection have been held at home and abroad, such as the International the Fingerprint Liveness

Detection Competition (FLDC) held every two years since 2009 [6,8], Fingerprint Verification Competition (FVC) and so on, indicating the importance of fingerprint liveness detection.

Among all kinds of FLD algorithms, deep learning achieves surprising performance, it is demanding on device performance. For mobile devices, the limited space and cost cannot meet such high requirements. In addition, deep network structures are complex and involve a large number of hyperparameters, and the training time is measured in hours or days. At the same time, the huge network structure leads to poor real-time performance. To solve the above problems, this paper proposes a novel fingerprint liveness detection based on broad learning with uniform local binary pattern (ULBP). Firstly, the region of interest (ROI) algorithm extracts the effective fingerprint region and removes redundant information. Secondly, ULBP is used to extract fingerprint texture features, which describe fingerprint information with few features. Finally, the extracted features are input as the BLS for subsequent training. Without multi-layer connections, the BLS does not need to use gradient descent to update the weights, so the computational speed is much better than that of deep learning. By increasing the width of the network, the network can adapt to different datasets and improve accuracy. Compared with increasing the number of layers in a deep network, the computational effort of increasing the width is negligible.

To sum up, the main contributions of this paper are enumerated as follows:

1) This paper proposes a novel fingerprint liveness detection method based on broad learning. Compared with common methods based on deep learning, our model has shorter training time. It has better detection performance at the same level of training time.
2) Different from deep learning methods, our method can exhibits outstanding real-time performance without GPU for training. Our model has lower requirements on device performance.
3) Without retraining the entire system, ours can update the model to scale with different datasets. Once a fake fingerprint is found to successfully deceive the FLD, the system can be quickly updated and put into use.

The rest of the paper is as follows. In Sect. 2, the related work on existing fingerprint liveness detection schemes are surveyed. After that, the relevant theories and techniques needed for this paper are provided in Sect. 3, including our model architecture and some basic theories. In Sect. 4, we will analyze the experimental results, and finally give the conclusion in Sect. 5.

2   Related Work

The concept of forged fingerprint attack was proposed in the late 20th century [9]. The researchers found that the fingerprint of the fake finger is extremely close to the human body, which has caused widespread concern in the research community. The FLD method is proposed to detect whether the input fingerprint to be tested comes from a real person. At present, fingerprint liveness detection methods are mainly divided into two categories: hardware-based FLD and software-based FLD [10].

2.1 Hardware-Based FLD

The hardware-based FLD integrates additional professional equipments into fingerprint biometric systems to identify the authenticity of fingerprint images by measuring skin temperature, conductivity, blood pressure, blood oxygen and other vital signs [11,12]. This allows for more accurate authentication and prevents fraud, increasing the false detection rate [13]. However, the system becomes more complex and expensive.

Although this method can achieve better detection accuracy, the equipment is expensive and the identification method is single, and it is easy for illegal users to find loopholes. The latter, software-based FLD, is more flexible and can save costs and simplify operations and minimize additional hardware [14–16]. Therefore, software-based method is the focus of current fingerprint liveness detection research.

2.2 Software-Based FLD

Compared with hardware-based FLD, software-based FLD only requires the input of a fingerprint. When the sample is captured by the scanner, true and fake fingerprint can be identified by using image processing techniques. According to the characteristics of different sensors and fake fingerprints, the method can flexibly use various algorithms to detect fake fingerprints. Existing software-based algorithms can be further divided into three categories: traditional, texture-based, and deep learning-based FLDs.

Traditional FLD. The traditional fingerprint liveness detection (FLD) methods usually utilize heuristic method to design appropriate discriminative features to distinguish the real and fake fingerprint images. The sweat pore-based detection method is the earliest proposed fingerprint liveness detection algorithm.

For high-resolution fingerprint images, the quality of artificial fingerprint images is often worse than that of real fingerprint images, since fake fingerprints are much weaker in level of detail and rougher than real fingerprints. Therefore, Moon [17] et al. proposed the idea of fingerprint liveness detection based on image quality. The fingerprint image is denoised and reconstructed by wavelet. The real and fake fingerprints are identified by calculating the noise residual between the reconstructed and original image. When the finger is pressed and rotated on the sensor, the real fingerprint can produce better elastic deformation than the fake fingerprint, which does not happen with fake fingerprints made of different materials. Therefore, researchers have proposed many fingerprint liveness detection algorithms based on the elastic deformation of fingerprint skin to identify the authenticity of fingerprints. Antonelli [18] et al. used the skin elastic deformation to verify the liveness of fingerprints for the first time. The researchers believe that in the process of imitating fingerprints, the sweat pores on the ridges of the finger epidermis are difficult to replicate. Manivanan [19] et al. used high-pass filter to extract effective sweat pore features and correlation filter to locate the position of sweat pores. Then extracted features are sent into classifier to distinguish forged fingerprint.

FLD Based on Texture Features. Texture feature refers to the changing trend of image grayscale and color, which is a common feature in fingerprint images. The difference in texture features between real and fake fingerprints is inconspicuous, which makes the eyes unable to identify fake fingerprints. However, by describing the overall information of the image, texture features are able to capture tiny differences, enabling effective fingerprint liveness detection.

Fingerprint texture features is usually represented by the grayscale distri bution of pixels and its adjacent pixels. Common texture feature descriptors include local binary pattern [20] (LBP), binarized statistical image feature [21] (BSIF), local phase quantization [22] (LPQ), Histogram of Gradient Orientation [23] (HOG), etc. Based on these texture feature descriptors, Xia [24] et al. proposed a weber local binary descriptor (WLBD) for FLD. WLBD consists of the local binary differential excitation component and the local binary gradient orientation component. FLD is implemented by feeding the combination of co-occurrence probability of intensity-variance and orientation features into a support vector machine (SVM) classifier. Mehboob [25] et al. proposed combined Shepard Magnitude and Orientation (SMOc), extracting global features of the fingerprint by computing the relation between perceived Shepard magnitude and initial pixel intensities in spatial domain. Firstly, considering the fingerprint as a 2-D vector, the descriptor combines the logarithmic function of the initial pixel intensity and Shepard magnitude to construct the perceived spatial stimulus. The phase information (CO) is then calculated in the frequency domain. Finally, SM and CO are binned and represented as a 2-D histogram.

FLD Based on Deep Learning. Deep learning technology not only shows excellent performance in the field of computer vision, but also has been successfully applied in forgery fingerprint detection. In the field of FLD, most methods segment the background or foreground of fingerprint images and extract regions of interest. Deep learning is a model architecture that learns the inherent laws and representation levels of sample data. Deep neural networks can extract high-level fingerprint features through self-learning and provide classification results, so CNNs are also used as feature extractors.

In 2012, Nogueira [26] et al. introduced CNN to FLD for the first time and compared four different CNNs, proving that pretrained CNNs achieve high accuracy in FLD. Kim [27] et al. proposed a deep belief network (DBN) based architecture that consists of multiple layers of restricted Boltzmann machines. Nevertheless, while achieving good accuracy, most CNN models have some shortcomings, such as fixed-scale input images. Although cropping or scaling can solve the scale problem, they easily lead to the loss of some key texture information and the degradation of image resolution, thus weakening the generalization performance of the classifier model. To solve this problem, Yuan [28] et al. proposed a scale-equalized deep convolutional neural network (DCNNISE) model to further improve the detection performance of forged fingerprints by using the retained subtle texture information. In performance evaluation, confusion matrix is first applied to FLD as a performance metric. Zhang [29] et al. found that the convolutional neural network model used for multi-classification cannot obtain satisfactory accuracy in FLD. This method ignores the difference between natural images and fingerprint images. Therefore, they propose a lightweight but powerful network structure Slim-ResCNN, suitable for fingerprint liveness detection. Since the texture features and the features extracted by the neural network do not coincide, the fusion features are more distinguishable between real and fake fingerprints. Anusha [30] et al. proposed a network structure that combines global and local patch features. LBP and Gabor operators are used for preprocessing, and DenseNet model is used to extract high-level semantic features. DenseNet with channel and spatial attention extracts local patch features. Aiming at the patch identification problem, a new patch attention network is proposed for feature fusion.

3   Methodology

3.1   Uniform Local Binary Pattern

The LBP feaure descriptor takes 3×3 pixel blocks as the basic unit. The difference between the central pixel and the neighboring 8 pixels is extracted as local texture features. When the pixel value of the adjacent pixel is less than the pixel value of the central pixel, the position of the adjacent pixel value is set to 1, otherwise it is set to 1. The pixel point in the upper left corner is the starting point, which is accumulated in binary form in a clockwise direction to obtain the LBP eigenvalue, as shown in Fig. 2.

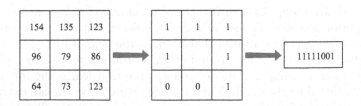

Fig. 2. An example of original Local Binary Pattern

The pixel value of the central pixel is x and the pixel value of the adjacent 8 pixels is $x_i (i = 0, ..., 7)$, respectively. The pixel block of is shown as

$$\begin{bmatrix} x_0 \ x_1 \ x_2 \\ x_7 \ x \ x_3 \\ x_6 \ x_5 \ x_4 \end{bmatrix} \tag{1}$$

The calculation formula of the LBP feature is as follow

$$x_i = \begin{cases} 1, x_i \geq x \\ 0, x_i < x \end{cases} \tag{2}$$

$$f(x) = \sum_{i=0}^{8} 2^i x_i \tag{3}$$

According to the corresponding position of the original pixel point, the LBP feature value is formed into a new image - LBP feature map, as shown in Fig. 3.

(a) Original sample (b) LBP feature map

Fig. 3. Comparison of the original sample and LBP feature map.

In addition, the adjacent pixels in the LBP operator are not only the surrounding 8, but can be adjusted according to the image. Obviously, the dimension of the lbp histogram (texture features) is 2^p, where p is the number of adjacent pixels. When the p is very large in the neighborhood, it is inconvenient to calculate in practice.

Most of the lbp features in the image contain at most two transitions, including "bright spots", "dark spots", 'flat areas', 'changed edges", etc. These transitions usually have the main content covered. When the binary number corresponding to a LBP has at most two transitions from 0 to 1 or from 1 to 0, the number is called an equivalence pattern, namely ULBP [31]. For instance, 00000000 (0 transition), 00000111 (only one transition from 0 to 1), 10001111 (two transitions) are equivalent modes. Patterns other than the equivalence pattern are classified into another, called the mixed pattern class. ULBP can be described as

$$ULBP = \begin{cases} \sum_{p=0}^{p-1} s(g_p - g_c) & \text{if } U(ULBP) \leq 2 \\ p(p-1) + 2 & \text{otherwise} \end{cases} \tag{4}$$

where g_c is the pixel of the center, and g_p is the pixel of the sample in the neighborhood. p is the number of adjacent pixels. $s()$ is signal function whose value is 0 or 1. $U()$ is number of transitions. The category of ULBP is $p(p-1)+2$. Thus the dimension of texture spectrum is shortened from 2^p to $p(p-1)+2$.

3.2 Broad Learning System Model

Fingerprint images usually have more than 40,000 pixels. There is vast redundant information in image. If the fingerprint compressed into a one-dimensional vector is fed into Broad Learning System (BLS), it not only incurs huge cost of training and testing, but also degrades the detection performance. Therefore, low dimensional features are extracted through feature descriptors. BLS is trained by these features. The structure of our method is shown as Fig. 4.

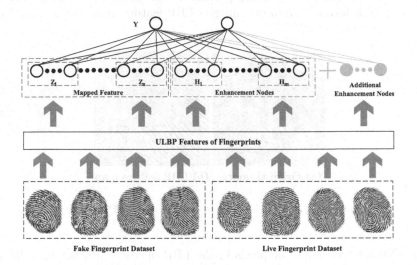

Fig. 4. The proposed structure of FLD based on Broad Learning. The fingerprint features are extracted by ULBP descriptors. Z_i and H_j denotes $i - th$ mapper feature and $j - th$ enhanced node severally.

Broad Learning System uses the features mapped by the input data as the feature nodes of the network [32]. The mapped features are augmented into augmented nodes with randomly generated weights. All mapped features and enhancement nodes are directly connected to the output, and the corresponding output coefficients can be derived by pseudo-inverse. The key to BLS is incremental learning, including augmentation node increments, feature node increments and input data increments. If the network structure needs to be expanded, there is no need to retrain the full network.

Through the feature node, input data generate mapped features, the equation is as follows

$$Z_i = f_i \left(\phi_i \left(XW_{e_i} + \beta_{e_i} \right) \right), i = 1, \dots, n \tag{5}$$

where Z_i denotes mapper feature nodes and X denotes the input data of BLS. W_{e_i} and β_{e_i} are the random weights coefficient and bias coefficient. n denotes the number of feature nodes. ϕ_i represents the transfer function, and f_i represents the normalization function. Denote $Z^i = [Z_1, \cdots, Z_i]$ as the concatenation of all the first i groups of mapping features.

Through the enhancement node, the mapped features are enhanced, the equation is as follows

$$H_m = \xi \left(Z^n W_{h_m} + \beta_{h_m} \right) \tag{6}$$

where, H_i denotes enhanced nodes. W_{h_m} and β_{h_m} are the random weights coefficient and bias coefficient. Denote $H^i = [H_1, \cdots, H_i]$ as the concatenation of all the first i groups of enhancement nodes.

All mapped and enhanced features are input into the output layer. The broad model can be represented as the equation of the form

$$\begin{aligned} Y &= [Z_1, \cdots, Z_n \mid H_1, \cdots, H_m] W^m \\ &= [Z^n \mid H^m] W \end{aligned} \tag{7}$$

where Y denotes the output of BLS. W denotes the connection weights and can be easily computed through the ridge regression. The solution process of the pseudo-inverse function is as follow

$$W = \left(A^T \times A + c \times I^{n+m} \right)^{-1} \times A^T \times Y \tag{8}$$

$$A^m = [Z^n \mid H^m] \tag{9}$$

where A is concatenation of mapped and enhanced features and c is a constant parameter.

If the model fails to achieve the desired accuracy, additional boost nodes can be inserted to achieve better performance. Denote A^{m+1} and pseudoinverse of are as

$$A^{m+1} \equiv \left[A^m \mid \xi \left(Z^n W_{h_{m+1}} + \beta_{h_{m+1}} \right) \right] \tag{10}$$

$$\left(A^{m+1} \right)^+ = \begin{bmatrix} (A^m)^+ - DB^T \\ B^T \end{bmatrix} \tag{11}$$

where $D = (A^m)^+ \xi \left(Z^n W_{h_{m+1}} + \beta_{h_{m+1}} \right)$

$$B^T = \begin{cases} (C)^+ & \text{if } C \neq 0 \\ \left(1 + D^T D \right)^{-1} B^T (A^m)^+ & \text{if } C = 0 \end{cases} \tag{12}$$

$$C = \xi \left(Z^n W_{h_{m+1}} + \beta_{h_{m+1}} \right) - A^m D \tag{13}$$

Finally, the new weights are as follow

$$W^{m+1} = \begin{bmatrix} W^m - DB^T Y \\ B^T Y \end{bmatrix} \tag{14}$$

Incremental learning only needs to compute the pseudo-inverse of the additional augmented nodes, instead of computing the entire (A^{m+1}), so the computation is fast.

4 Experiments and Results

In order to verify the effectiveness of the proposed system, we performed related experiments. The experimental equipments and environment: Intel(R) Core(TM) i7-10700 CPU @ 2.90 GHz, GTX 2070 (8 Gb) and 16 GB RAM. The operating system is Windows 10 professional 64 bit. The model training and evaluation are tested on Python3.7 platform.

Based on LivDet 2011 [7], 2013 [33], we conduct experiments on the dataset publicly provided by Liveness Detection Competitions (LivDet) to evaluate the performance of the proposed framework. The details of datasets are recorded in Table 1, including real and fake fingerprints collected by different sensors. The fingerprints collected by different sensors are quite different, and the fingerprints of different materials also have subtle differences. Fingerprint images collected by different sensors have different sizes, ranging from 208 to 1000 in width and from 324 to 1500 in height.

Table 1. Details of the LivDet datasets used in this paper. R is real fingerprint; F is fake fingerprint

Dataset	Sensor	Train(R/F)	Test(R/F)
LivDet2011	Biometrika(Bio)	1000/1000	1000/1000
	DigitalPersona(Dig)	1004/1000	1000/1000
	Italdata(Ita)	1000/1000	1000/1000
	Sagem(Sag)	1008/1008	1000/1036
LivDet2013	Biometrika(Bio)	1000/1000	1000/1000
	CrossMatch(Cro)	1250/1000	1250/1000
	Italdata(Ita)	1000/1000	1000/1000
	Swipe(Swi)	1221/979	1153/1000

In the field of fingerprint liveness detection, the common performance evaluation metric is Average Classification Error (ACE). The ACE is defined as the average value of False Reject rate (FRR) and False Accept rate (FAR), calculated as Eq. 15. FRR is the proportion of real fingerprints incorrectly identified as fake fingerprints. FAR is the proportion of fake fingerprints that are incorrectly identified as real fingerprints.

$$ACE = \frac{FRR + FAR}{2} \tag{15}$$

We will compare the classification ability and training time of our method with deep learning, including VGG16 [34], AlexNet [35] and ResNet18 [36] respectively. The ACE and training time for one round of different convolutional neural network are recorded in the table. The relevant parameters of our method are: 30 nodes and 30 windows of feature mapping layer, 300 nodes and 30 windows of enhance layer. The number of incremental learning steps is set to 10, and the best result is taken as the ACE in the table. Compared with classical CNNs, the training time of our method is drastically reduced. Furthermore, there is a large gap between the computing performance of CPU and GPU. The classification results of the above methods are as shown in Table 2.

Table 2. Classification resuls on LivDet datasets

Dataset	Model	ACE(%)				Training time
		Bio	Dig	Ita	Sag	
LivDet2011	Ours	1.8	15.6	21.3	16	1.22 s
	VGG16	9.8	13.8	26.2	13.4	147.34 s
	AlexNet	11	13.7	26.6	12.2	31.05 s
	ResNet18	9.4	13	21	15.9	83.42 s
		Bio	Ita	Swi		
LivDet2013	Ours	2.2	1.3	14.9		1.06 s
	VGG16	10.6	20.6	12.4		140.71 s
	AlexNet	14.1	26.6	13.9		27.87 s
	ResNet18	13.4	13	11.1		75.66 s

Deep learning requires powerful GPU, but our method requires only CPU for training and testing. With similar performance, training time is even much lower than deep learning. What's more, the number of parameters is also much lower than that of deep learning, as shown in the Table 3. In brief, our proposed method is more suitable for mobile devices with weak computing performance.

Table 3. Parameters quantity of different models

Model	Ours	VGG16	AlexNet	ResNet18
Parameter quantity	100k	138M	56M	30M

5 Conclusion

This paper proposes a FLD method based on broad learning with ULBP. The ULBP operator is used to extract the texture features of fingerprints, and the extracted features are fed into the BLS. The proposed method has short training time and small amount of parameters. At the same time, it has better detection performance at the same level of training time. Furthermore, our method does not require computationally powerful GPU and exhibits excellent real-time performance. This model has low requirements on device performance, which is in line with the needs of mobile devices and FLD embedded development. In the face of unknown fake fingerprint attack, our system allows the model to scale to different datasets without retraining. However, although the hardware equipment requirements are reduced, the detection accuracy is lower. In the future, we need to further enhance the detection performance.

Acknowledgement. This work is supported by the National Natural Science Foundation of China under grant 62102189; by the Jiangsu Basic Research Programs-Natural Science Foundation under grant BK20200807; by the Research Startup Foundation of NUIST under grant 2020r015; by the Public Welfare Technology and Industry Project of Zhejiang Provincial Science Technology Department under grant LGF21F020006; by the Key Laboratory of Public Security Information Application Based on Big-Data Architecture, Ministry of Public Security under grant 2021DSJSYS006; by NUDT Scientific Research Program under grant JS21-4; by the 2022 Excellent Undergraduate Graduation Design (Paper) support program of NUIST under grant 201983290123.

References

1. Yuan, C., Yu, P., Xia, Z., Sun, X., Wu, Q.M.J.: FLD-SRC: fingerprint liveness detection for AFIS based on spatial ridges continuity. IEEE J. Selected Topics Signal Process. **16**, 817–827 (2022). https://doi.org/10.1109/JSTSP.2022.3174655
2. Maltoni, D., Maio, D., Jain, A., Prabhakar, S.: Handbook of fingerprint recognition. Ch Synth. Fingerprint Gener. **33**(5–6), 1314 (2005). https://doi.org/10.1007/978-1-84882-254-2
3. Jia, X., et al.: Multi-scale local binary pattern with filters for spoof fingerprint detection. Inf. Sci. **268**, 91–102 (2014). https://doi.org/10.1016/j.ins.2013.06.041
4. Sousedik, C., Busch, C.: Presentation attack detection methods for fingerprint recognition systems: a survey. Iet Biometrics **3**(4), 219–233 (2014). https://doi.org/10.1049/iet-bmt.2013.0020
5. Schuckers, S.A.: Spoofing and anti-spoofing measures. Inf. Secur. Tech. Rep. **7**(4), 56–62 (2002). https://doi.org/10.1016/S1363-4127(02)
6. Marcialis, G.L., et al.: First international fingerprint liveness detection competition—LivDet 2009. In: Foggia, P., Sansone, C., Vento, M. (eds.) ICIAP 2009. LNCS, vol. 5716, pp. 12–23. Springer, Heidelberg (2009). https://doi.org/10.1007/978-3-642-04146-4_4
7. Yambay, D., Ghiani, L., P. Denti, P., Marcialis, G.L., Roli, F., Schuckers, S.: Livdet 2011 - fingerprint liveness detection competition 2011. In: 2012 5th IAPR International Conference on Biometrics (ICB), pp. 208–215 (2012). https://doi.org/10.1109/ICB.2012.6199810
8. Cappelli, R., Ferrara, M., Franco, A., Maltoni, D.: Fingerprint verification competition 2006. Biometric Technol. Today **15**(7–8), 7–9 (2007). https://doi.org/10.1016/S0969-4765(07)70140-6
9. Meyer, H.: Six biometric devices point the finger at security. Comput.Secur. **17**(5), 410–411 (1998). https://doi.org/10.1016/S0167-4048(98)80063-1
10. Nikam, S.B., Agarwal, S.: Texture and wavelet-based spoof fingerprint detection for fingerprint biometric systems. In: First International Conference on Emerging Trends in Engineering and Technology 2008, pp. 675–680 (2008). https://doi.org/10.1109/ICETET.2008.134

11. Putte, T., Keuning, J.: Biometrical fingerprint recognition: don't get your fingers burned. In: Smart Card Research and Advanced Applications, pp. 289–303 (2000)
12. Drahanský, M., Nötzel, R., Wolfgang, F.: Liveness detection based on fine movements of the fingertip surface. In: Proceedings of the 2006 IEEE Workshop on Information Assurance, pp. 42–47 (2006). https://doi.org/10.1109/iaw.2006.1652075
13. Kallo, P., Kiss, I., Podmaniczky, A., Talosi, J.: Detector for recognizing the living character of a finger in a fingerprint recognizing apparatus. US6175641B1 (2001)
14. 4 Abhyankar, A.S., Schuckers, S.C.: A wavelet-based approach to detecting liveness in fingerprint scanners. In: Proceedings of SPIE - The International Society for Optical Engineering, vol. 5404 (2004). https://doi.org/10.1117/12.542939
15. Schuckers, S., Abhyankar, A.: Detecting liveness in fingerprint scanners using wavelets: Results of the test dataset. In: Biometric Authentication, ECCV International Workshop, Bioaw, Prague, Czech Republic, May, vol. 3087 (2004). https://doi.org/10.1007/978-3-540-25976-3_10
16. Zhang, Y., Tian, J., Chen, X.: Fake finger detection based on thin-plate spline distortion model. in: Advances in Biometrics, International Conference, ICB 2007, Seoul, Korea, August 27–29, 2007, Proceedings, vol. 4642 (2007). https://doi.org/10.1007/978-3-540-74549-5_78
17. Moon, Y.S., Chen, J.S., Chan, K.C., So, K., Woo, K.C.: Wavelet based 545 fingerprint liveness detection. Electron. Lett. 41(20), 1112–1113 (2005). https://doi.org/10.1049/el:20052577
18. Antonelli, A., Cappelli, R., Maio, D., Maltoni, D.: Fake finger detection by skin distortion analysis. IEEE Trans. Inf. Forensics Secur. 1(3), 360–373 (2006). https://doi.org/10.1109/TIFS.2006.879289
19. Manivanan, N., Memon, S., Balachandran, W.: Automatic detection of active sweat pores of fingerprint using highpass and correlation filtering. Electron. Lett. 46(18), 1268–1269 (2010). https://doi.org/10.1049/el.2010.1549
20. Nikam, S.B., Agarwal, S.: Local binary pattern and wavelet-based spoof fingerprint detection. Int. J. Biometrics 1(2), 141–159 (2008). https://doi.org/10.1504/IJBM.2008.020141
21. Kannala, J., Rahtu, E.: BSIF: binarized statistical image features. In: 2012 21st International Conference on Pattern Recognition (ICPR). IEEE (2012)
22. Ghiani, L., Marcialis, G.L., Roli, F.: Fingerprint liveness detection by local phase quantization. In: Proceedings of the 21st International Conference on Pattern Recognition (ICPR2012), pp. 537–540 (2012)
23. Mohan, L.S., James, J.: Fingerprint spoofing detection using hog and local binary pattern (2017). https://doi.org/10.17148/IJARCCE.2017.64111
24. Xia, Z., Yuan, C., Lv, R., Sun, X., Xiong, N.N., Shi, Y.-Q.: A novel weber local binary descriptor for fingerprint liveness detection. IEEE Trans. Syst. Man Cybern. Syst. 50(4), 1526–1536 (2020). https://doi.org/10.1109/TSMC.2018.2874281
25. Mehboob, R., Dawood, H., Dawood, H., Ilyas, M.U., Guo, P., Banjar, A.: Live fingerprint detection using magnitude of perceived spatial stimuli and local phase information. J. Electron. Imaging 27(05), 053038 (2018). https://doi.org/10.1117/1.JEI.27.5.053038
26. Nogueira, R.F., de Alencar Lotufo, R., Machado, R.C.: Fingerprint liveness detection using convolutional neural networks. IEEE Trans. Inf. Forensics Secur. 11(6), 1206–1213 (2016). https://doi.org/10.1109/TIFS.2016.2520880
27. Kim, S., Park, B., Song, B.S., Yang, S.: Deep belief network based statistical feature learning for fingerprint liveness detection. Pattern Recogn. Lett. 77, 58–65 (2016). https://doi.org/10.1016/j.patrec.2016.03.015

28. Yuan, C., Xia, Z., Jiang, L., Wu, J., Sun, X.: Fingerprint liveness detection using an improved CNN with image scale equalization. IEEE Access **7**, 26953–26966 (2019)
29. Zhang, Y., Shi, D., Zhan, X., et al.: Slim-ResCNN: a deep residual convolutional neural network for fingerprint liveness detection. IEEE Access **7**, 91476–91487 (2019)
30. Banerjee, S., Chaudhuri, S.: DeFraudNet: End2End fingerprint spoof detection using patch level attention. In: Proceedings - 2020 IEEE Winter Conference on Applications of Computer Vision, WACV 2020, pp. 2684–2693 (2020). https://doi.org/10.1109/WACV45572.2020.9093397
31. Wang, Y., Mu, Z. Zeng, H.: Block-based and multi-resolution methods for ear recognition using wavelet transform and uniform local binary patterns. In: 2008 19th International Conference on Pattern Recognition, pp. 1–4 (2008). https://doi.org/10.1109/ICPR.2008.4761854
32. Chen, C.L.P., Liu, Z.: Broad learning system: an effective and efficient incremental learning system without the need for deep architecture. IEEE Trans. Neural Netw. Learn. Syst. **29**(1), 10–24 (2018). https://doi.org/10.1109/TNNLS.2017.2716952
33. Ghiani, L., et al.: Livdet 2013 fingerprint liveness detection competition 2013. In: International Conference on Biometrics (ICB) 2013, pp. 1–6 (2013). https://doi.org/10.1109/ICB.2013.6613027
34. Simonyan, K., Zisserman, A.: Very deep convolutional networks for large-scale image recognition. Computer Science (2014)
35. Krizhevsky, A., Sutskever, I., Hinton, G.E.: Imagenet classification with deep convolutional neural networks. Commun. ACM **60**(6), 84–90 (2017). https://doi.org/10.1145/3065386
36. He, K., Zhang, X., Ren, S., Sun, J.: Deep residual learning for image recognition. In: IEEE Conference on Computer Vision and Pattern Recognition (CVPR) 2016, pp. 770–778 (2016). https://doi.org/10.1109/CVPR.2016.90

A Novel Trajectory Tracking Controller for UAV with Uncertainty Based on RBF and Prescribed Performance Function

Xuelei Qi[1] , Chen Li[1] , and Hongjun Ma[2,3](✉)

[1] College of Information Science and Engineering, Northeastern University, Shenyang 110819, People's Republic of China
[2] School of Automation Science and Engineering, South China University of Technology, Guangzhou 510641, People's Republic of China
mahongjun@scut.edu.cn
[3] Key Laboratory of Autonomous Systems and Networked Control, Ministry of Education, Unmanned Aerial Vehicle Systems Engineering Technology Research Center of Guangdong, Guangzhou, People's Republic of China

Abstract. This paper proposes a novel trajectory tracking controller based on RBF neural network and fractional-order sliding mode control (FO-SMC). First, the prescribed performance control (PPC) is introduced into the system to make the tracking error converge to the predefined set. Then, the fractional-order calculus is introduced into SMC to alleviate the chattering of the system. Considering that RBF neural network can compensate for the uncertainty of the UAV motion model, RBF is introduced into the design of the controller. Besides, the Lyapunov theorem proves the stability of the system, and all signals in the closed-loop system are stable. Finally, a case study is carried out through simulation.

Keywords: Trajectory tracking · UAV · Fractional-order sliding mode control (FO-SMC) · RBF neural network · Prescribed performance control (PPC)

1 Introduction

With the development of intelligent technology, the unmanned aerial vehicle (UAV) has been widely used in monitoring, traffic control, disaster surveillance [1,2]. There are some study areas relative to UAV; however, the trajectory tracking problem is the most important for UAV. Generally, the path planning can be divided into two categories. One is path planning with local sensing in the complex and unknown environment, and the other is path planning based on global geographic information [3,4].

Nowadays, there are some researchers focus on how to make an accurate trajectory tracking for UAV. Some control (e.g. PID [5], sliding mode control [6],

H. Zhang et al. (Eds.): NCAA 2022, CCIS 1637, pp. 341–352, 2022.
https://doi.org/10.1007/978-981-19-6142-7_26

backstepping theory [7], neural network [8–10]) can play an important role in UAV. Among them, RBF neural network has the advantages of good flexibility, strong robustness, and not easy to fall into local optimization, and it has achieved well-designed results in solving the obstacle avoidance path planning problem of UAV. However, other scholars have done some improved research work and achieved some research results. Jose used an evolutionary algorithm and discrete-time recurrent neural network to act on the motion of robot [11]. A. L. Nelson evolved maze exploration behavior using recursive networks, including time delay and evolutionary strategy based on competition mechanism and discussed the evolution of more advanced behavior [12,13]. Pang et al. analyzed the advantages and disadvantages of genetic algorithm and overcame the chatting problem of the robot [14]. At present, there are still some complex problems, and a complete theoretical proof has not been formed. The steady-state performance needs to meet the requirements of the control, but it also needs to be considered the transient performance of the system. Therefore, the prescribed performance function needs to introduce into the system to make the tracking error converge to the predefined set [15,16].

Based on the above analysis, this paper proposes a novel trajectory tracking controller for UAV with uncertainty based on RBF and prescribed performance control (PPC). These two major contributions are as follow:

1) A novel trajectory tracking controller is proposed for UAV based on the fraction-order sliding mode control (FO-SMC) to reduce chattering caused by traditional SMC.
2) The prescribed performance control (PPC) is introduced into UAV system to make tracking error of position converge to the predefined set.

The rest of this paper can be organized as follows. Section 2 designs the system model and states the problem. In addition, Sect. 3 describes the main results, including: PPC, FO-SMC and RBF neural network. Section 4 verifies the effectiveness of the proposed method via MATLAB. Finally, Sect. 5 gives the conclusions.

2   System Model and Problem Statement

2.1   System Model

The unmanned aerial vehicle has a 6-DOF system with 4 rotors in the three-dimensional space, and it is also as the multi-input-multi-output (MIMO) system with strong coupling under-actuated states. Moreover, Fig. 1 shows the dynamic model of UAV system.

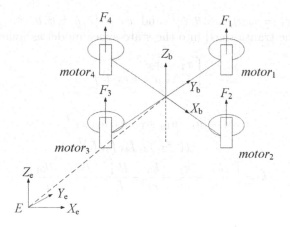

Fig. 1. The dynamic model of UAV.

The dynamic Lagrange model can be expressed

$$
\begin{cases}
\ddot{x} = \frac{u_1}{m}(cos\phi sin\theta cos\psi + sin\phi sin\psi) - \frac{k_1}{m}\dot{x} \\
\ddot{y} = \frac{u_1}{m}(cos\phi sin\theta sin\psi - sin\phi cos\psi) - \frac{k_2}{m}\dot{y} \\
\ddot{z} = \frac{u_1}{m}cos\phi cos\theta - g - \frac{k_3}{m}\dot{z} \\
\ddot{\theta} = u_2 - \frac{lk_4\dot{\theta}}{I_y} \\
\ddot{\psi} = u_3 - \frac{lk_5\dot{\psi}}{I_z} \\
\ddot{\phi} = u_4 - \frac{lk_6\dot{\phi}}{I_x}
\end{cases}
\tag{1}
$$

where (x,y,z) is the position of UAV, (ϕ,θ,ψ) can represent the roll angle, pitch angle, yaw angle, respectively. g is the gravity, m is the mass of UAV, l is the arm length of UAV, $k_i(i=1,2,3,4,5,6)$ is the air drag coefficient, I_x, I_y, I_z are the moment of inertia of UAV relative to the body coordinate system, $u_{1x}, u_{1y}, u_{1z}, u_2, u_3, u_4$ are the control inputs.

In order to simplify the design of the controller, the double closed-loop control structure is used in this paper. The position-loop subsystem of UAV is as the outer-loop and the attitude-loop subsystem is as the inner-loop. As simplify the system model, it can be transformed into the following systems:

$$
\begin{cases}
\ddot{x} = u_{1x} - \frac{k_1}{m}\dot{x} + f_1 \\
\ddot{y} = u_{1y} - \frac{k_2}{m}\dot{y} + f_2 \\
\ddot{z} = u_{1z} - \frac{k_3}{m}\dot{z} + f_3 \\
\ddot{\theta} = u_2 - \frac{lk_4\theta}{I_y} + f_4 \\
\ddot{\psi} = u_3 - \frac{lk_5\dot{\psi}}{I_z} + f_5 \\
\ddot{\phi} = u_4 - \frac{lk_6\phi}{I_x} + f_6
\end{cases}
\tag{2}
$$

where $u_{1x} = \frac{u_1}{m}(cos\phi sin\theta cos\psi + sin\phi sin\psi)$, $u_{1y} = \frac{u_1}{m}(cos\phi sin\theta sin\psi - sin\phi cos\psi)$, $u_{1z} = \frac{u_1}{m}cos\phi cos\theta - g$ and f represents the sum of the unmodeled dynamics and the model uncertainties of the UAV.

Let define $x_1 = [x, y, z, \phi, \theta, \psi]^T$ and $x_2 = [\dot{x}, \dot{y}, \dot{z}, \dot{\phi}, \dot{\theta}, \dot{\psi}]^T$, so the system model (2) can be transformed into the state-space model as follow:

$$\begin{cases} \dot{x}_1 = x_2 \\ \dot{x}_2 = u + kx_2 + f \end{cases} \tag{3}$$

where

$$u = [u_{1x}, u_{1y}, u_{1z}, u_2, u_3, u_4]^T$$

$$f = [f_1, f_2, f_3, f_4, f_5, f_6]^T$$

$$k = \left[-\frac{k_1}{m}, -\frac{k_2}{m}, -\frac{k_3}{m}, -\frac{lk_4}{I_y}, -\frac{lk_5}{I_z}, -\frac{lk_6}{I_x} \right]$$

2.2 Problem Statement

Considering the dynamic model (3), the objective of this paper is to track the desired position (x_d, y_d, z_d) and the tracking error of x, y, z-axis needs to convergence to the predefined set.

3 Main Results

In order to improve the accuracy of the position tracking, overcome the chattering problem in the traditional sliding mode controller and handle the uncertainty of the UAV model, so the adaptive fractional-order sliding mode controller (FO-SMC) is designed in this paper. First, we introduce the PPC into the design of controller so that the tracking error can be convergence to the predefined set. Then, the fractional calculus is introduced and replaced the traditional sliding mode approaching law, which can alleviate the chattering of system. Moreover, the system's uncertainty (e.g. the UAV dynamic model, uncertain external interference) is compensated by using RBF neural network. The overall diagram of the system is shown in Fig. 2.

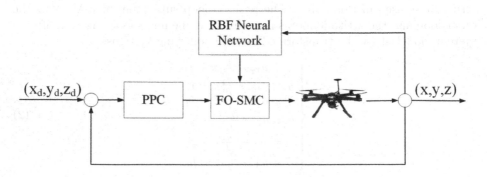

Fig. 2. The overall structure of the system.

3.1 Prescribed Performance Control

Assume that the prescribed performance function $\varphi(t) : R_+ \to R_+$ is positive and decreasing, and satisfies the following properties:

$$\begin{cases} -\delta\varphi(t) < e(t) < \varphi(t), e\,(0) \geq 0 \\ -\varphi(t) < e(t) < \delta\varphi(t), e\,(0) < 0 \end{cases} \tag{4}$$

where $0 \leq \delta \leq 1$ is design parameter, and $e(t)$ is error.

Let define the prescribed performance function

$$\varphi_i(t) = (\varphi_{i,0} - \varphi_{i,\infty})e^{-\mu_i t} + \varphi_{i,\infty} \tag{5}$$

where $\varphi_{i,0}, \varphi_{i,\infty}, \mu_i > 0$ represents the design parameters.

In order to convert the constraint form in (4) into the unconstrained form through the error transformation function:

$$e(t) = \varphi(t)\Phi(\varsigma) \tag{6}$$

where ς denotes the error after transformation, and ς is smooth, strictly increasing and reversible, which needs to meet the following properties:

$$\begin{cases} -\delta < \Phi(\varsigma) < 1, e\,(0) \geq 0 \\ -1 < \Phi(\varsigma) < \delta, e\,(0) < 0 \end{cases} \tag{7}$$

$$\begin{cases} \lim_{\varsigma \to -\infty} \Phi(\varsigma) = -\delta, \lim_{\varsigma \to \infty} \Phi(\varsigma) = 1, e(0) \geq 0 \\ \lim_{\varsigma \to -\infty} \Phi(\varsigma) = -1, \lim_{\varsigma \to \infty} \Phi(\varsigma) = \delta, e(0) < 0 \end{cases} \tag{8}$$

where, if ς is bounded, (7) establishes. When $e(0) \geq 0$, $-\delta\varphi(t) < \varphi(t)\Phi(\varsigma) = e(t) < \varphi(t)$. When $e(0) < 0$, $-\varphi(t) < \varphi(t)\Phi(\varsigma) = e(t) < \delta\varphi(t)$, resulting that (4) establishes.

In order to introduce the prescribed performance control into the design of controller, the derivative of (6) as

$$\dot{e}(t) = \dot{\varphi}(t)\Phi(\varsigma) + \varphi(t)\frac{\partial\Phi(\varsigma)}{\partial\varsigma}\dot{\varsigma} \tag{9}$$

From (9), it can obtain that the derivative of the transformation error ς

$$\dot{\varsigma} = \frac{\dot{e}(t) - \dot{\varphi}(t)\Phi(\varsigma)}{\varphi(t)\frac{\partial\Phi(\varsigma)}{\partial\varsigma}} \tag{10}$$

$$= F(\varsigma, \varphi) + G(\varsigma, \varphi)\dot{e}(t)$$

where

$$F(\varsigma, \varphi) = \frac{-\dot{\varphi}(t)\Phi(\varsigma)}{\varphi(t)\frac{\partial\Phi(\varsigma)}{\partial\varsigma}}$$

$$G(\varsigma, \varphi) = \frac{1}{\varphi(t)\frac{\partial\Phi(\varsigma)}{\partial\varsigma}}$$

Let define the tracking error as $e_{1x}(t) = x_d - x$, and (10) can be rewritten as

$$\dot{\varsigma} = F(\varsigma, \varphi) + G(\varsigma, \varphi)(\dot{x}_d - \dot{x}) \tag{11}$$

where x denotes the actual trajectory, x_d denotes the expected trajectory.

The coordinate transformation can be defined as

$$\eta_1 = \varsigma \tag{12}$$

Select the Lyapunov function as

$$\dot{V}_1 = \frac{1}{2}\eta_1^2 \tag{13}$$

Then, the derivative of (13) as

$$\begin{aligned}
\dot{V}_1 &= \eta_1 \dot{\eta}_1 \\
&= \eta_1(F + G(\dot{x}_d - \dot{x}))
\end{aligned} \tag{14}$$

Let $\dot{x}$ design as

$$\dot{x} = \frac{1}{G}(F + G\dot{x}_d + c_1\eta_1 - \eta_2), c_1 > 0 \tag{15}$$

where η_2 is the virtual control quantity.

Then, (15) is substituted into (14) as

$$\dot{V}_1 = -c_1\eta_1^2 + \eta_1\eta_2 \tag{16}$$

3.2 Fractional Order Sliding Mode Control

The Caputo fractional-order calculus is defined as

$$_\sigma D_t^r f(t) = \frac{1}{\Gamma(n-r)} \int_\sigma^t (t-\tau)^{n-r-1} f(\tau) d\tau \tag{17}$$

where σ and t represent the upper and lower limitation of calculus. Let r be the order of integration, and $\Gamma(.)$ is the gamma function.

Lemma 1. If $x = 0$ is the equilibrium point in (18), then

$$D^r f(t) = f(x, t) \tag{18}$$

where $f(x, t)$ satisfies Lipschitz conditions.

The sliding surface is designed as

$$s = \eta_2 + a \int_0^t \eta_2 dt \tag{19}$$

Then, its derivation can be obtained

$$\dot{s} = \dot{\eta}_2 + a\eta_2 \tag{20}$$

The reaching law can improve the dynamic quality (e.g. steady-state error and stability time) in the designed FO-SMC controller, but the traditional SMC exists the serious chattering and long convergence time problems. Therefore, the fractional calculus is introduced into the system

$$D^r s = -k_c \text{sgn}(s) \tag{21}$$

Then, it can be obtained

$$\dot{s} = D^{1-r}(-k_c \text{sgn}(s)) \tag{22}$$

The Lyapunov function can be selected as:

$$V_2 = \frac{1}{2}s^T s \tag{23}$$

where the function in (23) satisfies the conditions in (24)

$$\begin{cases} \omega_1\left(\|x\|\right) \le V(t, x(t)) \le \omega_2\left(\|x\|\right) \\ D^r V(t, x(t)) \le -\omega_3\left(\|x\|\right) \end{cases} \tag{24}$$

where $\omega_1, \omega_2, \omega_3$ are positive values, then the system (18) is stable.

The derivative of (23) is obtained as

$$\begin{aligned} \dot{V}_2 &= s^T \dot{s} \\ &= s^T D^{1-r}(-k_c \text{sgn}(s)) \end{aligned} \tag{25}$$

Due to $\text{sgn}(D^{1-r}(-k_c \text{sgn}(s))) = -k_c \text{sgn}(s)$, then

$$\begin{aligned} \text{sgn}(\dot{V}_2) &= \text{sgn}(s^T \dot{s}) \\ &= \text{sgn}(s^T)\text{sgn}(D^{1-r}(-k_c \text{sgn}(s))) \\ &= -k_c \end{aligned} \tag{26}$$

Then, according to Lemma 1, (26) is stable.

3.3   The FO-SMC with PPC and RBF

The RBF is introduced to approximate the nonlinear function with a higher convergence rate. It is a three-layer network structure, as shown in Fig. 3, including input, hidden, and output. In RBF, the number of hidden nodes determines the complexity of neural networks.

The uncertainty of UAV dynamic model is combined into a unified uncertainty term $f(\dot{x}_1, t)$, the RBF neural network can approximate $f(\dot{x}_1, t)$

$$b_j = e^{\frac{\|\bar{x} - c_j\|^2}{2d_j^2}} \tag{27}$$

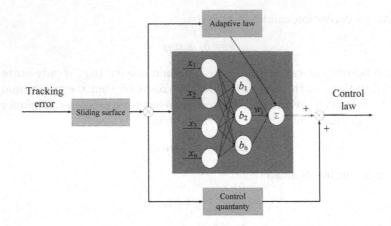

Fig. 3. The structure of RBF neural network.

$$f(\dot{x}_1, t) = W^{*T}g(x) + o \tag{28}$$

where b_j denotes the output of Gaussian function of RBF, let $\bar{x}$ be the network input, j-th denotes the node of the hidden layer, W^* represents the ideal weight values, and o is the approximation error.

Let $\tilde{f}(\dot{x}_1, t) = f(\dot{x}_1, t) - \hat{f}(\dot{x}_1, t)$, the uncertainty can be approximated by RBF as $\tilde{f}(\dot{x}, t) \overset{\Delta}{=} \tilde{W}^T g(x) + o$.

Then, the Lyapunov function can be selected as

$$V = \frac{1}{2}\eta_1^2 + \frac{1}{2}s^T s + \frac{1}{2\vartheta_1}\tilde{\varpi}^T\tilde{\varpi} \tag{29}$$

where $\vartheta_1 > 0, \tilde{W} = W - \hat{W}$.

Then, (30) can be obtained by calculating the derivative of (29)

$$\dot{V} = -c_1\eta_1^2 + \eta_1\eta_2 + s^T\dot{s} - \frac{1}{\vartheta_1}\tilde{\varpi}^T\dot{\hat{\varpi}} \tag{30}$$

Then, the control input u_{1x} and the adaption law $\dot{\hat{\varpi}}$ are designed as

$$u_{1x} = \frac{1}{G}\left(Q_x + G\frac{k_1}{m}\dot{x} + D^{1-r}(k_c\mathrm{sgn}(s)) - \hat{f}_x + a(F + c_1\eta_1 + G\dot{x}_d - G\dot{x}) + c_2 s\right) \tag{31}$$

$$\dot{\hat{\varpi}} = \vartheta_1 sh(\bar{x}) \tag{32}$$

where $Q_x = \dot{F} + G\dot{e}_{1x} + G\ddot{x}_d + c_1\dot{\eta}_1$.

One has

$$\dot{s} = \dot{\eta}_2 + a\eta_2$$
$$= Q_x - G\ddot{x} + a(F + c_1\eta_1 + G\dot{x}_d - G\dot{x})$$
$$= Q_x - G(u_{1x} - \frac{k_1}{m}\dot{x} + f_1) + a(F + c_1\eta_1 + G\dot{x}_d - G\dot{x})$$
$$= Q_x - Gu_{1x} + G\frac{k_1}{m}\dot{x} - f_x + a(F + c_1\eta_1 + G\dot{x}_d - G\dot{x})$$
$$= Q_x + G\frac{k_1}{m}\dot{x} - f_x + a(F + c_1\eta_1 + G\dot{x}_d - G\dot{x})$$
$$- (Q_x + G\frac{k_1}{m}\dot{x} + D^{1-r}(k_c\mathrm{sgn}(s)) - \hat{f}_x + a(F + c_1\eta_1 + G\dot{x}_d - G\dot{x}) + c_2s)$$
$$= -\tilde{f}_x - D^{1-r}(k_c\mathrm{sgn}(s)) \quad o_2o$$

(33)

where $f_x = Gf_1$.

Then, the derivative of the system can be obtained

$$\dot{V} = -c_1\eta_1^2 + \eta_1\eta_2 - s^T\tilde{f}_x - s^TD^{1-r}(k_c\mathrm{sgn}(s)) - c_2s^Ts - \frac{1}{\vartheta_1}\tilde{\varpi}^T\dot{\tilde{\varpi}}$$
$$\leq -(c_1 - \frac{1}{2})\eta_1^2 - (c_2 - \frac{1}{2})s^Ts - s^T(o + D^{1-r}(k_c\mathrm{sgn}(s)))$$
$$\leq -s^To - D^{1-r}k_c|s^T|$$

(34)

If $D^{1-r}k_c > |o|$, $\dot{V} < 0$, so the designed control law meets the requirements of Lyapunov stability.

Similar to the controller design in x-axis, the control law in y and z can be obtained as

$$u_{1y} = \frac{1}{G}(Q_y + G\frac{k_2}{m}\dot{y} + D^{1-r}(k_c\mathrm{sgn}(s)) - \hat{f}_y + a(F + c_1\eta_1 + G\dot{y}_d - G\dot{y}) + c_2s) \quad (35)$$

$$\dot{\tilde{\varpi}} = \vartheta_1sh(\bar{y}) \quad (36)$$

$$u_{1z} = \frac{1}{G}(Q_z + G\frac{k_3}{m}\dot{z} + D^{1-r}(k_c\mathrm{sgn}(s)) - \hat{f}_z + a(F + c_1\eta_1 + G\dot{z}_d - G\dot{z}) + c_2s) \quad (37)$$

$$\dot{\tilde{\varpi}} = \vartheta_1sh(\bar{z}) \quad (38)$$

The stability analysis is similar to the x-axis. The designed control laws u_{1y} and u_{1z} make the y-axis and z-axis Lyapunov stable, and meet the prescribed performance constrained conditions.

4   Experiments Simulation

In this section, a case study is carried out to simulate the designed FO-SMC controller via MATLAB for tracking the desired trajectory of UAV with the following parameters:

$$g = 9.8\,\mathrm{m/s}^2, m = 1.6\,\mathrm{kg}, l = 0.2\,\mathrm{m},$$

$$k_i = 0.02\,\text{N} \cdot \text{s}^2 (i = 1, 2, 3, 4, 5, 6)$$

$$I_x = 0.01\,\text{kg} \cdot \text{m}^2, I_y = 0.005\,\text{kg} \cdot \text{m}^2, I_z = 0.01\,\text{kg} \cdot \text{m}^2$$

For prescribed performance function

$$\varphi_{x,0} = 1, \varphi_{x,\infty} = 0.1, \mu_x = 1$$

$$\varphi_{y,0} = 0.2, \varphi_{y,\infty} = 0.02, \mu_y = 1$$

$$\varphi_{z,0} = 0.2, \varphi_{z,\infty} = 0.02, \mu_z = 1$$

Figure 4 shows the results between the actual position conducted by the proposed controller (FO-SMC) and the desired position. In addition, Fig. 5 shows the tracking errors of x, y, z-axis. More details are specific described as follow.

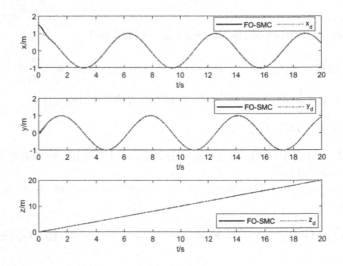

Fig. 4. The trajectory tracking curves under FO-SMC.

In Fig. 4, the black line shows the real trajectory of UAV; meanwhile, the red line demonstrates the desired trajectory. The simulation time is around 20 s. As for the x-axis, the initial value begins at 1.5 m, it tracks the desired trajectory within 1.13 s; furthermore, from 1.13 s to the end, it has no obvious bias between the designed trajectory and the desired trajectory so that the system controller looks good. As for the y-axis, it begins at 0 m, it shows that it can always track the designed trajectory during the simulation time 20 s . Finally, as for the z-axis, the initial value is 0 m at the beginning, and it rises up to 20 m in the final time. Moreover, the trajectories of y and z have no obvious fluctuation which can further prove that the designed FO-SMC controller has a better tracking performance in the UAV system.

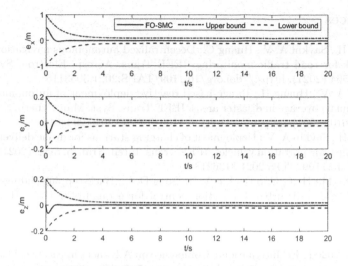

Fig. 5. The tracking errors of position.

In Fig. 5, it shows that position tracking with prescribed performance function, the solid line represents the trajectories of x, y, z-axis, the dash dot line shows the upper boundness, and the dotted line shows the lower boundness. The predefined set of x, y, z-axis is $[-0.1, 0.1]$, $[-0.02, 0.02]$, $[-0.02, 0.02]$, respectively. As for the x-axis, the initial value is 0.5, and it declines until it convergences to the predefined set around 1.581 s. But, the initial value of y-axis and z-axis is the same value with 0. As for the y-axis, the tracking error goes down at the beginning until around 0.712 s, then it fluctuates and convergences to the predefined set around 1.767 s. As for the z-axis, the trajectory of tracking error can converge to the predefined set around 1 s. Therefore, from these three trajectories of the tracking error, the system has a better convergence performance, so that the FO-SMC controller with prescribed performance function is well-designed.

5   Conclusions

Aiming at obtaining an accurate tracking trajectory, the FO-SMC controller based on RBF neural network and prescribed performance function is proposed in this paper. The FO-SMC controller can eliminate the system chattering. The RBF neural network compensates for the system uncertainty. Then, we introduce the prescribed performance function to make the tracking error convergence to the defined set and we also give a specific stability proofing by the Lyapunov theory. However, the future work considers to improve the steady-state and transient performance of the system, which is necessary.

Acknowledgement. This work was supported in part by the funds of the National Natural Science Foundation of China under Grant 61873306.

References

1. Huang H., Savkin A. V., Huang C.: Decentralized autonomous navigation of a uav network for road traffic monitoring. IEEE Trans. Aerosp. Electron. Syst. **57**(4), 2558–2564 (2021). https://doi.org/10.1109/TAES.2021.3053115
2. Savkin A. V., Huang H.: Range-based reactive deployment of autonomous drones for optimal coverage in disaster areas. IEEE Trans. Syst. Man Cybern. Syst. **51**(7), 4606–4610 (2021). https://doi.org/10.1109/TSMC.2019.2944010
3. Huang H., Savkin A. V.: Deployment of charging stations for drone delivery assisted by public transportation vehicles. IEEE Trans. Intell. Transp. Syst. (2021). https://doi.org/10.1109/TITS.2021.3136218
4. Xu L. X., Ma H. J., Guo D., Xie A. H., Song D. L.: Backstepping sliding-mode and cascade active disturbance rejection control for a quadrotor UAV. IEEE/ASME Trans. Mechatron. **25**(6), 2743–2753 (2020). https://doi.org/10.1109/TMECH.2020.2990582
5. Hu X., Liu J.: Research on UAV balance control based on expert-fuzzy adaptive PID. In: 2020 IEEE International Conference on Advances in Electrical Engineering and Computer Applications (AEECA), pp. 787–789 (2020). https://doi.org/10.1109/AEECA49918.2020.9213511
6. Li C., Luo Z., Xu D., Wu W., Sheng Z.: Online trajectory optimization for UAV in uncertain environment. In: 2020 39th Chinese Control Conference (CCC), pp. 6996–7001 (2020). https://doi.org/10.23919/CCC50068.2020.9189144
7. Jiao R., Chou W., Rong Y.: Disturbance observer-based backstepping control for quadrotor UAV manipulator attitude system. In: 2020 Chinese Automation Congress (CAC), pp. 2523–2526 (2020). https://doi.org/10.1109/CAC51589.2020.9327203
8. Xu M., Shi W.: RBF neural network PID trajectory tracking based on 6-PSS parallel robot. In: 2019 Chinese Automation Congress (CAC), pp. 5674–5678 (2019). https://doi.org/10.1109/CAC48633.2019.8996255
9. Li, X., Zhu, J.: Weighted average consensus in directed networks of multi-agents with time-varying delay. In: Zhang, H., Yang, Z., Zhang, Z., Wu, Z., Hao, T. (eds.) NCAA 2021. CCIS, vol. 1449, pp. 71–82. Springer, Singapore (2021). https://doi.org/10.1007/978-981-16-5188-5_6
10. Zhang, H., Yang, Z., Zhang, Z., Wu, Z., Hao, T. (eds.): NCAA 2021. CCIS, vol. 1449. Springer, Singapore (2021). https://doi.org/10.1007/978-981-16-5188-5
11. Jose, A.F., Marcelo, T., Gaardo, G.A.: Evolutionary reactive behavior for mobile robots navigation. Cybern. Intell. Syst. **1**, 532–537 (2004)
12. Nelson, A.L., Grant, E., Gatwtti, J.M.: Maze exploration behaviors using an integrated evolutionary robotic environment. Robot. Auton. Syst. **46**, 159–173 (2004)
13. Nelson, A.L., Grant, E., Barlow, G.: Evolution of complex autonomous robot behaviors using competitive fitness. Integr. Knowl. Intensive Multi-Agent Syst. **1**, 145–150 (2003)
14. Pang, K.K., Prahlad, V.: Evolution of control systems for mobile robots. Evol. Comput. **1**, 617–622 (2002)
15. Qiu J., Wang T., Sun K., Rudas I.J., Gao H: Disturbance observer-based adaptive fuzzy control for strict-feedback nonlinear systems with finite-time prescribed performance. IEEE Trans. Fuzzy Syst. **PP**(99), 1 (2021)
16. Sui, S., Chen, C.L.P., Tong, S.: A novel adaptive NN prescribed performance control for stochastic nonlinear systems. IEEE Trans. Neural Netw. Learn. Syst. **32**(7), 3196–3205 (2021)

Item-Behavior Sequence Session-Based Recommendation

Wangshanyin Zhao, Xiaohong Xiang$^{(\boxtimes)}$, Xin Deng, Wenxing Zheng,
and Hao Zhang

Chongqing University of Posts and Telecommunications,
Nan'an District Chongqing 400065, China
s200201042@stu.cqupt.edu.cn, {xiangxh,dengxin,zhanghao}@cqupt.edu.cn,
1766127449@qq.com

Abstract. Session-based Recommendation has become important and
popular in Recommendation System, which focuses on leveraging the
historic records to predict the next interactive item(s). The previous works only employ the item sequence or process item sequence
and users' behavior respectively. To model session with item sequence
and behavior sequence simultaneously, we propose the **Item-Behavior**
Sequence Session-based Recommendation (**IBSSR** for abbreviation).
The Hadamard product of item sequence and behavior sequence is
fed into an improved Transformer and behavior sequence is fed into a
Gated Recurrent Unit. Representation generated by concatenating these
embeddings with soft attention mechanism will be used to predicts the
next item. To verify the bound of our model, we conducted experiments
on two datasets where there are large differences in ability of behavior
contacting context. Experiment proofs that our model can effectively
model item sequence whose behavior sequence has strong ability in contacting context based on the session.

Keywords: Sequential recommendation · Fine-grained behavior ·
Residual · Transformer · GRU

1 Introduction

Recommendation system increasingly has become an important measure to solve
information overload problem which has appeared in many internet applications by providing item(s) which users may be interested in according to users'
or items' multifarious information. Conventional recommendation systems (e.g.,
collaborative filtering, CF [16]; factorization machine, FM [13] etc.) employ users'
or items' profiles and historic interaction items to recommend item to users
whose recommendation performance highly relies on the completeness of profiles
and they often prefer to the newer interactive recording, thus these methods are
often severely trapped in some real-world applications. Hence, several innovative
orientations are proposed to overcome these problems.

H. Zhang et al. (Eds.): NCAA 2022, CCIS 1637, pp. 353–366, 2022.
https://doi.org/10.1007/978-981-19-6142-7_27

Varying from traditional recommendation systems who conduct the recommendation task employing users' preferences in a statistic way, while sequential recommendation(SR for abbreviation) or session-based recommendation attempts to capture users' dynamic preferences to preform recommendation task exploiting the historical information which will be naturally produced by users' interactions in the most of all online platforms. Therefore, modeling users' preferences with their interactive items is ubiquitous method to capture the dynamic inclination.

Several modeling orientations have been put forward to establish sequential relation to capture the dynamic inclination. Markov Chain(MC for abbreviation) and Neural Network(NN for abbreviation) based method are two main orientations in SR. The basic MC method is FPMC [15], it will calculate the transition probability to predict which item user will interact. Recently, NN-based method gets more researchers' attentions due to the increasing mighty computing power. [5] is the first work introducing Recurrent Neural Network(RNN for abbreviation) into SR. Further research is carried out on the RNN-based approach by introducing novel ranking loss function to tailor RNN for SR [4] and personalizing RNN SR models with cross-session information transfer [12]. Except RNN-based method, more innovative methods are proposed to model sequential relationship. A hybrid encoder with an attention mechanism is proposed to capture the user's main purpose [7] and a memory model is proposed to capture users' general interests [9]. Furthermore, extensive non-sequential models come to the fore to capture complex item relation such as Knowledge Graph(KG for abbreviation) based method [21], Reinforcement Learning(RL for abbreviation) based method [19] and Graph Neural Network(GNN for abbreviation) based method [20] etc. [20] is a prominent work in divert sequential modeling to graph model.

Inspired by [10], the behaviors in items' production process also should be emphasized. Different from [10] taking behaviors to distinguish different sequences (i.e. distinguishing different users), we utilize behaviors to simulate the realistic interaction scenario.

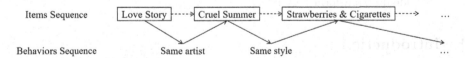

Fig. 1. A music example to describe how a song divert to another song.(The dotted line indicates observed item sequence, while the solid line indicates the real interaction sequence). For 'Love Story' and 'Cruel Summer', the transfer occurs for they are own to the same artist. While the transfer occurs between 'Cruel Summer' and 'Strawberries & Cigarettes' for the similar style.

In Fig. 1, an item sequence can be seen as a user. The user listens 'Love Story' and he/she approves the artist's savour, so a click happened on the artist's name to get some songs of same artist. Then 'Cruel Summer' in the list of the artist appears in the field of vision, so it is listened. Thirdly, 'Cruel Summer' also suits the taste, therefore, he/she clicks similar music for seeking more songs with similar style. Thus, we get two sequences, i.e. the dotted line and the solid line. The dotted line indicates the item sequence and the solid line indicate the realistic path user interacts.

To simulate the realistic interaction scenario, we propose a model named **Item-Behavior Sequence Session-based Recommendation (IBSSR** for abbreviation). In IBSSR, an item sequence and relevant behavior sequence are considered simultaneously. Specifically, a session consists of two parts and we need to learn item embeddings and behavior embeddings, based on which representation of a session is generated. To achieve this goal, the item sequence after Hadamard product with behavior sequence is fed into a Transformer [17] and the behavior sequence is fed into a Gated Recurrent Unit(GRU for abbreviation) [1] respectively. Two different learning mechanisms are adopted in light of different characteristics of behaviors and items following [10].

In summary, our contributions in this paper are as follows:

(1) Fine-grained behaviors are introduced to simulate the real scenario to improve SR performance.
(2) We propose a novel model to model item sequence and corresponding behavior sequence to improve SR performance.
(3) We conduct extensive experiment in our model to evaluate the result on two distinct behavior modes over two real world datasets(KKBOX and Movie-Lens).

In the rest of this paper, we introduce related work in Sect. 2. Details of our model will be given in Sect. 3. Experimental setting and analysis will follow in Sect. 4 and concluding our work is in Sect. 5.

2   Related Work

In this section, a brief research overview will be provided associated with our works.

2.1   Session-Based and Sequential Recommendation

The session-based recommendation is a subtask of sequential recommendation, which is defined as leveraging users' historic sequence to predict the next item(s) which user will interact and the divergence between sequential recommendation and session-based recommendation is the later more focuses on the recent users' interest. The earliest work is FPMC [15], which uses Markov chain to model item transition. For better performance, NN-based work is developed [5,7,9,20]. [5] is the first work using GRU to capture sequential feature by leveraging item sequence, and the output embeddings of the last GRU layer are regarded as the representation of users. [7] adds attention mechanism to eliminate accidental noise in item sequence to get the real users' interests. [9] adopts attention net and stresses on the last interactive item to get better result. To understand intricate item sequence, GNN-based methods are devised to capture expressive session embeddings [10,20] unlike Markov chain based method and RNN based method taking the follow item depending on the nearest as hypothesis. [20] employs GNN to learn items' embeddings and feeds them into an attention framework to obtain the session embeddings by converting sequence to directed graph. [10] utilizes out-degree and in-degree to construct undirected graph to obtain items' embeddings to recommend.

When taking account users' interactions with items into SR, additional effort is essential. [2,8,10,18,24] take users' behavior into consideration for better result. [10] feeds item sequence and behavior sequence into two processors respectively to obtain two embeddings, and then concatenating them as the embeddings of the session. [8] utilizes behavior-specific transition matrices to model users' behaviors and using a location specific transformation matrix to model the short-term context. [18] introduces a strong hypothesis to model monotonic behavior chains. [24] maps Triplet(item, behavior, duration) into embeddings by Word2Vec [11] and feeds it into RNN layer and attention to predict item. [2] organizes behaviors, items, categories as pyramid to obtain their embeddings and feeds them into Long Short-Term Memory(LSTM for abbreviation) to obtain the session embeddings.

3   Methodology

In this section, we introduce the details of IBSSR including the related algorithms in the model. We firstly formulize the problem addressed in this paper, and summarize the pipeline of IBSSR. Then, components of IBSSR will be given in following. In the following introductions, bold lowercase indicates a vector and a bold uppercase indicates a set, matrix or tensor.

3.1   Problem Formulation

In this paper, we focus on sequential model with item sequence and behavior sequence. We first define session set $S = \{s_1, s_2, \cdots, s_N\}$, where S denotes the set containing all sessions, the s_i denotes the i-th session in S, and the N denotes the number of session (i.e. $\|S\|$). For every session s_i, $s_i = \{\mathcal{I}_i, \mathcal{B}_i\}$, where the $\mathcal{I}_i$ denotes the item sequence for i-th session and the $\mathcal{B}_i$ denotes the behavior sequence for i-th session. For every item sequence, $\mathcal{I}_j = \{i_1^j, i_2^j, \cdots i_n^j\}$, where i_k^j denotes the k-th item for j-th session and n denotes the length of j-th item sequence(i.e. $\|\mathcal{I}_j\|$) and $i_k^j \in I$, where I is the set of all items. For every behavior sequence, $\mathcal{B}_j = \{b_1^j, b_2^j, \cdots, b_n^j\}$, where b_k^j denotes the k-th behavior for j-th session and n denotes the length of j-th behavior sequence(i.e. $\|\mathcal{B}_j\|$, obviously $\|\mathcal{I}_j\| = \|\mathcal{B}_j\| = n$), and $b_k^j \in B$, where B is the set of all behaviors.

The goal of our model is predicting the next interactive item i_{n+1} given the session s_j. To achieve this goal, a session s_j is fed into our model and a candidate item i_{n+1}, $(i_{n+1} \in I)$ is generated accord a matching score $\hat{y}_{s_j}$. According to $\hat{y}_{s_j}$, a top-k list will be obtained by the given session, and the item with highest score will be recommended to user.

3.2   Model Overview

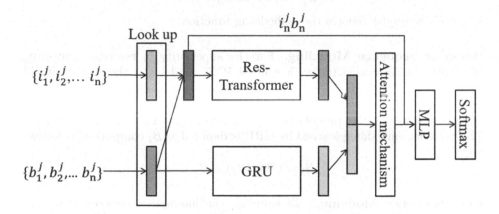

Fig. 2. The overall framework of our proposed IBSSR. The arrows in the figure denote the data flow. The colored blocks with different colors denote embeddings and the uncolored blocks denote different model components. The item sequence embeddings fusing with a behavior sequence by Hadamard product is fed into ResTransformer and the behavior sequence embeddings is fed into GRU to learning embeddings. These embeddings are concatenated and processed by attention mechanism and then fed into MLP with the last item-behavior interaction embedding to get the embedding of the session. The probability $\hat{y}_{s_j}$ is obtained by softmax function.

The framework of IBSSR is illustrated in Fig. 2.

To improve SR performance, we take account into the interactive behaviors. The embeddings of items and behaviors are obtained firstly(i.e. we get the blue block and the red block). Then interaction operation is conducted by Hadamard product with item embedding and sequence embedding(i.e. the item-behavior embedding, the purple block). Then the item-behavior embedding and the behavior embedding are fed into ResTransformer and GRU respectively considering the different properties of items and behaviors. Thus we obtain the processed embeddings of item-behavior and behavior(i.e. the orange block and the yellow block). Then, the concatenating embedding is fed into attention mechanism. After concatenating with the last item-behavior embedding, multi-layer perceptron(MLP for abbreviation) is utilized to calculate the session embedding and softmax function is utilized to obtain the probability.

3.3   Session Modeling

To convert sparse items and behaviors to dense vector, embedding block is adopted, the embedding outputs of $\mathcal{I}_j = \{i_1^j, i_2^j, \cdots, i_n^j\}$ and $\mathcal{B}_j = \{b_1^j, b_2^j, \cdots, b_n^j\}$ are denoted as $\mathcal{I}_j'$ and $\mathcal{B}_j'$, computing as following:

$$\begin{aligned} \mathcal{I}_j' &= Embedding(\mathcal{I}_j) \\ \mathcal{B}_j' &= Embedding(\mathcal{B}_j) \end{aligned} \tag{1}$$

where $Embeding(\cdot)$ denotes the embedding function.

Behavior Sequence Modeling. For the superiority in resolving vanishing gradient problem comparing with general RNN model and the superiority of time comparing with LSTM, GRU is adopted by our model to learning behavior embedding. Inspired by [3], residual blocks are adopted for the excellent performance in accelerating model convergence and improving model performance. The behavior embedding learned by GRU is denoted as $\tilde{\mathcal{B}}_j$ computing as following:

$$\tilde{\mathcal{B}}_j = GRU(\mathcal{B}_j) + \mathcal{B}_j \tag{2}$$

Item Sequence Modeling. To simulate the interaction between item and behavior, the Hadamard product is utilized between the embedding of items and behaviors. The output of Hadamard product is denoted as $\dot{\mathcal{I}}_j$ calculating as following:

$$\dot{\mathcal{I}}_j = \mathcal{I}_j' \odot \mathcal{B}_j' \tag{3}$$

where the $\odot$ denotes the Hadamard product.

For the superior performance of Transformer [17] in sequential modeling, it is also adopted by our works. Out of the same reason adding residual block in GRU component, residual is also added in every encoder layer and decoder layer of Transformer as follow:

$$\begin{aligned} Out &= EncoderLayer(In) + In \\ Out &= DecoderLayer(In) + In \end{aligned} \tag{4}$$

where the $EncoderLayer(\cdot)$ and the $DecoderLayer(\cdot)$ are basic encoder function and decoder function in Transformer. Note that the In and Out in Eq. 4 just indicate the input and output of encoder layer and decoder layer but not specific component in this paper.

The output of our Transformer with more residual block is named as ResTransformer in Fig. 2. And residual is also adopted by our ResTransformer. The embedding of item-behavior learned by ResTransformer is denoted as $\tilde{\mathcal{I}}_j$ computing as following:

$$\begin{aligned} \tilde{\mathcal{I}}_j' &= Encoder(\dot{\mathcal{I}}_j) \\ \tilde{\mathcal{I}}_j &= Decoder(\tilde{\mathcal{I}}_j') + \tilde{\mathcal{I}}_j' \end{aligned} \tag{5}$$

where the $\tilde{\mathcal{I}}_j'$ is the middle result, and the $Encoder(\cdot)$ and the $Decoder(\cdot)$ are stacked by $EncoderLayer(\cdot)$ and $DecoderLayer(\cdot)$ in Eq. 4.

3.4   Session Representation Generating

To obtain the session embedding, we need to polymerize all generated embeddings (i.e. item-behavior embedding $\tilde{\mathcal{I}}_j$ and behavior embedding $\tilde{\mathcal{B}}_j$). Firstly, concatenating both two embeddings is necessary and the result is denoted as $\mathcal{C}_j$

$$
\begin{aligned}
\mathcal{C}_j &= \tilde{\mathcal{I}}_j \oplus \tilde{\mathcal{B}}_j \\
&= \{\tilde{i}_1^j \oplus \tilde{b}_1^j, \tilde{i}_2^j \oplus \tilde{b}_2^j, \cdots, \tilde{i}_n^j \oplus \tilde{b}_n^j\} \\
&= \{\tilde{c}_1^j, \tilde{c}_2^j, \cdots, \tilde{c}_n^j\}
\end{aligned}
\tag{6}
$$

where $\oplus$ denotes the concatenation operation.

Inspired by [10], the local representation and global representation should be taken into account. The local representation is just the last term of $\mathcal{C}_j$ (i.e. $\tilde{c}_n^j$). For the global representation, we conduct following [10] by utilizing soft-attention mechanism [22] to weight every vector in $\mathcal{C}_j$ as following:

$$
\alpha_k = \beta^\top \sigma(W_1 \tilde{c}_k^j + W_2 \tilde{c}_n^j + b_\alpha)
\tag{7}
$$

where the α denotes the attention weight, the β denotes the weight for the after activate function $\sigma(\cdot)$, k denotes the k-th vector in $\mathcal{C}_j$, the W_1 and the W_2 denote weight for $\tilde{c}_k^j$ and $\tilde{c}_n^j$, and the b_α denotes the bias.

The global representation $\tilde{s}^j$ is calculated as following:

$$
\tilde{s}^j = \sum_{k=1}^{n} \alpha_k \tilde{c}_k^j
\tag{8}
$$

In the end, the final representation of the the session is as following:

$$
\tilde{s} = W_3[\tilde{s}^j; \tilde{c}_n^j]
\tag{9}
$$

where the W_3 is the weight.

After obtaining the session representation $\tilde{s}$, MLP and softmax function are adopted to obtain the probability of item $\hat{y}_{s_j}$ as follow:

$$
\hat{y}_{s_j} = softmax(MLP(\tilde{s} \oplus i))
\tag{10}
$$

To train IBSSR model, we choose binary cross-entropy as loss function of SR as following:

$$
\mathcal{L} = -\sum_{s_j \in S} \{y_{s_j} \log \hat{y}_{s_j} + (1 - y_{s_j} \log(1 - \hat{y}_{s_j}))\}
\tag{11}
$$

Our object is training all parameters in our model by minimizing the loss function $\mathcal{L}$ as following:

$$
\Theta = \arg\min \mathcal{L}
\tag{12}
$$

where the Θ denotes all parameters in our model.

4   Experiment

4.1   Experiment Settings

All our experiments are supported by RecBole [23], which provides supports in datasets, baselines, and evaluation metrics.

Datasets. We evaluate our model and baselines on the following two realistic datasets. The detail statistic information is shown in Table 1.

KKBOX[1]: This dataset is provided by a wellknown music service KKBOX. It contains users' information, items' information and interaction recordings. The behaviors we take in our paper is the 'source_system_tab', which records the users' interactive behaviors with tab such as accessing local library, search, discover etc. The number of the behavior is 9. Note that the null value exists in the original data.

MovieLens-1m[2]: This is a movie dataset that recommends movies for users creating in 1997 by GroupLens Research. We choose the version containing about 1m interaction recordings to experiment. The behavior is 'rating' record which does not contact context, while it reflects the users' preference.

For both two datasets, we conduct the k-core filter(i.e. keeping every item will be interacted at least k times), and we set the k as 3. And the max length of session is set as 50. Data augmentation is conducted. We set the 80% data for training, 10% for validation and 10% for testing. For KKBOX, we divide different sessions by the index following [10].

Table 1. Dataset statistics. The $\|\mathcal{S}\|$ denotes how much sessions in this dataset; $\|I\|$ denotes how much items in the dataset; the 'avg item' denotes how much interaction for every item; the 'Interaction' denotes how much interaction in this dataset per item and the 'Sparsity' is to describe the sparsity between items and users

	$\|\mathcal{S}\|$	$\|I\|$	Avg item	Interaction	Sparsity
KKBOX	966616	130005	46.57	6054514	99.995%
MovieLens-1m	17694	3493	283	988027	98.40%

Baselines. Five types of baseline are adopted to conduct the compare to measure the strengths and weaknesses of our model. An overview of every baseline is as following:

FPMC [15]: It is the first work on sequential recommendation adopting personalized Markov Chain, which is a general baseline in sequential recommendation and session-based recommendation.

---

[1] https://www.kaggle.com/c/kkbox-music-recommendation-challenge/data.
[2] http://grouplens.org/datasets/movielens.

GRU4Rec+BPR/CE [4,12]: These are masterpieces of the same working flow. [4] is an improved vision of [12]. GRU4Rec + BPR is another variant of GRU4Rec, which utilizes Bayes personalized ranking [14] as loss function and GRU4Rec + CE ultilizes cross-entropy as loss function.

NARM [7]: It is a GRU-based model with adding attention mechanism to eliminate accidental noise in item sequence to get the real user's interest.

STAMP [9]: It is a model taking account into time factor by adopting attention net and stresses on the last interactive item.

SR-GNN [20]: This is a work on session-based adopting GNN to capture non-continuous features to recommend.

Evaluation Metrics. Four types of evaluation metrics are adopted to compare performance comprehensively. An overview of every metric is as following:

Hit@K: It evaluates the recommendation system by whether the item appears or not. It is a general evaluation metric for SR. The higher the value is, the better the result is.

MRR@K: It is also called Mean Reciprocal Rank(MRR for abbreviation), which considers the position of the correct item. The higher the value is, the better the result is.

NDCG@K: It is also called Normalized Discounted Cumulative Gain (NDCG for abbreviation), which is a measure of ranking quality. It normalizes the result by taking account into the length of the session. The higher the value is, the better the result is.

Precision@K: It is also called positive predictive value, which evaluates how many recommendation items are correct. The higher the value is, the better the result is.

Hyper-Parameter Setup. For fair comparisons, all baseline set the default Hyper-parameter in RecBole. For our model, we set the item and behavior embedding size as 100. The hidden size of GRU is set as 100 and the layer number of GRU is set as 1. The number of head in multi-head mechanism of ResTransformer is set as 5 in encoder layer, and the number of the encoder layer is set as 6. For the decoder in ResTransformer has the same setting. The learning rate of Adam [6] optimizer is set as 0.001. In addition, all parameters are initialized by a Gaussian distribution with a mean of 0 and a standard deviation of 0.1. The batch size of every model is set as 256. The early stop step for every model is set as 10 epochs.

4.2 Global Performance Comparisons

First, we conpare the performance of all models in KKBOX to explore the toplimit as behavior in this dataset can concatenate the context. The Hit@20, MRR@20, NDCG@20 and the Precision@20 score (percentage value) are listed in Table 2.

The comparison shows that our model outperforms all baseline in KKBOX. The best Hit@20 and Precision@20 result of baseline are obtained by NARM and the best MRR@20 and NDCG@20 result of baseline are obtained by SR-GNN. This shows that the former is better at listing the interaction item in the top-20 list while it overlooks the location of occurrence. While the latter is the opposite. Focusing on our model, our model obtains 1.38%, 9.60%, 8.56%, 1.42% in relative promotion in all four metrics, and 0.39%,0.96%,1.2%,0.02% in absolute promotion in all metrics comparing with the best baseline. The success in KKBOX certificates that our model experts in the scenario that behavior can concatenate context.

Table 2. SR performance scores (percentage value) of all models in KKBOX show that our model obtains the best result in all four evaluation metrics.

	Hit@20	MRR@20	NDCG@20	Precision@20
FPMC	17.53	3.69	6.68	0.88
GRU4Rec+CE	12.61	2.68	4.8	0.63
GRU4Rec+BPR	28.17	8.44	12.77	1.37
NARM	28.18	7.94	12.38	1.41
STAMP	24.3	6.9	10.71	1.22
SR-GNN	26.58	10.0	13.67	1.33
IBSSR	**28.57**	**10.96**	**14.87**	**1.43**

Then, we compare the results in MovieLens-1m to explore the floor of our model for large difference between two datasets. The percentage scores of four metrics are listed in Table 3.

The comparison shows that our model connot obtain the best result. The difference between GRU4Rec with CE and BPR shows the importance of loss function. And the success of GNN-based method shows the importance of complex relational dependency. NARM obtains the best result validates that the next item is closer to last item, which may due to the duration of movie and music, the former is longer, thus user is more affected by the last movie.

To explore what causes this result, we perform ablation experiments on MovieLens-1m.

4.3 Ablation Study

To explore which portion leads to the bad result in MovieLens-1m, we adopt ablation study.

Table 3. SR performance scores (percentage value) of all models in MovieLens-1m show that some defects are in the scenario that behavior cannot concatenate context.

	Hit@20	MRR@20	NDCG@20	Precision@20
FPMC	18.06	4.84	7.7	0.9
GRU4Rec + CE	32.29	8	13.23	1.61
GRU4Rec + BPR	24.43	5.09	9.21	1.2
NARM	**33.66**	**8.14**	**13.64**	**1.68**
STAMP	25.8	5.97	10.21	1.29
SR-GNN	28.78	7.37	11.98	1.44
Ours	27.3	6.96	11.34	1.37

Without Interaction Between Item and Behavior. In this section, we want to explore the impact on interaction with item and behavior. Thus, we remove the interaction between item sequence and behavior sequence before feeding into ResTransformer. The percentage scores of 4 metrics are listed on Table 4. The result shows that we obtain the 36.16%, 17.57%, 26.14%, 37% in relative promotion in all four metrics, and obtain the 7.25%, 1.04%, 2.35%, 0.37% in absolute promotion. This shows that the interaction operation great improve the performance.

Table 4. SR performance scores (percentage value) of our model in MovieLens-1m show that our interaction operation make huge progress.

	Hit@20	MRR@20	NDCG@20	Precision@20
Ours	**27.3**	**6.96**	**11.34**	**1.37**
Ours w/o interaction	20.05	5.92	8.99	1.00

Without Behavior. In this section, we want to explore the impact on behavior. Thus, GRU module and behavior sequence are detached, and the item sequence is the unique input of our model correspondingly, which leads to smaller input embedding size in our attention mechanism as it concatenates two embedding vectors. The percentage scores of four metrics are listed on Table 5. The result shows that reductions occur in the processed model and demonstrates behavior is crucial for improving performance.

Table 5. SR performance scores (percentage value) of our models in MovieLens-1m show that model after eliminating behavior has worse performance

	Hit@20	MRR@20	NDCG@20	Precision@20
Ours	**27.3**	**6.96**	**11.34**	**1.37**
Ours w/o behavior	19.79	4.43	7.71	0.99

Without Local or Global Representation. In this section, we want to explore the impact on local and global representation in our attention mechanism. In our model detached either representation, attention mechanism will alter. The discrepancy in both is the lacked of global representation will invalidate the attention mechanism while the lacked of local representation will not. Table 6 reveals the result of absenting representation and the decline testify the significance of local and global representation. However, There is no significant difference between the absence of one of the two representations.

Table 6. SR performance scores (percentage value) of our models in MovieLens-1m show that model after eliminating local or global representation has worse performance

	Hit@20	MRR@20	NDCG@20	Precision@20
Ours	**27.3**	**6.96**	**11.34**	**1.37**
Ours w/o local	17.73	3.82	6.79	0.89
Ours w/o global	17.85	3.9	6.9	0.89

5 Conclusion

In this paper, we take account into the transformation between items to simulate the realistic users' interaction. A novel model is proposed to model both item sequence and behavior sequence. Our experiments validate our model's superiority over the state-of-the-art recommendation models in conventional users' behavior(i.e. our experiment in KKBOX); however, we cannot get the best performance in unconventional users' behavior(i.e. our experiment in MovieLens-1m), which can validate that tremendous differences appear in users' behavior types. In future studies, we will further explore the effects of different behavioral patterns on sequence modeling and the relationship between behaviors.

Ackonwledgement. This work was supported in part by the Natural Science Foundation of Chongqing(No. cstc2020jcyj-msxmX0284, cstc2019jcyj-msxmX0021), the Scientific and Technological Research Program of Chongqing Municipal Education Commission (No. KJQN202000625, KJCXZD2020027, KJQN202100637), National Natural Science Foundation of China (61806033).

References

1. Cho, K., Van Merriënboer, B., Bahdanau, D., Bengio, Y.: On the properties of neural machine translation: encoder-decoder approaches. arXiv preprint arXiv:1409.1259 (2014)
2. Gu, Y., Ding, Z., Wang, S., Yin, D.: Hierarchical user profiling for e-commerce recommender systems. In: Proceedings of the 13th International Conference on Web Search and Data Mining, pp. 223–231 (2020)

3. He, K., Zhang, X., Ren, S., Sun, J.: Deep residual learning for image recognition. In: Proceedings of the IEEE Conference on Computer Vision and Pattern Recognition, pp. 770–778 (2016)
4. Hidasi, B., Karatzoglou, A.: Recurrent neural networks with top-k gains for session-based recommendations. In: Proceedings of the 27th ACM International Conference on Information and Knowledge Management, pp. 843–852 (2018)
5. Hidasi, B., Karatzoglou, A., Baltrunas, L., Tikk, D.: Session-based recommendations with recurrent neural networks. arXiv preprint arXiv:1511.06939 (2015)
6. Kingma, D.P., Ba, J.: Adam: A method for stochastic optimization. arXiv preprint arXiv:1412.6980 (2014)
7. Li, J., Ren, P., Chen, Z., Ren, Z., Lian, T., Ma, J.: Neural attentive session-based recommendation. In: Proceedings of the 2017 ACM on Conference on Information and Knowledge Management, pp. 1419–1428 (2017)
8. Liu, Q., Wu, S., Wang, L.: Multi-behavioral sequential prediction with recurrent log-bilinear model. IEEE Trans. Knowl. Data Eng. 29(6), 1254–1267 (2017)
9. Liu, Q., Zeng, Y., Mokhosi, R., Zhang, H.: Stamp: short-term attention/memory priority model for session-based recommendation. In: Proceedings of the 24th ACM SIGKDD International Conference on Knowledge Discovery & Data Mining, pp. 1831–1839 (2018)
10. Meng, W., Yang, D., Xiao, Y.: Incorporating user micro-behaviors and item knowledge into multi-task learning for session-based recommendation. In: Proceedings of the 43rd International ACM SIGIR Conference on Research and Development in Information Retrieval, pp. 1091–1100 (2020)
11. Mikolov, T., Chen, K., Corrado, G., Dean, J.: Efficient estimation of word representations in vector space. arXiv preprint arXiv:1301.3781 (2013)
12. Quadrana, M., Karatzoglou, A., Hidasi, B., Cremonesi, P.: Personalizing session-based recommendations with hierarchical recurrent neural networks. In: Proceedings of the Eleventh ACM Conference on Recommender Systems, pp. 130–137 (2017)
13. Rendle, S.: Factorization machines. In: 2010 IEEE International Conference on Data Mining, pp. 995–1000. IEEE (2010)
14. Rendle, S., Freudenthaler, C., Gantner, Z., Schmidt-Thieme, L.: BPR: Bayesian personalized ranking from implicit feedback. arXiv preprint arXiv:1205.2618 (2012)
15. Rendle, S., Freudenthaler, C., Schmidt-Thieme, L.: Factorizing personalized Markov chains for next-basket recommendation. In: Proceedings of the 19th International Conference on World Wide Web, pp. 811–820 (2010)
16. Sarwar, B., Karypis, G., Konstan, J., Riedl, J.: Item-based collaborative filtering recommendation algorithms. In: Proceedings of the 10th International Conference on World Wide Web, pp. 285–295 (2001)
17. Vaswani, A., et al.: Attention is all you need. In: Advances in Neural Information Processing Systems 30 (2017)
18. Wan, M., McAuley, J.: Item recommendation on monotonic behavior chains. In: Proceedings of the 12th ACM Conference on Recommender Systems, pp. 86–94 (2018)
19. Wang, P., Fan, Y., Xia, L., Zhao, W.X., Niu, S., Huang, J.: KERL: a knowledge-guided reinforcement learning model for sequential recommendation. In: Proceedings of the 43rd International ACM SIGIR Conference on Research and Development in Information Retrieval, pp. 209–218 (2020)
20. Wu, S., Tang, Y., Zhu, Y., Wang, L., Xie, X., Tan, T.: Session-based recommendation with graph neural networks. In: Proceedings of the AAAI Conference on Artificial Intelligence, vol. 33, pp. 346–353 (2019)

21. Xian, Y., Fu, Z., Muthukrishnan, S., De Melo, G., Zhang, Y.: Reinforcement knowledge graph reasoning for explainable recommendation. In: Proceedings of the 42nd international ACM SIGIR Conference on Research and Development in Information Retrieval, pp. 285–294 (2019)
22. Xu, K., et al.: Show, attend and tell: neural image caption generation with visual attention. In: International Conference on Machine Learning, pp. 2048–2057. PMLR (2015)
23. Zhao, W.X., et al.: RecBole: towards a unified, comprehensive and efficient framework for recommendation algorithms. In: Proceedings of the 30th ACM International Conference on Information & Knowledge Management, pp. 4653–4664 (2021)
24. Zhou, M., Ding, Z., Tang, J., Yin, D.: Micro behaviors: a new perspective in e-commerce recommender systems. In: Proceedings of the Eleventh ACM International Conference on Web Search and Data Mining, pp. 727–735 (2018)

Human-Centered Real-Time Instance Segmentation with Integration with Data Association and SOLO

Lu Cheng[1], Mingbo Zhao[1(✉)], and Jicong Fan[2,3(✉)]

[1] Donghua University, Shanghai, China
2201840@mail.dhu.edu.cn, mzhao4@dhu.edu.cn
[2] The Chinese University of Hong Kong, Shenzhen, China
fanjicong@cuhk.edu.cn
[3] Shenzhen Research Institute of Big Data, Shenzhen, China

Abstract. Video instance segmentation (VIS) mainly has two methods in human-targeted tasks, one is based on mask propagation method, and the other is based on detection tracking mode method. Existing datasets pay less attention to all the characters appearing in the video, but focus on the key characters appearing in the video. Therefore, in this paper, an online joint detection and tracking model is proposed. Based on the single-stage segmentation method SOLO, appearance feature extraction and data association are added to realize the segmentation and tracking of human body in video. At the same time, in order to better evaluate the method, we extract a single person type on the basis of existing multi-type data sets, filter out high-quality annotations and various video scenes, and obtain the challenging data set PVIS. As a result, the method based on the PVIS dataset can achieve satisfied performance both for segmentation accuracy and tracking accuracy.

Keywords: SOLO · Video instance segmentation · Person Re-ID · Multi-object tracking

1 Background

Instance segmentation is a common task that is to achieve pixel-level segmentation of different instances in the research area of computer vision. Among different learning tasks, human instance segmentation is a key technology to automatically understand the human activities, and has been extensively utilized in different learning tasks such as video surveillance, virtual reality, et al. On the other hand, the efficiency and effectiveness of human instance segmentation methods have been developed fast, making the videos gradually occupying a more important position in human life, the human body segmentation task is not only limited to image-level instance segmentation, but proposed Instance segmentation requirements for continuous video.

Human video instance segmentation is essentially an extension of the instance segmentation task [31], that is, it is not only necessary to achieve pixel-level segmentation for each instance, but also to assign the same label to the same person in the video

H. Zhang et al. (Eds.): NCAA 2022, CCIS 1637, pp. 367–377, 2022.
https://doi.org/10.1007/978-981-19-6142-7_28

to realize the video frame tracking task [22]. Since the data set of video instance segmentation is obtained from dynamic video at a certain frame rate, the image quality will seriously affected caused by the extensive problems of motion deformation and blue [4,25,30]. Therefore, it is necessary to achieve high accuracy of segmentation, correlation effect, and efficient reference procedure. The task requirements still have high difficulty. The current mainstream video instance segmentation methods can be generally categorized as mask propagation methods and detection-and-tracking-mode methods [4,25,30].

The method based on mask propagation is suitable for the situation that target objects in the video of the first frame are labeled [1,18,19,24]. First, the target mask of the first frame of video is annotated by some certain instance segmentation method, and based on this, the masks of corresponding instances in following frames of videos is realized [5,20,28]. However, this method need high accuracy of mask in the first frame [2,32]. If higher-quality instance mask is not obtained [29], the subsequent masks cannot be propagated well as the segmentation error of instance masks will be aggravated, which will in turn seriously affect the following segmentation results of video frames [16,33–39].

The method based on detection and tracking mode is mainly divided into two steps SOLO [26,30]. First, the instance mask of each frame of video image is obtained by the instance segmentation method, and then the instance mask in the frame of video that belong to the same object is matched based on certain strategy of data association, and then the continuous motion trajectory of the instance is formed. The cross-frame data association method ranges from simple IOU matching conditions to directly learning character feature expressions for matching, and with the development of multi-task learning, joint learning of detection and matching features has begun to be implemented in a single network. Although the network complexity is reduced to a certain extent, it still increases the amount of computation and affects the real-time performance of segmentation as it is to integrate motion information with target appearance information for data matching.

Considering the fact that the segmentation masks of target in the first frame of video tend not to being provided and the efficiency are required, this paper proposes a SOLO-based human video instance segmentation framework.

The novelties in this work are given as:

1) On the basis of the single-stage segmentation network SOLO [26], the feature extraction and data association parts are added to realize the video instance segmentation task that integrates various tasks of target detection, segmentation and tracking;
2) Using the Siamese network structure and extracting the embedding of each target from the parameter-sharing backbone network as the appearance feature associated with the data, it improves the occlusion problem in the tracking process and the existence of ids in different scales for the same instance target. switch (id-switch) problem;
3) Assemble multiple video instance segmentation datasets to construct a PVIS dataset for specially handling human video segmentation tasks. Simulations results via the proposed PVIS dataset are conducted to evaluate the effectiveness of the proposed work.

This paper is structured with several sections: the related work for reviewing video instance segmentation is firstly given in Sect. 2; in Sect. 3, we give detail description of the proposed work; in Sect. 4, we will conduct extensive simulations to show the effectiveness of the proposed work on human video instance segmentation and final conclusion is drawn in Sect. 5.

2   Brief Review of Related Work

As an extension of instance segmentation technology, video instance segmentation is to handle both instance-level and video-level prediction problems. The classic method of the two-stage approach, Mask-RCNN [9], directly extends the segmentation branch on Faster-RCNN [21], through which object instances are segmented within the prior bounding box. In recent years, single-stage approaches that de-feature their parts have begun to emerge. YOLACT [3] generates image-sized prototype masks via a full convolutional network and then to predict a set of weight coefficients to formulate a linear combination of prototypes to calculate the masks of target instances. In addition, the SOLO and its variants [26] is to treat the mask of each instance as a output channel, so that predicted mask is global as the resolution is equal to the size of image. In addition, it does not require anchor boxes or bounding boxes, and combines a simple framework with better performance.

Compared with instance segmentation, video instance segmentation not only needs to segment a single frame of images, but also needs to consider the instance correspondence within continuous frames in video. For example, MaskTrack-RCNN [30] is the first work for handling video instance segmentation task, which is to add an Re-ID branch to Mask-RCNN [9] for handling data association, but only uses appearance features instead of temporal information of video sequences. CompFeat [8] further handle object-level and video-level appearance features by using spatial and temporal information. In addition, VisTR [28] further extends DETR [6] to implement a query-based end-to-end approach.

3   The Proposed Method

3.1   Overview

The overall goal of the article is to precisely segment the human instance in each frame of videos under the complex environment with changing scenes as well as to track the instance that belongs the same human in the frames of video. In general, the proposed method has adopted the global segmentation strategy developed by SOLO, where the total structure of the proposed method can be illustrated in Fig. 1. Here, SOLO [27] is to treat the mask of each instance as a output channel, so that predicted mask is global as the resolution is equal to the size of image. In detail, we first divide the size of input image as $S \times S$. In practice, each of $S \times S$ grid is used for predicting the classification label of instance as well as the mask given the center of a certain instance is fallen in a grid. At the same time, we then add a Re-ID embedding module to the backbone for extracting appearance feature of the instance, and we also involve another

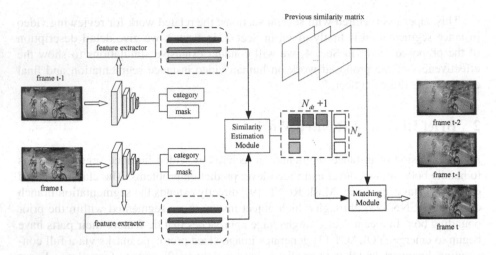

Fig. 1. Overall structure for illustrating how the proposed work handles human instance segmentation task. First, images are feed into model and is divided by $S \times S$. Each of $S \times S$ grid is used for predicting the classification label of instance as well as the mask given the center of an certain instance is fallen in a grid. At the same time, we then add a Re-ID embedding module to the backbone for extracting appearance feature of the instance, and we also involve another similarity estimation module for data association within different video subsequences, so as to implement instance segmentation and inter-frame segmentation in each frame for tracking.

similarity estimation module for data association within different video subsequences, so as to implement instance segmentation and inter-frame segmentation in each frame for tracking.

3.2   Feature Extraction Module

The motivations for this module are focusing on obtain features of a certain length to represent instances, and to achieve data association between different frames through appearance embeddings. As shown in the structure diagram, a pair of video frames separated by n timestamps is selected as input. Through the parameter sharing backbone network ResNet-101 [10], the feature of the instance center position is extracted from the feature maps of different levels and connected to obtain the instance embeddings. Since the video instance segmentation process will lead to large scale changes in the instances due to camera or personal movements, the scale invariance requirement is imposed on the embedding features of human instances. The instance segmentation method SOLO [26] divides the target into corresponding mask prediction grids according to different target sizes. Under this specification, the same instance of different sizes in adjacent frames will be assigned different identity information. If the feature connection of the FPN [14] structure is directly obtained, it will cause the problem of id change. To handle it, we in this work sample instance features via feature maps of different scales on the selected backbone network and connect them with a certain strategy. In this process, the feature maps generated by shallow convolution and deep convolution are fused, which have higher spatial resolution and more semantic information, respectively, which further improves the expressiveness of instance embedding. At the same time, the Siamese network structure [13] also superimposes the features corresponding to the same instance in different frames, which enhances the robustness of the embedding. In view of the test that the embedding vector is too large, the 1×1 convolution kernel is utilized for reducing the dimension before the feature embeddings in corresponding position.

3.3 Similarity Estimation Module

A data association matrix of pairs of input frames is built in this module, associating objects of a given frame with objects of previous frames. The main structure of this module is based on DAN [23] network. Since the number of instances in each frame in the video is uncertain, which brings difficulty to training, the maximum number of recognized instances in a single frame of pictures is fixed as N_m. The instance embedding features are matched in pairs to form a tensor $T_{t,t-n} \in \mathbb{R}^{(e \times 2) \times N_m \times N_m}$. After 5 layers of convolution, the number of channels is reduced to 1, and an $N_m \times N_m$ matrix is obtained to represent the possibility of association between any two targets. Additional rows or columns are added to the existing association matrix after considering instances that leave or emerge in later frames. When constructing the training target, the last row or column contains 1 to represent multiple instances of departure or change, and a binary association matrix is obtained. After that, the row and column SoftMax is performed on the row and column expansion matrix respectively to obtain the similarity matrix.

3.4 Tracking Trajectory Generation

The similarity matrix of each frame and the timestamp of the related trajectory are recorded in this module in order to correlate the current trajectory with all prior trajectories, and the Hungarian method is used to solve the instance matching problem. We use the accumulator matrix to cumulatively sum the similarity matrices of the instance object in the current frame and multiple historical frame objects. By maximizing the similarity relationship between things within current video frame and the targets with assigned identities during prior ones, the Hungarian algorithm creates a unique identity assignment. At the same time, because the Hungarian algorithm is a one-to-one constraint, and the last row or column added to the similarity matrix has several targets. Therefore, the one-to-one matching procedure must be performed many times until the matching is complete. The number of frames before storage must be limited to a specified range, considering the amount of computation and memory capacity.

3.5 Training

In the training process, two frames of video image pairs are selected for input, and the twin structure is used to extract features. According to the task requirements, the loss is divided into the instance segmentation part and the associated part. The calculation process is as follows:

$$L=\frac{1}{2}(\frac{1}{e^{s_i}}L_{seg} + s_i + \frac{1}{e^{s_j}}L_{match} + s_j) \tag{1}$$

The calculation process of L_{seg} is the same as that in SOLO, and due to the symmetrical structure of the network, the loss calculation process of the two frames of images is also required to be the same, and the loss function of the network is formed by summation. Since the learning objectives and difficulty of segmentation and identity matching are different, direct summation will affect the training effect of both, so parameters are added to balance the joint learning of tasks and avoid a certain task dominating. s_i, s_j are the parameters learned for handling the uncertainty in the loss of different tasks and for reducing the impact of different multi-tasks [11].

4 Experiments

4.1 Simulation Settings and Dataset

The PVIS Dataset

Currently, the video instance segmentation datasets specifically for humans are limited, the amount of data required for the experiment is large and the workload of self-labeling is too large, so a new benchmark dataset is constructed based on the existing video instance segmentation dataset [4,7,30]. The proposed method for training and evaluation. The constructed dataset has the following characteristics: high annotation accuracy; video scenes are complex and contain a dramatic change of different scenes including motion blur, occlusion, etc.; There are also some fuzzy annotations, especially when only a few prominent people are annotated in a crowded crowd.

According to the above data set requirements, we use the existing common data sets YouTube-VIS [30], DAVIS [4], DAVSOD [7], etc., to extract the human part and re-label and filter to obtain a new data set that are used for training and evaluating the model. In detail, DAVSOD dataset annotate salient targets in the frame of video, while DAVIS dataset also only pays attention to the salient people that may in the first frame appeared without annotating the other people appeared in the following frames of video. Different from these datasets, YouTube-VIS is generally to annotate all characters appeared in the whole time of the video. Therefore, new dataset is constructed motivated by YouTube-VIS, while we delete certain videos which cannot satisfy the requirement. Instead, we have also added the other two data Collect partial sequences that meet the requirements, crop and reassign instance numbers. As a results, 2.9k video sequences and 40 categories are selected from the YouTube-VIS dataset, which have annotated 4883 objects and corresponding 131k instance masks.

1117 video sequences are contained in the new PVIS dataset, which has 2025 targets and annotate approximately 60k instance masks. Here, we choose 128 videos to evaluate the proposed methods. Note the PVIS dataset is consist with a variety of different datasets, we set adjacent frame interval as 1-5FPS, so in some fast-moving videos, large inter-frame motion can be realized, and it also proposes a better matching of instances.

Measurement

Note the video instance segmentation is to integrate detection, tracking and segmentation into the unified framework. The evaluation metric method also adopts the HOTA [17] metric. This can be used to evaluate the effectiveness of how the method can handle detection and trajectory association simultaneously; on the other hand, HOTA aims to achieve a balance of multi-task learning based on the average mean of the detection and association results. If an instance is missed in a detector or false detection occurs, a lower detection accuracy score (DetA) will be obtained. However, dividing multiple instance masks in a continuous motion will results in low Association Accuracy Score (AssA). There are also some other measurements, such as detection recall rate DetRe, detection accuracy rate DetPr, association recall rate AssRe, association precision rate AssPr, positioning accuracy score LocA, and the number of identity transitions IDS.

4.2 Implementation Details

The network uses ResNet-101 as the backbone network and is a pre-trained model based on COCO [15]. Then, the model is fine-tuned through the PVIS dataset. Based on Pytorch, trained on 4 NVIDIA Titan X GPUs for 12 generations. During this process, we initialize the Adam optimizer [12] learning rate as 0.0002, use warmup the model by approximately 500 iterations, and then decay the parameter learning rate with a factor of 10 at the 9th and 11th generations, respectively. Through subsequent experimental verification, 11 layers are selected from the model for extracting features, and the length of embedded feature is set as 352. The longest trajectory storage time τ, the maximum interval of video frame pairs T_m and the maximum for recognizing people in a single video sequence N_m, are respectively 30, 10, and 50.

4.3 Comparison Results

In this subsection, we will evaluate the proposed work and compare the performance with other existing video instance segmentation methods, and the comparison object is MaskTrack-RCNN [30], the first classical method proposed for this task. The simulation results for visualization with different compared methods are illustrated in Fig. 2, and different measurements for evaluate the proposed method quantitatively are given in Table 1. From the simulation results, we can see the proposed work can well handle instance segmentation as well as tracking task illustrating its superiority to other state-of-the-art methods from many aspects.

The below results are the visualization results after using our method and the MaskTrack-RCNN method. Each row is a 5-frame segmentation result selected from the test video, where the same instance is represented by the same color. The first row, third row, and fifth row are the results of MaskTrack-RCNN running, and the second row, fourth row, and sixth row are the results of the method in this paper running on the same video sequence. Since this method utilize the multi-scale feature map to sample the corresponding instance features in the feature extraction procedure, and the matching includes two stages of feature association and IOU association, it can better cope with partial occlusion in the second row or two instances in the fourth row. When the

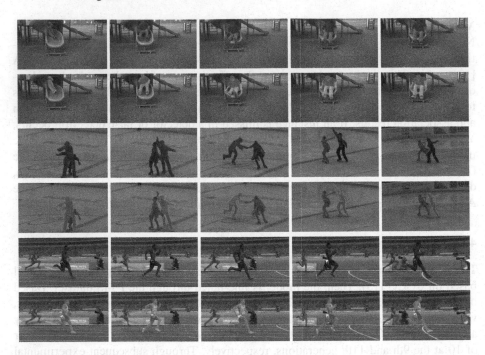

Fig. 2. The simulation results of segmentation compared with the proposed work with MaskTrack-RCNN, where every row shows five frames selected from a video, using the same color to represent the same person instance.

distance is short, the identity of the instance cannot be judged. However, MaskTrack-RCNN will be affected in the occlusion video sequence and cannot accurately segment the instance, and after the two instances are interlaced, their instance identities will be exchanged with each other, and the identity association in the video sequence cannot be guaranteed.

Through the quantitative evaluation results in Table 1, the overall video instance segmentation effect is improved by 3.4% in HOTA, 3.5% in DetA and 2.5% in AssA compared with MaskTrack-RCNN, and the id conversion is reduced. Combining the results presented by the two methods, we can see there are little difference in performance between two methods given the frame in the slow video having one character; however, while in some videos with high-speed motion and lots of characters, the example features used in the method proposed in the paper can grasp good semantic information and achieve satisfied tracking results. In addition, the identify conversion phenomenon can be alleviated due to the fast change for the instance size during high-speed motion.

Table 1. Quantitative Results between the proposed work with other state-of-the-art methods based on PVIS dataset.

Methods	HOTA	DetA	AssA	IDS
SipMask [5]	51.8	38.6	72.1	76.0
MaskTrack-RCNN [30]	54.2	42.1	72.7	91.2
Our method	58.6	45.7	77.1	65.0

5   Conclusion

In this paper, feature extraction and data association are added to the single-stage segmentation network SOLO, and a video instance segmentation method in complex situations is proposed. This method has good performance in terms of accuracy. On the basis of the single-stage segmentation network SOLO, the feature extraction and data association parts are added to realize the video instance segmentation task that integrates various tasks of target detection, segmentation and tracking; for the occlusion problem in the video or the problem of concentration of sampling points caused by the overlapping of multiple characters, the maximum contour centroid sampling strategy is adopted to eliminate the influence of the occluded part or the overlapping part on the instance feature representation, which can better cope with such video scenes. Meanwhile, we have assembled multiple video instance segmentation datasets to construct a PVIS dataset for specially handling human video segmentation tasks. Simulations results via the proposed PVIS dataset are conducted to evaluate the effectiveness of the proposed work.

Acknowledgement. This work is supported by National Natural Science Foundation of China (61971121, 62106211).

References

1. Bertasius, G., Torresani, L.: Classifying, segmenting, and tracking object instances in video with mask propagation. In: Proceedings of the IEEE/CVF Conference on Computer Vision and Pattern Recognition, pp. 9739–9748 (2020)
2. Bochinski, E., Eiselein, V., Sikora, T.: High-speed tracking-by-detection without using image information. In: 2017 14th IEEE International Conference on Advanced Video and Signal Based Surveillance (AVSS), pp. 1–6. IEEE (2017)
3. Bolya, D., Zhou, C., Xiao, F., Lee, Y.J.: Yolact: real-time instance segmentation. In: Proceedings of the IEEE/CVF International Conference on Computer Vision, pp. 9157–9166 (2019)
4. Caelles, S., Pont-Tuset, J., Perazzi, F., Montes, A., Maninis, K.K., Van Gool, L.: The 2019 davis challenge on vos: Unsupervised multi-object segmentation. arXiv preprint arXiv:1905.00737 (2019)
5. Cao, J., Anwer, R.M., Cholakkal, H., Khan, F.S., Pang, Y., Shao, L.: SipMask: spatial information preservation for fast image and video instance segmentation. In: Vedaldi, A., Bischof, H., Brox, T., Frahm, J.-M. (eds.) ECCV 2020. LNCS, vol. 12359, pp. 1–18. Springer, Cham (2020). https://doi.org/10.1007/978-3-030-58568-6_1

6. Carion, N., Massa, F., Synnaeve, G., Usunier, N., Kirillov, A., Zagoruyko, S.: End-to-end object detection with transformers. In: Vedaldi, A., Bischof, H., Brox, T., Frahm, J.-M. (eds.) ECCV 2020. LNCS, vol. 12346, pp. 213–229. Springer, Cham (2020). https://doi.org/10. 1007/978-3-030-58452-8_13

7. Fan, D.P., Wang, W., Cheng, M.M., Shen, J.: Shifting more attention to video salient object detection. In: Proceedings of the IEEE/CVF Conference on Computer Vision and Pattern Recognition, pp. 8554–8564 (2019)

8. Fu, Y., Yang, L., Liu, D., Huang, T.S., Shi, H.: Compfeat: comprehensive feature aggregation for video instance segmentation. arXiv preprint arXiv:2012.03400 (2020)

9. He, K., Gkioxari, G., Dollár, P., Girshick, R.: Mask r-cnn. In: Proceedings of the IEEE International Conference on Computer Vision, pp. 2961–2969 (2017)

10. He, K., Zhang, X., Ren, S., Sun, J.: Deep residual learning for image recognition. In: Proceedings of the IEEE Conference on Computer Vision and Pattern Recognition, pp. 770–778 (2016)

11. Kendall, A., Gal, Y., Cipolla, R.: Multi-task learning using uncertainty to weigh losses for scene geometry and semantics. In: Proceedings of the IEEE Conference on Computer Vision and Pattern Recognition, pp. 7482–7491 (2018)

12. Kingma, D.P., Ba, J.: Adam: a method for stochastic optimization. arXiv preprint arXiv:1412.6980 (2014)

13. Leal-Taixé, L., Canton-Ferrer, C., Schindler, K.: Learning by tracking: siamese CNN for robust target association. In: Proceedings of the IEEE Conference on Computer Vision and Pattern Recognition Workshops, pp. 33–40 (2016)

14. Lin, T.Y., Dollár, P., Girshick, R., He, K., Hariharan, B., Belongie, S.: Feature pyramid networks for object detection. In: Proceedings of the IEEE Conference on Computer Vision and Pattern Recognition, pp. 2117–2125 (2017)

15. Lin, T.-Y., et al.: Microsoft COCO: common objects in context. In: Fleet, D., Pajdla, T., Schiele, B., Tuytelaars, T. (eds.) ECCV 2014. LNCS, vol. 8693, pp. 740–755. Springer, Cham (2014). https://doi.org/10.1007/978-3-319-10602-1_48

16. Liu, J., Lin, M., Zhao, M., Zhan, C., Li, B., Chui, J.K.T.: Person re-identification via semi-supervised adaptive graph embedding. Appl. Intell. 1–17 (2022)

17. Luiten, J., et al.: Hota: a higher order metric for evaluating multi-object tracking. Int. J. Comput. Vision **129**(2), 548–578 (2021)

18. Oh, S.W., Lee, J.Y., Sunkavalli, K., Kim, S.J.: Fast video object segmentation by reference-guided mask propagation. In: Proceedings of the IEEE Conference on Computer Vision and Pattern Recognition, pp. 7376–7385 (2018)

19. Oh, S.W., Lee, J.Y., Xu, N., Kim, S.J.: Video object segmentation using space-time memory networks. In: Proceedings of the IEEE/CVF International Conference on Computer Vision, pp. 9226–9235 (2019)

20. Peng, J., et al.: Chained-tracker: chaining paired attentive regression results for end-to-end joint multiple-object detection and tracking. In: Vedaldi, A., Bischof, H., Brox, T., Frahm, J.-M. (eds.) ECCV 2020. LNCS, vol. 12349, pp. 145–161. Springer, Cham (2020). https:// doi.org/10.1007/978-3-030-58548-8_9

21. Ren, S., He, K., Girshick, R., Sun, J.: Faster r-cnn: Towards real-time object detection with region proposal networks. Adv. Neural Inf. Process. Syst. **28**, 91–99 (2015)

22. Smeulders, A.W., Chu, D.M., Cucchiara, R., Calderara, S., Dehghan, A., Shah, M.: Visual tracking: an experimental survey. IEEE Trans. Pattern Anal. Mach. Intell. **36**(7), 1442–1468 (2013)

23. Sun, S., Akhtar, N., Song, H., Mian, A., Shah, M.: Deep affinity network for multiple object tracking. IEEE Trans. Pattern Anal. Mach. Intell. **43**(1), 104–119 (2019)

24. Voigtlaender, P., Chai, Y., Schroff, F., Adam, H., Leibe, B., Chen, L.C.: Feelvos: fast end-to-end embedding learning for video object segmentation. In: Proceedings of the IEEE/CVF Conference on Computer Vision and Pattern Recognition, pp. 9481–9490 (2019)
25. Voigtlaender, P., et al.: Mots: multi-object tracking and segmentation. In: Proceedings of the IEEE/CVF Conference on Computer Vision and Pattern Recognition, pp. 7942–7951 (2019)
26. Wang, X., Kong, T., Shen, C., Jiang, Y., Li, L.: SOLO: segmenting objects by locations. In: Vedaldi, A., Bischof, H., Brox, T., Frahm, J.-M. (eds.) ECCV 2020. LNCS, vol. 12363, pp. 649–665. Springer, Cham (2020). https://doi.org/10.1007/978-3-030-58523-5_38
27. Wang, X., Zhang, R., Kong, T., Li, L., Shen, C.: Solov2: dynamic and fast instance segmentation. arXiv preprint arXiv:2003.10152 (2020)
28. Wang, Y., et al.: End-to-end video instance segmentation with transformers. In: Proceedings of the IEEE/CVF Conference on Computer Vision and Pattern Recognition, pp. 8741–8750 (2021)
29. Xu, Y., Osep, A., Ban, Y., Horaud, R., Leal-Taixé, L., Alameda-Pineda, X.: How to train your deep multi-object tracker. In: Proceedings of the IEEE/CVF Conference on Computer Vision and Pattern Recognition, pp. 6787–6796 (2020)
30. Yang, L., Fan, Y., Xu, N.: Video instance segmentation. In: Proceedings of the IEEE/CVF International Conference on Computer Vision, pp. 5188–5197 (2019)
31. Yu, R., Tian, C., Xia, W., Zhao, X., Wang, H., Yang, Y.: Real-time human-centric segmentation for complex video scenes. arXiv preprint arXiv:2108.07199 (2021)
32. Zhang, Y., Yang, Q.: A survey on multi-task learning. arXiv preprint arXiv:1707.08114 (2017)
33. Zhao, M., Chow, T.W., Wu, Z., Zhang, Z., Li, B.: Learning from normalized local and global discriminative information for semi-supervised regression and dimensionality reduction. Inf. Sci. **324**, 286–309 (2015)
34. Zhao, M., Chow, T.W., Zhang, Z., Li, B.: Automatic image annotation via compact graph based semi-supervised learning. Knowl.-Based Syst. **76**, 148–165 (2015)
35. Zhao, M., Lin, M., Chiu, B., Zhang, Z., Tang, X.S.: Trace ratio criterion based discriminative feature selection via l2, p-norm regularization for supervised learning. Neurocomputing **321**, 1–16 (2018)
36. Zhao, M., Liu, J., Zhang, Z., Fan, J.: A scalable sub-graph regularization for efficient content based image retrieval with long-term relevance feedback enhancement. Knowl.-Based Syst. **212**, 106505 (2021)
37. Zhao, M., Zhang, Y., Zhang, Z., Liu, J., Kong, W.: ALG: adaptive low-rank graph regularization for scalable semi-supervised and unsupervised learning. Neurocomputing **370**, 16–27 (2019)
38. Zhao, M., Zhang, Z., Chow, T.W.: Trace ratio criterion based generalized discriminative learning for semi-supervised dimensionality reduction. Pattern Recogn. **45**(4), 1482–1499 (2012)
39. Zhao, M., Zhang, Z., Chow, T.W., Li, B.: A general soft label based linear discriminant analysis for semi-supervised dimensionality reduction. Neural Netw. **55**, 83–97 (2014)

Extracting Key Information from Shopping Receipts by Using Bayesian Deep Learning via Multi-modal Features

Jiaqi Chen[1], Lujiao Shao[1], Haibin Zhou[1], Jianghong Ma[1], Weizhi Meng[2],
Zenghui Wang[3], and Haijun Zhang[1(✉)]

[1] Departmant of Computer Science, Harbin Institudte of Technology,
Shenzhen, China
hjzhang@hit.edu.cn
[2] Department of Applied Mathematics and Computer Science,
Technical University of Denmark, Kongens Lyngby, Denmark
[3] Department of Electrical and Mining Engineering, University of South Africa,
Florida 1710, South Africa

Abstract. This research presents a new key information extraction algorithm from shopping receipts. Specifically, we train semantic, visual and structural features through three deep learning methods, respectively, and formulate rule features according to the characteristics of shopping receipts. Then we propose a multi-class text classification algorithm based on multi-modal features using Bayesian deep learning. After post-processing the output of the classification algorithm, the key information we seek for can be obtained. Our algorithm was trained on a self-labeled Chinese shopping receipt dataset and compared with several baseline methods. Extensive experimental results demonstrate that the proposed method achieves optimal results on our Chinese receipt dataset.

Keywords: Information extraction · Multi-modal feature fusion ·
Bayesian deep learning · Shopping receipt

1 Introduction

Shopping receipts, as a proof of customers' offline consumption, usually contain a large amount of information about stores, products and customers. Specifically, for customers, these kinds of information can reflect their own consumption habits and record the details of the purchase of products. For stores, the information can be used for potential market planning by building customer points systems. However, these shopping receipts are often shot by mobile phones of customers when they upload them to the points systems. As a result, it is challengeable to accurately extract the key information conveyed by these shopping receipt images due to unbalanced shooting quality, diverse expression forms of synonymous text and inconsistent template forms, even though these images have been successfully digitalized by a well-performed optical character recognition (OCR) system.

Actually, there exist many key information extraction methods, which can be roughly categorized into rule, machine learning and deep learning-based methods. Specifically, the rule-based approaches are usually heavily dependent on experienced developers who work in the related business domain, and they need to take a lot of efforts to construct the domain-specific rules which typically cannot generalize well to other business domains. To overcome these issues, researchers have given considerable attention to machine learning-based methods, in which the information extraction problem is formulated as a classification process implemented by well-known classifiers such as naive Bayes, support vector machines or maximum entropy models. At present, methods based on deep learning become more and more common for document image understanding. For example, Gao et al. [1] designed a hierarchical attention network method to automatically extract information from unstructured pathological reports. Luo et al. [2] proposed a domain-specific language, which allows users to describe the values and layout of text data in images, and achieves the extraction of structured medical text information from images of medical documents. Liu et al. [3] proposed an information extraction algorithm which combines textual and visual information in visually rich documents (VRDs). This algorithm uses graph convolution network to calculate the graph feature embeddings of each text segment in a document. After graph and text feature embeddings are combined, the algorithm uses Bi-LSTM and conditional random field (CRF) to extract key information. Furthermore, LayoutLM [4] was developed to model the text and layout information in scanned document images. In LayoutLM, image features were used to integrate the visual information of words. LayoutLM achieves promising results in the tasks of form understanding, receipt understanding and document image classification. However, the document images targeted by the above method were all scanned. For the document images taken by cameras in different environments such as mobile phones, we need to process them by considering more intrinsic features hidden in these target images.

In this research, we extract features from a shopping receipt through different modalities by using deep neural networks. Moreover, we propose a key information extraction algorithm to extract the information which can be used in customer points business. Specifically, main contributions of this paper are summarized into two folds: 1) we achieve the feature representations of shopping receipts in terms of rules, semantics, visual information and structural information, and standardize them to better represent the fields of a receipt; and 2) we propose a multi-modal feature-based Bayesian deep learning algorithm for key information extraction. The algorithm enhances the model's capability of uncertainty reasoning so as to highly improve the robustness of the extracted results.

The rest of this article is organized as follows. Section 2 briefly reviews the related work. Section 3 introduces the extraction methods of different modality features and the framework of key information extraction associated with implementation details. Section 4 describes the verification experiment and ablation study. This article concludes with future work propositions in Sect. 5.

2   Related Work

In this section, we review related state-of-the-arts for extraction methods and Bayesian deep learning, and highlight the features of our method finally.

1) Extraction Methods: Text representation can be formulated based on word bag models or topic models. A common way to generate word embeddings is to use the Word2Vec [5], which has high speed and versatility and is effective in most cases. Another word embedding model fastText [6] also has a fast-training speed and uses N-grams for character-level words. It can solve the problem of polysemy and construct word vectors of unregistered words. At present, most methods to obtain text representations are mainly based on pre-trained model. For example, BERT [7] can obtain dynamic representations for words that are literally the same but have different meanings. It uses a deeper network and massive corpora to get the final text vectors. On the other hand, convolutional neural network (CNN) is utilized to extract image features. Generally, shallow networks extract texture and fine-grained features, while deep networks extract contours, shapes, and the most prominent features. However, deepening the network depth usually tends to lead to rapid degradation of network performance. The network structure of residual block in ResNet [8] addresses well the degradation problem of CNN after stacking many layers. Graph neural network (GCN) [9] is another type of CNN that can work directly on graphs and make use of their structural information, aiming at solving the problem of node classification in graphs under semi-supervised conditions.

2) Bayesian Deep Learning: Ordinary feedforward neural networks are prone to overfitting, and when applied to supervised or reinforcement learning problems, these networks are often unable to correctly evaluate the uncertainty in training data. As a result, feedforward neural networks make overconfident decisions about the right category, prediction, or action. Bayes by Backprop [10] uses variational Bayesian learning to introduce uncertainty in network weight parameters to solve aforementioned problems. In Bayesian deep learning, the ownership value of neural network is represented by the probability distribution of possible values, rather than a single fixed value of weights. Attributed to the generalization capability of Bayesian deep learning, it has been widely used in a variety of fields, such as computer vision, natural language processing and recommendation systems [11]. For example, Kendall et al. [12] proposed a unified Bayesian deep learning framework, which allows computer vision models to learn mappings from input data to arbitrary uncertainties and combine these mappings with cognitive uncertainty approximations. Siddhant and Lipton [13] performed a great deal of deep active learning on three natural language processing tasks (i.e., sentiment analysis, named entity recognition and semantic role labeling), each of which considers a variety of datasets, models, and acquisition functions. Teng et al. [14] proposed Bayesian deep collaborative matrix factorization (BDCMF) for collaborative filtering. It considers various item content information generated by item potential content factors through generation network for recommendation systems.

3) Features of Our Method: There exist several studies focusing on key information extraction from document image text. For example, the work in [3] uses a GCN to calculate the graph feature embeddings of each text segment in a document. The work in [4] adds 2-D position embedding and image embedding on the basis of the existing pre-trained models so that the document structure and visual information can be combined effectively. Both works [3] and [4] introduce features from more than one modality of document images. Our algorithm also uses more multi-modal features, and adopts Bayesian deep learning to realize the key information extraction module. We extract rules, semantics, visual information and structural features from different shopping receipts. We then propose a multi-class text classification model using Bayesian deep LSTM to categorize a receipt's text fields. In this way, we extract the key information from the text fields which are classified as key categories.

3   Key Information Extraction Based on Bayesian Deep Learning

This section introduces a key information extraction algorithm. First, the framework of key information extraction from receipts is introduced. Feature extraction, Bayesian deep network for classification, and implementation details are then described sequentially.

3.1   Problem Formulation and Framework Overview

The textbox of a receipt image obtained by OCR mostly contains independent semantic information and can be regarded as a key entity. In this research, the key information extraction task of a receipt image is modeled by multi-class classification. Formally, a shopping receipt can be defined as follows:

$$field_i = (t_i, p_i, v_i), i = 1, \ldots, n_{r_2}, \tag{1}$$

$$receipt = \{field_i \mid i = 1, \ldots, n_r\}, \tag{2}$$

where, $receipt \in RD$ is a specific receipt in a shopping receipt dataset RD, $field_i$ represents the ith textbox of the receipt, t_i, p_i and v_i are the text, coordinate and image information of $field_i$ respectively, and n_r is the number of textboxes of the receipt. To sum up, the task of this research can be described as follows: classify each $field_i$, and the $field_i$ classified as key information categories will undergo some post-processing to realize the extraction of key information of shopping receipts. For clarity, the overall framework of key information extraction from receipts is shown in Fig. 1. We elaborate the implementation details of each module in the following context.

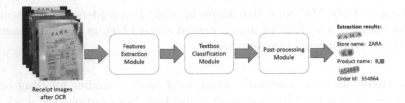

Receipt images
after OCR

Fig. 1. Framework of key information extraction from receipts

3.2 Feature Extraction

1) Rule Features: Shoppxing receipts contain a lot of intrinsic rules and prominent structures in terms of the contents and distributions over the texts. Obviously, adding rules features for key information extraction is highly helpful to improve the accuracy of the overall model. Table 1 defines the text content rules based on the observations of the receipt texts. Here, the first eight categories of text rule parameters are character-level statistics of texts, and the other categories are based on the content analysis and judgment of texts.

Table 1. Examples of text content rules for receipts

Rule parameters	Examples
number_of_letters	/
number_of_digits	/
number_of_chinese	/
contain_order_num_describe_words	order; order id; serial number
contain_order_num_pattern	JS01230001707260059
contain_date_describe_words	time; order time
contain_date	2017-03-17; 17/03/17
contain_time	14:10:22; 2:10PM
contain_price_describe_words	price; total
contain_price	29CNY; ¥29; 29.00

According to the observations of receipt images, the defined structural rules are summarized in Table 2. The "relative_positon_of_text_box" parameter is given by the regional position of a receipt image. In this paper, the region division method is formulated according to the distributions of receipt text contents. A receipt can be clustered into four areas from top to bottom according to text semantics: "shop, order details area", "commodity details area", "amount settlement area" and "rest information area". In order to automatically determine the receipt regions, we adopt a K-means clustering method. The number of clustering centers was set to 4, and the initial clustering center was selected from the textbox which has the largest word frequency and largest coordinate distance from other textboxes in the dataset.

Table 2. Examples of structure rules for receipts

Rule parameters	Explain
text_box_height	Height of a textbox
text_box_width	Width of a textbox
average_char_font_size	Average character size in a textbox
relative_positon_of_text_box	The region of a textbox in a receipt
sort_position_of_text_box	The position sequence of a textbox sorted by top to bottom, left to right

2) Semantic Features: In the corpus of Chinese shopping receipt set, the existence of polysemy rarely appears. However, fields which contain categories like "store name" and "mall name" often have unlogged words and low-frequency words. And the feature generation part should not take up most of the time of the entire extraction system. Therefore, fastText [6] was employed to generate the semantic features of receipt fields in this paper.

3) Visual Features: The image of a shopping receipt conveys many details that cannot be reflected by textual features such as the font, thickness and color of texts, the material of the paper, and the clarity of printing, etc. Therefore, to get better extraction results, we used ResNet [8] to learn the image visual features of receipt images. Specifically, an image is firstly cut according to the coordinate information of a textbox identified in the receipt image. The subgraphs after cutting need to be scaled to obtain a uniform size, and then fed into ResNet for representation learning. Finally, the output is processed through a Max pooling layer to obtain the visual features of the fields.

4) Structural Features: Based on the semi-structured form of a receipt image, we constructed a receipt graph, and used GCN to calculate the structural features of receipt images. In the step of constructing the receipt graph $G_{receipt} = (N_{receipt}, E)$, each $field_i$ is defined as a node $n_i \in N_{receipt}$ of the graph, and the edge $e_{i,j} \in E$ between nodes is established according to the coordinate relationship of the textbox, as shown in Fig. 2, where ΔW and ΔH are the width offset and the height offset of the target text box, respectively; $\Delta x_{i,t}$ and $\Delta y_{i,t}$ are the horizontal and vertical distance between the center of the contrast textbox and the target textbox center (x_{center}, y_{center}), respectively; And Top, Bottom, Left and Right are the top, bottom, left, and right coordinates of the textbox, respectively. When the contrast textbox and the target textbox are in the same line (column) and within a certain vertical (horizontal) offset range, the two textboxes are considered to have a horizontal (vertical) structural relationship, indicating that there is an edge connecting them, as shown in the form of:

$$e_{i,j} = \begin{cases} 1, \text{Top}(t_i) \leq \text{Bottom}(t_j), \text{Bottom}(t_i) \geq \text{Top}(t_j), \Delta y_{i,j} \leq \Delta H \\ 1, \text{Left}(t_i) \leq \text{Right}(t_j), \text{Right}(t_i) \geq \text{Left}(t_j), \Delta x_{i,j} \leq \Delta W \\ 0, \text{ else} \end{cases} \quad (3)$$

The setting of the horizontal and vertical offset range prevents connections between unrelated nodes that are far apart. A receipt image with large angle transformation can also generate the corresponding graph network well. Finally, the constructed graph consists of 376,443 nodes and 618,824 edges in our dataset (see Sect. 4.1).

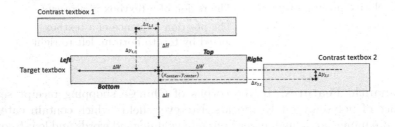

Fig. 2. Diagram of relationships between textboxes

The input of GCN consists of an eigenmatrix X with dimension $N \times D$, where N is the number of nodes in the graph and D is the number of input features of each node, and a graph adjacency matrix A with dimension $N \times N$. Its output is an eigenmatrix with dimension $N \times F$, where F is the number of output features of each node. The propagation rule of GCN is formulated as follows:

$$H^{(l+1)} = \sigma \left(\widetilde{D}^{-\frac{1}{2}} \widetilde{A} \widetilde{D}^{-\frac{1}{2}} H^{(l)} W^{(l)} \right), \tag{4}$$

where, $W^{(l)}$ and $H^{(l)}$ are the weight matrix and activation matrix of the lth neural network layer, respectively; $\sigma(\cdot)$ is the nonlinear activation function ReLU; and $\widetilde{A} = A + I_N$ is the matrix obtained by adding a self-loop to each node in the graph. The complete feature acquisition module is shown in Fig. 3. The output F_{out} is the learned feature from the receipt's structural features.

3.3   Bayesian Deep Network for Classification

We add Bayesian estimation to the recurrent neural network (RNN). Its weight matrix is calculated by Bayesian back propagation of a variational posteriori distribution. The objective function of Bayesian deep recurrent neural network on the input sequence of length T is shown in the form of:

$$\mathcal{F}(\mathcal{D}, \theta) = \mathrm{KL}[q(\mathbf{w} \mid \theta) \| P(\mathbf{w})] - \mathbb{E}_{q(\mathbf{w}|\theta)} \left[\log P\left(\mathbf{y}_{1:\mathbf{T}} \mid \theta, \mathbf{x}_{1:\mathbf{T}}\right)\right], \tag{5}$$

where $P\left(\mathbf{y}_{1:\mathbf{T}} \mid \theta, \mathbf{x}_{1:\mathbf{T}}\right)$ is the likelihood probability of a generated sequence when the states of RNN are input into the appropriate probability distribution. Then, back propagation is carried out in small batches, as shown in the form of:

$$\mathcal{F}(\mathcal{D}, \theta) = \mathrm{KL}[q(w \mid \theta) \| P(w)] - \mathbb{E}_{q(w|\theta)} \left[\log \sum_{b=1}^{B} \sum_{c=1}^{C} p\left(y^{(b,c)} \mid \theta, x^{(b,c)}\right)\right], \tag{6}$$

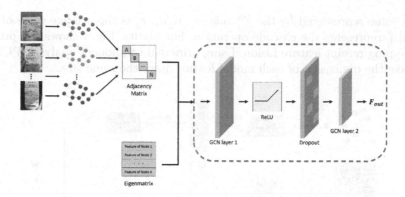

Fig. 3. Structural feature acquisition module

where, B is the batch number, C is the number of sequences in one batch, and (b, c) represents the cth sequence in batch b. The loss function is described as follows:

$$\mathcal{L}_{(b,c)}(\theta) = w_{\text{KL}}^{(b,c)} \text{KL}[q(w \mid \theta) \| P(w)] - \mathbb{E}_{q(w|\theta)} \left[\log p \left(y^{(b,c)} \mid \theta, x^{(b,c)}, s_{prev}^{(b,c)} \right) \right], \tag{7}$$

where, $w_{\text{KL}}^{(b,c)}$ allocates KL divergence to each batch of sequences, and $s_{prev}^{(b,c)}$ represents the initial state of a batch in the RNN.

3.4 Implementation Details

By combining the features of receipts in different modalities, the performance of the deep learning model can be improved and the heterogeneous difference between modalities can be compensated. In this paper, the fusion of multi-modal features is carried out by direct concatenation of four types of features. The structural, semantic, visual and rule features are realized as follows:

$$F_{st} = \text{GCN}_{2-layers}(A, H), \tag{8}$$

$$F_{se} = \text{Bi-LSTM}\,(Wt_i), \tag{9}$$

$$F_{vi} = \text{ResNet}\,18\,(v_i), \tag{10}$$

$$F_{ru} = \text{Concat}\,(r_j^i - \bar{r}_j), \tag{11}$$

and the specific feature fusion method is defined as follows:

$$F_{\text{fused}} = F_{st} \| F_{se} \| F_{vi} \| F_{ru}, \tag{12}$$

where $A \in \mathbb{R}^{N \times N}$ is the adjacency matrix of the constructed receipt graph, $H \in \mathbb{R}^{N} \times D_G$ is the eigenmatrix of the nodes in the graph, N is the number of nodes, and D_G is the number of nodes; $t_i \in \mathbb{R}^{D_t}$ is the vector representation of $field_i$, D_t is the length of the vector, W is the projection matrix of word vector obtained by fastText; v_i is the tensor representation of the $field_i$'s image; $S_j^i \in \mathbb{R}$

is the value represented by the jth rule of $field_i$, $\bar{r}_j$ is the average value of rule j; and $\|$ represents the cascade operation. For clarity, Fig. 4 shows the process of shopping receipt feature fusion. Using principal component analysis (PCA) is to make the dimension of each modal feature have the same size.

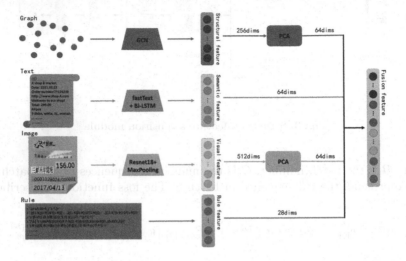

Fig. 4. The process of shopping receipt feature fusion

The gradient dispersion of RNN makes it have only short-term memory, so it is easy to lose information when faced with long sequence data. For this reason, LSTM is used in our model instead of RNN structure. The overall ReceiptBDL text classification algorithm with two LSTM layers is shown in Fig. 5. The input part of the algorithm is the sequence $x_{1:T}$ formed after word segmentation of the text t_i. Each x_t is calculated to get word embeddings. The word embeddings are combined with the fusion features F_{fused}^i of the text t_i to generate a new sequence. The sequence is trained by double-layer Bayes LSTM, and then calculated by Softmax. Afterwards, the classification results of input texts can be obtained.

Most of the texts classified as key information can be directly used as the final results. However, there exist some exceptions: (1) the distances between the description text of a key field and the actual information in the image are too close, and they are in the same textbox after OCR, as shown in Fig. 6(a); (2) the key field is too long, resulting in cross-lines in the image. OCR will divide it into upper and lower textboxes, as shown in Fig. 6(b); (3) there are gaps in the key fields, and OCR will divide them into left and right textboxes, as shown in Fig. 6(c). For case 1, the text was divided into two parts according to punctuation by regular expression, and the text in the last part was treated as the final result. For cases 2 and 3, we used the representation of inter-field relations as shown in Fig. 2 to merge fields that are close to each other and of the same category, and the merged text was regarded as the final result.

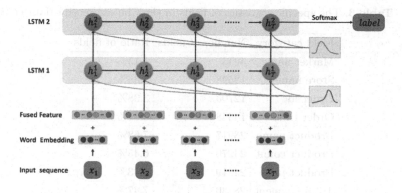

Fig. 5. Diagram of text classification model with Bayesian deep learning

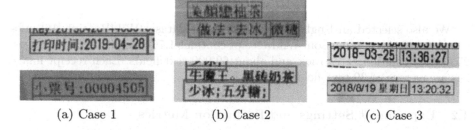

(a) Case 1   (b) Case 2   (c) Case 3

Fig. 6. Diagram of text classification model with Bayesian deep learning

4   Experiment

In order to evaluate the performance of our proposed algorithm ReceiptBDL, we conduct experiments on key information extraction on a large-scale dataset built by ourselves. In this section, we firstly introduce our dataset and experiment settings. Then, we compare our method with other baseline models to exhibit the superiority of our method. Finally, an ablation study on different modal features is illustrated.

4.1   Dataset

The dataset used in this paper comes from Chinese shopping receipts images. The Chinese shopping receipt dataset is composed of receipt images taken by customers of shopping malls in Beijing, Shanghai, Wuxi and Changsha [15]. It contains eight key information categories. Each receipt image has an average of 35 text fields. At present, 376,443 text fields in 10,731 shopping receipt images have been manually marked. The specific statistics of fields are shown in Table 3. The ratio of key fields in the dataset is 39.70%.

Table 3. The specific statistics of fields in the Chinese receipt dataset

Field category	Number of fields	Ratio of fields
Market name	4258	1.13%
Store name	12598	3.35%
Order id	12705	3.38%
Order time	14348	3.81%
Product name	28917	7.68%
Product count	24270	6.45%
Product price	24219	6.43%
Total payment	28130	7.47%
Others	226998	60.30%

We also selected an English shopping receipt dataset WildReceipt [16] published in 2020 for comparison. WildReceipt contains 1,768 English receipt images, 25 key information categories, and about 69,000 text fields. Each receipt image has an average of 39 text fields.

4.2 Experiment Settings and Evaluation Metrics

The input of Bayesian deep LSTM is the text sequence obtained by field segmentation. The learning rate of the training model was 1e-2, the number of training cycles epoch was set to 30, and the dropout value was 0.3. The proportion of training, validation and test set to total data set is 80%, 10% and 10%, respectively. During model training, the model parameters with the best performance were saved. Then we used the best model parameters we saved to test the models on test set. Our experiment was performed on a machine with a two-way Geforce GTX 1080 GPU with 128-GB memory.

In order to obtain a fair comparison, standard F1-Score was used as a quantitative evaluation metric for each text category. Specifically, we take the data of one category as the positive sample and the data of the other categories as the negative sample for multiple evaluations. Meanwhile, Macro-F1 and Weighted-F1 were employed to better evaluate the overall multi-class classification performance of a model. Macro-F1 and Weighted-F1 were obtained after the macro-average and weighted macro-average of F1-Score.

4.3 Results and Comparison

We selected seven baseline text classification algorithms to compare with the algorithm we proposed. Among them, TextCNN [17], TextRNN [18], TextRCNN [19] and TCN [20] learn the features of fields through neural network and conduct text classification. BERT [7] and ERNIE [21] perform text classification through pre-training and fine-tuning. Bayesian deep LSTM (BD-LSTM) conducts text classification based on an LSTM with distributed parameters. Table 4 and Table 5 show the F1-score results of different models on Chinese and English shopping receipt datasets, respectively.

Table 4. Comparative results of F1-Score in the Chinese receipt dataset

	Market name	Store name	Order id	Order time	Product name	Product count	Product price	Total payment	Others
TextCNN	0.4754	0.7070	0.5064	0.8389	0.7468	0.8657	0.4017	0.5047	0.8620
TextRNN	0.5414	0.7092	0.5288	0.8506	0.7604	0.8798	0.4795	0.5914	0.8732
TextRCNN	0.5270	0.7126	0.5464	0.8573	0.7454	0.8784	0.4910	0.6122	0.8754
TCN	0.4684	0.7626	0.2726	0.8241	0.8444	0.8755	0.4772	0.6195	0.8747
BERT	0.5796	0.8273	0.6868	0.8883	**0.8776**	0.8939	0.5257	0.6710	0.9012
ERNIE	**0.5935**	**0.8336**	0.7008	**0.8890**	0.8542	0.8976	0.5169	0.6696	0.8992
BD-LSTM	0.4432	0.7744	0.5968	0.8688	0.8435	0.8926	0.5262	0.6195	0.8863
ReceiptBDL (ours)	0.4677	0.8081	**0.7294**	0.8860	0.8575	**0.8989**	**0.6337**	**0.7479**	**0.9066**

Table 5. Comparative results of F1-Score in the English receipt dataset

	Store name	Order date	Order time	Product name	Product count	Product price	Total payment	Others
TextCNN	0.0186	0.7317	0.1918	0.1220	0.8620	0.5225	0.0000	0.7856
TextRNN	0.0370	0.7749	0.6226	0.2902	0.8819	0.6530	0.1239	0.8048
TextRCNN	0.0000	0.8353	0.6788	0.4370	0.8745	0.6772	0.1129	0.7843
TCN	0.2500	0.8421	0.7368	0.3364	0.8982	0.7070	0.0000	0.8210
BERT	0.3672	**0.9847**	0.9069	0.7746	0.9231	0.7400	0.3353	0.8485
ERNIE	0.0000	0.9404	0.9283	0.6976	0.9199	0.7186	0.4717	0.8248
BD-LSTM	0.2319	0.8810	0.7821	0.3343	0.8926	0.7201	0.0000	0.8220
ReceiptBDL (ours)	**0.7928**	0.9395	**0.9789**	**0.9555**	**0.9610**	**0.9324**	**0.5751**	**0.9508**

The results show that, in the experiment of Chinese dataset, our ReceiptBDL algorithm based on multi-modal features proposed in this paper has a higher F1-score over five categories than other algorithms, and the results of the other four categories are only slightly lower than BERT and ERNIE's results. In the experiment of English dataset, the F1-score of seven categories produced by our method is higher than those achieved by other algorithms. Especially for the "Store Name" column, it can be seen that the F1-Score of the BD-LSTM is very low in comparison with other baseline algorithms. However, after adding multi-modal features, the F1-Score value has been greatly improved, which is ascribed to the impact of semantic and rule features. In terms of the rule features, the external knowledge base is used to determine whether a field contains a store name. In the part of semantic features, the field features pre-trained by fastText have achieved promising results in the English set.

Table 6 shows the results with respect to the multi-class classification evaluation metrics. Our proposed algorithm outperforms the compared methods in both Chinese and English datasets. In the Chinese dataset, the Accuracy, Macro-F1 and Weighted-F1 were 1.69%, 0.62% and 1.51% higher than the optimal baseline method ERNIE, respectively. In the English dataset, the evaluation indicators were 6.99%, 9.16% and 6.52% higher than the optimal baseline method BERT. Both ERNIE and BERT need to fine-tune the pre-trained models so as to achieve better results in downstream tasks. However, ReceiptBDL does not need the process of pre-training. So the model has a smaller number of parameters as well as faster training speed. At the same time, multi-modal features can also make the model pay special attention to the features of receipts in other modalities except for texts.

Table 6. Comparative results in the Chinese and English receipt datasets

	Accuracy		Macro-F1		Weighted-F1	
	Chinese	English	Chinese	English	Chinese	English
TextCNN	0.7861	0.6894	0.6565	0.4042	0.7740	0.6210
TextRNN	0.8030	0.7293	0.6905	0.5234	0.7962	0.6838
TextRCNN	0.8030	0.7132	0.6940	0.5500	0.7994	0.6938
TCN	0.8060	0.7509	0.6688	0.5740	0.7965	0.7081
BERT	0.8421	0.8077	0.7613	0.7101	0.8434	0.8013
ERNIE	0.8423	0.7767	0.6876	0.6762	0.8432	0.7662
BD-LSTM	0.8225	0.7546	0.7168	0.5830	0.8211	0.7108
ReceiptBDL (ours)	**0.8592**	**0.9347**	**0.7711**	**0.8857**	**0.8583**	**0.9332**

4.4 Ablation Study

In this subsection, we ablate experiments on the multi-modal features. Figure 7 and Fig. 8 show the results of ablation experiments on Chinese and English shopping receipt datasets, respectively. Among them, blue, orange and gray columns are listed as a group.

The experimental results of Fig. 7 show that the Weighted-F1 scores of the BD-LSTM model with rules, semantic, structures and visual features were 2.31%, 0.28%, 1.62% and 2.56% higher than those of the original model without any extra features. The Accuracy, Macro-F1 and Weighted-F1 scores of the model with all features are the best. To sum up, the features of different modalities in Chinese shopping receipts have a positive impact on the performance improvement of the model in this paper.

The experimental results of Fig. 8 show that when semantic features are added to the original model, its performance is improved the most, and its Weighted-F1 score increases by 20.7%. Moreover, adding rules or structural features can also improve the classification performance of the model. However, since the receipt images selected from the WildReceipt dataset are of good quality, i.e., without curl, blur or other quality issues, the model with adding visual features has basically the same effect as the original model. When all of modality features are included in the model, the results turn out to be suboptimal. To sum up, the language model can generate effective semantic features for receipt

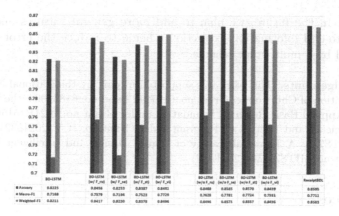

Fig. 7. Comparative results of different feature combinations in the Chinese receipt dataset

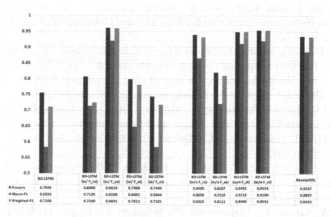

Fig. 8. Comparative results of different feature combinations in the English receipt dataset

texts in English context. Therefore, we can only use semantic features in the English receipt dataset to improve the overall performance of the model.

5 Conclusion

In this paper, we propose a key information extraction algorithm for shopping receipts using Bayesian deep learning. Specifically, we formulate the input of the LSTM model with multi-modal features, and change the weight parameters in the network from a fixed value to a learnable probability distribution. Experimental results show that our proposed algorithm achieves significantly better performance in terms of evaluation metrics on a Chinese shopping receipt dataset in comparison to several baseline methods. Moreover, the ablation study shows that adding features of different modalities, indeed, can improve the results of our model. In the future, we plan to add more general features and consider using end-to-end information extraction scheme to reduce the error accumulation caused by a multi-stage scheme.

Acknowledgements. This work was supported in part by the National Natural Science Foundation of China under Grant no. 61972112 and no. 61832004, the Guangdong Basic and Applied Basic Research Foundation under Grant no. 2021B1515020088, the Shenzhen Science and Technology Program under Grant no. JCYJ20210324131203009, and the HITSZ-J&A Joint Laboratory of Digital Design and Intelligent Fabrication under Grant no. HITSZ-J&A-2021A01.

References

1. Gao, S., et al.: Hierarchical attention networks for information extraction from cancer pathology reports. J. Am. Med. Inf. Assoc. **25**(3), 321–330 (2018)
2. Luo, K., Lu, J., Zhu, K.Q., Gao, W., Wei, J., Zhang, M.: Layout-aware information extraction from semi-structured medical images. Comput. Biol. Med. **107**, 235–247 (2019)
3. Liu, X., Gao, F., Zhang, Q., Zhao, H.: Graph convolution for multimodal information extraction from visually rich documents. In: Proceedings of the 2019 Conference of the North American Chapter of the Association for Computational Linguistics: Human Language Technologies, vol. 2, pp. 32–39 (2019)
4. Xu, Y., Li, M., Cui, L., Huang, S., Wei, F., Zhou, M.: LayoutLM: pre-training of text and layout for document image understanding. In: Proceedings of the 26th ACM SIGKDD International Conference on Knowledge Discovery & Data Mining, pp. 1192–1200 (2020)
5. Mikolov, T., Chen, K., Corrado, G., Dean, J.: Efficient estimation of word representations in vector space. In: Proceedings of the International Conference on Learning Representations (2013)
6. Joulin, A., Grave, É., Bojanowski, P., Mikolov, T.: Bag of tricks for efficient text classification. In: Proceedings of the Conference of the European Chapter of the Association for Computational Linguistics, pp. 427–431 (2017)
7. Kenton, J.D.M.W.C., Toutanova, L.K.: BERT: pre-training of deep bidirectional transformers for language understanding. In: Proceedings of the Conference of the North American Chapter of the Association for Computational Linguistics: Human Language Technologies, pp. 4171–4186 (2019)
8. He, K., Zhang, X., Ren, S., Sun, J.: Deep residual learning for image recognition. In: Proceedings of the IEEE Conference on Computer Vision and Pattern Recognition, pp. 770–778 (2016)
9. Kipf, T.N., Welling, M.: Semi-supervised classification with graph convolutional networks. In: Proceedings of the International Conference on Learning Representations (2017)
10. Blundell, C., Cornebise, J., Kavukcuoglu, K., Wierstra, D.: Weight uncertainty in neural network. In: Proceedings of the International Conference on Machine Learning, pp. 1613–1622 (2015)
11. Wang, H., Yeung, D.Y.: A survey on bayesian deep learning. ACM Comput. Surv. **53**(5), 1–37 (2020)
12. Kendall, A., Gal, Y.: What uncertainties do we need in bayesian deep learning for computer vision? Adv. Neural Inf. Process. Syst. **30**, 5574–5584 (2017)

13. Siddhant, A., Lipton, Z.C.: Deep bayesian active learning for natural language processing: results of a large-scale empirical study. In: Proceedings of the Conference on Empirical Methods in Natural Language Processing, pp. 2904–2909 (2018)
14. Xiao, T., Liang, S., Shen, W., Meng, Z.: Bayesian deep collaborative matrix factorization. In: Proceedings of the AAAI Conference on Artificial Intelligence, pp. 5474–5481 (2019)
15. Ren, L., Zhou, H., Chen, J., Shao, L., Wu, Y., Zhang, H.: A transformer-based decoupled attention network for text recognition in shopping receipt images. In: Zhang, H., Yang, Z., Zhang, Z., Wu, Z., Hao, T. (eds.) NCAA 2021. CCIS, vol. 1449, pp. 563–577. Springer, Singapore (2021). https://doi.org/10.1007/978-981-16-5188-5_40
16. Sun, H., Kuang, Z., Yue, X., Lin, C., Zhang, W.: Spatial dual-modality graph reasoning for key information extraction. arXiv preprint arXiv:2103.14470 (2021)
17. Kim, Y.: Convolutional neural networks for sentence classification. In: Proceedings of the Conference on Empirical Methods in Natural Language Processing, pp. 1746–1751 (2014)
18. Liu, P., Qiu, X., Huang, X.: Recurrent neural network for text classification with multi-task learning. In: Proceedings of the 25th International Joint Conference on Artificial Intelligence, pp. 2873–2879 (2016)
19. Lai, S., Xu, L., Liu, K., Zhao, J.: Recurrent convolutional neural networks for text classification. In: Proceedings of the 29th AAAI Conference on Artificial Intelligence, pp. 2267–2273 (2015)
20. Bai, S., Kolter, J.Z., Koltun, V.: An empirical evaluation of generic convolutional and recurrent networks for sequence modeling. arXiv preprint arXiv:1803.01271 (2018)
21. Zhang, Z., Han, X., Liu, Z., Jiang, X., Sun, M., Liu, Q.: ERNIE: enhanced language representation with informative entities. In: Proceedings of the 57th Annual Meeting of the Association for Computational Linguistics, pp. 1441–1451 (2019)

A Multi-Surrogate-Assisted Artificial Bee Colony Algorithm for Computationally Expensive Problems

Tao Zeng[1] , Hui Wang[1(✉)], Tingyu Ye[1] , Wenjun Wang[2], and Hai Zhang[1]

[1] School of Information Engineering, Nanchang Institute of Technology,
Nanchang 330099, China
huiwang@whu.edu.cn
[2] School of Business Administration, Nanchang Institute of Technology,
Nanchang 330099, China

Abstract. Artificial bee colony (ABC) is a popular swarm intelligence algorithm, which has shown excellent performance on many optimization problems. However, it is rarely used to solve computationally expensive problems. In this paper, a multi-surrogate-assisted ABC (called MSABC) algorithm is proposed to solve computationally expensive problems. Multiple surrogates cannot only improve the prediction performance and estimate the degree of prediction uncertainty, but also capture both global and local features of the fitness landscape. In the employed bee phase, Radial Basis Function (RBF) network is used as a global surrogate model to assist ABC to quickly find the region where the global optimum might be located. In the onlooker bee phase, Kriging model is employed as a local surrogate built around some top best data points. To speed up the convergence, multiple dimensions for each solution are updated simultaneously. Experimental studies on CEC 2014 expensive optimization benchmark set show that the proposed approach can effectively solve those expensive problems under a limited computational budget.

Keywords: Surrogate model · Artificial bee colony · Kriging model · RBF network · Dimension perturbation

1 Introduction

Engineering and scientific fields has a variety of optimization problems, e.g., path planning, scheduling problem, investment optimization, etc. Traditional algorithms are hard to solve some of the complicated problems. Therefore, the evolutionary algorithms (EAs) are proposed to solve such problems. The commonly used EAs are: genetic algorithm (GA) [1,2], artificial bee colony (ABC) [3,4], ant colony optimization (ACO) [5,6], differential evolution (DE) [7,8], particle swarm optimization (PSO) [9,10], and so on.

ABC has a simple structure, few control parameters, and powerful search abilities [11]. Therefore, ABC has received much attention since its introduction.

H. Zhang et al. (Eds.): NCAA 2022, CCIS 1637, pp. 394–405, 2022.
https://doi.org/10.1007/978-981-19-6142-7_30

However, ABC has a poor exploitation ability. The search strategy determines the search capability, to balance exploration and exploitation, researchers proposed various improved strategies. Inspired by DE algorithm, Gao and Liu [12] added mutation operator to the strategy. According to the idea of PSO, Zhu and Kwong [13] use the best individual *gbest* to guide the population, and finally improves convergence speed. Enlightened by GA, Gao et al. [14] incorporated the crossover operator into the search strategy to solve the oscillation phenomenon [13]. There are a variety of search strategies, each of them has its own search characteristics. The combination of multiple strategies may significantly increase performance. The difficulty in using multiple search strategies is how to select appropriate one for present search phase. A strategy pool that contains three search strategies was built by Wang et al. [15]. When the search strategy cannot get a better offspring, another strategy is chosen at random in the remaining strategies. Ye et al. [16] defined a evaluating indicator to adaptively choose the suitable search manner. According to the success rate, Zeng et al. [17] adaptively select an appropriate search strategy. Literature [18] adopted five different strategies, which are selected according to their success rates. In [19], a random neighborhood structure is used to design the effective search strategy.

Despite the fact that ABC has performed well on optimization tasks, it faced challenges for solving computationally expensive problems. It is unaffordable in these cases since huge evaluations are required. In order to extend ABC to address such problems, a multi-surrogate-assisted ABC (called MSABC) algorithm is proposed. It is essential for surrogate-assisted algorithms to select an appropriate surrogate model [20]. However, each surrogate model has its advantages and disadvantages. Choosing an appropriate surrogate model is not a simple task. To address this issue, multiple surrogate models are used. Lim et al. [21] applied both global and local surrogate model when conduct the local search. It is helpful to gain from the 'less of uncertainty' and alleviate the 'curse of uncertainty'. In the literature [22], the local optima were smoothed out using a global surrogate model, and the estimate accuracy was improved using a local surrogate model. Wang et al. [23] combined three kinds of surrogate models for fitness prediction. In our approach, two surrogate models are utilized. In the employed bee phase, Radial Basis Function network is utilized to help ABC swiftly locate the global optimum as a global model. In the onlooker bee phase, Kriging model is employed as a local surrogate built around some top best data points. To validate the efficacy of MSABC, five ABC variants are compared with MSABC on CEC 2014 expensive optimization benchmark set.

This paper is structured as follows: Sect. 2 provides a short overview of the original ABC. Detailed descriptions of the proposed MSABC are given in Sect. 3. In Sect. 4, both the result and the experiments are provided. Finally, conclusions is given in last part.

2   Artificial Bee Colony Algorithm

ABC is an efficiency evolutionary algorithm proposed by Karaboga [24]. It is inspired by the behaviors of bees. According to their duty, there are three distinct classifications of bees. Each kind of bees plays an important role. Including initialization phase, there are four different phases in ABC:

(1) Initialization phase

Supposing that D is the dimensions of the test problems, SN is the population size. Each solution X_i will get an initial position between lower and upper bound (x_{min} and x_{max}), as shown below:

$$x_{i,j} = x_{min,j} + rand(0,1) \cdot (x_{max,j} - x_{min,j}) \tag{1}$$

where $i = 1, 2, \cdots, SN$, $j = 1, 2, \cdots, D$, $rand(0,1)$ is a random number in the range $[0, 1]$.

(2) Employed bee phase

In order to find a better nectar source, the employed bees will search around the current position. The search process can be abstracted into the following formula:

$$v_{i,j} = x_{i,j} + \phi_{i,j} \cdot (x_{i,j} - x_{k,j}) \tag{2}$$

where V_i is the offspring of X_i, j is a randomly selected dimension, note that only $j - th$ dimension will be updated, $j \in \{1, 2, \cdots, D\}$, X_k is another individual selected from the swarm, and $i \neq k$. V_i will substitute for X_i only when the quality of V_i is better than X_i.

(3) Onlooker bee phase

The onlooker bees only search around the nectar source which has a high quality. The quality of nectar source, called fitness, can be defined as follows:

$$fit(X_i) = \begin{cases} \frac{1}{1+f(X_i)}, & \text{if } f(X_i) \geq 0 \\ 1 + |f(X_i)|, & \text{if } f(X_i) < 0 \end{cases} \tag{3}$$

where $f(X_i)$ is function value of X_i. In the minimization optimization problem, a lower $f(X_i)$ can get a higher $fit(X_i)$, which means the individual has a higher quality. The roulette is applied to choose the individuals which has a high quality, the selection probability can be calculated as follows:

$$p_i = \frac{fit(X_i)}{\sum_{i=1}^{SN} fit(X_i)} \tag{4}$$

(4) Scout bee phase

In this stage, the nectar source with a poor quality will be abandoned. Such individuals may fall into local minimization, slow down the convergence speed. Here we set a counter, marked as $trail_i$, to record the number of consecutive non updates for $i - th$ individual. If $trail_i$ exceeds the threshold $limit$, X_i will reinitialize according to Eq. (1), in the meantime, $trail_i$ will set to 0.

3   Proposed Approach

As described in Sect. 2, ABC can be divided into four stages based on different search characters. Each individual has the same opportunity to be updated throughout the employed bee stage. However, in onlooker bee stage, outstanding individuals have more opportunities to generate the offspring than the inferior solutions. According to the features of different phases, the Radial Basis Function (RBF) network serves as a global model in the employed bee phase. It assists ABC to swiftly locate the area where the global optimum is most probable to be. The Kriging model is applied as a local surrogate model in the onlooker bee phase, built around some top best data points. The overall framework of our approach can be found in Algorithm 1. How to update the parent solution X_i is given in Algorithm 2.

3.1 Modified Search Strategy

ABC excels at exploration but weak in exploitation. To balance exploration and exploitation, the search equation is adjusted. Equation (5) is a search strategy selected from CABC [14], and it has a strong exploration capability. In our approach, the employed bee stage adopts Eq. (5) as the search strategy. This is helpful for the swarm to find a better search region. In [15], the new search strategy (Eq. (6)) used the information of best solution to improve exploitation. In MSABC, Eq. (6) is used during the onlooker bee stage to assist the outstanding solutions quickly converge.

$$v_{i,j} = x_{r1,j} + \phi_{i,j}(x_{r1,j} - x_{r2,j}) \tag{5}$$

$$v_{i,j} = x_{\text{best},j} + \phi_{i,j}(x_{\text{best},j} - x_{k,j}) \tag{6}$$

where $\phi_{i,j}$ is a uniform random number in the range of $[-1, 1]$, X_{r1}, X_{r2}, X_{r3} are solutions selected from the swarm, and $i \neq k$, $r1 \neq r2 \neq i$. X_{best} is the best individual.

3.2 Random Dimension Perturbation

There are modest differences between parent and offspring since ABC just changes one dimension at a time. This may lead to two problems. Firstly, the convergence speed is slow; secondly, because the information carried by the new solution is insufficient, the reliability of the surrogate model cannot be significantly improved by using it. In our approach, the random dimension perturbation is applied in this paper to tackle the issues aforementioned.

It is essential to update the values in different dimensions simultaneously. However, too many updated dimensions may affect the convergence. Because it is difficult for excellent solutions to find better offspring in a large search space. Assume that the upper limit of updated dimensions is $m*D$, where m is the scaling factor. For i-th individual, the number of dimensions that need to be updated is $J_{num}^i = rand(1, m*D)$, where $rand(1, m*D)$ means randomly selecting a positive integer from $[1, m*D]$. We pick J_{num}^i numbers from $\{1, 2, \cdots, D\}$. The set of selected numbers is denoted as S_i, which contains the dimensions that need to be updated. Then, the neighborhood search is used to update each dimension in S_i.

3.3   Radial Basis Function Network

Hardy [25] initially suggested the concept of RBF to describe irregular surfaces. Broomhead and Lowe [26] created a three-layer structured RBF neural network. Powell [27] introduced multivariate interpolation using the RBF. Literature [28] showed that the RBF network can be used to get close to any continuous function. RBF network is used as global surrogates in this paper to find the optimal region. Below is a description of the RBF network:

$$\varphi(\boldsymbol{x}) = \sum_{i=1}^{CN} w_i \rho\left(\boldsymbol{x}, \boldsymbol{c}_i\right) \tag{7}$$

where $\rho\left(\boldsymbol{x}, \boldsymbol{c}_i\right)$ is the RBF kernel, c_i is center of hidden layer, input vector, marked as x, has a total of d dimensions, there are CN hidden neurons, and the weight of the hidden layer is w_i. The Gaussian kernel is applied in this work, which is detailed below:

$$\rho(\boldsymbol{x}, \boldsymbol{c}_i) = e^{\left(\frac{-||x - c_i||^2}{\sigma_i^2}\right)} \tag{8}$$

where σ_i is a positive value and represents the width of the hidden layer.

3.4   Kriging Model

The Kriging model [29, 30], commonly known as the Gaussian process regression model, is a form of interpolation regulated by prior covariances. The Kriging model can be defined as follows by assuming that x follows a joint normal distribution:

$$y = \mu(x) + \epsilon(x) \tag{9}$$

where the mean of the Kriging model is denoted by $\mu(x)$, and the deviation is denoted by $\epsilon(x)$. Training samples $\mathcal{D} = \left(\mathbf{x}_k^d, y_k^d\right)$ are applied to determine $\mu(x)$ and $\epsilon(x)$. According to maximum likelihood estimation, the posterior list below can be got:

$$p(\theta \mid \mathbf{x}, y) = \frac{p(y \mid \mathbf{x}, \theta) p(\theta)}{p(y \mid \mathbf{x})} \tag{10}$$

$\epsilon(x^*)$ and $\mu(x^*)$ could acquire for x^* as follows after optimizing the hyperparameter θ:

$$\varepsilon\left(\mathbf{x}^*\right) = k\left(\mathbf{x}^*, \mathbf{x}^*\right) - k\left(\mathbf{x}^*, \mathbf{x}\right) \left(\mathcal{K} + \sigma_n^2 I\right)^{-1} k\left(\mathbf{x}, \mathbf{x}^*\right) \tag{11}$$

$$\mu\left(\mathbf{x}^*\right) = k\left(\mathbf{x}^*, \mathbf{x}\right) \left(\mathcal{K} + \sigma_n^2 I\right)^{-1} y \tag{12}$$

where $\mathcal{K}$ stands for kernel matrix, k represents the covariance function. The Kriging models have included a wide variety of well-known kernel functions, including RBF, multiquadric, and exponential kernels.

3.5   Multi-Surrogate-Assisted ABC

3.5.1   RBF Network Assisted the Employed Bee Stage

The new position (V_i) of each solution is generated by Eq. (5)in employed bee phase. Then, RBF network is applied to estimate the function value $(\hat{f}_{RBF}(V_i))$ of each offspring. The individual will be updated if the estimated function value is greater than $f(X_i)$. However, procedure is a little different for the best individual. Suppose that X_i is the best individual, if $\hat{f}_{RBF}(V_i) < f(X_i)$, then X_i will be evaluated by the expensive function and stored in archive AR. Only when it gets a truly better position will it be updated. In this way, the search process will not be misled by false optimums of the surrogate models [31]. At the end of this phase, After being evaluated by the expensive function, a solution (x_R^b) that has the best estimated function value will be stored in archive AR. Following equation shows how to select x_R^b:

$$x_R^b = \arg\min_x \hat{f}_{RBF}(x) \tag{13}$$

3.5.2   Kriging Assisted the Onlooker Bee Phase

Roulette is employed to choose the corresponding solution for the neighborhood search during the onlooker bee phase. Then, the estimated function value for each solution is obtained by the Kriging model. The process of the greedy selection is consistent with the employed bee phase.

The reliability of the surrogate model can be improved by evaluating some of the potential solutions. In general, the data with the best estimated function value (x_K^b) as well as the largest uncertainty (x^u) are potential solutions [23]. The following equations show how to select them:

$$x_K^b = \arg\min_x \hat{f}_{Kri}(x) \tag{14}$$

$$x^u = \arg\max_x U_{Kri}(x) \tag{15}$$

where $\hat{f}_{Kri}(x)$ is the predict function value given by Kriging model, $U_{Kri}(x)$ represents the confidence level calculated by the Kriging model.

3.6   Surrogate Model Management

The number of fitness evaluations is limited when solving expensive problems. So, the data is very precious. Whenever a real objective function is used for evaluation, the corresponding data will save in the archive AR. The training samples for the RBF network and the Kriging model are stored in TS_{RBF} and TS_{Kri}, respectively. To reduce the complexity of constructing the surrogate model, we restrict the size of AR.

Employed bees just search around parent individual to find the offspring. Therefore, the training data should not be selected far away from the current swarm. Equation 16 shows how to calculate the distance from a individual AR_i to the current swarm:

$$dist_i = \min\{d(AR_i, X_j) \mid j \in \{1, 2, \cdots, SN\}\} \tag{16}$$

where AR_i and X_j are i-th, j-th individual in AR and swarm, respectively. $d(AR_i, X_j)$ stands for the distance between AR_i and X_j. $dist_i$ is the minimum distance between AR_i and the individual within current population.

Before starting the employed bee phase, clear the data in TS_{RBF}, calculate $dist_i$ for every sample in AR. Selecting the sample with the minimum distance and put it into TS_{RBF}, repeat this operation until TS_{RBF} is full or AR is empty. Then, the updated training set TS_{RBF} could be applied to construct the RBF Network.

Kriging model is used to model the local area near the outstanding solutions. The elite samples in archive AR are a good choice to train the model. At the end of employed bee phase, clearing the data in TS_{Kri}, choosing the sample with minimum function value in AR, then put it into TS_{Kri}, repeat this operation until TS_{Kri} is full or AR is empty. Then, the Kriging model is built with the data in TS_{Kri}.

4   Experimental Study

4.1   Test Problems and Parameter Settings

This paper perform an experimental investigation on CEC 2014 expensive optimization benchmark set [32] to validate efficacy of MSABC. There are 8 different problems, each with a different number of dimensions (10, 20, 30). The function name and search range can be found in Table 1, where f_1-f_4 are unimodal problems, f_5-f_6 are typical multimodal problems, f_7-f_8 are very complex multimodal problems. The optimum function value is zero for all test problems. Error values $\leq 10^{-8}$ are taken as 0.

For 10, 20 and 30 dimensional function the maximum number of evaluations are 500, 1000, 1500, respectively. The size of swarm is set to 50, $limit$ is 100. In MSABC, $MaxTSSize = maxnode * D + maxnode$, where $maxnode$ represents the number of hidden nodes, which takes 8. σ can be calculated by referring to this literature [33].

4.2   Computational Results

In this study, we select five classic ABC variants to compare with MEABC: ABC, GABC [13], CABC [14], MPEABC [34] and MABC [35]. A single strategy is used in the first three algorithms, whereas MABC has two and MPEABC has three search strategies. MPEABC is the only algorithm that uses dimensional perturbation in comparison algorithms.

Table 2 lists the results on CEC 14 expensive optimization test problems for six ABC variants. MSABC achieves the best results whether the test problem is

---

Algorithm 1: Proposed Approach (MSABC)

1 Population initialization;
2 **while** $FEs < MaxFEs$ **do**
3   Update the train set TS_{RBF};
4   Build RBF network with the samples in TS_{RBF};
5   **for** $i = 1$ to SN **do**
6   Generate offspring V_i by Eq. (5);
7   Approximate the function value $\hat{f}_{RBF}(V_i)$ using the RBF network;
8   Greedy Select;
9   **end**
10   Search x_R^b by Eq. (13);
11   Evaluate x_R^b using the expensive function, and add it into archive AR;
12     $FEs + +$;
13   Update the train set TS_{Kri};
14   Build the Kriging model with the samples in TS_{Kri};
15   Generate selection probability p_i according to Eq. (4);
16   Set $i = 1$ and $counter = 1$;
17   **while** $counter \leftarrow SN$ **do**
18   **if** $rand(0,1) < p_i$ **then**
19   Generate offspring V_i according to Eq. (6);
20   Approximate the function value $\hat{f}_{Kri}(V_i)$ using Kriging model;
21           $counter = counter + 1$;
22   Greedy Select;
23   **end**
24        $i = (i+1)\%SN$;
25   **end**
26   Search x_K^b and x^u by Eq. (14) and Eq. (15), respectively;
27   Evaluate x_K^b and x^u using the expensive function, then add it into archive AR;
28     $FEs+ = 2$;
29   **if** max $\{trial_i\} > limit$ **then**
30   Initialize X_i by Eq. (1);
31   Calculate $f(X_i)$, set $FEs + +$ and $trial_i = 0$;
32   **end**
33 **end**

Algorithm 2: Greedy Select

1 Suppose that $\hat{f}(V_i)$ is the approximate function value using the surrogate model;
2 **if** $\hat{f}(V_i) < f(X_i)$ **then**
3   **if** X_i is the $best$ $individual$ **then**
4   Evaluate V_i using the expensive function, and add it into archive AR;
5        $FEs + +$;
6   **if** $f(V_i) < f(X_i)$ **then**
7   Update parent with offspring and set $trial_i = 0$;
8   **end**
9   **else**
10           $trial_i + +$;
11   **end**
12   **end**
13   **else**
14   Update parent with offspring and set $trial_i = 0$;
15   **end**
16 **end**
17 **else**
18     $trial_i + +$;
19 **end**

Table 1. CEC 2014 expensive optimization test problems.

Functions	D	Search range	Functions	D	Search range
f_1: Shifted Sphere	10	$[-20, 20]$	f_5: Shifted Ackley	10	$[-32, 32]$
	20			20	
	30			30	
f_2: Shifted Ellipsoid	10	$[-20, 20]$	f_6: Shifted Griewank	10	$[-20, 20]$
	20			20	
	30			30	
f_3: Shifted Rotated Ellipsoid	10	$[-20, 20]$	f_7: Shifted Rotated Rosenbrock	10	$[-20, 20]$
	20			20	
	30			30	
f_4: Shifted Step	10	$[-20, 20]$	f_8: Shifted Rotated Rastrigin	10	$[-20, 20]$
	20			20	
	30			30	

Table 2. Results on CEC 2014 expensive optimization test problems.

Functions	D	ABC	CABC	MABC	GABC	MPEABC	MSABC
f_1	10	3.32E+02	1.78E+02	5.01E+02	1.54E+02	1.90E+02	**5.85E-03**
	20	1.00E+03	5.94E+02	1.52E+03	7.16E+02	5.04E+02	**0.00E+00**
	30	1.92E+03	1.11E+03	2.67E+03	1.42E+03	8.80E+02	**1.53E-08**
f_2	10	1.39E+03	5.84E+02	2.48E+03	7.67E+02	9.00E+02	**3.87E-07**
	20	9.09E+03	5.09E+03	1.41E+04	6.49E+03	4.62E+03	**7.26E-06**
	30	2.68E+04	1.46E+04	3.54E+04	1.96E+04	1.05E+04	**3.39E-05**
f_3	10	1.53E+03	8.88E+02	3.24E+03	8.73E+02	1.41E+03	**1.54E-04**
	20	9.46E+03	5.94E+03	1.42E+04	7.20E+03	5.93E+03	**1.29E-03**
	30	3.89E+04	2.55E+04	5.53E+04	3.08E+04	2.30E+04	**9.13E+00**
f_4	10	2.90E+02	1.76E+02	4.33E+02	1.65E+02	1.80E+02	**2.87E+00**
	20	1.05E+03	5.40E+02	1.49E+03	7.16E+02	5.22E+02	**3.42E+01**
	30	1.97E+03	1.16E+03	2.59E+03	1.41E+03	8.26E+02	**8.49E+01**
f_5	10	1.81E+01	1.72E+01	1.97E+01	1.78E+01	1.74E+01	**1.14E+01**
	20	1.93E+01	1.88E+01	2.04E+01	1.93E+01	1.80E+01	**1.24E+01**
	30	2.00E+01	1.90E+01	2.05E+01	1.96E+01	1.84E+01	**1.37E+01**
f_6	10	5.92E+01	3.51E+01	1.36E+02	3.98E+01	4.43E+01	**1.72E-02**
	20	2.09E+02	1.19E+02	3.64E+02	1.81E+02	1.06E+02	**6.44E-04**
	30	3.89E+02	2.40E+02	5.97E+02	3.22E+02	1.85E+02	**4.28E-02**
f_7	10	4.27E+02	2.59E+02	1.58E+03	1.35E+02	3.95E+02	**1.20E+02**
	20	1.86E+03	9.84E+02	4.68E+03	1.17E+03	1.21E+03	**2.16E+02**
	30	9.42E+03	3.20E+03	2.05E+04	5.03E+03	4.67E+03	**1.10E+03**
f_8	10	8.90E+01	8.01E+01	1.12E+02	8.33E+01	8.36E+01	**4.43E+01**
	20	2.19E+02	1.90E+02	2.67E+02	2.02E+02	1.96E+02	**1.16E+02**
	30	4.21E+02	3.69E+02	5.15E+02	4.03E+02	3.79E+02	**2.23E+02**

Table 3. Friedman test.

Algorithms	Mean ranking ($D = 10$)	Mean ranking ($D = 20$)	Mean ranking ($D = 30$)
ABC	5.00	5.00	5.00
CABC	2.50	2.75	2.75
MABC	6.00	6.00	6.00
GABC	2.63	3.88	4.00
MPEABC	3.88	2.38	2.25
MSABC	**1.00**	**1.00**	**1.00**

Table 4. Wilcoxon test.

MSABC vs.	ρ-value ($D - 10$)	ρ-value ($D = 20$)	ρ-value ($D = 30$)
ABC	**1.17E-02**	**1.17E-02**	**1.17E-02**
CABC	**1.17E-02**	**1.17E-02**	**1.17E-02**
MABC	**1.17E-02**	**1.17E-02**	**1.17E-02**
GABC	**1.17E-02**	**1.17E-02**	**1.17E-02**
MPEABC	**1.17E-02**	**1.17E-02**	**1.17E-02**

in 10, 20, or 30 dimensions. Table 3 shows the Friedman test results. MSABC has the top mean ranking value among all comparison algorithms. At the same time, we can find that CABC obtains better results than other single-search strategy algorithms, or even the multi-search strategy algorithm MABC. It shows that CABC has great search capability in addressing such expensive and complex problems. MPEABC is the only algorithm that uses dimension perturbation in the comparison algorithm. It has the best results among the five comparison algorithms. To some extent, it also indicates that using dimension perturbation can speed up the convergence when solving expensive and complex problems. The Wilcoxon results are listed in Table 4, the rho-values demonstrate that MSABC surpasses all comparison algorithms for all dimensions. The above results show that an effective search strategy, dimension perturbation, and multi-surrogate model can significantly increase the capacity of ABC algorithm to solve expensive and complex problems.

5   Conclusion

This study presents a multi-surrogate-assisted ABC (MSABC)algorithm to reduce fitness evaluations. Based on the characteristics of different search phases, in the employed bee phase, RBF network is selected as a global model. In onlooker bee phase, Kriging model is employed as a local surrogate built around some top best data points. In addition, random dimension perturbation is utilized to speed up the search.

To test the effectiveness of MSABC, the CEC 2014 expensive optimization benchmark set is chosen. MSABC is compared with five ABC variants. Under limited evaluations, MSABC can outperform the comparison algorithms for all test functions, based on the results obtained. Though the proposed approach can effectively save computational budget on the benchmark set, it would be fascinating to see how it performs in real-world problems.

Acknowledgment. This work was supported by National Natural Science Foundation of China (No. 62166027), and Jiangxi Provincial Natural Science Foundation (Nos. 20212ACB212004, 20212BAB202023, and 20212BAB202022).

References

1. Whitley, D.: A genetic algorithm tutorial. Stat. Comput. **4**(2), 65–85 (1994)
2. Metawa, N., Hassan, M.K., Elhoseny, M.: Genetic algorithm based model for optimizing bank lending decisions. Expert Syst. Appl. **80**, 75–82 (2017)
3. Ye, T.Y., Zeng, T., Zhang, L.Q., Xu, M.Y., Wang, H., Hu, M.: Artificial bee colony algorithm with an adaptive search manner. In: Neural Computing for Advanced Applications, pp. 486–497. Springer, Singapore (2021). https://doi.org/10.1007/s00521-022-06981-4
4. Zeng, T., Ye, T., Zhang, L., Xu, M., Wang, H., Hu, M.: Population diversity guided dimension perturbation for artificial bee colony algorithm. In: Zhang, H., Yang, Z., Zhang, Z., Wu, Z., Hao, T. (eds.) NCAA 2021. CCIS, vol. 1449, pp. 473–485. Springer, Singapore (2021). https://doi.org/10.1007/978-981-16-5188-5_34
5. Dorigo, M., Birattari, M., Stutzle, T.: Ant colony optimization. IEEE Comput. Intell. Maga. **1**(4), 28–39 (2006)
6. Dorigo, M., Stützle, T.: Ant colony optimization: overview and recent advances. In: Handbook of Metaheuristics, pp. 311–351 (2019)
7. Price, K., Storn, R.M., Lampinen, J.A.: Differential Evolution: A Practical Approach to Global Optimization. Springer, Heidelberg (2006). https://doi.org/10.1007/3-540-31306-0
8. Wu, G.H., Shen, X., Li, H.F., Chen, H.K., Lin, A.P., Suganthan, P.N.: Ensemble of differential evolution variants. Inf. Sci. **423**, 172–186 (2018)
9. Kennedy, J., Eberhart, R.: Particle swarm optimization. In: Proceedings of ICNN 1995 - International Conference on Neural Networks, vol. 4, pp. 1942–1948 (1995)
10. Tian, D.P., Shi, Z.Z.: MPSO: modified particle swarm optimization and its applications. Swarm Evol. Comput. **41**, 49–68 (2018)
11. Karaboga, D., Basturk, B.: A powerful and efficient algorithm for numerical function optimization: artificial bee colony (abc) algorithm. J. Glob. Optim. **39**(3), 459–471 (2007)
12. Gao, W.F., Liu, S.Y.: Improved artificial bee colony algorithm for global optimization. Inf. Process. Lett. **111**(17), 871–882 (2011)
13. Zhu, G., Kwong, S.: Gbest-guided artificial bee colony algorithm for numerical function optimization. Appl. Math. Comput. **217**(7), 3166–3173 (2010)
14. Gao, W.F., Liu, S.Y., Huang, L.L.: A novel artificial bee colony algorithm based on modified search equation and orthogonal learning. IEEE Trans. Cybern. **43**(3), 1011–1024 (2013)
15. Wang, H., Wu, Z., Rahnamayan, S., Sun, H., Liu, Y., Pan, J.S.: Multi-strategy ensemble artificial bee colony algorithm. Inf. Sci. **279**, 587–603 (2014)

16. Ye, T.Y., Wang, H., Wang, W.J., Zeng, T., Zhang, L.Q., Huang, Z.K.: Artificial bee colony algorithm with an adaptive search manner and dimension perturbation. Neural Comput. Appl., 1–15 (2022)

17. Zeng, T., et al.: Artificial bee colony based on adaptive search strategy and random grouping mechanism. Expert Syst. Appl. **192**, 116332 (2022)

18. Kiran, M.S., Hakli, H., Gunduz, M., Uguz, H.: Artificial bee colony algorithm with variable search strategy for continuous optimization. Inf. Sci. **300**, 140–157 (2015)

19. Ye, T.Y., et al.: Artificial bee colony algorithm with efficient search strategy based on random neighborhood structure. Knowl.-Based Syst. **241**, 108306 (2022)

20. Jin, Y.C., Wang, H.D., Sun, C.L.: Data-Driven Evolutionary Optimization. Springer, Heidelberg (2021). https://doi.org/10.1007/978-3-030-74640-7

21. Lim, D., Jin, Y.C., Ong, Y.S., Sendhoff, B.: Generalizing surrogate-assisted evolutionary computation. IEEE Trans. Evol. Comput. **14**(3), 329–355 (2009)

22. Sun, C., Jin, Y., Zeng, J., Yu, Y.: A two-layer surrogate-assisted particle swarm optimization algorithm. Soft Comput. **10**(6), 1461–1475 (2014). https://doi.org/10.1007/s00500-014-1283-z

23. Wang, H.D., Jin, Y.C., Doherty, J.: Committee-based active learning for surrogate-assisted particle swarm optimization of expensive problems. IEEE Trans. Cybern. **47**(9), 2664–2677 (2017)

24. Karaboga, D.: An idea based on honey bee swarm for numerical optimization. Technical report (2005)

25. Hardy, R.L.: Multiquadric equations of topography and other irregular surfaces. J. Geophys. Res. **76**(8), 1905–1915 (1971)

26. Broomhead, D.S., Lowe, D.: Multivariable functional interpolation and adaptive networks. Complex Syst. **2**(3), 321–355 (1988)

27. Powell, M.J.D.: Radial Basis Functions for Multivariable Interpolation: A Review, pp. 143–167. Clarendon Press, USA (1987)

28. Park, J., Sandberg, I.W.: Universal approximation using radial-basis-function networks. Neural Comput. **3**(2), 246–257 (1991)

29. Matheron, G.: Principles of geostatistics. Econ. Geol. **58**(8), 1246–1266 (1963)

30. Emmerich, M.: Single-and multi-objective evolutionary design optimization assisted by gaussian random field metamodels. Ph.D. thesis, Dortmund, University, Dissertation (2005)

31. Jin, Y.C., Olhofer, M., Sendhoff, B.: A framework for evolutionary optimization with approximate fitness functions. IEEE Trans. Evol. Comput. **6**(5), 481–494 (2002)

32. Liu, B., Chen, Q., Zhang, Q., Liang, J., Suganthan, P., Qu, B.: Problem definitions and evaluation criteria for computational expensive optimization. Technical Report (2013)

33. Sun, C.L., Jin, Y.C., Cheng, R., Ding, J.L., Zeng, J.C.: Surrogate-assisted cooperative swarm optimization of high-dimensional expensive problems. IEEE Trans. Evol. Comput. **21**(4), 644–660 (2017)

34. Wang, H., et al.: Multi-strategy and dimension perturbation ensemble of artificial bee colony. In: 2019 IEEE Congress on Evolutionary Computation (CEC), pp. 697–704. IEEE (2019)

35. Gao, W.F., Liu, S.Y.: A modified artificial bee colony algorithm. Comput. Oper. Res. **39**(3), 687–697 (2012)

Multi-view Spectral Clustering with High-order Similarity Learning

Yanying Mei[1] , Zhenwen Ren[2,3](✉) , Bin Wu[1], and Yanhua Shao[1]

[1] School of Information Engineering, Southwest University of Science and Technology, Mianyang 621010, China
{myy930,wubin}@swust.edu.cn, syh@cqu.edu.cn
[2] State Key Laboratory for Novel Software Technology, Nanjing University, Nanjing 210008, China
rzw@njust.edu.cn
[3] School of National Defence Science and Technology, Southwest University of Science and Technology, Mianyang 621010, China

Abstract. Clustering objects with diverse attributes obtained from multiple views is full of challenges in fusing the multi-view information. Many of the present multi-view clustering (MVC) methods concentrate on direct similarity learning among data points and fail to excavate the hidden high-order similarity among different views. Therefore, it is difficult to obtain a dependable clustering assignments. To address this problem, we propose the high-order similarity (HOS) learning model for multi-view spectral clustering (MCHSL). The proposed MCHSL learns the first-order similarity (FOS), second-order similarity (SOS), and the HOS collaboratively to excavate the local structure relations, proximity structure relations of paired data points and the interactive-view relations among different views instead of the common similarity learning. Then spectral clustering is performed to obtain the final clustering assignments. Extensive experiments performed on some public datasets indicate that the proposed MCHSL has better clustering performance than benchmark methods in most cases and is able to reveal a dependable underlying similarity structure hidden in multiple views.

Keywords: First-order similarity · Second-order similarity · High-order similarity · Multi-view clustering

1 Introduction

Clustering with multi-view data from multiple sources has been extensively used in data processing, machine vision, and statistics [3,7]. For instance, in image clustering, images or videos are usually described by multiple features, such as pixel features, SIFT features and HOG features, which can be regarded as a classic multi-view clustering (MVC) problem. The main idea for solving this problem is to acquire the consistent representation learned from multi-view data.

MVC is an important clustering method that can completely excavate the multiple information learned from multi-view data. The current MVC methods are mainly divided into five types: graph-based clustering, collaborative training clustering, kernel-based clustering, deep clustering, and subspace clustering. For graph-based clustering methods, they usually focus on learning an optimal affinity graph, followed by a spectral algorithm or graph cut algorithm used for the graph to obtain the final clustering result. Despite their excellent results, most existing MVC approaches still have the following shortcomings. (1) These methods fail to fully consider the local structure relations and proximity structure relations of paired data points; (2) these methods only concentrate on exploiting consensus structure among multiple views while neglect intrinsic independent structure within each view.

This work proposes a multi-view spectral clustering (MVSC) with HOS learning (MCHSL) model which explores the local structure relations, the proximity structure relations of paired data points, and the interactive-view relations of different views for clustering. MCHSL reveals the latent consensus structure among different data points by utilizing the FOS, SOS and HOS simultaneous learning. Therefore, the latent similarity learning model is optimized by regularizing the FOS, SOS, and HOS simultaneously, and the spectral clustering (SPC) is used on the learned latent similarity matrix to achieve the final clustering assignments. Experiments are carried on four public datasets, and the clustering results indicate that MCHSL has superior performance in MVC. The main contributions of our work are presented as follows:

- In order to explore true clustering affiliations of data points for MVC, the FOS and SOS of paired data points are learned collaboratively to excavate their local structure relations and proximity structure relations.
- In order to explore each view-specific information and the consensus information of multiple views for MVC, the HOS is learned to excavate their interactive-view relations of different views.
- Compared with the present most advanced methods, the proposed MCHSL has better performance on multiple benchmark datasets in diversified applications.

The rest of this work is arranged as follows: Sect. 2 generally presents the related work. The proposed method, optimization, and complexity analysis are presented in Sect. 3. The experiments and convergence analysis are discussed in Sect. 4. In the end, we obtain the conclusion in Sect. 5.

2   Related Work

Many graph-based clustering methods related to our work have been proposed [8,9,15,19,20,22]. Most of which are about learning a k-nearest neighbor affinity matrix [12], an adaptive neighbor affinity matrix [11], or a subspace affinity matrix [17], and the final clustering assignments are decided by the learned affinity matrix. For instance, Nie et al. [12] assigned the optimal neighbors for each data point to learn the affinity matrix. Later, Nie et al. [11] extended this model on multi-view graph clustering. Wang et al. [18] proposed a MVC method that automatically weighted each graph matrix to derive the consensus graph matrix. Chen et al. [2] proposed a latent embedding space learning approach for MVC which excavated the underlying structure of multiple views by introducing a latent embedding representation. Peng et al. [13] proposed a HOS learning model for multi-dimensional data clustering. Although the above methods have obtained better clustering performance, they can not fully learn the proximity structure relations among different data points and the interactive-view relations among different views. Our work can solve these limitations.

In our work, MVSC [3,6,10] is also used, which learns the intrinsic structure relations among data points by excavate the attributes of affinity graph for clustering. The core thought of SPC is to construct a affinity graph and achieve normalized cut on the affinity graph for clustering. MVC based on the graph cutting will cluster the data by cutting the fusion graph directly such that the clustering results are related to the construction of fusion graph. However, the construction of the fusion graph is decided by the k nearest neighbors, so the SPC is used in our method.

3   Proposed Method

In this part, we firstly propose the FOS and SOS collaborative learning model , and then the MVSC with HOS learning model is proposed, namely MCHSL.

3.1   Main Notations

The nouns are described in this part. The matrices are denoted in bold uppercase letters (e.g., $\mathbf{Y}$), vectors are represented by bold lowercase letters (e.g., $\mathbf{y}$), and scalars are represented by Greek letters (e.g., α, β, γ). The j-th column vector and the (i, j)-th entry of matrix $\mathbf{Y}$ are represented by $\mathbf{y}_j$ and y_{ij}, respectively. The $\mathrm{Tr}(\mathbf{Y})$ denotes the trace of the matrix $\mathbf{Y}$, the matrix $\mathbf{I}$ represents a identity matrix, and the Frobenius norm of the matrix $\mathbf{Y}$ are represented by $\|\mathbf{Y}\|_F$.

3.2   First and Second Order Similarity Learning

Although current MVC methods have good clustering performance, they can not fully consider the proximity structure relation among data points, which is helpful for data clustering. In order to fully excavate the relations among data points, we come up with learning the FOS and SOS simultaneously to achieve an optimal similarity matrix for clustering assignments.

Given the data points set $\mathbf{Y} = \{\mathbf{y}_1, \cdots, \mathbf{y}_N\}$, which is defined as the nodes of the affinity graph. According to [4], we have

Definition 1 [4] *(FOS:First order similarity). The edge value of the graph is defined as a_{ij} , which denote the FOS between data points $\mathbf{y}_i$ and $\mathbf{y}_j$, they are considered as the primary metric of similarity between data points.*

According to Definition 1, the second order similarity between nodes is defined as

Definition 2 [4] *(SOS:Second order similarity). The SOS denotes the similarity of neighbors of paired data points. Given* $\mathbf{a}_i = [a_{i1}, \cdots, a_{iN}]$ *, which denotes the FOS between* $\mathbf{y}_i$ *and the other data points. Thus, the SOS between* $\mathbf{y}_i$ *and* $\mathbf{y}_j$ *is decided by the similarity of* $\mathbf{a}_i$ *and* $\mathbf{a}_j$ *.*

Although the FOS is the primary measure of similarity for clustering, in the real-world scenario, there are often all kinds of outliers and noise which are detrimental to data clustering, so SOS learning is proposed. The SOS is learned to excavate the similarity of the proximity structures of paired data points and the data points are considered as similar if they have the same proximity structures. In this paper, we explore learning the FOS and SOS collaboratively for data clustering. According to [4], the model is given by

$$
\min_{\mathbf{A}, \mathbf{A}^{(v)}} \sum_{v=1}^{V} \left(\sum_{i,j=1}^{N} a_{ij}^{(v)} \left\| \mathbf{y}_i^{(v)} - \mathbf{y}_j^{(v)} \right\|_2^2 + \sum_{i=1}^{N} \left\| \mathbf{y}_i^{(v)} - \sum_{j=1}^{N} a_{ij}^{(v)} \mathbf{y}_j^{(v)} \right\|_2^2 + \alpha \|\mathbf{A}^{(v)}\|_F^2 \right) + \beta \sum_{v=1}^{V} \left\| \mathbf{A} - \mathbf{A}^{(v)} \right\|_F^2 ,
$$
$$
\text{s.t. } 0 \leq \mathbf{A}^{(v)} \leq 1, (\mathbf{A}^{(v)})^\top = \mathbf{A}^{(v)}.
$$
(1)

in which $\mathbf{Y}^{(v)} = \{\mathbf{y}_1^{(v)}, \cdots, \mathbf{y}_N^{(v)}\}$ represents the data points in the v-th view, N represents the number of data points, V represents the number of views, $a_{ij}^{(v)}$ in similarity matrix $\mathbf{A}^{(v)} \in \mathbb{R}^{N \times N}$ denotes the affinity between $\mathbf{y}_i^{(v)}$ and $\mathbf{y}_j^{(v)}$, and $\alpha > 0$, $\beta > 0$ represent the trade-off parameters. According to [14], we introduce matrix $\mathbf{M}$, which represent a complete graph, thus we have

$$
\min_{\mathbf{A}^{(v)}} \sum_{i,j=1}^{N} \left\| \mathbf{y}_i^{(v)} - \mathbf{y}_j^{(v)} \right\|_2^2 a_{ij} = \min_{\mathbf{A}^{(v)}} \text{Tr}(\mathbf{M}^{(v)\top} \mathbf{A}^{(v)}),
$$
(2)

in which $\mathbf{M}^{(v)}$ represents the v-th complete graph with $m_{ij}^{(v)} = \left\| \mathbf{y}_i^{(v)} - \mathbf{y}_j^{(v)} \right\|_2^2$. If $m_{ij}^{(v)}$ in $\mathbf{M}^{(v)}$ is smaller, the similarity $a_{ij}^{(v)}$ is greater.

According to [4], we have

$$
\min_{\mathbf{A}^{(v)}} \sum_{i=1}^{N} \left\| \mathbf{y}_i^{(v)} - \sum_{j=1}^{N} a_{ij}^{(v)} \mathbf{y}_j^{(v)} \right\|_2^2 = \min_{\mathbf{A}^{(v)}} \text{Tr}\left(\mathbf{Y}^{(v)}(\mathbf{I} - \mathbf{A}^{(v)})(\mathbf{I} - \mathbf{A}^{(v)})^\top \mathbf{Y}^{(v)\top} \right).
$$
(3)

Therefore, Eq. (1) can be rewritten as

$$
\min_{\mathbf{A}, \mathbf{A}^{(v)}} \sum_{v=1}^{V} \left(\text{Tr}\left(\mathbf{Y}^{(v)}(\mathbf{I} - \mathbf{A}^{(v)})(\mathbf{I} - \mathbf{A}^{(v)})^\top \mathbf{Y}^{(v)\top} \right) + \text{Tr}(\mathbf{M}^{(v)\top} \mathbf{A}^{(v)}) + \alpha \|\mathbf{A}^{(v)}\|_F^2 \right) + \beta \sum_{v=1}^{V} \left\| \mathbf{A} - \mathbf{A}^{(v)} \right\|_F^2
$$
$$
\text{s.t. } 0 \leq \mathbf{A}^{(v)} \leq 1, (\mathbf{A}^{(v)})^\top = \mathbf{A}^{(v)}.
$$
(4)

3.3 Multi-view Spectral Clustering with High-order Similarity Learning

We propose to learn the HOS among different views which can capture the consensus information and view-dependent information in each view and offer guidance for learning a robust latent similarity matrix.

Given multi-view matrix $\mathbf{Y} = \left[\mathbf{Y}^{(1)}, \mathbf{Y}^{(2)}, \cdots, \mathbf{Y}^{(V)}\right]$, which have V similarity matrices $\mathbf{A}^{(v)}$, $v = 1, 2, ..., V$. To overcome the shortcoming of direct similarity, we fuse the consensus similarity matrix with the view-dependent similarity matrix to learn a reliable HOS among different views.

Definition 3 [13] *(HOS:High-order similarity). Given $\mathbf{A}$ and $\mathbf{A}^{(v)}$, which are defined as the consensus latent similarity matrix and the v-th similarity matrix. The HOS in the v-th view is defined as $\mathbf{A}\mathbf{A}^{(v)}$, in which the element is defined as $(\mathbf{A}\mathbf{A}^{(v)})_{ij} = \sum_{k=1}^{N} \mathbf{A}_{ik}\mathbf{A}_{kj}^{(v)}$, and N represents the number of data points.*

Unlike the common similarity matrix $\mathbf{A}^{(v)}$, which mainly relies on the paired data points affinity, the similarity expression $\mathbf{A}\mathbf{A}^{(v)}$ learns the HOS among multiple views. Therefore, the HOS can be adaptively updated in the latent similarity learning process by the supplements of the predefined similarity learning.

To facilitate consensus learning across different views, one can constrain the error term $\left\|\mathbf{A}\mathbf{A}^{(v)} - \mathbf{A}\mathbf{A}^{(w)}\right\|_{F}^{2}$, $v = 1, 2, ..., V$, $w = 1, 2, ..., V$. Comparing with direct similarity learning $\left\|\mathbf{A} - \mathbf{A}^{(v)}\right\|_{F}^{2}$, the HOS learns both each view-dependent information and the consensus information among multiple views, thereby it is robust to outliers and noises. Besides, the consensus information brought by the HOS learning makes the structure in each view similar to the learned consensus latent similarity matrix, which would be beneficial for the latent similarity learning. The HOS learning model can be written as

$$
\min_{\mathbf{A},\mathbf{A}^{(v)}} \sum_{v=1}^{V} \left(\mathrm{Tr}\left(\mathbf{Y}^{(v)}(\mathbf{I} - \mathbf{A}^{(v)})(\mathbf{I} - \mathbf{A}^{(v)})^{\top}\mathbf{Y}^{(v)\top} \right) + \mathrm{Tr}(\mathbf{M}^{(v)\top}\mathbf{A}^{(v)}) + \alpha\|\mathbf{A}^{(v)}\|_{F}^{2} \right)
$$
$$
+ \lambda_1 \sum_{v=1}^{V} \sum_{w=1}^{V} \left\|\mathbf{A}\mathbf{A}^{(v)} - \mathbf{A}\mathbf{A}^{(w)}\right\|_{F}^{2} + \beta \sum_{v=1}^{V} \left\|\mathbf{A} - \mathbf{A}^{(v)}\right\|_{F}^{2}
$$
$$
\text{s.t. } \mathbf{0} \leq \mathbf{A}^{(v)} \leq \mathbf{1}, (\mathbf{A}^{(v)})^{\top} = \mathbf{A}^{(v)}.
$$

$$(5)$$

Obviously latent similarity $\mathbf{A}$ is decided by the similarity $\mathbf{A}^{(v)}$ of different views, which may have bad structure due to the noises and outliers. As we all know, the strict block diagonal property of the similarity matrix is important for similarity learning clustering, and the trace constraint for the similarity matrix which is usually determined by the true number of clusters can promote the affinity graph to be block diagonal. Therefore, according to [16], we have

$$\min_{\mathbf{A},\mathbf{A}^{(v)}} \sum_{v=1}^{V} \left(\mathrm{Tr}\left(\mathbf{Y}^{(v)}(\mathbf{I} - \mathbf{A}^{(v)})(\mathbf{I} - \mathbf{A}^{(v)})^{\top} \mathbf{Y}^{(v)\top} \right) + \mathrm{Tr}(\mathbf{M}^{(v)\top} \mathbf{A}^{(v)}) + \alpha \|\mathbf{A}^{(v)}\|_F^2 \right)$$

$$+ \lambda_1 \sum_{v=1}^{V} \sum_{w=1}^{V} \left\| \mathbf{A}\mathbf{A}^{(v)} - \mathbf{A}\mathbf{A}^{(w)} \right\|_F^2 + \beta \sum_{v=1}^{V} \left\| \mathbf{A} - \mathbf{A}^{(v)} \right\|_F^2$$

$$\text{s.t. } \mathrm{Tr}(\mathbf{A}^{(v)}) = c, 0 \le \mathbf{A}^{(v)} \le 1, (\mathbf{A}^{(v)})^{\top} = \mathbf{A}^{(v)}.$$

$$(6)$$

where $\mathbf{Y}^{(v)}$ denotes the data points in the v-th view, V represents the number of views, $\mathbf{A}^{(v)} \in \mathbb{R}^{N \times N}$ represents the similarity matrix, c represents the number of clusters, and $\beta > 0$, $\alpha > 0$, $\lambda_1 > 0$ represent the trade-off parameters.

Considering the above discussion, we propose a high order similarity learning model MCHSL. First, MCHSL learns FOS and SOS among data points to excavate their local structure relations and proximity structure relations to avoid noise interference. Second, the HOS is learned to excavate the interactive-view relations of different views, so that the consensus information learned from multiple views can be efficiently excavated. Moreover, a trace constraint is used on the learned similarity matrix. Finally, SPC is used to obtain the real clustering assignments.

3.4 Optimization

The problem in Eq. (6) has the solution by iteratively update one variable and fixing the others until convergence. The specific updated steps are presented as follows:

Step 1: By fixing $\mathbf{A}^{(v)}$ to solve variable $\mathbf{A}$, Eq. (6) can be rewritten as

$$\min_{\mathbf{A}} \lambda_1 \sum_{v=1}^{V} \sum_{w=1}^{V} \left\| \mathbf{A}\mathbf{A}^{(v)} - \mathbf{A}\mathbf{A}^{(w)} \right\|_F^2 + \beta \sum_{v=1}^{V} \left\| \mathbf{A} - \mathbf{A}^{(v)} \right\|_F^2 \qquad (7)$$

$$\text{s.t. } \mathrm{Tr}(\mathbf{A}^{(v)}) = c, 0 \le \mathbf{A}^{(v)} \le 1, (\mathbf{A}^{(v)})^{\top} = \mathbf{A}^{(v)}.$$

The derivative of Eq. (7) is set to zero, and the closed-form solution of $\mathbf{A}$ can be given by

$$\mathbf{A} = 2\beta \sum_{v=1}^{V} \mathbf{A}^{(v)} (4\lambda_1 V \sum_{v=1}^{V} \mathbf{A}^{(v)2} - \sum_{v=1}^{V} \sum_{w=1}^{V} \mathbf{A}^{(v)} \mathbf{A}^{(w)} + 2\beta V \mathbf{I})^{-1} \qquad (8)$$

Step 2: Fixing $\mathbf{A}$ and $\mathbf{A}^{(w)}$, $w = 1, ..., V$, $w \ne v$ to update $\mathbf{A}^{(v)}$, Eq. (6) refer to $\mathbf{A}^{(v)}$ can be rewritten as

$$\min_{\mathbf{A}^{(v)}} \sum_{v=1}^{V} \left(\mathrm{Tr}(\mathbf{Y}^{(v)}(\mathbf{I} - \mathbf{A}^{(v)})(\mathbf{I} - \mathbf{A}^{(v)})^{\top} \mathbf{Y}^{(v)^{\top}}) + \mathrm{Tr}(\mathbf{M}^{(v)^{\top}} \mathbf{A}^{(v)}) + \alpha \|\mathbf{A}^{(v)}\|_F^2 \right)$$

$$+ \lambda_1 \sum_{v=1}^{V} \sum_{w=1}^{V} \left\| \mathbf{A}\mathbf{A}^{(v)} - \mathbf{A}\mathbf{A}^{(w)} \right\|_F^2 + \beta \sum_{v=1}^{V} \left\| \mathbf{A} - \mathbf{A}^{(v)} \right\|_F^2$$

$$\text{s.t. } \mathrm{Tr}(\mathbf{A}^{(v)}) = c, 0 \le \mathbf{A}^{(v)} \le 1, (\mathbf{A}^{(v)})^{\top} = \mathbf{A}^{(v)}.$$

$$(9)$$

Then, Eq. (9) can be written as

$$\min_{\mathbf{A}^{(v)}} \frac{1}{2} \left\| \mathbf{A}^{(v)} - \mathbf{B}^{(v)} \right\|_F^2 \text{ s.t. } \mathrm{Tr}(\mathbf{A}^{(v)}) = c, 0 \le \mathbf{A}^{(v)} \le 1, (\mathbf{A}^{(v)})^{\top} = \mathbf{A}^{(v)},$$

$$(10)$$

where

$$\mathbf{B}^{(v)} = \left(4\mathbf{Y}\mathbf{Y}^{\top} - \mathbf{M}^{\top} + 2\beta\mathbf{A} + 2\lambda_1 \mathbf{A}^2 \sum_{w \ne v}^{V} \mathbf{A}^{(w)} \right) \left(2\lambda_1 V \mathbf{A}^2 + (2\alpha + 2\beta)\mathbf{I} + 4\mathbf{Y}\mathbf{Y}^{\top} \right)^{-1}$$

$$(11)$$

Such a problem can be solved according to [16].

Algorithm 1: Optimization for MCHSL

Input: Multi-view data $\mathbf{Y} = [\mathbf{Y}^{(1)}, \ldots, \mathbf{Y}^{(V)}]$, parameters α, β and λ_1.

Output: The clustering results.

1 Initialize each similarity matrix $\mathbf{A}^{(v)}$ and $maxIter = 200$.

2 **while** *not convergence* **do**

3 | Update consensus similarity matrix $\mathbf{A}$ via (7);

4 | Update each similarity matrix $\mathbf{A}^{(v)}$ via (9);

5 **end**

6 Perform spectral clustering on $\mathbf{A}$.

Up to now, we update $\mathbf{A}$, $\mathbf{A}^{(v)}$ alternatively, and this work is repeated until the convergence criteria or the maximum iteration is reached. The whole optimization of MCHSL is summarised in algorithm 1. After obtaining $\mathbf{A}$, the similarity matrix is represented by $\mathbf{A} = \frac{\mathbf{A}^{\top} + \mathbf{A}}{2}$, then we use the SPC [10] to get the final clustering assignments.

3.5   Complexity Analysis

The computation steps of the solution for MCHSL is summarized in Algorithm 1. The computation cost of MCHSL includes solving Eq. (7) and (9). Thus, the complexity of MCHSL is $\mathcal{O}(N^3)$, in which N represents the number of data points.

4   Experiment

Our experimental platform is Windows 10(x64) operating system with Intel Core i5 (2.9 GHz) CPU, 16 GB memory, and the simulation software is MATLAB 2019b.

4.1   Datasets

Four real public datasets are adopted in our work which include images and text. Then the detailed descriptions about these datasets are given as follows.

ORL[1] is a general face dataset which is consist of 400 face images including 40 distinct objects, which were collected at different lights, times, facial expression, and facial details. These images are described by intensity feature, LBP features, and Gabor features. We use these three features data as three views in our experiments.

Yale[2] includes 165 face images from 15 people, which are about facial expressions or configuration images. These images are described by raw pixel values, Gabor features, and LBP features. We use these three features data as three views in our experiments.

BBCSport [21] includes 544 documents, which are collected from the BBC Sports website. All these documents are extracted in two different features about sports news from 2004 to 2005 including business, politics, entertainment, technology and sports.

Webkb [13] contains 230 course web pages and 821 non-course web pages, which are collected from the computer science department websites in four countries. Similar to [13], we use the two features data as two views in our experiments.

4.2   Baselines and Evaluation Metrics

We compare our work with eight most outstanding clustering methods, such as SC [10], Co-Reg [5], RMSC [21], DIMSC [1], LMSC [24], MVGL [23], MCLES [2], HSC [13]. We use three indicators to evaluate the clustering performance: Accuracy (ACC), Purity, and Normalized Mutual Information (NMI). The specific parameter setting and result analysis are given as below. In the experiments, each method is performed in 10 times in order to avoid the random values, and then the mean and standard deviation are presented, where the higher values show the better performance.

4.3   Results Analysis

We show the experimental results in Tables 1, 2, 3 and 4, and mark the best values in bold. Observing these results, the following conclusions can be obtained:

---

[1] http://www.cl.cam.ac.uk/research/dtg/.
[2] http://cvc.yale.edu/projects/yalefaces/yalefaces.html.

Table 1. Clustering performance of MCHSL and comparison methods on the ORL dataset.

Method	ACC	NMI	Purity
SC1 [10]	0.657(0.024)	0.805(0.010)	0.693(0.020)
SC2 [10]	0.774(0.026)	0.891(0.010)	0.803(0.021)
SC3 [10]	0.697(0.032)	0.841(0.017)	0.733(0.026)
Co-Reg [5]	0.692(0.004)	0.838(0.002)	0.729(0.003)
RMSC [21]	0.760(0.026)	0.720(0.021)	0.739(0.017)
DIMSC [1]	0.428(0.000)	0.634(0.000)	0.478(0.000)
LMSC [24]	0.801(0.033)	0.907(0.020)	0.838(0.029)
MVGL [23]	0.735(0.000)	0.865(0.000)	0.795(0.000)
MCLES [2]	0.797(0.023)	0.902(0.012)	0.840(0.018)
HSC [13]	0.783(0.000)	0.869(0.000)	0.800(0.000)
MCHSL	**0.820(0.035)**	**0.916(0.014)**	**0.863(0.027)**

- The single-view clustering (SVC) approaches have acquired good clustering performance. But in general, MVC approaches perform better than SVC approaches. This shows that the complementary information from multiple views helps improve clustering performance.
- The proposed MCHSL method obtains the best performance in purity, NMI, and ACC on all datasets, and is significantly superior to all other MVC methods.
- Compared with MCLES and MVGL, MCHSL can learn the high-order affinity from multiple views, so as to capture more important structural information of data points for clustering.
- Compared with HSC, MCHSL performs better in most instances. This indicates that FOS and SOS learning simultaneously is beneficial for excavating the local structure relations among data points and their proximity structure relations, thus improving the clustering performance.

4.4   Parameters Analysis

In our experiments, three parameters α, β and λ_1 are adopted to balance the regularization term on each affinity graph, the common similarity term and the HOS learning term. Each parameter is verified by fixing the other parameters. Taking the ORL datasets for example, from Fig. 1, the performance of our MCHSL is pretty stable when α, β and λ_1 vary in $[10^{-2}, 10^4]$. In general, when these parameters change, our algorithm is insensitive on different datasets.

4.5   Convergence Analysis

The convergence analysis of MCHSL is given in this section. The each iteration error is presented in Fig. 2 on the four real-word dataset. In MCHSL, the error

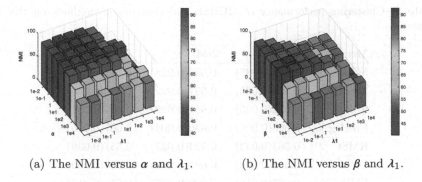

(a) The NMI versus α and λ_1.   (b) The NMI versus β and λ_1.

Fig. 1. The NMI of MCHSL on the ORL dataset with different parameter settings.

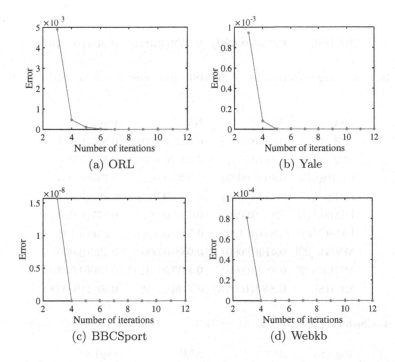

Fig. 2. Convergence curve of MCHSL on the four real-world datasets.

Table 2. Clustering performance of MCHSL and comparison methods on the Yale dataset.

Method	ACC	NMI	Purity
SC1 [10]	0.550(0.035)	0.589(0.028)	0.561(0.035)
SC2 [10]	0.563(0.035)	0.597(0.022)	0.572(0.030)
SC3 [10]	0.632(0.035)	0.651(0.026)	0.637(0.034)
Co-Reg [5]	0.596(0.006)	0.636(0.004)	0.607(0.005)
RMSC [21]	0.563(0.043)	0.524(0.037)	0.551(0.036)
DIMSC [1]	0.655(0.000)	0.691(0.000)	0.655(0.000)
LMSC [24]	0.667(0.018)	0.690(0.016)	0.671(0.017)
MVGL [23]	0.630(0.000)	0.638(0.000)	0.642(0.000)
MCLES [2]	0.705(0.012)	0.725(0.015)	0.706(0.013)
HSC [13]	0.751(0.003)	0.734(0.004)	0.752(0.003)
MCHSL	**0.764(0.029)**	**0.772(0.016)**	**0.801(0.016)**

Table 3. Clustering performance of MCHSL and comparison methods on the BBC-Sport.

Method	ACC	NMI	Purity
SC1 [10]	0.845(0.001)	0.672(0.003)	0.845(0.001)
SC2 [10]	0.511(0.001)	0.235(0.000)	0.572(0.000)
Co-Reg [5]	0.693(0.007)	0.538(0.002)	0.735(0.003)
RMSC [21]	0.774(0.010)	0.765(0.012)	0.760(0.011)
DIMSC [1]	0.877(0.000)	0.749(0.001)	0.877(0.000)
LMSC [24]	0.851(0.120)	0.745(0.136)	0.856(0.105)
MVGL [23]	0.419(0.000)	0.088(0.000)	0.423(0.000)
MCLES [2]	0.880(0.003)	**0.807(0.011)**	0.880(0.003)
MCHSL	**0.883(0.078)**	0.716(0.141)	**0.883(0.078)**

Table 4. Clustering performance of MCHSL and comparison methods on the Webkb.

Method	ACC	NMI	Purity
SC1 [10]	0.776(0.000)	0.001(0.000)	0.781(0.000)
SC2 [10]	0.768(0.000)	0.006(0.000)	0.781(0.000)
Co-Reg [5]	0.578(0.000)	0.003(0.000)	0.781(0.000)
DIMSC [1]	0.615(0.000)	0.001(0.000)	0.781(0.000)
MVGL [23]	0.751(0.000)	0.010(0.000)	0.781(0.000)
HSC [13]	0.848(0.000)	0.398(0.000)	0.848(0.000)
MCHSL	**0.909(0.000)**	**0.443(0.000)**	**0.909(0.000)**

that is used to control the terminal condition is set to $\xi = 1.0 \times 10^{-4}$, which is defined as: error $= \left\| \mathbf{A}^{t+1} - \mathbf{A}^t \right\|_F^2$, where $\mathbf{A}^t$ represents the obtained similarity matrix in the t-th iteration. According to Fig. 2, with increase of iterations, the error decreases and converges within 10 iterations. This indicates that the optimization for MCHSL is convergent. The similar phenomenon can also be observed in the other datasets.

5   Conclusion

In this paper, a high-order similarity learning model for MVSC, namely MCHSL, is proposed, which can comprehensively explore the local structure relations, proximity structure relations of paired data points and the interactive-view relations among different views. The proposed model learns the HOS among different views and the common similarity among different data points collaboratively, instead of the directly similarity learning, enabling to learn an optimal similarity matrix. Experimentally, MCHSL has better clustering performance compared with the other MVC methods.

Due to the impact of computational complexity, the proposed MCHSL is mainly applied to small-sized and medium-sized datasets. In the future work, we wish to use it in the large-scale datasets.

Acknowledgement. This research was supported by the National Natural Science Foundation of China (Grant nos. 62106209).

References

1. Cao, X., Zhang, C., Fu, H., Liu, S., Zhang, H.: Diversity-induced multi-view subspace clustering. In: IEEE Conference on Computer Vision and Pattern Recognition, CVPR, pp. 586–594. IEEE Computer Society (2015)
2. Chen, M., Huang, L., Wang, C., Huang, D.: Multi-view clustering in latent embedding space. In: The Thirty-Fourth AAAI Conference on Artificial Intelligence, pp. 3513–3520. AAAI Press (2020)
3. Elhamifar, E., Vidal, R.: Sparse subspace clustering: algorithm, theory, and applications. IEEE Trans. Pattern Anal. Mach. Intell. **35**(11), 2765–2781 (2013)
4. Goyal, P., Ferrara, E.: Graph embedding techniques, applications, and performance: a survey. Knowl.-Based Syst. **151**, 78–94 (2018)
5. Kumar, A., Rai, P., Daumé, H.: Co-regularized multi-view spectral clustering. In: Advances in Neural Information Processing Systems, pp. 1413–1421 (2011)
6. Liang, W., et al.: Multi-view spectral clustering with high-order optimal neighborhood laplacian matrix. IEEE Trans. Knowl. Data Eng. 1 (2020)
7. Liu, G., Lin, Z., Yan, S., Sun, J., Yu, Y., Ma, Y.: Robust recovery of subspace structures by low-rank representation. IEEE Trans. Pattern Anal. Mach. Intell. **35**(1), 171–184 (2013)
8. Ma, J., Zhang, Y., Zhang, L.: Discriminative subspace matrix factorization for multiview data clustering. Pattern Recogn. **111**, 107676 (2021)

9. Mei, Y., Ren, Z., Wu, B., Shao, Y., Yang, T.: Robust graph-based multi-view clustering in latent embedding space. Int. J. Mach. Learn. Cybern. **13**(2), 497–508 (2021). https://doi.org/10.1007/s13042-021-01421-6

10. Ng, A.Y., Jordan, M.I., Weiss, Y.: On spectral clustering: analysis and an algorithm. In: Advances in Neural Information Processing Systems, vol. 14, pp. 849–856. MIT Press (2001)

11. Nie, F., Cai, G., Li, X.: Multi-view clustering and semi-supervised classification with adaptive neighbours. In: Proceedings of the Thirty-First AAAI Conference on Artificial Intelligence, pp. 2408–2414. AAAI Press (2017)

12. Nie, F., Wang, X., Huang, H.: Clustering and projected clustering with adaptive neighbors. In: The 20th ACM SIGKDD International Conference on Knowledge Discovery and Data Mining, pp. 977–986. ACM (2014)

13. Peng, H., Wang, H., Hu, Y., Zhou, W., Cai, H.: Multi-dimensional clustering through fusion of high-order similarities. Pattern Recogn. **121**, 108108 (2022)

14. Ren, Z., Li, H., Yang, C., Sun, Q.: Multiple kernel subspace clustering with local structural graph and low-rank consensus kernel learning. Knowl.-Based Syst. **188**, 105040 (2020)

15. Ren, Z., Mukherjee, M., Lloret, J., Venu, P.: Multiple kernel driven clustering with locally consistent and selfish graph in industrial IoT. IEEE Trans. Industr. Inf. **17**(4), 2956–2963 (2020)

16. Ren, Z., Sun, Q., Wei, D.: Multiple kernel clustering with kernel k-means coupled graph tensor learning. In: Proceedings of the Thirty-Fifth AAAI Conference on Artificial Intelligence, pp. 9411–9418. AAAI Press (2021)

17. Vidal, R., Favaro, P.: Low rank subspace clustering (LRSC). Pattern Recogn. Lett. **43**, 47–61 (2014)

18. Wang, H., Yang, Y., Liu, B.: GMC: graph-based multi-view clustering. IEEE Trans. Knowl. Data Eng. **32**(6), 1116–1129 (2020)

19. Wang, H., et al.: Kernelized multiview subspace analysis by self-weighted learning. IEEE Trans. Multimedia **23**, 3828–3840 (2020)

20. Wen, J., et al.: Unified tensor framework for incomplete multi-view clustering and missing-view inferring. In: Thirty-Fifth AAAI Conference on Artificial Intelligence, AAAI, pp. 10273–10281. AAAI Press (2021)

21. Xia, R., Pan, Y., Du, L., Yin, J.: Robust multi-view spectral clustering via low-rank and sparse decomposition. In: Proceedings of the Twenty-Eighth AAAI Conference on Artificial Intelligence, pp. 2149–2155. AAAI Press (2014)

22. Ye, Q., Huang, P., Zhang, Z., Zheng, Y., Fu, L., Yang, W.: Multiview learning with robust double-sided twin SVM. IEEE Trans. Cybern. (2021)

23. Zhan, K., Zhang, C., Guan, J., Wang, J.: Graph learning for multiview clustering. IEEE Trans. Cybern. **48**(10), 2887–2895 (2018)

24. Zhang, C., Hu, Q., Fu, H., Zhu, P., Cao, X.: Latent multi-view subspace clustering. In: 2017 IEEE Conference on Computer Vision and Pattern Recognition, CVPR. pp. 4333–4341. IEEE Computer Society (2017)

An Improved Convolutional Neural Network Model by Multiwavelets for Rolling Bearing Fault Diagnosis

Gangxing Ren, Jing Yuan[✉], Fengxian Su, Huiming Jiang, and Qian Zhao

School of Mechanical Engineering, University of Shanghai for Science and Technology, 516 Jun Gong Road, Shanghai 200093, China

yuanjing@usst.edu.cn

Abstract. Convolutional neural network (CNN) is increasingly applied to data-driven fault diagnosis of mechanical equipment spare parts. However, CNN training network parameters need a large amount of fault data, and better training effect can be obtained by updating network parameters repeatedly. In this paper, an improved CNN by multiwavelets is introduced multiwavelets into convolution layer, the natural convolution attribute of multiwavelets is fused with convolution layer to fully release the two-channel feature extraction ability of multiwavelets transform. At the same time, we change the parameters of the multiwavelets convolution kernel to discuss the overall diagnostic performance of the network in the same dataset. Thus, different multiwavelets kernel parameters are customized according to different signal characteristics. The feasibility and effectiveness of improved CNN by multiwavelets for case Western Reserve University fault-bearing data are verified.

Keyword: Deeping learning · CNN · Multiwavelets · Signal processing · Fault identification

1 Introduction

Rotating machinery is an indispensable part of the equipment manufacturing industry, and rolling bearing is an important part of rotating machinery. Its small defects (such as bearing crack and bearing abrasion damage) may lead to catastrophic accidents of the whole mechanical structure. Therefore, the timely and accurate classification of rolling bearing fault types has attracted many scholars [1–3].

The development of data-driven mechanical fault diagnosis benefits from the rapid development of sensing technology, computing systems and information storage technology in recent years. These technologies provide technical guarantees for data acquisition, transmission, and storage in manufacturing systems [4, 5]. However, the abnormal response of rolling bearing caused by fault is usually irrelevant. Due to the complex nonlinear behaviors e.g., friction contact between components, radial clearance of bearing and small vibration, the original time-domain signal show not only transient phenomenon but also nonlinear dynamic effects. Noise and various uncertainties further

H. Zhang et al. (Eds.): NCAA 2022, CCIS 1637, pp. 419–433, 2022.
https://doi.org/10.1007/978-981-19-6142-7_32

exacerbate this situation [6]. Traditional fault diagnosis methods consist of three main phases: Original signal acquisition, feature extraction and selection, fault classification and fault prediction. For example, Yuan et al. proposed high-fidelity noise-reconstructed empirical mode decomposition for mechanical multiple and weak fault extractions [7]. Qiao et al. used empirical mode decomposition, fuzzy feature extraction, and support vector machines to diagnose and verify the faults of steam turbine generator sets under three different working conditions [8]. Chen combined wavelet packet feature extraction technology and machine learning technology, an online monitoring model of logistic regression is proposed. The effectiveness of the scheme is verified by analyzing the tool wear vibration signal [9]. However, the traditional fault diagnosis methods listed above have their limitations. First, traditional methods rely on manual experience selection, and the quality of extracted features directly affects the performance of the final classification algorithm. Therefore, the stability of the diagnosis effect is not reliable. Second, In the current era of big data, a large amount of data needs to be processed in time and quickly. It takes time and effort to construct features manually. Even if the extracted features are used to train the model, the efficiency of the final model prediction or classification still needs to be improved.

Meanwhile, in order to meet the challenges of the big data era, people have done a lot of research on intelligent fault diagnosis methods in recent years. Deep learning models automatically find nonlinear features and realize classification by superimposing multiple network layers, which has achieved promising results in many tasks of artificial intelligence. Along with the development of the internet industry, lots of sensors are used in the operation detection system of mechanical equipment. The large explosion of data has made the traditional fault diagnosis methods difficult to meet the needs of the market, which also aroused people's interest in exploring the application of deep learning methods in fault diagnosis. For example: stacked Auto-Encoders (SAEs) [10], Deep Belief Networks (DBNs) [11], Recurrent Neural Networks (RNNS) [12], Generative Adversarial Networks (GAN) [13]. Due to deep learning's powerful function of nonlinear feature mapping and advantage of end-to-end learning advantages, the results of bearing diagnosis based on deep learning method have been significantly improved.

Convolutional neural network (CNN) is one of the representative deep learning algorithms, it contains convolution layer, pooling layer, and complete connection layer. With the help of these, CNN can directly train the original time-domain vibration signal and identify the fault features hidden in the signal, so as to diagnose and classify different types of faults, this effectively avoids subjective feature selection and human intervention. As such, fault diagnosis technology driven by original data set has become attractive. Weimer et al. adopt deep convolutional neural network, which overcomes the difficulty of redefining manual fault characteristics for each new situation in the production process and improves the automation and accuracy of monitoring [14]. Ince and Abdeljaber et al. use 1DCNN to detect motor faults and this method has higher accuracy than the model-based method [15, 16]. However, on one hand, the accuracy of CNN is extremely dependent on large-scale datasets, for the research of rolling bearing diagnosis, the fault data set is usually limited, and the cost of obtaining a large amount of training data is very high. On the other hand, the performance of the diagnosis method will decline when the training data set contains multiple fault features. The signal is caused by the coupled composite fault characteristics in the dynamic signals of complex systems. Besides, when the data source distribution in the test set deviates from the training set data source in the target domain, the diagnostic performance of the network will be observed to decline. The signal differences are caused by different monitoring environments in which vibration signals are collected, e.g., different working loads, sensor positions, and rotation speeds. CNN network model is difficult to train the network parameters of multiple characteristic signals, but the actual working conditions are always changing, and the same fault type does not exist in a single form.

Therefore, this paper by introducing the combination of multiwavelets transform and CNN, the ability of multiwavelets containing multiple different frequency domain basis functions to extract multiple faults at the same time, multiwavelets also fully inherit the properties of single wavelets such as the property of orthogonality, symmetry, compact support [17]. Besides, the features extracted from the first layer of the convolutional neural network will affect the overall diagnostic performance of the network, the quality of feature extraction directly affects the accuracy of whole network fault identification and multiwavelets transformation is a natural convolution process based on multiple wavelet basis functions. Therefore, replacing the first convolution layer of convolution neural network with multiwavelets transform can not only give full play to the excellent attributes of multiwavelets transform but also have better compatibility.

Besides, the multiwavelets basic functions are similar to fault features could extract fault features in dynamic signals, this paper constructs customized multiwavelets layer parameters according to different input signals, customized multiwavelets construct the most matching multiwavelets basis function according to different signal characteristics. The improved CNN by multiwavelets reduces the network training parameters, reduces the problem of big data driving, and it improves the problem of poor fitting effect of classical CNN in the case of multiple fault classification.

The rest of the paper is organized as follows. Section 2 mainly expounds on the basic theory of multiwavelets and the basic structure of convolution neural networks. Section 3 describes the basic structure of the improved convolution neural network by replacing the multiwavelets. The simulation results are presented and compared with the existing scheme in Sect. 4. Finally, we conclude this paper.

2 Basic Theory of Method

2.1 Multiwavelets

Multiwavelets Multiscales Analysis
Multiwavelets refer to wavelets generated by two or more functions as scaling functions. The basic theory is to expand the multi-resolution analysis space generated by a single wavelet with multiscales functions [18].

The vector valued function $F(x) = (f_1(x), f_2(x), \cdots f_r(x))^T$, if there is $f_j(x) \in L^2(R), j = 1, 2, \cdots r$, it is recorded as $F(x) \in L^2(R)^r(\frac{\pi}{2} - \theta)$, if $\Phi = (\phi_1, \cdots \phi_r)^T \in L^2(R)^2$ satisfy the two-scale relationship:

$$\Phi(t) = \sum_{k=0}^{N} H_k \Phi(2t - k) \tag{1}$$

where $\{H_K\}$, $K = 0, 1, \cdots, M$ is r × r two-scales matrix sequence, which is r-order scale function. The r multiresolution analysis generated by $\Phi(x)$ is defined as:

$$V_j = clos_{L^2(R)} \left\{ 2^{j/2} \varphi_i \left(2^j x - k \right) : 1 \leq i \leq r, k \in Z \right\} \tag{2}$$

V_j is subspace with a resolution of 2^j.

If W_j is the complement subspace of V_j space in V_{j+1}, the vector function $\Psi(x) = (\psi_1, \psi_2, \cdots, \psi_r)^T \in L^2(R)^r$, the expansion and translation components construct a Rise basis of W_j subspace

$$W_j = clos_{L^2(R)} \left\{ 2^{j/2} \psi_i \left(2^j x - k \right) : 1 \leq i \leq r, k \in Z \right\} \tag{3}$$

There is a matrix sequence $\{G_k\}_{k \in Z}$, make $\Psi(x) = (\psi_1, \psi_2, \cdots, \psi_r)^T$ satisfy the following two-scale relationship:

$$\Psi(x) = \sum_{k=0}^{N} G_k \Phi(2x - k) \tag{4}$$

where $\{G_k\}$, $k = 0, 1, \cdots, N$ is $r \times r$ two-scales matrix sequence, which is r-order wavelet function. Studies on multiwavelets with multiplicity $r > 2$ are rare. Hence, $r = 2$ is studied in this paper.

By the dilations of Eq. (1) and Eq. (4), the following recursive relationship between the coefficients $(c_{1,j,k}, c_{2,j,k})^T$ can be obtained.

$$\begin{pmatrix} c_{1,j-1,k} \\ c_{2,j-1,k} \end{pmatrix} = \sqrt{2} \sum_{n=0}^{K} H_n \begin{pmatrix} c_{1,j,2k+n} \\ c_{2,j,2k+n} \end{pmatrix}, \quad j, k \in Z \tag{5}$$

$$\begin{pmatrix} d_{1,j-1,k} \\ d_{2,j-1,k} \end{pmatrix} = \sqrt{2} \sum_{n=0}^{K} G_n \begin{pmatrix} c_{1,j,2k+n} \\ c_{2,j,2k+n} \end{pmatrix}, \quad j, k \in Z \tag{6}$$

Hermite Spline Multiwavelets

Hermite spline multiwavelets is the orthogonal multiwavelets constructed by using third-order spline function, which have 2-order continuous differentiability. The multiscales support interval [0, 2], and they have 4-order approximation order. The support interval of the mulitwavelets function is [0, 3] and they have 2-order approximation order. The two scale relationship is shown in Eq. (7) and Eq. (8).

$$\Phi(x) = \begin{bmatrix} \varphi_1(x) \\ \varphi_2(x) \end{bmatrix} = H_0 \Phi(2x) + H_1 \Phi(2x - 1) + H_2 \Phi(2x - 2) \tag{7}$$

where $H_0 = \begin{bmatrix} \frac{1}{2} & \frac{3}{4} \\ -\frac{1}{8} & -\frac{1}{8} \end{bmatrix}$, $H_1 = \begin{bmatrix} 1 & 0 \\ \frac{1}{2} & \frac{1}{8} \end{bmatrix}$, $H_2 = \begin{bmatrix} \frac{1}{2} & -\frac{3}{4} \\ \frac{1}{8} & -\frac{1}{8} \end{bmatrix}$.

$$\Psi(x) = \begin{bmatrix} \psi_1 \\ \psi_2 \end{bmatrix} = G_0 \Psi(2x) + G_1 \Psi(2x - 1) + G_2 \Psi(2x - 2)$$

$$+ G_3 \Psi(2x - 3) + G_4 \Psi(2x - 4) \tag{8}$$

where

$$G_0 = \begin{bmatrix} \frac{67}{240} & \frac{7}{240} \\ -\frac{95}{972} & -\frac{1}{162} \end{bmatrix}, G_1 = \begin{bmatrix} -1 & \frac{187}{60} \\ \frac{89}{243} & \frac{91}{81} \end{bmatrix}, G_2 = \begin{bmatrix} \frac{173}{120} & 0 \\ 0 & \frac{26}{9} \end{bmatrix}, G_3 = \begin{bmatrix} -1 & \frac{187}{60} \\ -\frac{89}{243} & \frac{91}{81} \end{bmatrix},$$

$$G_4 = \begin{bmatrix} \frac{67}{240} & -\frac{7}{240} \\ \frac{95}{972} & -\frac{1}{162} \end{bmatrix}$$

2.2 CNN

Convolutional Layer

Convolution layer extracts data features by convoluting the convolution layer parameters with the input data. Usually, a convolution layer has multiple convolution kernels. Because the same convolution kernel shares parameters in the process of convolution, one convolution kernel learns a class of features, which is called mapping graph. To calculate the output y^i, first the inputs $x^1, x^2, ..., x^d$ are convoluted with convolutional kernels $W^{i,d}$. Then, add all the convolutional operation results, the sum of convolution results is added to scalar offset value b^i. The output of the convolutional layer Z^i can be obtained as.

$$Z^i = \mathbf{W}^i \otimes \mathbf{X} + b^i = \sum_1^d W^{i,d} \otimes x^d + b^i \tag{9}$$

where $\otimes$ represents the convolutional operation and $W^i \in \mathbb{R}^{m \times n \times d}$ is the convolution kernel. Based on the nonlinear activation function, the output feature map y^i can be represented as.

$$y^i = g\left(Z^i\right) \tag{10}$$

where $g(\cdot)$ represents the nonlinear activation function. In this research, convolution layer is adopt the rectified linear unit (ReLU) function as the activation function.

Polling Layer

The pooled layer performs subsampling by checking the input data and extracts features while reducing the dimension of the data. Pooling includes maximum pooling and average pooling, among which maximum pooling has the best effect, which can be described as the following.

$$p^{1(i,j)} = \max_{(j-1)w < t < jw}\{a^{1(i,t)}\}j = 1, 2, \cdots, q \tag{11}$$

where $p^{1(i,j)}$ denotes t-th neuron of the i-th feature map in layer 1, w represent width od convolutional kernel, j represent j-th pooling kernel.

Fully Connected and Output Layers

The signal features extracted from the upper network are input to the first full connection layer for one-dimensional sequence expansion. Each λ_c matrix by the finally output layer will be input into a softmax function $\varphi(\cdot)$, which is defined by

$$\varphi(\lambda_c) = \frac{e^{\lambda_c}}{\sum_{c=1}^{C} e^{\lambda_c}}, c = 1, \cdots C \tag{12}$$

The $\varphi = [\varphi(\nu_1), \cdots, \varphi(\nu_C)]$ represent a C-dimensional probability vector. Represents the probability distribution under C kinds of test conditions. The output value of the softmax function represents the probability distribution of the input signal in each tag.

3   The Presented Network Structure

Aiming at bearing fault, this paper improves the network framework based on CNN is illustrated in Fig. 1. Original time-domain signals from sensor are directly extracted features by multiwavelets layer of improved CNN, the feature signal extracted by multiwavelets is transmitted to deep network composed of one-dimensional convolutional and pooling layers. Since the characteristic signal after multiwavelets transform is a two-dimensional signal, the convolution kernel of the next convolution layer is also a two-dimensional matrix, and the characteristic signal after passing through the first convolution layer will be the one-dimensional characteristic signal, the dimension of convolution kernel in other convolution layers is one-dimensional convolution kernel. Finally, a fully connected (FC) layer and a multi category output layer composed of softmax function are used as the bottom architecture of the network.

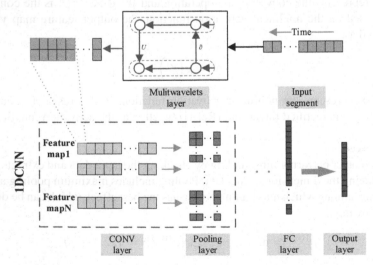

Fig. 1. The structure of the improved CNN by multiwavelets.

3.1 Multiwavelets Layer

The specific implementation steps of the multiwavelets layers are as follows.

Split sublayer: the signal f is divided into two new signal according to the odd and even bits of the data sequence, even sequence samples are p and odd sequence samples are q.

$$p(x) = f(2x) \tag{13}$$

$$q(x) = f(2x+1)x \in z \tag{14}$$

Predict sublayer: we use the optimizer matrix ∂ to convolute even samples q to predict odd samples p. The error between the predicted value and real value is Δ that defined as detail coefficients.

$$\Delta = p - \partial * q \tag{15}$$

where ∂ is the matrix vector of prediction operator, the symbol $*$ represent matrix vector convolution operation.

Update sublayer: the detail coefficients obtained by the predictor are transferred into the update sublayer, it convoluted with updater U composed of the parameter matrix to perform the operation and adding the result to p. The update sequence p^U represent the vector of approximation coefficients.

$$p^U = p + U * \Delta \tag{16}$$

where U represents the vector of two dimensionals matrix as updater.

The custom multiwavelet layer is based on the evolution of adaptive multiwavelet theory, and also integrates the biorthogonal perfect reconstruction multi filter bank to process the input signal. The transformation of parameter k control the matrix low-pass and highpass filters $\{H, G, \tilde{H}, \tilde{G}\}$ as well as predict sublayer operation and update sublayer operation $\{\partial, U\}$.

$$G(k) = k\left(I - \partial\left(k^2\right)/k\right) \tag{17}$$

$$H(k) = I + U\left(k^2\right)G(k) \tag{18}$$

$$\tilde{H}(k) = I + \partial\left(k^2\right)/k \tag{19}$$

$$\tilde{G}(k) = \left(I - H(k)kU\left(k^2\right)\right)/k \tag{20}$$

In order to ensure the linear phase of predictor filters, it require symmetry, the operator ∂ of predict sublayer is subjected to the symmetry condition.

$$\partial(0) = \begin{pmatrix} \frac{1}{2} & \frac{1}{4} \\ c & -\frac{1}{4} \end{pmatrix}, \quad \partial(-1) = \begin{pmatrix} \frac{1}{2} & -\frac{1}{4} \\ -c & -\frac{1}{4} \end{pmatrix} \tag{21}$$

Next, operator U of update sublayer closely related to ∂ can be calculated as

$$U = \{U(0), U(1)\}, \quad U(0) = \partial(-1)/2 = \begin{pmatrix} \frac{1}{4} & -\frac{1}{8} \\ -\frac{c}{2} & -\frac{1}{8} \end{pmatrix}, \quad U(1) = \partial(0)/2 = \begin{pmatrix} \frac{1}{4} & \frac{1}{8} \\ \frac{c}{2} & -\frac{1}{8} \end{pmatrix} \quad (22)$$

Equation (22) show that the parameter c affects not only the matrix vector of predictor ∂ but also the matrix vector updater U. In other words, free parameter c will change the multiscales and multiwavelets functions due to Eq. (21).

3.2 Multiwavelets Layer Parameters

The customized multiwavelets layer relies on the basic function to change its kernel function with different input signals, to match different fault features more accurately. Table 1 shows the important parameters of improved CNN by multiwavelets, here each data sample of the input network contains 1024 sampling points. The shallow convolution layer is replaced by a multiwavelets layer with a kernel size of 10, and the number of output channels is set to 1, which greatly reduces the shallow training parameters of the network and improves the overall training speed and convergence speed of the network. The fault features are extracted by multiwavelets layer and transferred into convolution layers, the fault feature information is further mined. Two convolutional layers are set, Conv1D represent one vector between convolution kernel and the input parameters. The next adaptive maximum pooling layer adopt to the kernel number 16. Then pass feature signal to the full connection layer. The final dimension of output data is m which is consistent with the type of input label.

Figure 2 shows the date flow framework structure of improved CNN by multi-wavelets, described as follows.

Step1: In the data preprocessing stage, the collected time-domain signals containing various fault types are divided into training set and test set according to the ratio of 6:4.
Step2: In the training stage, two different fault timing signals are packaged into the same two-dimensional matrix, the improved CNN is initialized.
Step3: According to different fault signal types, set different parameter c constraint matrices ∂ and U, therefore, the kernel parameters of the multiwavelet layer are determined.
Step4: The data is processed by the split sublayer, predict sublayer, and update sublayer to extract fault features.
Step5: The fault features extracted from the shallow layer are introduced into two one-dimensional convolutional layers and adaptive maximum pooling layer, the fault features characteristics of the input signals will be fully mined.
Step6: The results are calculated by the softmax function get the probability of each feature label. At the same time, the gradient of each layer in the network model is calculated by using the back propagation algorithm to continuously modify the network parameters and improve the training accuracy.
Step7: The loss rate is calculated according to the cross entropy loss function to judge whether the network accuracy has completed the training.
Step8: The well-trained network model is applied to the testing data.
Step9: Identify the fault type label for the input signal.

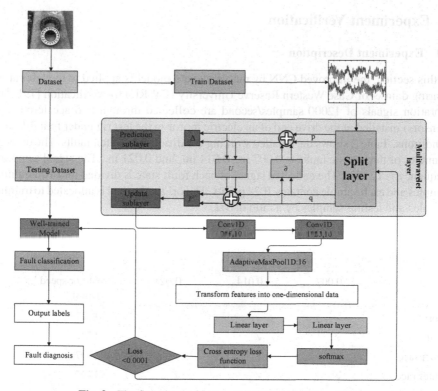

Fig. 2. The flowchart of improved CNN by multiwavelets.

Table 1. The parameters of improve CNN model.

	Name	Kernel size	Channel	Data size
	Data size			1×1024
	Preprocessing			2×512
L1	Multiwavelets layer	2×5	1	$2 \times 1 \times 256$
L2	Conv1D	2×5	10	$1 \times 10 \times 52$
L3	Conv1D	1×25	16	$1 \times 16 \times 227$
L4	AdaptiveMaxPool1D		16	$1 \times 16 \times 16$
L5	Linear layer		1	$1 \times 1 \times 100$
L6	Softmax layer		1	m

4 Experiment Verification

4.1 Experiment Description

In this section, the improved CNN by multiwavelets model is applied to the laboratory bearing dataset of Case Western Reserve University (CWRU) for verification [19]. The vibration signals of 12000 samples/second are collected through two accelerometers (sensors) installed at the drive end of an electric motor in the test rig under four different conditions. Table 2 shows the detailed working condition data of ten faults. The damage diameter of three basic faults is 0.007 in., 0.014 in., and 0.021 in.. The signal sampling frequency is 12 kHz. The collected signal of each fault state is divided into 100 training samples and each sample contains 1024 points. The original signal is allocated to training samples and testing samples by a ratio of 3:2.

Table 2. Rolling bearing operation state

	0.007	0.014	0.021	Motor speed (rpm)
Normal	–	–	–	1797
Ball	✓	✓	✓	1797
Inner race	✓	✓	✓	1797
Outer race	✓	✓	✓	1797

4.2 Parameter Optimization of Multiwavelets Layer

In the process of training parameters, the network will initialize parameters, but the value of random initialization parameters is uncertain, so it is difficult to fully show the feature extraction ability of the multiwavelets layer, Fig. 3 shows the loss function decreases during the iterative training of network parameters, which compares the network training speed and fitting effect of parameters under different values, it can be observed from the figure that when the $c = -2$, the network training speed and the stability of parameter fitting reach the best, Table 3 shows fault classification average accuracy of multiwavelets layer under different parameter values, and when multiwavelets parameter $c = -2$, the network model full play to performance. Therefore, in the case of parameter $c = -2$, the network model can achieve the best speed, stability, and diagnostic accuracy of multi-fault classification.

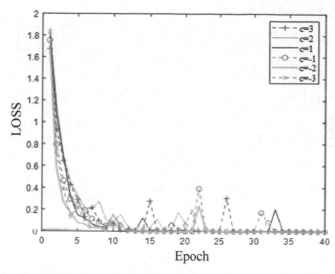

Fig. 3. Changes of cross-entropy loss in different parameters.

Table 3. The accuracy of the dataset.

Dataset	$C = -3$	$C = -2$	$C = -1$	$C = 1$	$C = 2$	$C = 3$
Accuracy (%)	99.67	99.83	99.63	99.87	99.67	99.33

4.3 The Performance of Bearing Fault Diagnosis

Multilayer Perceptron (MLP), and 1DCNN are taken for comparison with the presented solution, improve CNN model $c = -2$ as the parameter of multiwavelet. MLP consists of four layers: input layer, output layer, and two hidden layers, 1DCNN has the same convolution layer parameters as the improve CNN, in which the convolution kernel size of the shallow network is 10. The two groups of experimental schemes are fully in line with the principle of controlling a single variable.

Figure 4 is shown the CNN and improved CNN by multiwavelets decreased rapidly in the first five iterations and converged in the seventh iteration, the MLP is difficult fully converge in the multi-fault classification experiment, and the Cross entropy loss function fluctuates greatly, this phenomenon proved that MLP model has obvious shortcoming in nonlinear fitting problem. CNN network shows a good training effect in multi-fault nonlinear fitting fault. However, CNN fluctuated greatly in the 14th, 37th and 38th iterations show that the training parameters are unstable in the process of multiple training. The improved CNN by multiwavelets model not only inherits the excellent nonlinear fitting ability of the CNN model in the face of multi-fault problems but also has the spatial-temporal features extracted by the multiwavelets, which makes the model maintain better network stability and reliability in multiple iterations compared with the classical CNN model.

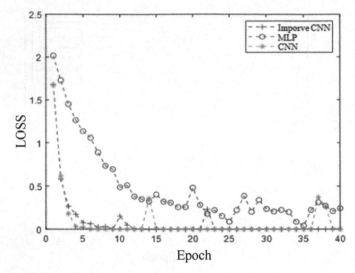

Fig. 4. Variation of loss value in different experimental schemes.

Table 4. Classification results in case.

Method	Max accuracy	Min accuracy	Mean accuracy
IMPROVE CNN	100.0%	99.3%	99.8%
MLP	59.0%	53.1%	54.0%
1DCNN	99.8%	97.4%	97.5%

The accuracy of these methods on common Dataset is listed in Table 4. After training these three schemes five times, this paper brings the training parameters of each time into the test set for verification, and the average accuracy shows that improved CNN has a stronger ability to extract fault features. Fig. 5 visualizes the verification results of 400 sets of test sets by three methods, Figure (a) shows the confusion matrix thermodynamic diagram of the improved CNN by multiwavelets. Figure(b) and Figure(c) show the confusion matrix thermodynamic diagram of CNN and MLP respectively. These figures indicate that the improved CNN has higher accuracy in fault classification.

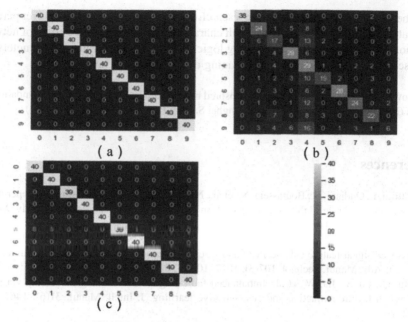

Fig. 5. Sample classification in validation set.

5 Conclusion

CNN method relies on training a large number of the same fault data and updating network parameters through backpropagation to establish the recognition ability of the same type of fault. However, under the actual working conditions, the types of bearing faults are constantly changing, and it is difficult to collect sufficient data for the same type of faults. When small datasets are employed, the first layer of CNN is difficult to effectively extract deep feature parameters and it influences the performances of the entire network. Therefore, an improved CNN by multiwavelets is proposed in this paper. The first layer of the model is the multiwavelets transform in signal processing, multiwavelets layer completely inherits the advantages of the fast extraction of features signals from multiple wavelets base of multiwavelets. Original signals in the time domain are directly input into the improved CNN, multiwavelets layer is used as a multi-channel filter to extract multi fault features at the same time, subsequently, the features extracted by the multiwavelets layer are fused as inputs to the next layer. By enhancing the ability of network shallow feature extraction, accurately extract fault features and reduce network parameters. By incorporating multiwavelets feature extraction, this proposed CNN method can accurately diagnose faults with smaller data set samples in practical bearing diagnosis, and by customizing multiwavelets layer parameters, the influence of shallow layer parameters on the overall network performance is discussed. The validity of this method is fully verified by publicly experimental datasets and comparsion with the traditional methods and classical methods.

Based on the CNN shallow layer replacement multiwavelets model, this paper mainly finds the multiwavelets layer parameters most suitable for fault characteristics through parameter optimization. In the future research direction, the multiwavelets layer parameters can be adaptively matched with fault characteristics to improve the effect of network diagnosis efficiency, and the underlying logic of CNN network training parameters can be discussed based on the physical meaning of multiwavelets layer.

Acknowledgements. This research is sponsored by the National Natural Science Foundations of China (No. 51975377 and 52005335), Shanghai Sailing Program (No. 21YF1430600).

References

1. Attoui, I., Oudjani, B., Boutasseta, N., et al.: Novel predictive features using a wrapper model for rolling bearing fault diagnosis based on vibration signal analysis. Int. J. Adv. Manuf. Technol. **106**(7), 3409–3435 (2020)
2. Wang, H., Chen, J., Zhou, Y., et al.: Early fault diagnosis of rolling bearing based on noise-assisted signal feature enhancement and stochastic resonance for intelligent manufacturing. Int. J. Adv. Manuf. Technol. **107**(3), 1017–1023 (2020)
3. Xu, Q., Lu, S., Jia, W., et al.: Imbalanced fault diagnosis of rotating machinery via multi-domain feature extraction and cost-sensitive learning. J. Intell. Manuf. **31**(6), 1467–1481 (2020)
4. Lund, D., MacGillivray, C., Turner, V., et al.: Worldwide and regional internet of things (IoT) 2014–2020 forecast: a virtuous circle of proven value and demand. Int. Data Corp. (IDC), Tech. Rep. **1**(9) (2014)
5. Chen, Z., Li, W.: Multisensor feature fusion for bearing fault diagnosis using sparse autoencoder and deep belief network. IEEE Trans. Instrum. Meas. **66**(7), 1693–1702 (2017)
6. Sousa, R., Antunes, J., Filipe, C., et al.: Robust cepstral-based features for anomaly detection in ball bearings. Int. J. Adv. Manuf. Technol. **103**(5–8), 2377–2390 (2019)
7. Yuan, J., Xu, C., Zhao, Q., Jiang, H., Weng, Y.: High-fidelity noise- reconstructed empirical mode decomposition for mechanical multiple and weak fault extractions. ISA Trans. (2022). https://doi.org/10.1016/j.isatra.2022.02.017
8. Hu, Q., He, Z.J., Zi, Y., et al.: Intelligent fault diagnosis in power plant using empirical mode decomposition, fuzzy feature extraction, and support vector machines. Key Eng. Mater. **293**, 373–382 (2005)
9. Chen, B., Chen, X., Li, B., et al.: Reliability estimation for cutting tools based on a logistic regression model using vibration signals. Mech. Syst. Signal Process. **25**(7), 2526–2537 (2011)
10. Haidong, S., Hongkai, J., Xingqiu, L., et al.: Intelligent fault diagnosis of rolling bearing using deep wavelet auto-encoder with extreme learning machine. Knowl. Based Syst. **140**, 1–14 (2018)
11. Shao, H., Jiang, H., Zhang, H., et al.: Electric locomotive bearing fault diagnosis using a novel convolutional deep belief network. IEEE Trans. Ind. Electron. **65**(3), 2727–2736 (2017)
12. Pan, H., He, X., Tang, S., et al.: An improved bearing fault diagnosis method using one-dimensional CNN and LSTM. Strojniski Vestnik/J. Mech. Eng., 64 (2018)
13. Liu, H., Zhou, J., Xu, Y., et al.: Unsupervised fault diagnosis of rolling bearings using a deep neural network based on generative adversarial networks. Neurocomputing **315**, 412–424 (2018)
14. Weimer, D., Scholz-Reiter, B., Shpitalni, M.: Design of deep convolutional neural network architectures for automated feature extraction in industrial inspection. CIRP Ann. **65**(1), 417–420 (2016)

15. Ince, T., Kiranyaz, S., Eren, L., et al.: Real-time motor fault detection by 1-D convolutional neural networks. IEEE Trans. Ind. Electron. **63**(11), 7067–7075 (2016)
16. Abdeljaber, O., Avci, O., Kiranyaz, S., et al.: Real-time vibration-based structural damage detection using one-dimensional convolutional neural networks. J. Sound Vib. **388**, 154–170 (2017)
17. Sun, H., He, Z., Zi, Y., et al.: Multiwavelet transform and its applications in mechanical fault diagnosis–a review. Mech. Syst. Signal Process. **43**(1–2), 1–24 (2014)
18. Keinert, F.: Wavelets and Multiwavelets. CRC Press, Boca Raton (2003)
19. Bearing Data Center, Case Western Reserve University, Cleve land, OH, USA (2004). http://csegroups.case.edu/bearingdatacenter/home

TextSMatch: Safe Semi-supervised Text Classification with Domain Adaption

Yibin Xu[1], Ge Lin[2], Nanli Zeng[3], Yingying Qu[4], and Kun Zeng[1(✉)]

[1] School of Computer Science and Engineering, Sun Yat-sen University,
Guangzhou, China
`xuyb37@mail2.sysu.edu.cn, zengkun2@mail.sysu.edu.cn`
[2] National Engineering Research Center of Digital Life, Sun Yat-sen University,
Guangzhou, China
`linge3@mail.sysu.edu.cn`
[3] Cloud Communication Division, China Mobile Internet Co., Ltd.,
Guangzhou, China
[4] School of Business, Guangdong University of Foreign Studies, Guangzhou, China
`jessie.qu@gdufs.edu.cn`

Abstract. The performance of many efficient Deep semi-supervised learning(SSL) is severely degraded when the distribution of unlabeled and labeled data does not match. Some recent approaches have chosen to weaken or even remove out-of-distribution (OOD) data, which can lose the potential value of OOD data. We propose TextSMatch to solve this issue, a simple, safe and effective SSL method for text classification, which recycles the OOD data near the labeled domain to make full use of the information in OOD data. Specifically, adversarial domain adaptation is applied to the OOD data to project it into the space of ID and labeled data, and its recoverability is assessed using the use of migration probabilities. Moreover, TextSMatch unifies the mainstream methods. In addition to consistency regularization training of class probabilities for unlabeled data and its augmented data, we also normalized the structure of embedding with contrastive learning based on pseudo-labeled. TextSMatch performs significantly better than other baseline methods on AG News and Yelp datasets in scenarios such as class mismatch and different amounts of labeled data.

Keywords: Semi-supervised learning · Out-of-distribution · Adversarial domain adaptation · Text classification

1 Introduction

Deep semi-supervised learning (SSL), which uses large amounts of inexpensive unlabeled data to improve deep neural network performance and thus reduce the need for labeled data, is a long-standing problem in natural language processing and computer vision [1–5]. Recent state-of-the-art approaches can be summarized in three main categories: (1) Generating low-entropy label predictions

H. Zhang et al. (Eds.): NCAA 2022, CCIS 1637, pp. 434–448, 2022.
https://doi.org/10.1007/978-981-19-6142-7_33

using sharpened prediction technique [6–8]; (2) unsupervised or self-supervised pre-training followed by fine-tuning [9–11]; (3) using consistency regularization for unlabeled data and its enhanced data for learning [5,12–14].

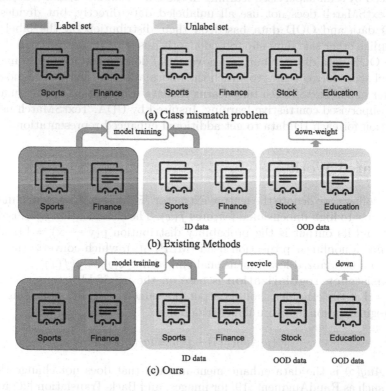

Fig. 1. Motivation for the study. (a) Demonstration of the class mismatch problem. (b) Existing solution strategies to identify the OOD data and then discard or down the weight. (c) Our approach, having separated ID data from OOD data, recycles valuable OOD data and discards non-recyclable OOD data.

However, the excellent performance of the above approaches is based on a assumption that the distribution of labeled data (C_l) and unlabeled data (C_u) is the same. Unfortunately, it is difficult to maintain this assumption in many practical scenarios. A common situation is that the label field of labeled data is usually smaller than that of unlabeled data, such problem for realistic SSL is called class mismatch [15]. For example, as shown in Fig. 1(a), in text classification, unlabeled text crawled from the Internet based on keywords generally contains unknown categories in the tagged data. The samples in the unlabeled set that belong to category C_l are referred to as in-distribution (ID) data, while the remaining samples that belong to classes C_u only are referred to as out-of-distributed as (OOD) data. Faced with this open set of training data, the model often classifies text of unknown categories into the known label space.

At this point, the performance of the deep semi-supervised model is usually not improved, or even worse than the supervised model [16].

To this end, TextSMatch is proposed to address the above-mentioned problems faced by semi-supervised learning in the NLP domain. Unlike existing deep SSL, TextSMatch does not use all unlabeled data directly, but divides them into ID data and OOD data based on their distribution on the label space. Also, unlike other safe semi-supervised algorithms [16,17], TextSMatch does not remove OOD data directly but uses adversarial domain adaptation to map it to the label space for recycling to enrich the information learned by the model. On the other hand, TextSMatch fully learns ID data using consistency regularity as well as supervised contrastive learning. Inspired by UDA, TextSMatch use back translation for each ID data to get additional enhanced representation.

2   Related Work

In what follows, we denote the feature extractor (BERT) as $f(\circ)$, which maps the input text x to high-dimensional features $f(x)$. The classification head is defined as $h(\circ)$ and its output is the probability distribution $p(y - x) = h(f(x))$. We also define a nonlinear projection head (MLP) $g(\circ)$, which converts the feature $f(x)$ into a normalized low-dimensional embedding $z(x) = g(f(x))$.

Consistency regularization is base on the manifold assumption [18]. It assumes that the classifier should output the same classificationollowing consistency regularization loss to unlabeled samples:

$$||p(y|Aug_1(x)) - p(y|Aug_2(x))||_2^2, \tag{1}$$

where $Aug(\circ)$ is the data enhancement method that does not change the text labels, such as RandAugment [19] for images, and Back-Translation [20] for text. Temporal Ensembling [21] reduces the overhead of forward computing the sample twice in each iteration by accumulating the historical predictions using exponential moving average (EMA). Mean Teacher [13] weights the model weights instead of weighting them as Temporal Ensembling weights the model predictions as well. UDA [5], ReMixMatch [8], and FixMatch [22] use cross-entropy loss instead of squared error and apply stronger augmentation, and UDA also performs confidence-based masking, below which confidence thresholds are not involved in the loss calculation.

Entropy minimization is a method that benefits from unlabeled data in a maximum posterior estimation framework. By minimizing the entropy of unlabeled data, the overlap of probability distributions between different categories can be reduced to avoid splitting dense sample data points to either side of the decision boundary. It can be achieved either explicitly by minimizing the entropy probability of unlabeled samples [4,21,23] as in Eq. (2), or implicitly by constructing low entropy pseudo-labels on unlabeled samples and using them as training targets for cross-entropy loss [6,22]. Some methods [5,7,8] also use sharpening functions to post-process "soft" pseudo-labels to reduce entropy, while FixMatch generates "hard" pseudo-labels for samples with maximum category probabilities below a threshold.

$$H(y|x) = -\sum_{i=1}^{C} p(y|x)logp(y|x) \tag{2}$$

Self-supervised learning uses unlabeled data to enable the model to learn intrinsic knowledge from the data. It has achieved great success in NLP in recent years, most typically with pre-training and fine-tuning frameworks and reaching SOTA in various NLP tasks [24]. Coupled with some novel techniques such as skewed delta learning rate and gradual thawing, with a small amount of labeled data, the pre-trained model can also have very good performance. It is worth mentioning that, pre-training approaches usually have different goals, such as language modeling [25] and masked language modeling [26]. Recently, contrastive learning methods [9,27] have received increasing attention as they have been shown to learn more robust feature representations for downstream tasks. The basic idea is to align similar samples in the feature space and push away dissimilar ones, so that the learned features are discriminative

$$-log\frac{exp(z(Aug(x_i)) * \frac{z(Aug(x_i))}{t})}{\sum_{j=1}^{N} exp(z(Aug(x_i)) * \frac{z(Aug(x_i))}{t})}, \tag{3}$$

where x_j includes x_i and $N-1$ other texts (i.e., negative samples).

Class mismatch problem is a common problem in semi-supervised learning, but to the best of our knowledge, the problem has not been well solved. The current work focuses on detecting OOD data and de-discarding it. OOD data detection is the problem of detecting observations belonging to different training data distributions [28]. An intuitive approach is to identify OOD samples based on confidence scores estimated as the maximum softmax probability. However, softmax-based confidence estimates for individual DNNs can be problematic, as DNNs are often overconfident [29]. To address this issue, one study focused on confidence calibration [30] to form smoother prediction distributions that incorporate uncertainty. Rooted in a similar idea, Chen et al. proposed USAD [16] to derive soft targets that can be used as OOD detection metrics and introduced a simple OOD filter to automatically discard OOD samples without require a large computational cost to train the OOD filter.

3 Method

3.1 Overview

Given a batch of labeled samples of size B, $X = \{(x_b, y_b)\}_{b=1}^{B}$, where $y_b \in Y \in \{1, 2, ..., c\}$, Y is the label space and c is the number of known categories, and a batch of unlabeled text $U = \{u_b\}_{b=1}^{\mu B}$, where μ represents the relative size of the number of samples in X and U. In particular, U is composed of ID data set U_{id}

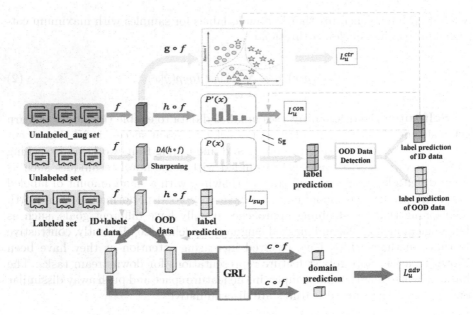

Fig. 2. Framework of the propose TextSMatch.

whose tags all belong to Y and OOD data set U_{ood} whose tags do not belong to Y, $U \in U_{id} \cup U_{ood}$.

First, TextSMatch trains the labeled text X, learns the feature distribution of each category, and obtains the model M_0. For unlabeled samples, we use M_0 for prediction. If the maximum category probability does not exceed the threshold τ, the sample belongs to D_{ood}. Based on the detection results of OOD data, we can not only calculate the consistency loss and comparison loss of ID data, but also obtain the characteristic representation of ID data and OOD data respectively. Then, the labeled data and the unlabeled ID data are regarded as one domain, and the OOD data is regarded as another domain, and Domain-adversarial learning is used to obtain the transferable score of the OOD data and recycle part of the data. On the whole, TextSMatch jointly optimizes four objective functions: (1) the supervised classification loss of labeled data L_{sup}, (2) the consistency loss of unlabeled ID data L_{con}^u, (3) Supervised comparison loss L_{ctr}^u for unlabeled ID data based on pseudo-labels, and (4) Adversarial loss L_u^{adv} for transferable OOD data. Specifically, L_{sup} is defined as the cross entropy between the model prediction on labeled sample and its ground-truth label.

$$L_{sup} = \frac{1}{Z} \sum_{x_b, y_b \in X} I(p(y|x_b) < \eta_t) * H(y_b, p(y|x_b)) \tag{4}$$

$$Z = \sum_{x_b, y_b \in X} I(p(y|x_b) < \eta_t) \tag{5}$$

In Eq. (4), the entropy of two distributions y and p can be expressed as $H(y, p)$, and $I(p(y|x_b) < \eta_t)$ represents TSA (Training Signal Annealing) technology [5]. Due to the large gap between unlabeled data and labeled data in quantity, the model can easily overfit the labeled data.

The consistency loss of ID data L_u^{con} is defined as the KL divergence loss of the original unlabeled data and the enhanced data output.

$$L_u^{con} = \frac{1}{2\mu B} \sum_{i=1}^{2} \sum_{u_b \in U} KL(p_{\tilde{\theta}}(y|u_b) \| p_{\theta}(y|Aug_i(u_b))) \tag{6}$$

According to UDA and ReFixMatch, for ID data, we also use confidence-based masking, sharpening prediction, and distribution alignment techniques. Specifically, for the predicted label distribution of unlabeled data, we first align it according to the label distribution of labeled data. Then, for samples whose classifiers are not so certain, we do not calculate their loss.

The contrastive loss L_u^{ctr} is realized through pseudo-labels and contrast learning, and we implicitly minimize entropy by optimizing the contrastive loss:

$$L_u^{ctr} = \frac{-1}{2N - 1} \sum_{i=1}^{N} \sum_{j=1}^{N} \Gamma_{[i \neq j]} * \Gamma_{[y_i^{id} = y_j^{id}]} *$$
$$log \frac{s(z_i, z_j)}{\sum_{k=1}^{N} \Gamma_{[i \neq k]} * s(z_i, z_k) + \epsilon}, \tag{7}$$

where $\Gamma_{[i \neq j]}$ and $\Gamma_{[y_i^{id} = y_j^{id}]}$ are indicator functions that take 1 when and only when $i \neq j$ and $y_i^{id} = y_j^{id}$, and 1 otherwise. $s(.)$ is used to represent the distance function of two embedding vectors, and cos similarity is used in the experiments. And $\epsilon \cong 0$ is a constant used to avoid dividing by 0.

The adversarial loss consists of the discriminative loss of two domains. The parameter update of the domain classifier adopts the Greedent reversal Layer (GRL), which can finally adapt the OOD data field to the ID data, that is, achieve the effect of recycling. Therefore, the overall training goal of the model is Eq. (8).

$$L = L_{sup} + \lambda_{con} L_u^{con} + \lambda_{ctr} L_u^{ctr} + \lambda_{adv} L_u^{adv} \tag{8}$$

Note that, the relative weights of the above four loss functions are controlled by the non-negative coefficients $lambda_{con}$, $lambda_{ctr}$ and $lambda_{adv}$. The complete model framework is shown in Fig. 2. From the above description, we can find that OOD Data Detection and Adversarial Domain Adaption are two relatively important modules for our method. Next we will introduce in detail in the following chapters.

3.2 OOD Data Detection

The objective of this module is to separate ID data from OOD data for unlabeled text. Many existing works treat OOD data as abnormal data and perform abnormal detection on unlabeled data. On the one hand, it can be achieved by delimiting the threshold by the softmax score of unlabeled data during network training [31,32]. On the other hand, it can also be judged by the reconstruction loss of unlabeled text [33], because the distribution of labeled categories is learned during network training. The OOD data distribution is not trained for the neural network, so the reconstruction error of the OOD data will be relatively large. Because the first method not only learns the text data, but also learns the category label information, it will use the data more fully than the other methods. In addition, the first type of method can also obtain pseudo-labels of ID data while performing OOD data detection, so as to prepare data for subsequent modules. So, we follow Liang et al. to get label prediction $P(x) = [p_1(x), p_2(x), ..., p_c(x)]^T$, where $p_i(x)$ is calculated as follows:

$$p_i(x) = \frac{exp(q_i/\tau)}{\sum_{j=1}^{c} exp(q_j/\tau)}, \tag{9}$$

where $q = h \circ f$ can be understood as the probability that the input belongs to a certain category, $\tau \in R^+$ is the temperature coefficient used to prevent the distribution from being too even [31], and *softmax score* refers to the maximum value in $P(x)$, the formula can be expressed as Eq. (10).

$$s(x) = \max_{i \in \{1,2,...,c\}} p_i(x) \tag{10}$$

However, in our TextsMatch method, as the neural network is trained, the softmax score of unlabeled data will oscillate, which will make them inconsistent in subsequent iterations. To alleviate this problem, we used exponential moving average (EMA) before calculating the softmax score [34]. By assigning more weight to the most recent prediction, and at the same time exponentially reducing the weight of early prediction, we can obtain more stable label prediction. Therefore, the improved label prediction of x is as follows:

$$\tilde{P}(x)^{(t)} = \eta \tilde{P}(x)^{(t-1)} + (1 - \eta)P(x)^{(t)}, \tag{11}$$

where $P(x)^{(t)}$ calculates the label prediction at the current moment by formula (11), and $\tilde{P}(x)^{(t-1)}$ and $\tilde{P}(x)^{(t)}$ represent the integrated label prediction at time $t - 1$ and t, respectively. $\eta \in [0, 1]$ is a momentum coefficient used to control the magnitude of the current impact of historical predictions. Inspired by [32], the softmax scores of different samples can be used to determine whether the unlabeled text is ID or OOD data, because the softmax scores of OOD data are obviously smaller than the softmax scores of ID data. After getting the stable $\tilde{s}(x)$, we use the threshold δ to distinguish OOD data or ID data in the unlabeled text D_u. The detection process of OOD data can be expressed as Eq. (12).

$$t(x; \delta) = \begin{cases} 0, if \ \tilde{s}(x) > \delta \\ 1, if \ \tilde{s}(x) \leq \delta \end{cases} \tag{12}$$

When $t(x; \delta) = 0$, it means that x belongs to OOD data, and they will be merged into D_{ood}, and the remaining unlabeled data is ID data. At the same time, we can obtain the corresponding pseudo-label $Y^{id} \in \{1, 2, ..., c\}$ of ID data.

$$Y^{id} = argmax \; (\tilde{s}(x)) \;\; if \; \tilde{s}(x) > \delta \tag{13}$$

3.3 Adversarial Domain Adaption

In addition to the learning of ID data above, in order to fully extract the information contained in U_{ood}, another major task of ours is to recycle the texts in U_{ood}. Specifically, the migration scores of U_{ood} data are first calculated to quantify the recyclable value of the texts in U_{ood}, and then these texts are migrated to the space of $X \cup U_{id}$ to train a semi-supervised classifier. Inspired by Yaroslav et al. [35] and the limited amount of X data, we treat U_{ood} and $X \cup U_{id}$ as two different domains and adopt an adversarial domain adaptation approach to solve this issue. Adversarial domain adaptation uses a domain classifier to classify whether the data points come from the source or target domain, to encourage domain confusion and to extract domain invariant features. The adversarial domain adaptation is divided into generative [36] and non-generative models [37] according to whether they belong to the generative model or not, and TextSMatch belongs to the second category.

In our case, the domain discriminator c minimizes the cross-entropy loss to classify the ID data from the OOD data and distinguish which domain the data comes from. And there is a gradient reversal layer (CRL) between the feature extractor f and the domain discriminator c, which will automatically invert the gradient direction during backpropagation, so the parameter θ_f of the feature extractor f is trained to maximize the cross-entropy loss to extract invariant features between different domains. As a result, "easily" migrated OOD data can potentially be recycled. Thus, the objective function of the above adversarial process can be expressed as Eq. (14).

$$\min_{\theta_f} \max_{\theta_c} = \frac{1}{|U_{ood}|} \sum_{x_i \in U_{ood}} w(x_i) \; log \; c(f(x_i))$$
$$+ \frac{1}{N+B} \sum_{x_i \in U_{id} \cup X} log(1 - c(f(x_i))) \tag{14}$$

$|U_{ood}|$ is the amount of data in the OOD data, $w(x_i)$ is the migratable score, which measures whether the text in U_{ood} is recyclable or not, and the calculation of this score will be described in detail later. By optimizing the above objective function, the encoder f maps the data to a specific feature space and enables the classifier h to determine the class of data from the source domain while the domain discriminator c is unable to distinguish which domain the data comes from. Also, we train the classifier h and the nonlinear projection head g by minimizing the contrast loss L_u^{ctr} with the supervised cross-entropy loss L_{sup}. Therefore, TextSMatch can extract useful knowledge from OOD data and use it for feature representation as well as text classification tasks on label space Y. In order to measure whether the data in OOD data has potential value for recycling, we propose two metrics, domain similarity score $w_s(x)$ and domain propensity score $w_t(x)$. First, the classifier c is trained to determine whether the unlabeled data belongs to the source or target domain, so we can get the probability of OOD sample x_i come from the domain $X \cup U_{id}$ from the output of c as:

$$c(f(x_i)) = p(x_i \in X \cup U_{id}|x_i), \tag{15}$$

where $p(.)$ represents the probability. Intuitively, if the output probability is small, we know that x_i is not similar to the known space $X \cup U_{id}$. Therefore, we should recover these samples correctly by giving them smaller weights. On the other hand, if the output probability is large, the equivalent x_i may be more related to the known space, so the similarity scores of these texts should be large so that both classifier c and classifier h will ignore them. Finally, the domain similarity score is formulated as Eq. (16).

$$w_s(x_i) = \frac{c(f(x_i))}{\frac{1}{|U_{ood}|} \sum_{j=1}^{|U_{ood}|} c(f(x_j))}, \ x_i \in U_{ood} \tag{16}$$

Second, for an OOD data sample, it is not enough to determine whether it is recyclable by the output probability of classifier c. Inspired by [38], we learn that the label prediction $\tilde{s}^{ood}(x)$ of the OOD data output by classifier h is also rich in transferable information. Specifically, we use the size of the interval between the maximum prediction probability and the second largest prediction probability to express the domain propensity score of the OOD samples.

$$\tilde{w}_t(x_i) = \max_{j \in \{1,...,c\}} \tilde{p}_j(x) - \max_{k \in \{1,...,c\}; k \neq j} \tilde{p}_k(x) \tag{17}$$

If the predicted difference of an OOD example is small, this means that the category propensity of this sample is not significant and such OOD data is not valuable to be recovered. If the difference is large, it means that this sample can extract valuable information and be recycled by category j. Finally, we also normalize the domain propensity score $w_t(x)$.

$$w_t(x_i) = \frac{\tilde{w}_t(x_i)}{\frac{1}{|U_{ood}|} \sum_{j=1}^{|U_{ood}|} \tilde{w}_t(x_j)}, \ x_i \in U_{ood} \tag{18}$$

So the migration score $w(x_i)$ of OOD data is obtained by summing the domain similarity score $w_c(x_i)$ and the domain propensity score $w_t(x_i)$.

$$w(x_i) = w_s(x_i) + w_t(x_i) \tag{19}$$

4   Experiment

4.1   Dataset

We constructed two class-mismatch datasets using two different tag types of English datasets, AG News[1] and Yelp-5[2], respectively. Specifically, AG News

---

[1] https://huggingface.co/datasets/ag_news.
[2] https://www.yelp.com/dataset/challenge.

contains 30,000/1,900 training/test texts in 4 classes (World, Sports, Business, Sci/Tec). Here, we randomly select 200 samples from each of the World and Sports categories in the training set to form the labeled set, and select 15,000 training texts among the four categories to form the unlabeled data-set. Thus, the texts in the categories Business and Sci/Tec are OOD data, and the texts in the other two categories are ID data. Yelp-5 is a fine-grained sentiment multi-categorization dataset, is composed of 130,000/10,000 training/test images from five classes "1" to "5". We randomly selected 200 samples from each of classes "2" to "4" as the labeled set, the sample 10,000 training texts from five classes to make up the unlabeled data-set. The statistics of the specific dataset and the composition information are presented in Table 1.

Table 1. Statistics of class mismatch datasets.

Dataset	Label Type	#Classes	#Train data	#ID data	#OOD data	#Test
AG News	News Topic	4	50 — 200 — 500	15000	15000	20000
Yelp-5	Review Sentiment	5	50 — 200 — 500	15000	10000	15000

4.2 Baseline Methods

We compare TextSMatch to the following models that have been experimented in text classification: BERT [39], UDA [5], MixText [12]. Note that in order to validate the effectiveness of the semi-supervised model, the BERT model only calculates the supervised loss to the detriment of the additional unlabeled data. With this in mind, TextSMatch as well as the other two methods use the same feature extractor as BERT.

BERT: The pre-trained model was considered as a special kind of deep semi-supervised model. In our experiments we used the BERT-base-uncased[3] pre-trained model and then fine-tuned it using labeled data. Specifically, we used [CLS] pooling after the output of the BERT encoder and two layers of MLP to predict the labels.

UDA: The algorithm was proposed to surpass all existing semi-supervised learning methods and had very vast application scenarios. Model-wise we uniformly used BERT-base-uncased, unlabeled data, batch size, and softmax sharpen temperature are all consistent with TextSMatch.

MixText: MixText is a text classification model based on TMix with consistent training. Inspired by the image classification algorithm Mixup, a new data enhancement algorithm TMix is proposed. TMix achieves data enhancement by linearly interpolating different training samples in the embedded space (hidden space). Specifically, we use the mixup layer set 7, 9, 12 because, this layer set contains most syntactic and semantic information and has the best model performance.

---

[3] https://pypi.org/project/pytorch-transformers/.

4.3 The Settings

We tag the text with a BERT-based-uncased tokenizer, the BERT-based-uncased model is used as encoder, and treated embedding of [CLS] markers as feature representation of the whole sentence. The classifier h consists of an MLP layer of 768 hidden size and tanh activation function. And the domain discriminator consists of a the layer of MLP with 768 hidden size as well as ReLU activation function. The maximum sentence length was set to 128. For sentences exceeding this limit, we keep the first 128 tokens. The learning rate of the BERT encoder is $2e-5$, and that of the MLP is $1e-4$. The threshold τ used to determine whether the data belongs to the OOD data was set to 0.57, and the Adversarial loss weight λ_{adv} was set to 0.005.

For methods other than BERT, the batch size is set to 4 for labeled data and 16 for unlabeled data. In MixText, we set $K = 2$, i.e. for each unlabeled data we performed the augmentation twice, especially for German and Russian. For UDA and TextSMatch, the threshold for Confidence-based masking is set to 0.45.

4.4 The Results

We compare TextSMatch not only with several popular SSL models related to text classification mentioned above, but also with supervised learning method. The proportion of OOD data in the unlabeled data set is varied to find the performance of various methods with different amounts of class mismatched data-set. Specifically, both SSL and supervised methods used all labeled training data, where AG News had 20,000 samples per class and the Yelp-5 dataset had 130,000 labeled data per class. Then we set the $\{0\%, 25\%, 50\%, 75\%\}$ of the OOD data set to infer the performance of the TextSMatch and baselines methods. The results are shown in Fig. 3. Based on these results, we can find that all SSL methods significantly outperform supervised learning methods when the class distribution of the data is consistent. However, the performance of all existing deep SSL methods decreases as the mismatch in class distribution increases. While models such as UDA approach the performance of baseline supervised learning methods when 50% of the classes of unlabeled data do not belong to the labeled domain of the labeled set, our proposed TextSMatch throws maintains the performance improvement when more than 75% of the unlabeled instances are from unseen classes.

The results on the two text classification datasets are shown in Table 2. TextS-Match performs significantly better than BERT when labeled data is 50 per class. In the case of data with 20 labels, the model accuracy improves from 35.7% to 47.3% on Yelp-5 and from 86.8% to 96.8% on AG News. It is worth mentioning that the performance of TextSMatch with 50 labeled data close to that of BERT with 500 labeled data. Because MixText not only incorporates unlabeled data, but also exploits the implicit relationship between labeled and unlabeled data through TMix, the performance is slightly better than UDA. Compared to UDA with MixText, our propose TextSMatch consistently shows the best performance on both datasets since TextSMatch not only makes fuller use of the unlabeled

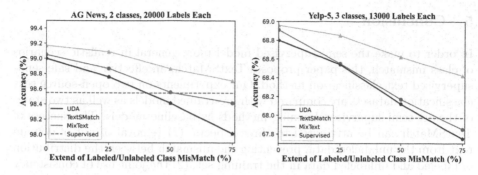

Fig. 3. Classification accuracy of TextSMatch on different datasets with varying class mismatch ratios between labeled and unlabeled data.

Table 2. Comparison of model performance in accuracy(%) with different amounts of labeled data

Dataset	Model	50	200	500	Dataset	Model	50	200	500
AG News	BERT	86.8	95.4	97.0	Yelp-5	BERT	35.7	52.9	59.6
	UDA	96.2	97.3	97.4		UDA	46.2	59.9	62.6
	MixText	96.3	97.5	97.6		MixText	46.5	60.1	63.3
	TextSMatch	**96.8**	**97.7**	**98.2**		TextSMatch	**47.3**	**60.7**	**64.1**

ID data through contrastive learning, but also project the OOD data in the unlabeled data to the label space through domain-adaptation, and recycles the samples near the labeled domain, which will finally enrich the information for text classification as well as feature extraction.

We measure the performance of TextSMatch by removing each component individually, and show the results in Table 3. We observe a drop in performance after dropping each component, which indicates each component in TextSMatch has an impact on the final performance. The performance of the model degrades more after removing the contrast learning compared with the consistency canonical when unlabeled data is removed, suggesting that contrast learning based on pseudo-labeled graphs make better use of unlabeled text. As we expected, the largest impact on model performance is from removing unlabeled data, and the smallest impact is from Contrastive learning.

Table 3. Performance in terms of accuracy (%) on Yelp-5 with 15000 unlabeled data and 200 labeled data for each category after removal various TextSMatch parts.

Model	Accuracy(%)
TextSMatch	**64.1**
- Unlabeled data	52.9
- Consistency regularization	61.5
- Metric learning	63.8
- Adversarial Domain Adaption	62.3

5   Conclusion

In order to make the semi-supervised model more general in realistic scenarios of class mismatch, this paper proposes TextSMatch, an effective and safe semi-supervised text classification method. Our experiments on two open-source text classification datasets are compared with pre-trained models as well as two state-of-the-art semi-supervised learning methods as baseline models. The success of TextSMatch can be attributed to three aspects: (1) removal of out-of-domain text from the unlabeled data, preventing the mismatch between the distribution of labeled and unlabeled data in the training set. (2) the joint use of consistency regularity and entropy minimization methods. (3) contrast learning based on pseudo-labele graph to better learn text embedding representations.

Funding. The publication of this paper is funded by NSFC (No. U1711266) and Key-Area Research and Development Program of Guangdong Province (No. 2019B0 10153001).

References

1. Jia, L., Zhang, Z., Wang, L., Jiang, W., Zhao, M.: Adaptive neighborhood propagation by joint l2, 1-norm regularized sparse coding for representation and classification. In: 2016 IEEE 16th International Conference on Data Mining (ICDM), pp. 201–210. IEEE (2016)
2. Zhang, H., Zhang, Z., Zhao, M., Ye, Q., Zhang, M., Wang, M.: Robust triple-matrix-recovery-based auto-weighted label propagation for classification. IEEE Trans. Neural Networks Learn. Syst. **31**(11), 4538–4552 (2020)
3. Zhang, Z., Li, F., Jia, L., Qin, J., Zhang, L., Yan, S.: Robust adaptive embedded label propagation with weight learning for inductive classification. IEEE Trans. Neural Networks Learn. Syst. **29**(8), 3388–3403 (2017)
4. Sajjadi, M., Javanmardi, M., Tasdizen, T.: Regularization with stochastic transformations and perturbations for deep semi-supervised learning. In: NIPS, pp. 1163–1171 (2016)
5. Xie, Q., Dai, Z., Hovy, E., Luong, M.-T., Le, Q.V.: Unsupervised data augmentation for consistency training. In: NIPS (2020)
6. Lee, D.-H., et al.: Pseudo-label: the simple and efficient semi-supervised learning method for deep neural networks. In: Workshop on Challenges in Representation Learning, ICML, volume 3 (2013)
7. Berthelot, D., et al.: Mixmatch: a holistic approach to semi-supervised learning. In: NIPS, pp. 5050–5060 (2019)
8. Berthelot, D., et al.: Semi-supervised learning with distribution alignment and augmentation anchoring. In: ICLR, Remixmatch (2020)
9. Chen, T., Kornblith, S., Norouzi, M., Hinton, G.: A simple framework for contrastive learning of visual representations. In: International Conference on Machine Learning, pp. 1597–1607. PMLR (2020)
10. He, K., Fan, H., Wu,, Y., Xie, S., Girshick, R.: Momentum contrast for unsupervised visual representation learning. In: Proceedings of the IEEE/CVF Conference on Computer Vision and Pattern Recognition, pp. 9729–9738 (2020)

11. Caron, M., Misra, I., Mairal, J., Goyal, P., Bojanowski, P., Joulin, A.: Unsupervised learning of visual features by contrasting cluster assignments. In: NIPS (2020)
12. Chen, J., Yang, Z., Yang, D.: Mixtext: linguistically-informed interpolation of hidden space for semi-supervised text classification. In: ACL, pp. 2147–2157 (2020)
13. Tarvainen, A., Valpola, H.: Mean teachers are better role models: weight-averaged consistency targets improve semi-supervised deep learning results. In: ICLR, Workshop Track Proceedings (2017)
14. Miyato, T., Maeda, S., Koyama, M., Ishii, S.: Virtual adversarial training: a regularization method for supervised and semi-supervised learning. IEEE Trans. Pattern Anal. Mach. Intell. **41**(8), 1979–1993 (2018)
15. Oliver, A., Odena, A., Raffel, C., Cubuk, E.D., Goodfellow, I.J.: Realistic evaluation of deep semi-supervised learning algorithms. arXiv preprint arXiv:1804.09170 (2018)
16. Chen, Y., Zhu, X., Li, W., Gong, S.: Semi-supervised learning under class distribution mismatch. In: Proceedings of the AAAI Conference on Artificial Intelligence, vol. 34, pp. 3569–3576 (2020)
17. Guo, L.-Z., Zhang, Z.-Y., Jiang, Y., Li, Y.-F., Zhou, Z.-H.: Safe deep semi-supervised learning for unseen-class unlabeled data. In: International Conference on Machine Learning, pp. 3897–3906. PMLR (2020)
18. Yang, X., Song, Z., King, I., Xu, Z.: A survey on deep semi-supervised learning. arXiv preprint arXiv:2103.00550 (2021)
19. Cubuk, E.D., Zoph, B., Shlens, J., Le, Q.V.: Randaugment: practical data augmentation with no separate search. arXiv preprint arXiv:1909.13719, **2**(4), 7 (2019)
20. Edunov, S., Ott, M., Auli, M., Grangier, D.: Understanding back-translation at scale. arXiv preprint arXiv:1808.09381 (2018)
21. Laine, S., Aila, T.: Temporal ensembling for semi-supervised learning. In: ICLR (2017)
22. Sohn, K., et al.: Simplifying semi-supervised learning with consistency and confidence. In: NIPS, Fixmatch (2020)
23. Grandvalet, Y., Bengio, Y., et al.: Semi-supervised learning by entropy minimization. In: CAP, pp. 281–296 (2005)
24. Radford, A., Narasimhan, K., Salimans, T., Sutskever, I.: Improving language understanding by generative pre-training (2018)
25. Peters, M.E., et al.: Deep contextualized word representations. In: NAACL-HLT, pages 2227–2237 (2018)
26. Liu, Y., et al.: Roberta: a robustly optimized bert pretraining approach. arXiv preprint arXiv:1907.11692 (2019)
27. Behrmann, N., Fayyaz, M., Gall, J., Noroozi, M.: Long short view feature decomposition via contrastive video representation learning. In: Proceedings of the IEEE/CVF International Conference on Computer Vision, pp. 9244–9253 (2021)
28. Hendrycks, D., Gimpel, K.: A baseline for detecting misclassified and out-of-distribution examples in neural networks. arXiv preprint arXiv:1610.02136 (2016)
29. Nguyen, A., Yosinski, J., Clune, J.: Deep neural networks are easily fooled: high confidence predictions for unrecognizable images. In: Proceedings of the IEEE Conference on Computer Vision and Pattern Recognition, pp. 427–436. IEEE Computer Society (2015)
30. Liang, S., Sun, R., Li, Y., Srikant, R.: Understanding the loss surface of neural networks for binary classification. In: Dy, J.G., Krause, A. (eds.) International Conference on Machine Learning, pp. 2835–2843. PMLR, PMLR (2018)
31. Hinton, G., Vinyals, O., Dean, J.: Distilling the knowledge in a neural network. arXiv preprint arXiv:1503.02531 (2015)

32. Liang, S., Li, Y., Srikant, R.: Enhancing the reliability of out-of-distribution image detection in neural networks. arXiv preprint arXiv:1706.02690 (2017)
33. Lotfollahi, M., Naghipourfar, M., Theis, F.J., Alexander Wolf, F.: Conditional out-of-distribution generation for unpaired data using transfer vae. Bioinformatics **36**(Supplement_2), i610–i617 (2020)
34. Cai, Z., Ravichandran, A., Maji, S., Fowlkes, C., Tu, Z., Soatto, S.: Exponential moving average normalization for self-supervised and semi-supervised learning. In: Proceedings of the IEEE/CVF Conference on Computer Vision and Pattern Recognition, pp. 194–203 (2021)
35. Ganin, Y., et al.: Domain-adversarial training of neural networks. J. Mach. Learn. Res. **17**(1), 2030–2096 (2016)
36. Isola, P., Zhu, J.-Y., Zhou, T., Efros, A.A.: Image-to-image translation with conditional adversarial networks. In: IEEE, pp. 1125–1134 (2017)
37. Ganin, Y., Lempitsky, V.: Unsupervised domain adaptation by backpropagation. In: International Conference on Machine Learning, pp. 1180–1189. PMLR (2015)
38. Yao, Y., Deng, J., Chen, X., Gong, C., Wu, J., Yang, J.: Deep discriminative CNN with temporal ensembling for ambiguously-labeled image classification. In: AAAI, vol. 34, pp. 12669–12676 (2020)
39. Devlin, J., Chang, M.-W., Lee, K., Toutanova, K.: Pre-training of deep bidirectional transformers for language understanding. In: NAACL-HLT, Bert (2019)

Recommendation Method of Cross-language Computer Courses

Jiajun Ou, Lin Zhou, Zhenzhen Li, and Shaohong Zhang(✉)

School of Computer Science and Cyber Engineering, Guangzhou University,
Guangzhou, Guangdong, China
2112106079@e.gzhu.edu.cn, zimzsh@qq.com

Abstract. In traditional course recommendation methods, courses are usually recommended through a single language content. However, learners can not be effectively recommended using current methods when they want to access cross-language course content. In computer science, foreign computer technology has first mover advantages, so cross-language recommendation of foreign courses is important for learners. But there is still comparatively little work on related recommendation methods, traditional recommendation methods are insufficient to address the above needs. To meet the above needs, we propose a cross-language computer course recommendation method in this paper. Our method can be effectively used in the cross-language course recommendation scenario to recommend the most relevant top-N English courses through the content description information of Chinese courses. Our experiment was conducted in computer course data from 15 universities and a number of Massive Open Online Courses (MOOCs). As a result, our experiment achieved the accuracy of 89%, the precision of 90%, the recall of 85% and the F1-score of 86%. The evaluation indexes reflect the effectiveness and feasibility of our method. Our method can also be applied to course recommendation in other subject fields.

Keywords: Course recommendation system · Cross-language · Content-based recommendation

1 Introduction

With the rapid development of big data and artificial intelligence, there is a lot of redundant information appearing online, so it is very important for users to select the information they are interested in. The same problem is encountered in online courses.

In recent years, researchers have conducted a lot of research on course recommendation systems. A large number of recommendation system algorithms and models have emerged to facilitate the development of this field. A common approach is to use collaborative filtering for course recommendation, which is shown to be effective in [1–4]. In addition, there are also research [5–8] that incorporate traditional machine learning approaches for course recommendation. Furthermore, some of interesting research [9,10] recently combine reinforcement learning for course recommendation. Due to the increase in computing power, several research [11,12] efforts combining deep learning methods have emerged in this field.

However, most of the current recommendation systems only recommend content based on a single language. When the domestic courses that users requirements are insufficient, they will go to foreign websites and spend a lot of time to find relevant courses for learning, which will cost a lot of labor and time. But no research on cross-language recommendation systems for course recommendation has been found. Therefore, a cross-language computer course recommendation system is proposed in this paper to address this aspect of the problem.

Our method is based on the course description content for recommendation. The deep learning models at the transformers [13] are fine-tuned for training to obtain a course content classification models, input the Chinese course data that needs to be recommended and get the predicted classification label of the course. After that, we continue to use the Chinese text data to translate the text by the Chinese-English translation model. The translated text is computed in the corresponding English category data using the text similarity model and the obtained similarity list is sorted to get the top-N list for English course recommendation providing users with convenient access to cross-language course information. The contributions of this paper are listed as follows:

1) We propose the task of cross-language course recommendation in computer field to solve the problem that current recommendation systems can only make recommendation in a single language.
2) We propose a cross-language recommendation method, which is proven to be effective for this task through the experiments.

The rest of the text is structured as follows. In the second section, we will detail other related work in the field of course recommendation. In the third section, we will introduce the specific implementation of the two different methods and models of cross-language recommendation. In the fourth section, we present the data of English and Chinese courses used in the experiments, the experimental setup to validate the recommendation method and the experimental results obtained from the comparison experiments. In the last section, a summary of our work will be presented.

2   Related Works

We introduce in detail the research related to course recommendation systems in this paragraph.

A course recommendation method combining collaborative filtering and knowledge graph is proposed in [1], which uses a knowledge graph to characterize course information embedded in a low-dimensional space, calculates course similarity and fuses the information into a collaborative filtering algorithm for course recommendation. A cross-user domain collaborative filtering recommendation method is proposed in [2], which performs course recommendation by predicting the grade scores of students elective courses. A personalized approach based on cross-platform and collaborative filtering algorithm is proposed in [3] for course recommendation. A hybrid recommendation method based on association rules, content filtering and collaborative filtering is proposed in [4], which performs personalized course recommendation for learners. The use of Latent Dirichlet Allocation (LDA) for topic classification of course content is proposed in [5], which combines student history data for online course recommendation. An Apriori algorithm-based Massive Open Online Courses (MOOCs) recommendation system is proposed in [6], which combines Spark framework of big data and multiple algorithms for course recommendation. A machine learning method that combines with student ratings and learning styles for personalized course recommendation is proposed in [7]. An approach based on a combination of K-Means algorithm and Apriori association rules for online course recommendation is proposed in [8]. A hierarchical reinforcement learning algorithm-based course recommendation system for MOOCs is proposed in [9]. A dynamic attention and hierarchical reinforcement learning adaptive framework for course recommendation is proposed in [10]. An end-to-end graph neural network approach to integrate heterogeneous contextual information of MOOCs into knowledge concepts for course recommendation is proposed in [11]. An approach based on a double-layer attention mechanism for course recommendation is proposed in [12]. A personalized online course model based on user features, item features and cross-cutting features for recommendation is proposed in [14]. An ontology-based hybrid filtering approach for course recommendation is proposed in [15]. A collaborative Bayesian variational network approach based on demand-awareness is proposed in [16] for online course recommendation. A context-aware approach incorporating Radio Frequency Identification (RFID) technology for online course recommendation is mentioned in [17]. A personalized recommendation approach based on mobile learning for learning resource recommendation is proposed in [18]. An online teaching method based on progressive structure for course recommendation is proposed in [19]. A recommendation method based on sentiment factor is proposed in [20] for course recommendation. An approach based on bio-inspired algorithm for course recommendation is proposed in [21].

The related works described above are all based on a single language for course recommendation. Our approach differs from others in that we focus on cross-language course recommendation scenario.

3   Recommendation Methods

3.1   Content-Based Recommendation Method

In the field of recommendation systems, common approaches include item-based recommendation, content-based recommendation, user-based recommendation, collaborative filtering-based recommendation, mixed-method recommendation and deep learning-based recommendation methods. The most widely used approach in industry is collaborative filtering. Our goal is to provide learners with cross-language course recommendation methods by recommending top-N English course from the Chinese course descriptions that learners are currently studying. In this task, what learners want most is to get course content information. In order to solve the above problem, we choose to use recommendation system based on course description content. The idea of the content-based recommendation method is to find the most relevant content to recommend to the learner. This method is highly feasible and can accurately recommend the information that the learners want to obtain. The most important technique of this method is to calculate the cosine similarity of the course description content. The cosine similarity formula is shown below. i, j denote two different sentence vectors.

$$sim(i,j) = cos(i,j) = \frac{i \cdot j}{||i|| * ||j||} \tag{1}$$

3.2   Cross-Language Recommendation Methods

In this section, we will discuss in detail the cross-language recommendation method proposed, which is divided into two different recommendation strategies, the Chinese-based course content classification recommendation method and the English-based course content classification recommendation method. The main difference is the data language type when classifying and labeling the text.

3.2.1   Recommendation Method Based on Chinese Course Classification

The first recommendation method proposed is based on the recommendation of Chinese course content classification model. Firstly, we input a Chinese course description data, the description content should meet certain length requirements, the maximum length can not exceed 512. The description content will be truncated if it is too long. Then, we classify and label the input course content by pre-training the Chinese course content classification model with fine-tuned parameters to get the label of the input course content. After that, we use the label to match the English course content data with the corresponding label in the data, so that the course data with the corresponding label can be used as the alternative recommendation course datasets. At the same time, we translate the Chinese course description by the Chinese-English translation model to get the translated English course content description. At this time, we combine the translated English course information with the obtained English recommendation course datasets list to perform one-to-many cosine similarity calculation using the text similarity model and get the relevance list of cosine similarity. Finally, we sort the similarity list in descending order, so that the top-N English courses are recommended as target courses. The specific implementation process of Method one described above is shown in Fig. 1.

3.2.2 Recommendation Method Based on English Course Classification

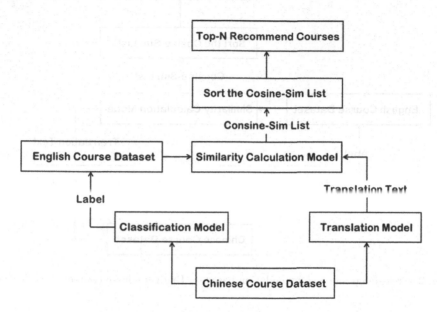

Fig. 1. Cross-language recommendation based on Chinese course content classification

The second course recommendation method proposed is based on the recommendation of English course content classification. The difference between two methods is that the process of obtaining course classification labels and performing Chinese-English translation is in a different order. The specific recommendation process is described as follows. Firstly, we input the Chinese course content and translate it using the Chinese-English model. Then, we get the translated English course description content and input it into the English course classification model for classifying and obtaining labels. After that, we use the obtained classification labels to match the datasets with the same labels in the English course recommendation data. At the same time, we input the translated English course descriptions and the labeled English courses datasets into the sentence cosine similarity model to obtain the list of recommendation English courses that are the most similar to the target courses. Finally, we sort the list in descending order and obtain the top-N recommended English courses. The specific implementation process of Method two described above is shown in Fig. 2.

3.3 Other Models

3.3.1 Classification Models
In order to implement our method, we use the current advanced deep learning models in each part of the recommendation system. In terms of classification models are introduced as described below. The model framework of BERT [22] is based

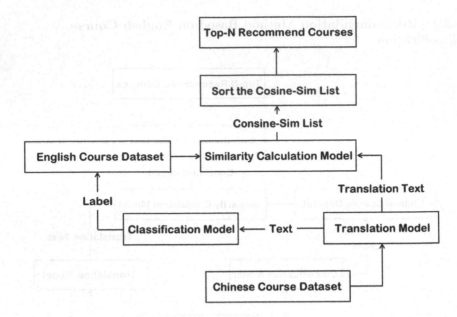

Fig. 2. Cross-language recommendation based on English course content classification

on a bidirectional transformer structure consisting of a Masked Language Model (MLM) and Next Sentence Prediction (NSP) as its main training tasks. The MLM task is to mask and replace part of the input data, and then perform word prediction of the modified part. The NSP task is to input two sentences to determine whether the two sentences are contextual. The ALBERT [23] model goes through two methods to reduce the number of parameters. Method one is a factorization of the input vector, which the vector is input to the lower dimensional space first before mapping to the hidden space. Method two is cross-layer parameter sharing, which allows each layer of the ALBERT model to share parameter updates. In terms of task, ALBERT uses Sentence Order Prediction (SOP) to replace NSP. The SOP task swaps the order of two sentences in the context and then goes to determine the coherence of these two sentences. RoBERTa [24] used a larger amount of data than BERT and used more GPU resources for pretraining. Structurally, RoBERTa makes a more detailed tuning of BERT. In the task, RoBERTa directly removes the NSP and uses a dynamic masking strategy that allows the model to learn different feature representations during training. The data use Byte-Pair Encoding (BPE), which can represent a larger vocabulary than WordPiece of BERT for solving out of vocabulary problem. XLNet [25] uses a Permutation Language Model (PLM) to replace the MLM task, which obtains different arrangements between words by rearranging the word sequences allowing the model to learn more information in different positions. Structurally, the transformer structure of BERT is replaced with the Transformer-XL structure gaining the ability to train long texts. ELETRAC [26] introduces an idea like Generative Adversarial Nets (GAN) by using a Generator-Discriminator structure, in which masked text data is first put into Generator to predict the word at [MASK] position and then Discriminator determines whether the word has been replaced. This method allows ELETRAC to achieve excellent results in multiple tasks with smaller parameters and less training time. DistilBERT [27] introduces a Student and Teacher strategy, which the Student model has half the number of parameters of the original BERT and the Teacher model has the same number of parameters as the BERT, by allowing the Student model to learn as much as possible from the Teacher model. The model achieves 97% of the BERT with fewer resources and less time. The main objective of MacBERT [28] is to train a model based on Chinese data that uses a [MASK] strategy to improve the MLM task using a combination of the advantages of whole word masking strategy and N-gram masking strategy. Word2Vec [29] are trained in the data, and then correlations between words are obtained. The original method of random word replacement is changed to use similar words for replacement in the MLM task.

3.3.2   Translation Models

In terms of translation models, we use pre-trained models from the Language Technologies Research Group at the University of Helsinki [30], which provides a large number of pre-trained models for translation. We use Chinese-English translation model Helsinki-NLP/opus-mt-zh-en and the English-Chinese translation model Helsinki-NLP/opus-mt-en-zh in transforms package to implement the translation module part of the cross-language recommender system task of this paper.

3.3.3   Text Similarity Model

In terms of text similarity, we use the Sentence-BERT [31] model to calculate the similarity of course content. The model uses the Siamese structure, which is divided into two parts, the training model and the inference model. First of all, the two sentences to be compared are converted into a sentence Embedding vector representation, then passed into the training model for training. The two vectors intersection features are extracted and merged together with the intersection features again for predicting the results. The structure of the inference model is similar to that of the training model, except that the cosine similarity is calculated for the two processed sentence vectors obtained after the input model. Finally, we use all-MiniLm-L6-v2 of the Sentence-Transformer package. This model is a light weight model, its effect of similarity calculation is better than most of the models and the speed of calculation is the fastest. So we use this model for the task.

4 Experimental Setups and Results

4.1 Datasets

The Chinese course description content data is partly obtained from 15 computer related college majors at domestic universities through downloading from the official websites of the universities. The other part is the data related to computer courses obtained from online MOOCs through web crawlers, with a total of 1074 Chinese course data. The English course data are partly computer courses obtained from the official websites of more than 70 foreign universities and the course contents crawled through English MOOCs websites, totaling 2007 data. Finally, the data corresponding to Chinese and English language courses were obtained from CS2013 [32] computer course specifications and MIT OpenCourseWare websites as validation data set, totaling 247 data.

To improve the search speed of the recommendation method, the data was divided into 10 categories of data in order to avoid calculating the similarity in all English data. Although we initially tried to use Word2Vec to extract word vectors and LDA for topic classification, the results were only 32% accuracy. Therefore, a total of 3328 Chinese and English courses were manually classified and labeled for each course. The 10 categories into which the categories are divided are programming, mathematics, physics, hardware, software engineering, algorithm, data, computer, artificial intelligence and network, as shown in the following Table 1.

Table 1. Description of data classification and percentage.

Category	Description	Percentage	Label
Programming	Java, C, C++, Python, etc.	16.2%	0
Mathematics	Related to Mathematics, etc.	7.3%	1
Physics	Signal, Information, Electricity, Optics, etc.	7.8%	2
Hardware	IoT, SCM, Controller, VR, etc.	10.8%	3
Software Engineering	Software Engineering, Software Testing, etc.	8.8%	4
Algorithm	Data Structures, etc.	5.1%	5
Data	Database, Data Analysis, Big Data, etc.	9.5%	6
Computer	Intorduction of Computer etc.	12.7%	7
Artificial Intelligence	AI, NLP, CV, Computer Graphics, etc.	6.5%	8
Network	Network Security, Cryptography, etc.	15.3%	9

4.2 Fine-Tuning of the Course Classification Model

In order to evaluate the recommendation methods more objectively, we use a variety of pre-trained models for course classification to predict the corresponding course categories. We fine-tune these models by using the models in the transformers package. The Chinese text models include bert-base-chinese and hfl/chinese-macbert-base. The English text models include google/electra-base-discriminator, bert-base-uncased, roberta-base, albert-base-v2, xlnet-base-cased and distilbert-base-uncased. The course data is trained on a Tesla K80 GPU with epochs set to 100 for each model. Each model took 5 h to train for a total of 8 models with using 40 h. Since the amount of used data was not large, the training data to test data ratio was appropriately increased to 9:1. The specific model training situation are shown in the following Table 2.

Table 2. Training and Testing Data Accuracy of Each Model.

Model	Training data accuracy	Testing data accuracy
BERT-CH	0.98	0.99
MacBERT-CH	0.99	0.98
BERT-EN	0.97	0.97
ALBERT-EN	0.97	0.93
DistilBERT-EN	0.98	0.98
RoBERTa-EN	0.96	0.96
XLNet-EN	0.97	0.96
ELECTRA-EN	0.96	0.98

4.3 Experimental Evaluation

To evaluate our cross-language recommendation system, we use accuracy, precision, recall and F1-score for the evaluation of the classification model. The 247 Chinese and English data are used for the validation data to evaluate the classification models. Since it is a multi-category text classification, the evaluation methods of macro, micro and evaluation with weighted are used for evaluation. The macro method averages the precision and recall of all categories before calculating the F1-score. The micro method calculates the total precision and recall for all categories first, and then calculates the total F1-score. The method with weights is a calculation that is based on the size of the various types of sample data of the assessment data and then adjusted proportionally. The effect of the three assessment methods is shown in the following Table 3,4 and 5.

Taking the evaluation index of macro method as an example, the best model for Chinese course classification is the Chinese BERT model. The accuracy achieves 89%, which is 9% higher than the MacBERT model. The precision achieves 90%, which is 9% higher than the MacBERT model. The recall achieves 85%, which is 7% higher than the MacBERT model. The F1-score achieves 86%, which is 7% higher than the MacBERT model. Therefore, the BERT model has the best performance among the Chinese models in terms of all evaluation metrics. For the English course classification task, the best model for accuracy is XLNet with 72%, 7% higher than the worst ALBERT. The best model for precision is RoBERTa with 70%, 7% higher than the worst ALBERT. The best

458   J. Ou et al.

Table 3. Accuracy, precision, recall, F1-score of macro method for each model.

Model	Accuracy	Precision	Recall	F1-Score
BERT-CH	0.89	0.90	0.85	0.86
MacBERT-CH	0.80	0.81	0.79	0.79
BERT-EN	0.71	0.69	0.70	0.68
ALBERT-EN	0.65	0.63	0.64	0.62
DistilBERT-EN	0.69	0.66	0.67	0.66
RoBERTa-EN	0.70	0.70	0.70	0.68
XLNet-EN	0.72	0.69	0.71	0.69
ELECTRA-EN	0.70	0.67	0.67	0.67

Table 4. Accuracy, precision, recall, F1-score of micro method for each model.

Model	Accuracy	Precision	Recall	F1-Score
BERT-CH	0.89	0.89	0.89	0.89
MacBERT-CH	0.80	0.80	0.80	0.80
BERT-EN	0.71	0.71	0.71	0.71
ALBERT-EN	0.65	0.65	0.65	0.65
DistilBERT-EN	0.69	0.69	0.69	0.69
RoBERTa-EN	0.70	0.70	0.70	0.70
XLNet-EN	0.72	0.72	0.72	0.72
ELECTRA-EN	0.70	0.70	0.70	0.70

Table 5. Accuracy, precision, recall, F1-score of weighted method for each model.

Model	Accuracy	Precision	Recall	F1-Score
BERT-CH	0.89	0.90	0.89	0.89
MacBERT-CH	0.80	0.81	0.80	0.80
BERT-EN	0.71	0.73	0.71	0.71
ALBERT-EN	0.65	0.66	0.65	0.65
DistilBERT-EN	0.69	0.70	0.69	0.69
RoBERTa-EN	0.70	0.74	0.70	0.70
XLNet-EN	0.72	0.73	0.72	0.72
ELECTRA-EN	0.70	0.70	0.70	0.69

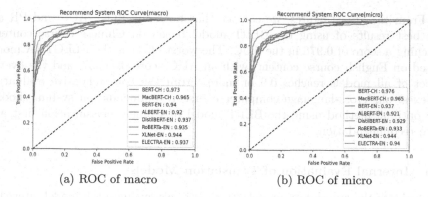

(a) ROC of macro   (b) ROC of micro

Fig. 3. Comparison of ROC curves and AUC scores of recommended methods for each model

model for recall is XLNet with 71%, the worst is ALBERT with 64%. The best F1-score is XLNet reaching 69%, 7% higher than the worst ALBERT. Overall it can be seen that the method of using Chinese courses for classification and labeling first performs better overall than the method of using Chinese courses for translation and then classification and labeling.

4.4 Recommendation Method Evaluation

The recommendation system evaluation method is evaluated using Receiver Operating Characteristic Curve (ROC), Area Under the Curve (AUC) and Precision Recall Curve (PR). Our evaluation is also based on comparative experiments between the macro and micro methods. The details are shown in the following Figs. 3 and 4.

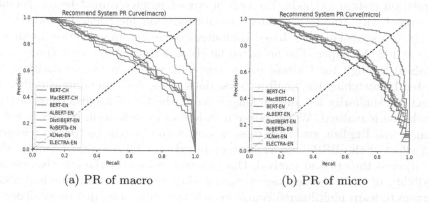

(a) PR of macro   (b) PR of micro

Fig. 4. Comparison of PR curves of recommended methods for each model

From the above figures, we can see that both the ROC, AUC and PR are the best results of using the BERT model based on Chinese course content, reaching a score of 0.976 in the AUC. The worst effect is the ELECTRA model based on English course content, with an AUC score of 0.937, and the overall effect of all models reaches 0.9 or more. From the comprehensive evaluation indexes, our cross-language computer course recommendation system chooses the optimal method using the BERT model of Chinese classification first and then recommendation.

4.5   Internal Evaluation of Translation Models

To evaluate the translation models we choose, we evaluate the model internally using two methods. The first evaluation method obtains the model of the Chinese translated short text by using the English short text with its English-Chinese translation Helsinki-NLP/opus-mt-en-zh model, and then inputs the Chinese translated short text with its Chinese translation model Helsinki-NLP/opus-mt-zh-en to obtain the English translated text. The original English short text and the translated English short text were put into Sentences-BERT for cosine similarity calculation, our experiment shows the average similarity is over 90%. The second method is to use the validation set course text to input into the Chinese-English translation model to get the English course text and then compare it with the manually translated data for cosine similarity. There is an average of more than 85% similarity. As shown above, Chinese-English translation model has the efficiency and feasibility in our method.

5   Conclusion

The current course recommendation system has a lot of recommendation methods for a single language, but there is less research on cross-language recommendation system methods. The task of cross-language cannot be competently performed with traditional course recommendation methods. To address this problem, a cross-language based computer course recommendation method is proposed in the paper. The predicted labels and translated course contents are obtained through the Chinese course contents, and then the similarity is calculated by matching the English course data of the corresponding classification to get the similarity list and ranking. Finally, the top-N English course recommendation is realized. We introduce a variety of classification models based on Chinese and English, evaluates these models with different metrics to select the best effect of the BERT model strategy based on Chinese course content classification as the optimal method. Our experiments reflect the effectiveness and feasibility of the cross-language recommendation method. Our method allows learners to learn multilingual course knowledge while providing convenience. In addition, our method is not only used in computer courses, but also can be migrated for learning to be used in other fields.

Acknowledgment. The work described in this paper was partially supported by grants from the funding of Guangzhou education scientific research project [No. 1201730714], and Guangdong Basic and Applied Basic Research Foundation [No. 2022A1515011697].

References

1. Xu, G., Jia, G., Shi, L., Zhang, Z.: Personalized course recommendation system fusing with knowledge graph and collaborative filtering. Comput. Intell. Neurosci. **2021** (2021)
2. Huang, L., Wang, C.D., Chao, H.Y., Lai, J.H., Philip, S.Y.: A score prediction approach for optional course recommendation via cross-user-domain collaborative filtering. IEEE Access **7**, 19550–19563 (2019)
3. Li, J., Ye, Z.: Course recommendations in online education based on collaborative filtering recommendation algorithm. Complexity 2020 (2020)
4. Xiao, J., Wang, M., Jiang, B., Li, J.: A personalized recommendation system with combinational algorithm for online learning. J. Ambient. Intell. Humaniz. Comput. **9**(3), 667–677 (2018)
5. Apaza, R.G., Cervantes, E.V., Quispe, L.C., Luna, J.O.: Online courses recommendation based on lda. In: SIMBig, pp. 42–48. Citeseer (2014)
6. Zhang, H., Huang, T., Lv, Z., Liu, S., Zhou, Z.: MCRS: a course recommendation system for MOOCS. Multimedia Tools Appl. **77**(6), 7051–7069 (2018)
7. Nafea, S.M., Siewe, F., He, Y.: On recommendation of learning objects using felder-silverman learning style model. IEEE Access **7**, 163034–163048 (2019)
8. Aher, S.B., Lobo, L.: Combination of machine learning algorithms for recommendation of courses in e-learning system based on historical data. Knowl.-Based Syst. **51**, 1–14 (2013)
9. Zhang, J., Hao, B., Chen, B., Li, C., Chen, H., Sun, J.: Hierarchical reinforcement learning for course recommendation in MOOCS. In: Proceedings of the AAAI Conference on Artificial Intelligence, vol. 33, pp. 435–442 (2019)
10. Lin, Y., Feng, S., Lin, F., Zeng, W., Liu, Y., Wu, P.: Adaptive course recommendation in MOOCS. Knowl.-Based Syst. **224**, 107085 (2021)
11. Gong, J., Wang, S., Wang, J., Feng, W., Peng, H., Tang, J., Yu, P.S.: Attentional graph convolutional networks for knowledge concept recommendation in MOOCS in a heterogeneous view. In: Proceedings of the 43rd International ACM SIGIR Conference on Research and Development in Information Retrieval, pp. 79–88 (2020)
12. Zhu, Q.: Network course recommendation system based on double-layer attention mechanism. Scientific Programming 2021 (2021)
13. Wolf, T., et al.: Transformers: state-of-the-art natural language processing. In: Proceedings of the 2020 Conference on Empirical Methods in Natural Language Processing: System Demonstrations, pp. 38–45 (2020)
14. Chen, Q., Yu, X., Liu, N., Yuan, X., Wang, Z.: Personalized course recommendation based on eye-tracking technology and deep learning. In: 2020 IEEE 7th International Conference on Data Science and Advanced Analytics (DSAA), pp. 692–968. IEEE (2020)
15. Ibrahim, M.E., Yang, Y., Ndzi, D.L., Yang, G., Al-Maliki, M.: Ontology-based personalized course recommendation framework. IEEE Access **7**, 5180–5199 (2018)

16. Wang, C., Zhu, H., Zhu, C., Zhang, X., Chen, E., Xiong, H.: Personalized employee training course recommendation with career development awareness. In: Proceedings of the Web Conference 2020, pp. 1648–1659 (2020)
17. Wang, S.-L., Wu, C.-Y.: Application of context-aware and personalized recommendation to implement an adaptive ubiquitous learning system. Expert Syst. Appl. **38**(9), 10831–10838 (2011). https://doi.org/10.1016/j.eswa.2011.02.083
18. Hsu, C.-K., Hwang, G.-J., Chang, C.-K.: A personalized recommendation-based mobile learning approach to improving the reading performance of EFL students. Comput. Educ. **63**, 327–336 (2013). https://doi.org/10.1016/j.compedu.2012.12.004
19. Rafiq, M.S., Jianshe, X., Arif, M., Barra, P.: Intelligent query optimization and course recommendation during online lectures in e-learning system. J. Ambient. Intell. Humaniz. Comput. **12**(11), 10375–10394 (2021)
20. Wang, Y.: Research on online learner modeling and course recommendation based on emotional factors. Sci. Program. **2022** (2022)
21. Gil, A.B., García-Peñalvo, F.J.: Learner course recommendation in e-learning based on swarm intelligence. J. Univers. Comput. Sci. **14**(16), 2737–2755 (2008)
22. Devlin, J., Chang, M.W., Lee, K., Toutanova, K.: Bert: pre-training of deep bidirectional transformers for language understanding. arXiv preprint arXiv:1810.04805 (2018)
23. Lan, Z., Chen, M., Goodman, S., Gimpel, K., Sharma, P., Soricut, R.: Albert: a lite Bert for self-supervised learning of language representations. arXiv preprint arXiv:1909.11942 (2019)
24. Liu, Y., et al.: Roberta: a robustly optimized Bert pretraining approach. arXiv preprint arXiv:1907.11692 (2019)
25. Yang, Z., Dai, Z., Yang, Y., Carbonell, J., Salakhutdinov, R.R., Le, Q.V.: Xlnet: generalized autoregressive pretraining for language understanding. In: Advances in Neural Information Processing Systems, vol. 32 (2019)
26. Clark, K., Luong, M.T., Le, Q.V., Manning, C.D.: Electra: pre-training text encoders as discriminators rather than generators. arXiv preprint arXiv:2003.10555 (2020)
27. Sanh, V., Debut, L., Chaumond, J., Wolf, T.: Distilbert, a distilled version of Bert: smaller, faster, cheaper and lighter. arXiv preprint arXiv:1910.01108 (2019)
28. Cui, Y., Che, W., Liu, T., Qin, B., Wang, S., Hu, G.: Revisiting pre-trained models for Chinese natural language processing. arXiv preprint arXiv:2004.13922 (2020)
29. Mikolov, T., Chen, K., Corrado, G., Dean, J.: Efficient estimation of word representations in vector space. arXiv preprint arXiv:1301.3781 (2013)
30. Tiedemann, J.: The tatoeba translation challenge-realistic data sets for low resource and multilingual MT. arXiv preprint arXiv:2010.06354 (2020)
31. Reimers, N., Gurevych, I.: Sentence-Bert: sentence embeddings using Siamese Bert-networks. arXiv preprint arXiv:1908.10084 (2019)
32. Draft, S.: Computer science curricula 2013. ACM and IEEE Computer Society, Incorporated, New York, NY, USA (2013)

Temperature Prediction of Medium Frequency Furnace Based on Transformer Model

Shifeng Ma, Yanping Li[✉], Dongyue Luo, and Taotao Song

School of Information and Electrical Engineering, Shandong Jianzhu University, Jinan 250101, China

Liyanping0531@126.com

Abstract. Aiming at the difficulty of continuous measurement of charge temperature in intermediate frequency furnace smelting process, this paper proposes a method of continuous measurement of charge temperature in intermediate fre quency furnace smelting process by using Transformer model, which makes use of the characteristic that the resistivity of metal changes nonlinearly with temperature in the melting process. Analyze the circuit structure of the medium frequency furnace, build the charge resistance data acquisition system on the medium frequency furnace, collect a large number of charge temperature, equivalent resistance and other data with the help of instruments, train one part of the collected data with the transformer model, and take another part of the collected data for performance evaluation. After the model is trained, import the resistance data inferred by the built resistance acquisition system into the transformer model, in this way, the temperature data of the current charge of the medium frequency furnace can be calculated in real time, and the temperature data can be transmitted to the upper computer for real-time display.

Keywords: Medium frequency furnace · Transformer model · Resistance data acquisition · Temperature measurement

1 Introduction

With the continuous development of metallurgical industry, medium frequency furnace has become an important means of metallurgy with its advantages of high smelting efficiency, less element burning loss, easy temperature control, convenient operation and low pollution [1, 2]. Temperature is not only an important thermodynamic parameter, but also an important kinetic parameter for the metal smelting process. It not only has a significant impact on the reaction direction and degree of each chemical reaction and the relative reaction speed between elements, but also has a significant impact on the mass transfer and heat transfer speed of the molten pool. In order to quickly and effectively remove the impurities in the metal, accelerate the melting of the charge, reduce splashing, improve the furnace life and save the energy consumed by smelting, the charge temperature must be controlled. If the temperature control of the whole process can be done well, stable production will come into being. In order to achieve effective temperature control, the charge temperature should be measured accurately first.

© The Author(s), under exclusive license to Springer Nature Singapore Pte Ltd. 2022
H. Zhang et al. (Eds.): NCAA 2022, CCIS 1637, pp. 463–476, 2022.
https://doi.org/10.1007/978-981-19-6142-7_35

In recent years, the temperature measurement methods used for medium frequency furnace charge in smelting plants mainly include thermocouple (thermal resistance), infrared radiation temperature measurement and so on [3], among them, the thermocouple can not carry out continuous temperature measurement and its service life is limited. Infrared radiation temperature measurement has high requirements for the environment. The research on charge temperature measurement at home and abroad is also making breakthroughs. SEI and ETX developed an acoustic temperature measurement system for utility boilers [4]; Soo young Lee et al. Proposed a deep learning model to predict the distribution of molten steel temperature by considering both transient and steady-state characteristics [5]. With the wide application of deep learning, more and more people try to predict the temperature in the smelting process through deep learning [6–16].

The medium frequency furnace is a power supply device that converts the power frequency 50 Hz alternating current into medium frequency (above 300 Hz to 1000 Hz) [17]. The three-phase power frequency alternating current is rectified and converted into direct current, and then the direct current is changed into adjustable medium frequency current to supply the medium frequency alternating current flowing through the capacitor and induction coil, and produce high-density magnetic lines in the induction coil, and cut the metal material contained in the induction ring, resulting in a large eddy current in the metal material. The medium frequency induction furnace is composed of furnace shell, fixed frame, furnace cover, furnace tilting mechanism, induction coil, yoke and other parts. Its internal structure is shown in Fig. 1. According to the different coupling modes, the medium frequency furnace is divided into parallel resonant medium frequency furnace and series resonant medium frequency furnace. Because the series medium frequency furnace has the advantages of good constant power output capacity, fast melting speed and low energy consumption, the series resonant medium frequency furnace is widely used.

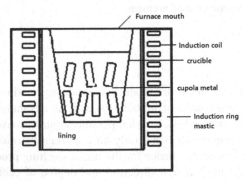

Fig. 1. Internal structure of medium frequency furnace.

The power consumption of medium frequency furnace is very huge. During smelting, it is necessary to continuously measure the charge temperature and adjust the working frequency of medium frequency furnace, so as to use the minimum electric energy and achieve the best smelting effect. If the temperature measurement is not accurate, the medium frequency furnace will not achieve the best smelting effect, and the waste of electric energy will be huge. Therefore, it is very necessary to measure the charge temperature in the medium frequency furnace.

Firstly, this paper introduces the circuit principle of series resonant medium frequency furnace, the influence of temperature on metal resistivity and Transformer model, and simulates and verifies the circuit model of medium frequency furnace. The resistance and temperature are collected through the design of circuit, and processed through transformer model to realize the function of temperature prediction.

2 Principle Introduction

2.1 Principle of Series Resonant Medium Frequency Furnace

The circuit model of the series resonant medium frequency furnace is shown in Fig. 2. The circuit model can be summarized as follows: firstly, the 50 Hz power frequency alternating current is rectified into direct current through a bridge rectifier circuit, and then filtered by the filter (reactor LDC), and then the direct current is inverted into single-phase medium frequency alternating current through the inverter (SCR1-4) to supply the load. The load circuit includes capacitance, inductance and resistance. The inductance is composed of induction coil. Here, the induction coil and charge can be equivalent to a mutual inductance coupling. The induction coil, resistance and capacitance form a primary loop, and the charge can be equivalent to a secondary loop with one turn of inductance.

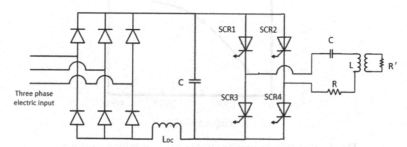

Fig. 2. Series resonant circuit model.

2.2 Effect of Temperature on Metal Resistivity

Resistivity is a physical quantity that reflects the conductivity of a conductor. When the resistivity is low, the conductivity of the material is good; when the resistivity is high, the conductivity of the material is poor. The resistivity is determined by the nature of the material itself, and the resistivity of the same material changes with the change of temperature. Generally, the temperature is high and the resistivity is large. This is also true in metal materials. In reference [16–21], the resistivity of metals presents a nonlinear change trend due to the continuous change of temperature during heating, and the resistivity of different types of metals varies with temperature. Figure 3 shows the resistivity temperature relationship curve of aluminum silicon alloy liquid. Through the curve, it can be seen that after the temperature is higher than 400 °C, The resistivity changes obviously with temperature. Figure 4 shows the change of resistivity of tungsten with temperature. Through the change of metal resistivity with temperature, this paper proposes to find the corresponding relationship between temperature and resistivity in the process of metal smelting, so as to calculate the temperature of charge in the process of smelting.

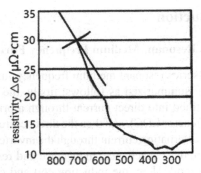

Fig. 3. Resistivity temperature relation curve of Al Si alloy liquid.

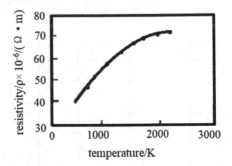

Fig. 4. Variation curve of tungsten resistivity with temperature.

2.3 Transformer Model

Transformer abandons CNN (Convolutional Neural Networks) and RNN (Recurrent Neural Network) used in previous deep learning tasks. At present, the popular Bert is built based on Transformer. This model is widely used in NLP fields, such as machine translation, question answering system, text summarization and speech recognition [22]. This paper uses Transformer model for data processing.

Transformer is essentially an Encoder-Decoder structure. The Encoder is composed of six coding blocks. Similarly, the Decoder is composed of six decoding blocks. Like all generation models, the output of the encoder will be used as the input of the Decoder, as shown in Fig. 5:

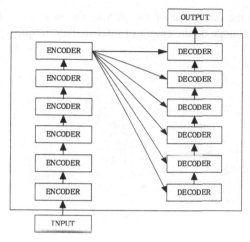

Fig. 5. Transformer model structure.

For the Encoder, there are two layers, a self-attention layer and a feedforward neural network. Self-attention can help the current node not only focus on the current word, but also obtain the semantics of the context. Firstly, the model needs to embed the input data. After embedding, it is input to the Encoder layer. After self-attention processes the data, it sends the data to the feedforward neural network. The calculation of the feedforward neural network can be parallel, and the output will be input to the next Encoder.

The Decoder also includes the two-layer network mentioned by the Encoder, but there is also an attention layer between the two layers to help the current node obtain the key contents that need to be paid attention to at present.

Transformer abandons the traditional CNN and RNN, and the whole network structure is completely composed of attention mechanism. More precisely, Transformer consists of and only consists of self-attention and feed forward neural network. A trainable neural network based on Transformer can be built in the form of stacked Transformer, by building Encoder-Decoder with 6 layers of Encoder and Decoder and 12 layers in total, and refreshing the Bleu value in machine translation. Firstly, it uses the attention mechanism to reduce the distance between any two positions in the sequence to a constant; Secondly, it is not a sequential structure similar to RNN, so it has better parallelism and conforms to the existing GPU framework.

2.4   Simulation Verification of Circuit Model of Medium Frequency Furnace

According to the principle that the metal resistance changes with the temperature under different temperatures, the hardware simulation circuit of the intermediate frequency furnace as shown in Fig. 6 is built on Multisim, in which T1 is the coil on the intermediate frequency furnace, the number of turns is 24, and the number of turns is the measured number of turns of a series resonant intermediate frequency furnace. Xfg1 is the signal source, XMM1 and XMM4 are ammeters, XMM2 and XMM3 are voltmeters. A sliding rheostat with a maximum resistance of 10 mΩ is connected in series with the secondary coil of T1. The change of metal resistance at different temperatures is simulated by artificially changing the resistance of the sliding rheostat, so as to observe the data impact of resistance change on the circuit of intermediate frequency furnace.

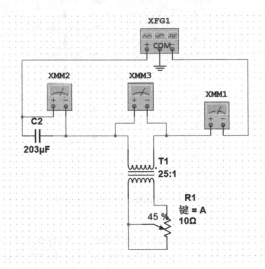

Fig. 6. Simulation circuit construction.

Figure 7 shows the data change after randomly changing the sliding rheostat resistance value of 0.001 mΩ in the secondary circuit. Through the data, it can be seen that after the resistance value of 0.001 mΩ, the current of the primary circuit changes by 0.4 A, the voltage changes by 1 V, and the circuit data changes significantly. Therefore, through the simulation, it can be seen that it is possible to reflect the temperature through the resistance.

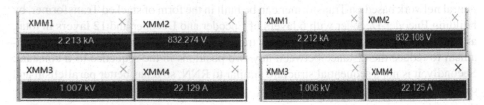

Fig. 7. Simulation data.

3   Acquisition of Training Data of Transformer Model

Through the above simulation, it can be clearly seen that the resistance will change significantly with the change of temperature. Due to the strong eddy current in the charge during the smelting process of medium frequency furnace, it will be very difficult to measure directly. Through the above circuit analysis, due to the influence of mutual inductance coupling, the resistance change of charge will affect the primary circuit, that is, the resistance of charge will produce an equivalent resistance associated with it in the primary circuit. Therefore, this paper calculates the temperature data by calculating and collecting the equivalent resistance.

3.1 Collection Principle of Equivalent Resistance of Charge Resistance in Primary Circuit

Because the temperature of the charge resistance is extremely high during the heating process, and there is a large current of thousands of amperes inside, the method of directly measuring the resistance is unrealistic, so it is necessary to calculate the resistance by other methods. The conditional formula of series resonance is known as formula (1):

$$\omega_0 L - \frac{1}{\omega_0 C} - 0 \qquad (1)$$

According to formula (1), the voltage $\dot{V}_{L_0}$ on L and C during resonance and $\dot{V}_{L_0}$ equal size, phase difference $180°$, $\dot{V}_{L_0} = -\dot{V}_{c_0}$, escribe the intermediate frequency voltage $\dot{V}_s$ input through the primary circuit equal to the voltage drop on $\dot{V}_R$, therefore, the period, frequency and duty cycle of the primary circuit power signal after passing through the inverter can be measured, and the discrete Fourier transform can be done. Take the fundamental component. Since the passband of the resonant circuit has the function of frequency selection for the external voltage, in fact, only the fundamental component is working through the input voltage of the inverter, and the other harmonics are filtered. Therefore, the fundamental component is the input intermediate frequency voltage $\dot{V}_s$ of the primary circuit, then measure the current of the primary circuit $\dot{I}_0$, through resistance value $R = \frac{\dot{V}_s}{\dot{I}_0}$, then R can be obtained, and the equivalent resistance Rr can be calculated through R.

3.2 Construction of Equivalent Resistance Data Acquisition System

Acquisition and Processing of Voltage Data After Inverter. It can be seen from the series resonant circuit model in Fig. 2 that since the input voltage of the primary circuit after passing through the inverter is very large, and it is the voltage mixed with the fundamental component and harmonic component, which cannot be measured directly, the voltage input into the primary circuit after passing through the inverter can be reduced in equal proportion of the order of magnitude through the parallel voltage Transformer, set the sampling rate through Field Programmable Gate (FPGA) for Analog to Digital (AD) sampling to obtain the amplitude, frequency and period of voltage. The specific flow chart is shown in Fig. 8:

Measure the period, frequency and duty cycle of the voltage signal of the primary circuit, and perform discrete Fourier transform on the sampled voltage signal. Due to the selectivity of the passband of the resonant circuit and the frequency selection of the external voltage, only the fundamental component A obtained by discrete Fourier transform acts on the resonant circuit, that is, $A = \dot{V}_s$, based on this, the input voltage $\dot{V}_s$ of the primary circuit can be obtained.

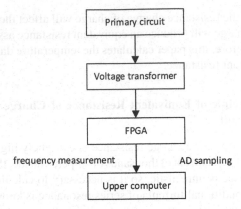

Fig. 8. Primary circuit voltage signal acquisition process.

Acquisition of Primary Circuit Current Data. Primary loop current $\dot{I}_0$ Rogowski coil plus integrator is selected as the measurement method. In the method of sampling the signal, as shown in Fig. 9, Rogowski coil is an AC current sensor, which is a hollow ring coil, which can be directly sleeved on the measured conductor to measure AC current. Rogowski coil is applicable to the measurement of AC current in a wide frequency range. It has no special requirements for conductor and size, and has fast instantaneous response ability. It is widely used in occasions where traditional current measurement devices such as current Transformer cannot be used, especially in high-frequency and large current measurement. Therefore, it is selected to measure the current $\dot{I}_0$ through Rogowski coil.

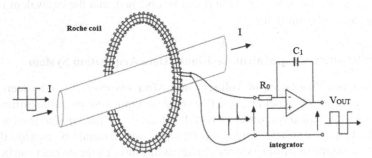

Fig. 9. Schematic diagram of measuring current with Rogowski coil.

The output voltage of the coil can be expressed by the formula $V_{OUT} = M\frac{di}{dt}$. Where m is the mutual inductance of the coil and $\frac{di}{dt}$ is the transformation ratio of the current. Another AC voltage signal can be obtained by integrating the voltage signal output by the coil with a special integrator, which can accurately reproduce the waveform of the measured current signal.

The digital output integrator completes the integration and AD sampling of the coil output voltage signal in the integrator, and uploads the AD sampling results to the secondary instrument or host computer with optical fiber as the medium. It can effectively avoid loss and interference in the transmission process.

Calculation of Equivalent Resistance. After collecting the voltage signal $\dot{V}_s$ and the current signal $\dot{I}_0$ of the primary circuit after passing through the inverter, the total resistance R of the circuit can be obtained through Eq. (2).

$$R = \frac{\dot{V}_s}{\dot{I}_0} \tag{2}$$

where, R = R0 + Rr. R0 represents the coil resistance of the primary circuit and Rr represents the equivalent resistance of the secondary circuit.

3.3 Collection of Charge Temperature Data

In order to ensure the accuracy of charge temperature data acquisition, the temperature acquisition in this paper applies thermocouple to collect. While collecting resistance data, collect its corresponding temperature data under the resistance value to obtain the corresponding data group of temperature and resistance.

4 Using Transformer Model to Process Data

For the data processing of this paper, the Transformer model is used. Using the attention mechanism of the model, the temperature and resistance are translated into two languages. In this paper, 9000 groups of equivalent resistance and temperature data are collected as training data.

The Transformer model consists of two parts: Encoder and Decoder. For the Encoder, there are two layers, a self-attention layer and a feedforward neural network. Self-attention can help the current node not only focus on the current word, but also obtain the semantics of the context. Firstly, the model needs to embed the input data. After embedding, it is input to the encoder layer. After self-attention processes the data, it sends the data to the feedforward neural network. The calculation of the feedforward neural network can be parallel, and the output will be input to the next Encoder. Like the seq2seq model, the Transformer model also adopts the encoder Decoder architecture. However, its structure is more complex than that of attention. The Encoder layer is stacked by six Encoders, and so is the Decoder layer (Fig. 10).

In the Encoder of Transformer, the data will first pass through a module called 'self-attention' to obtain a weighted eigenvector Z, namely Attention (Q, K, V). Self-attention accountant calculates three new vectors. The dimension of the vector can be 512. We call these three vectors query, key and value respectively. These three vectors are the result of multiplying the embedding vector with a matrix. The matrix is randomly initialized and the dimension is (64, 512). Note that the second dimension needs to be the same as

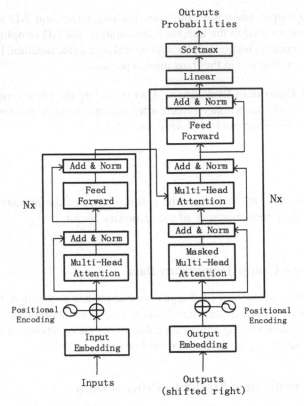

Fig. 10. Transformer overall structure.

the embedding dimension, Its value will be updated all the time in the process of BP. The dimension of these three vectors is 64 lower than the embedding dimension.

The input consists of query vector q and key vector k with dimension d_k and value v with dimension d_v. Then we calculate the dot product of q and all the bonds and divide by $d_k^{1/2}$ and apply the Soft-max function to obtain the weight of these values. In practical application, a set of attention functions of q will be calculated at the same time. These q are packaged into a matrix Q, and k and v are also packaged into matrices K and V. The calculation output matrix is expressed as:

$$Attention(Q, K, V) = soft\max(\frac{QK^T}{d_k^{1/2}})V \tag{3}$$

After obtaining Z, it will be sent to the next module of the Encoder, namely Feed Forward Neural Network. This full connection has two layers. The activation function of the first layer is ReLU, and the second layer is a linear activation function, which can be expressed as:

$$FNN = \max(0, ZW_1 + b_1)W_2 + b_2 \tag{4}$$

The difference between Decoder and Encoder is that the Decoder has an Encoder - Decoder attention, which is used to calculate the weight of input and output respectively. In addition to the main Encoder and Decoder, there is also the part of data preprocessing. Transformer abandons RNN, and the biggest advantage of RNN is the abstraction of data in time series. It proposes two methods of positional Encoding, sums the Encoded data with embedded data, and adds relative position information. After the model is trained, take another part of the data as the data set for performance evaluation.

Transformer model has the following advantages: 1 It breaks through the limitation that RNN model can not be parallel computing, and the amount of parallel computing can be measured by the minimum number of sequence operations required. 2. Compared with CNN, the number of operations required to calculate the correlation between two locations does not increase with distance. Self-attention mechanism can produce more interpretable models.

5 Construction of Burden Smelting Temperature Calculation System

After the above collected temperature and resistance data are trained by Transformer model, the data collected by the resistance acquisition system built on the medium frequency furnace can be imported into the model, so as to build the temperature calculation system. The specific flow chart is as follows (Fig. 11):

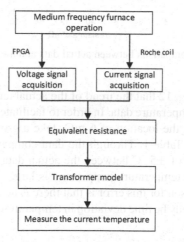

Fig. 11. Build the working process of temperature measurement system.

Firstly, the equivalent resistance acquisition system for voltage signal acquisition and current signal acquisition built on the circuit conductor of the primary circuit is used. After the operation of the medium frequency furnace, the voltage and current data are collected and processed in real time to infer the current equivalent resistance, and the inferred equivalent resistance value is introduced into the Transformer model for operation to calculate the current temperature value, the host computer is used to display the real-time on the screen.

6   Result Analysis

In order to verify the advantages and disadvantages of the temperature measurement system, while measuring the temperature of the system, use thermocouples to measure the temperature of the medium frequency furnace, record the measured temperature data and the actual temperature data, and make a curve, as shown in Fig. 12:

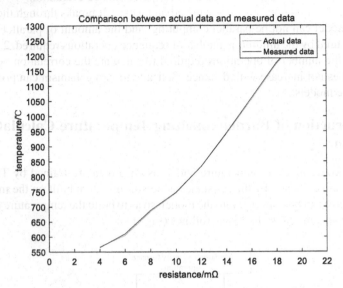

Fig. 12. Comparison between actual data and measured data.

It can be seen from Fig. 12 that the trend of the actual temperature data is the same as that of the measured temperature data. In order to facilitate observation, starting from the two groups of data at the location of 6 mΩ, take a group of data every 2 mΩ to obtain the data shown in Table 1. Through the data comparison in Table 1, it can be seen that there is an error of ±5 °C between the actual data and the measured data. In the smelting industry, the temperature jump range is large, and the error is within the acceptable range. The reason for this error is that there is no error compensation for the collected data, and there may be time deviation in temperature acquisition and resistance acquisition.

Table 1. Actual data and measured data.

Equivalent resistance value/mΩ	Actual temperature/°C	Measured temperature/°C
6	605	610
8	684	688
10	746	748
12	834	834
14	940	943
16	1053	1058
18	1170	1174

7 Conclusion

This paper uses the characteristic that the resistivity of metal will change with the change of temperature in the melting process. A method for continuously measuring the charge temperature of medium frequency furnace is designed. In order to collect the equivalent resistance data, a charge resistance data acquisition system is built. Firstly, the system collects the voltage data and primary circuit current data after passing through the inverter, so as to infer the equivalent resistance of the medium frequency furnace charge in the primary circuit, and collect the corresponding temperature data of the charge while collecting the equivalent resistance data, then use the Transformer model to train the collected data, collect the equivalent resistance of the charge in the primary circuit through the built charge resistance data acquisition system, upload the collected resistance data to the transformer model, calculate the charge temperature data of the current medium frequency furnace, and use the host computer, the measured temperature data of medium frequency furnace charge is displayed in real time.

References

1. Shi, M., Chen, R.: Application of medium frequency furnace in cast steel production. Casting **67**(1), 49–50, 54 (2018). https://doi.org/10.3969/j.issn.1001-4977.2018.01.012
2. Xi, W.: Intermediate Frequency Electric Stove Energy-Saving Principle and Method of Research. Shandong University of Technology, Shandong (2012). https://doi.org/10.7666/d.D586340
3. Saxena, M.R., Maurya, R.K., Mishra, P.: Assessment of performance, combustion and emissions characteristics of methanol-diesel dual-fuel compression ignition engine: a review. J. Traffic Transp. Eng. (Engl. Ed.) **8**(05), 638–680 (2021)
4. Lu, Q.: Application and maintenance of medium frequency furnace. Equip. Manag. Maint. (18), 64–65 (2018). https://doi.org/10.16621/j.cnki.issn1001-0599.208.09D.37
5. Tang, L., et al.: Unravelling the precipitation evolutions of AZ80 magnesium alloy during non-isothermal and isothermal processes. J. Mater. Sci. Technol. **75**(16), 184–195 (2021)
6. Zhou, D., Cheng, S.: A new method to detect the high temperature distribution in the iron-making and steelmaking industry. In: Hwang, J.-Y., et al. (eds.) 8th International Symposium on High-Temperature Metallurgical Processing. The Minerals, Metals & Materials Series, pp. 497–505. Springer, Cham (2017). https://doi.org/10.1007/978-3-319-51340-9_49

7. Sa'id Waladin, K., Jasim Omar, F., Raafat Omar, F.: Estimation of induction furnace charge temperature using multiple model adaptive estimator (MMAE). In: 2019 16th International Multi-Conference on Systems, Signals & Devices (SSD), pp. 207–212. IEEE (2019)

8. Soares Fabio, M., Oliveira Roberto, C.L.: Modelling of temperature in the aluminium smelting process using neural networks. In: International Joint Conference on Neural Networks. IEEE (2010)

9. Li, J., Ma, B.: Parameters adjustment for VOD endpoint carbon content and endpoint temperature prediction model. In: International Symposium on Instrumentation & Measurement. IEEE (2014)

10. Zhai, N., Zhou, X.: Temperature prediction of heating furnace based on deep transfer learning. SENSORS **20**(17), 1–27 (2020)

11. Zhou, P., Guo, D., Wang, H., Chai, T.: Data-driven robust M-LS-SVR-based NARX modeling for estimation and control of molten iron quality indices in blast furnace ironmaking. IEEE Trans. Neural Netw. Learn. Syst. **29**, 4007–4021 (2018)

12. Wang, X.: Ladle furnace temperature prediction model based on large-scale data with random forest. IEEE/CAA J. Autom. Sinica **4**, 770–774 (2017)

13. Lee, S.Y., Tama, B.A., Choi, C., et al.: Spatial and sequential deep learning approach for predicting temperature distribution in a steel-making continuous casting process. IEEE Access **PP**(99), 1 (2020)

14. Zhang, X., Kano, M., Matsuzaki, S.: A comparative study of deep and shallow predictive techniques for hot metal temperature prediction in blast furnace ironmaking. Comput. Chem. Eng. **130**(1), 106575 (2019)

15. Hua, B., Xu, H.: Development and application of non—contact thermometry in combustion process. Instrum. Anal. Monit. (2) (2021). https://doi.org/10.3969/j.issn.1002-3720.2021.02.005

16. Leon-Medina, J.X., Camacho, J., et al.: Temperature prediction using multivariate time series deep learning in the lining of an electric arc furnace for ferronickel production. Sensors, 21, 6894 (2021)

17. Roy, S., Ramana, C.V.: Effect of sintering temperature on the chemical bonding, electronic structure and electrical transport properties of β-Ga_(1.9)Fe_(0.1)O_3 compounds. J. Mater. Sci. Technol. **67**(08), 135–144 (2021)

18. Wang, W., et al.: Microstructure and properties of novel Al-Ce-Sc, Al-Ce-Y, Al-Ce-Zr and Al-Ce-Sc-Y alloy conductors processed by die casting, hot extrusion and cold drawing. J. Mater. Sci. Technol. **58**(23), 155–170 (2020)

19. Cheng, Y., Ma, D., Guo, C., Yang, F., Mu, T., Gao, Z.: An experimental study on the conductivity changes in coal during methane adsorption-desorption and their influencing factors. Acta Geologica Sinica (Engl. Ed.) **93**(03), 704–717 (2019)

20. Bajorek, A., Chekowska, G.: Microstructure and electrical resistivity in the GdNi_(5–x)Cu_x intermetallic series. J. Rare Earths **35**(01), 71–78 (2017)

21. Xu, W., Hou, Y., Song, W., Zhoum Y., Yin, T.: Resistivity and thermal infrared precursors associated with cemented backfill mass. J. Cent. South Univ. **23**(09), 2329–2335 (2016)

22. Ashish, V., Noam, S., Niki, P., et al.: Attention Is All You Need. arXiv (2017)

LQR Optimal Control Method Based on Two-Degree-of Freedom Manipulator

Taotao Song[⊠], Yanping Li[⊠], Shifeng Ma, and Honggang Li

School of Information and Electrical Engineering, Shandong Jianzhu University, Jinan 25010, China

614227901@qq.com, Liyanping0531@126.com

Abstract. The robotic arm is a very complex multi-input, multi-output nonlinear system, and it needs to have high trajectory tracking accuracy. In this paper, a mathematical modeling and optimal trajectory tracking control of a two axis robotic arm is adopted to address the problem of unstable and large input control of the robotic arm. Firstly, the dynamic characteristics of the robot arm are analyzed, and then the dynamic model of the robot arm is established by Lagrange's equation method. Secondly, the control model of the system is derived, and the linear quadratic optimal control is introduced, and the weight matrices R and Q in the quadratic performance index function are optimized by genetic algorithm. The experimental results through simulation show that the optimized control method can effectively improve the convergence speed and stability of the system and obtain a better trajectory tracking effect compared with the traditional empirical values, which verifies the feasibility of the method.

Keywords: Dynamic model · LQR (Linear Quadratic Regulator) · Genetic algorithm · Trajectory tracking

1 Introduction

Robotics is an extremely important engineering discipline in industrialized societies, and in recent years, with the technological development in the field of robotics, the application of robotic arms in life will become possible. At the same time, the need for high precision, high stability and low consumption will become higher and higher. Robots are also used in a wide range of applications, all over the home robots [1] to medical robots [2] and industrial robots [3–5]. And with the development of the times, robots will become more and more common in people's lives. Many scholars are engaged in the study of control methods for robotic arms, and excellent trajectory tracking [6] performance will become extremely important, and the analysis of robotic arms is an inevitable requirement. One of the analysis process contains robot mathematical modeling and control theory analysis parts.

Fundamentally, the robotic arm needs to be described by a complex nonlinear model, while the control of such a model is often tedious and difficult. Therefore, a simplified, appropriate and effective dynamics model can be established for specific analytical

H. Zhang et al. (Eds.): NCAA 2022, CCIS 1637, pp. 477–488, 2022.
https://doi.org/10.1007/978-981-19-6142-7_36

control methods. Therefore, in this paper, a two-degree-of-freedom R-R robotic arm will be modeled and its optimal control will be investigated.

Since the robot is a system with complexity and coupling, and its mathematical model has obvious nonlinear characteristics, the study of the robotic arm necessarily involves solving the dynamics and control problems. The widely used robotic arm control methods include PID (Proportion-Integral-Derivative) control, LQR (Linear Quadratic Regulator) control [7], and robust control [8]. Among them, LQR control is widely used and the fundamental factor is that it can obtain higher control performance at a lower cost.

The literature [9] used LQR to compare PD (Proportional- Derivative) control, but used empirical weighting parameters, which were not effective enough. Literature [10] uses LQR and PID control for comparison, which shows the superiority of LQR over traditional control methods, but the weight parameters have not reached the optimal parameters. The literature [11] used GWO (Gray Wolf algorithm) optimization seeking for the parameters in the LQR controller.

Since the traditional LQR control is based on the weight values Q and R, and Q and R are often determined based on engineering experience. Although, the results can be basically achieved satisfactorily under over by LQR control, it is often difficult to achieve accurate guaranteed weight values when searching for the best weights, meanwhile, genetic algorithm [12] has the feature of global optimization, which can find better weights to optimize the LQR controller to obtain better control effect. Therefore, in this paper, the parameters of the LQR controller are rectified by genetic algorithm and a two-axis robotic arm model is established for comparison, and the method can be obtained to have a better control effect.

2 Dynamical Modeling

The derivation of robot dynamics models [13] plays an important role in robot simulation, structural analysis of robotic arms and the design of control algorithms. And the Lagrange method [14] is a more commonly used energy-based robot dynamics modeling method. In the Lagrange expression, the variables in the equations of motion can be independent of the reference coordinate system of the robotic arm, and these variables will describe the robotic arm relationship more effectively. In this paper, a simplified two-link robotic arm is chosen as the object of study and the second-class Lagrange equation is invoked to derive the robotic arm dynamics equations. When using the Lagrange equations, only the kinetic and potential energies of the system are required, and based on the principle of energy conservation, a model of the robot arm dynamics can be derived in (1):

$$\tau = M(q)\ddot{q} + C(q, \dot{q}) + G(q) \tag{1}$$

M is the inertia matrix, C is the Coriolis and the centrifugal matrix, G is the gravity matrix, τ is the moment.

Dynamical properties:

The inertia matrix is symmetric and positive definite.

The reciprocal of the inertia matrix and the centrifugal force matrix can form an antisymmetric matrix relationship:

$$\dot{M}(q) - 2C(q, \dot{q}) \tag{2}$$

Lagrange function is the difference between the kinetic K and potential P of the system, as shown in (3):

$$L = K - P \tag{3}$$

K and P can be represented with any coordinate system, the Lagrange equation for the dynamic state of the system described by the Lagrange function is:

$$F_i = \frac{d}{dt}\frac{\partial L}{\partial \dot{q}_i} - \frac{\partial L}{\partial q_i} \quad i = 1, 2, \ldots, n \tag{4}$$

where:

n is the number of links in the formula;

q_i is the defined generalized coordinate;

$\dot{q}_i$ is the defined generalized velocity;

F_i is the set of forces defined in generalized coordinates;

The dimension of generalized force F_i is the dimension n of moment and the dimension N/M of moment, which should be determined by $\dot{q}_i$ as linear coordinate and angular coordinate. Whether the dimension of generalized coordinate is m or rad depends on whether is linear coordinate or angular coordinate. Similarly, whether the dimension of generalized velocity $\dot{q}_i$ is m/s or rad/s depends on whether $\dot{q}_i$ is linear velocity or angular velocity.

The model of planar two degree of freedom manipulator is a simplified object, which is conducive to calculation and molecular design. The vertical plane manipulator is easier to handle than the horizontal one. The vertical manipulator model used in this paper is shown in Fig. 1.

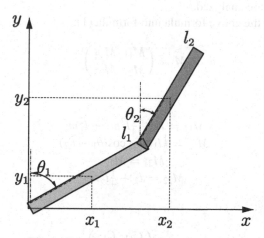

Fig. 1. Two degree of freedom manipulator

Assuming that each manipulator is a homogeneous connecting rod and the center of mass is in the center of the rod, a dynamic model is established for the particle in X-Y, where m_1 is the mass of connecting rod 1, m_2 is the mass of connecting rod 2, l_1 is the length of connecting rod 1, l_2 is the length of connecting rod 2, θ_1 is the angle between connecting rod 1 and Y axis, and θ_2 is the angle between connecting rod 2 and Y axis.

The formula is as follows:

$$x_1 = \frac{1}{2}l_1 \sin \theta_1$$

$$y_1 = \frac{1}{2}l_1 \cos \theta_1$$

$$x_2 = l_1 \sin \theta_1 + \frac{1}{2}l_2 \sin \theta_2$$

$$y_2 = l_1 \cos \theta_1 + \frac{1}{2}l_2 \cos \theta_2 \tag{5}$$

where x_1 and y_1 are the abscissa and ordinate of the centroid of connecting rod 1, and x_2 and y_2 are the abscissa and ordinate of the centroid of connecting rod 2.

According to the fact that the kinetic energy of the manipulator is the sum of translational kinetic energy and rotational kinetic energy, as shown in (6), (7):

$$T_i = \frac{1}{2}m_i\left(\dot{x}_i^2 + \dot{y}_i^2\right) + \left(\frac{1}{2}I_i\dot{\theta}_i^2\right) \tag{6}$$

$$P_1 = m_1 g l_1 \cos \theta_1, \; P_2 = m_2 g \left(l_1 \cos \theta_1 + \frac{1}{2}l_2 \cos(\theta_1 + \theta_2)\right) \tag{7}$$

where T_i is the kinetic energy of each connecting rod, I_i is the Inertia, the kinetic energy is $T = T_1 + T_2$ and the potential energy of the whole is $P = P_1 + P_2$.

According to the formula (6), (7) and (4). The dynamic equations of the two links of the manipulator are analyzed.

By substituting the above formula into formula (1):

$$M = \begin{pmatrix} M_{11} & M_{12} \\ M_{21} & M_{22} \end{pmatrix}$$

where:

$$\begin{aligned} M_{11} &= J_1 + \tfrac{1}{4}l_1^2 m_1 + l_1^2 m_2 \\ M_{12} &= \tfrac{1}{2}l_1 l_2 m_2 \cos(\theta_1 - \theta_2) \\ M_{21} &= M_{12} \\ M_{22} &= J_2 + \tfrac{1}{4}l_2^2 m_2 \end{aligned} \tag{8}$$

C which is given by:

$$C = \begin{pmatrix} C_{11} & C_{12} \\ C_{21} & C_{22} \end{pmatrix}$$

$$\begin{aligned} C_{11} &= -\tfrac{1}{2}l_1 l_2 m_2 \sin(\theta_1 - \theta_2)\dot{\theta}_2^2 \\ C_{12} &= C_{21} = 0 \\ C_{22} &= \tfrac{1}{2}l_1 l_2 m_2 \sin(\theta_1 - \theta_2)\dot{\theta}_1^2 \end{aligned} \tag{9}$$

The gravity vector $G = \begin{bmatrix} G_1 & G_2 \end{bmatrix}^T$ is given by:

$$G_1 = \left(\tfrac{1}{2}gl_1m_1 + gl_1m_2\right) \sin \theta_1$$
$$G_2 = \tfrac{1}{2}gl_2m_2 \sin \theta_2 \tag{10}$$

Finally, F is torque exerted by actuators at each joint.

3 LQR Controller

Linear quadratic controller is an optimal design technique that can simplify the direct model by using state space input. It is a typical method in linear system theory. In essence, it is a feedback coupling control system, as shown in Fig. 2.

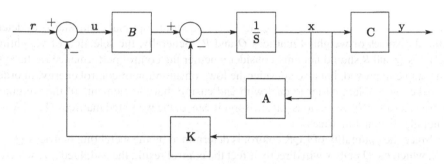

Fig. 2. LQR control block diagram

According to the design requirements, it seeks the optimal control rate $u(t)$. Then looking for the feedback gain value K can make the feedback control system achieve the balance and optimization between output error, system stability and dynamic performance according to the design requirements.

State space equation of linear time invariant continuous system, as shown in (11):

$$\begin{cases} \dot{x}(t) = Ax(t) + Bu(t) \\ y = Cx(t) \end{cases} \tag{11}$$

Error vector $e(t)$ is:

$$e(t) = y_r(t) - y(t) \tag{12}$$

where is the ideal input vector. Then the optimal control problem based on LQR is to find a set of optimal solutions, that is, control quantities. Minimize formula (10).

$$J = \frac{1}{2} \int_0^\infty \left[e^T(t)Qe(t) + u^T(t)Ru(t) \right] dt \tag{13}$$

where: J is the cost function, Q [15] is the error weight matrix, and R is the control weight matrix.

For the LQR optimal tracking problem, under the condition that the state space is controllable and observable, assuming that Q and R are known, the extreme value of the cost function can be obtained, which has the following form

$$u(t) = -R^{-1}B^T P x(t) + R^{-1} B^T g(t) \tag{14}$$

The matrix P is the solution of Riccati algebraic equation, as shown in (15)

$$g(t) = \left(P B R^{-1} B^T - A^T \right)^{-1} C^T Q y_r(t) \tag{15}$$

The feedback gain matrix can be obtained by solving the LQR optimal control, as shown in (17)

$$K = R^{-1} B^T P \tag{16}$$

One of the main difficulties in designing linear quadratic optimal controller is to determine the values of weighted matrices Q and R. Generally, the selection of weighting matrices Q and R should not only consider whether the control performance of the system can be improved, but also consider the low consumption of control energy. In order to make the problem easier to deal with and ensure that the elements of the weighted matrices Q and R have obvious physical significance, the weighted matrices Q and R are generally diagonal matrices.

Since the optimality of LQR control is determined by the weighting matrices Q and R, in which the Q matrix will directly affect the control result, the weighted q is selected to find the best. Therefore, Q can be selected as a four-dimensional positive semi definite diagonal matrix and R as a one-dimensional positive definite matrix. The details are as follows:

$$Q = \begin{bmatrix} q_{11} & 0 & 0 & 0 \\ 0 & q_{22} & 0 & 0 \\ 0 & 0 & q_{33} & 0 \\ 0 & 0 & 0 & q_{44} \end{bmatrix}, R = 1 \tag{17}$$

According to empirical selection, the ranges of the four parameter variables in the Q matrix are set to $q_{11} \in [1 \ 100]$, $q_{22} \in [1 \ 100]$, $q_{33} \in [1 \ 100]$, $q_{44} \in [1 \ 100]$. Through MATLAB, the feedback gain can be obtained according to the given weighting matrix.

4   Genetic Algorithm

Genetic algorithm is a general algorithm to solve the search problem. It has great advantages for selecting the output of the optimal solution. The process is generally to determine parameters, parameter coding, generate population, calculate fitness, generate new population, mutation, crossover and so on. The specific flow chart of genetic algorithm is shown in Fig. 2.

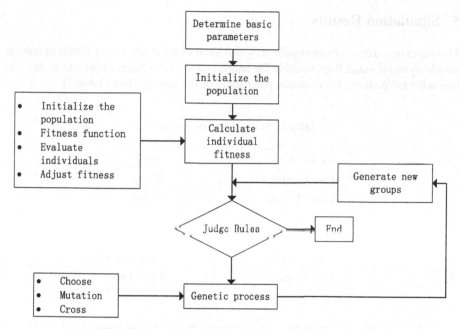

Fig. 3. Flow chart of genetic algorithm optimization

Step1: Determine the constraint range of the control variables to be optimized and generate the initial population;

Step2: The values of each individual in the initial population are successively assigned to the variables, q_{11}, q_{22}, q_{33}, and q_{44} of the weighting matrix Q, and the corresponding optimal control feedback gain matrix K is obtained according to the formula (10). The optimal control input $u(t)$ is obtained, and the control model of the system is simulated in the environment of MATLAB [16] software SIMULINK;

Step3: Combining each individual in the initial population with constraints, the dynamic response of the system under each parameter is calculated respectively, and the objective function value J of each individual is output after calculation;

Step4: Take the individual fitness as the corresponding objective function and judge whether the termination conditions of genetic algorithm are met according to the calculation results. If not, carry out genetic operation, otherwise directly skip to step 6;

Step5: Carry out genetic operation, set the initial population size $M = 20$, the termination evolution algebra $G = 100$, the crossover probability $PC = 0.5$, and the mutation probability $PM = 0.002$. According to the objective function value of each individual, select, crossover and mutation genetic operators according to the given crossover probability and mutation probability, generate new population individuals, jump to step 2, and cycle the optimization process of genetic algorithm;

Step6: Output the individual value of the population corresponding to the minimum value of the objective function, and calculate the optimal control of the manipulator according to the optimal result obtained by the algorithm.

5 Simulation Results

This paper uses a two-axis manipulator, which has the characteristics of simple operation, simple dynamic modeling, suitable for analysis and control algorithm and so on. The following table shows the dynamic parameters of the manipulator (Table 1).

Table 1. Manipulator specifications

No	Parameter	Value
1	Link 1 length	50 cm
2	Link 2 length	50 cm
3	Link 1 mass	0.4 kg
4	Link 2 mass	0.4 kg
5	Link 1 inertia	0.027 kg/m^2
6	Link 1 inertia	0.027 kg/m^2

In order to verify the trajectory tracking problem of manipulator under the control of linear quadratic controller optimized by genetic algorithm, numerical analysis and simulation experiments are carried out with MATLAB software. Firstly, LQR is simulated according to the empirical value, and $q_{11} = 100$, $q_{22} = 100$, $q_{33} = 1$, $q_{44} = 1$, $r = 1$ is selected according to the traditional experience. The optimized feedback gain matrix is:

$$K = [-28.8100 \quad -163.3252 \quad -24.6687 \quad -26.2270]$$

In the process of genetic algorithm optimization, the optimal individual fitness function value can be obtained. As shown in Fig. 3, with the continuous change of the population, and with the continuous decrease of the individual fitness function value, until the optimal individual is obtained after 10 generations. The optimal individual value is $q_{11} = 98.8$, $q_{22} = 100$, $q_{33} = 19.2751$, $q_{44} = 1$, $r = 1$. The optimal feedback gain matrix obtained by optimization is:

$$K = [-28.8141 \quad -95.6368 \quad -15.8670 \quad -14.9760]$$

The parameters of LQR controller are optimized by genetic algorithm, and the actual effect can be reflected by the change of cost function value. As shown in Fig. 4, the evolution process of genetic algorithm.

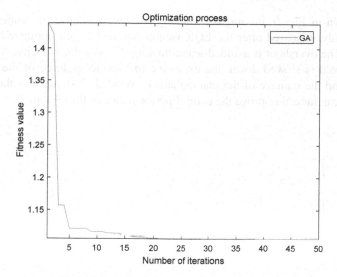

Fig. 4. Genetic algorithm optimization curve

As shown in Fig. 5, the angle tracking response of connecting rod 1 is significantly improved after the LQR weight parameters are optimized by genetic algorithm. Reach the specified position faster and reduce the error faster.

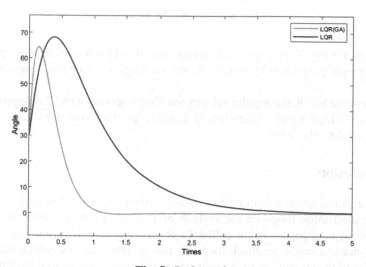

Fig. 5. Rod 1 angle

As shown in Fig. 6, the angular velocity tracking response of connecting rod 1 is significantly improved after the LQR weight parameters are optimized by genetic algorithm. The overshoot is avoided when tracking the angular velocity. Although the response speed is slowed down, the excessive instantaneous force of the moment is prevented and the damage of the manipulator is avoided. It shows that the optimized parameters can indeed improve the control performance of the system.

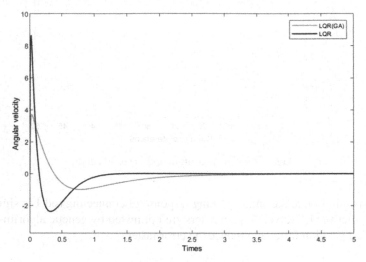

Fig. 6. Rod 1 angular velocity

As shown in Fig. 7, the angle tracking response of rod 2 is improved after optimizing the LQR weight parameters by genetic algorithm. Angle tracking response is relatively faster.

As shown in Fig. 8, the angular velocity tracking response of rod 2 is improved after optimizing the LQR weight parameters by genetic algorithm. Angular velocity tracking response is relatively faster.

6   Conclusion

The two-axis manipulator with uniform mass distribution is studied and modeled by Lagrange mechanics. Based on the analysis of a typical LQR controller, the position motion of the manipulator is controlled. LQR has only a small amount of overshoot and only needs a single feedback loop and two weight matrices, which reduces the complexity of the system. At the same time, genetic algorithm is used to optimize the weight matrix parameters of linear quadratic form in trajectory tracking control. The experimental results show that the weight matrix optimized by genetic algorithm can better balance the control effect and input. Compared with traditional experience, the optimized results can effectively shorten the response time and avoid excessive input impact. It is expected that LQR control optimized by Particle swarm optimization and genetic algorithm may lead to a better response which is left for future work.

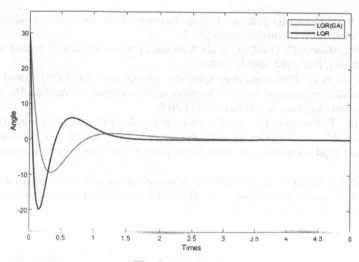

Fig. 7. Rod 2 angle

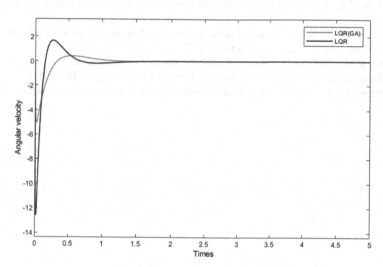

Fig. 8. Rod 2 angular velocity

References

1. Kidd, C.D., Breazeal, C.: Robots at home: Understanding long-term human-robot interaction. In: IEEE/RSJ International Conference on Intelligent Robots and Systems. IEEE (2008)
2. Beasley, R.A.: Medical robots: current systems and research directions. J. Robot. **2012**(1), 401613.1-401613.14 (2012)
3. Li, L.I., Shang, J., Feng, Y., et al.: Research of trajectory planning for articulated industrial robot: a review. Comput. Eng. Appl. **54**, 36–50 (2018)
4. Kim, S.H., Nam, E., Ha, T.I., et al.: Robotic machining: a review of recent progress. Int. J. Precis. Eng. Manuf. **20**(9–12), 1629–1642 (2019). https://doi.org/10.1007/s12541-019-001 87-w

5. Dauth, W., Findeisen, S., Südekum, J., et al.: German robots - the impact of industrial robots on workers. CEPR Discussion Papers (2017)
6. Yuan, M., Manzie, C., Good, M., et al.: A review of industrial tracking control algorithms. Control. Eng. Pract. **102**, 104536 (2020)
7. Dumlu, A., et al.: Design and linear quadratic optimal regulator (LQR) control of 6-DOF RSS parallel manipulator. In: 2015 International Conference on Applied Mechanics and Mechatronics Engineering (AMME 2015) (2015)
8. Zhou, J.L.: Robust optimal control for robotic manipulator. Comput. Eng. Appl. (2007)
9. Fabian, J., Monterrey, C, Canahuire, R.: Trajectory tracking control of a 3 DOF delta robot: a PD and LQR comparison. In: IEEE XXIII International Congress on Electronics. IEEE (2017)
10. Saraf, P., Ponnalagu, R.N.: Modeling and simulation of a point to point spherical articulated manipulator using optimal control. In: 2021 7th International Conference on Automation, Robotics and Applications (ICARA) (2021)
11. Choubey, C., Ohri, J.: Gwo-based tuning of LQR–PID controller for 3-DOF parallel manipulator (2021)
12. Vose, M.D.: The Simple Genetic Algorithm, vol. 1, pp. 31–57. MIT Press, Cambridge (1999)
13. Guo, L., Liao, Q., Wei, S., et al.: Design of linear quadratic optimal controller for bicycle robot. In: IEEE International Conference on Automation and Logistics, pp. 1968–1972. IEEE (2009)
14. Li, X., Wang, X., et al.: A kind of Lagrange dynamic simplified modeling method for multi-DOF robot. J. Intell. Fuzzy Syst. Appl. Eng. Technol. **31**, 2393–2401 (2016)
15. Doina, Z.M.: LQG/LQR optimal control for flexible joint manipulator. In: International Conference and Exposition on Electrical and Power Engineering. IEEE (2013)
16. Coleman, T.F., Zhang, Y.: Optimization toolbox for use with MATLAB. Rice University (1999)

Multi-layer Echo State Network with Nonlinear Vector Autoregression Reservoir for Time Series Prediction

Heshan Wang[✉], Yuxi Liu, Dongshu Wang[✉], Yong Luo, and Jianbin Xin

College of Electrical Engineering, Zhengzhou University, Zhengzhou 450001, People's Republic of China
{h.s.wang,wangdongshu,luoyong,j.xin}@zzu.edu.cn

Abstract. Reservoir Computing (RC) defines a Recurrent Neural Network (RNN) method that optimally processes transient information that generated based on recurrent tasks utilizing sequential datasets. As a typical actualization of the RC classic, Echo State Network (ESN) can train the transient data by linear optimization using very small training datasets and make the network operations fairly fast. However, classical ESNs use a completely randomized reservoir matrix to delineate the fundamental recursive operations and are accompanied by a large number of hyper-parameters to be optimized, such as input weights, reservoir weights, warm-up time steps, and activation functions. The recently proposed Nonlinear Vector Autoregressive (NVAR) algorithm which does not require random inputs and reservoir weights is equated with the traditional ESN reservoir. As a future direction of RC development, NVAR requires few hyper-parameters to be optimized. In this study, a multi-layer ESN model based on the NVAR (NVAR-MLESN) is constructed and applied to time series prediction. Three sequential benchmarks which are synthetic Mackey-Glass chaotic sequence, Sunspot sequence, and real-world Santa Fe Laser sequential dataset are implemented to show the advantage of NVAR-MLESN. Exhaustive simulated outcomes exhibit that our NVAR-MLESN is efficient to develop the performance of classical MLESN and single layer NVAR-ESN. Meanwhile, NVAR-MLESN has fewer parameters to be optimized, faster computing speed, and better performance comparing with the traditional deep ESNs or MLESNs.

Keywords: Multi-layer echo state network · Reservoir computing · Time series forecasting · Nonlinear vector autoregression reservoir

1 Introduction

Reservoir Computing (RC) [1–4] is a special variation of the classical Dynamic or Recurrent Neural Network (RNN) network which is ground on the concept of dividing the nonlinear time-dependent coefficient of the RNN. RC is especially suitable for learning time series datasets due to its dynamic reservoir which is initially generated completely randomly in the condition of stability constraints. Echo state network (ESN) [5, 6] is a typical actualization of the RC paradigm. The weight metrics of the hidden layer and input layer are constructed through a fully randomized manner under a stability constraint but keep unchanged during the training process, which can make the computational power of training process to focus simply on the output layer of the ESN algorithm, thus causing for an astonishing fast training speed. ESNs have been variously utilized in numerous areas, such as chaotic sequence forecasting [7], automatic speech or voice recognition [8, 9], motion identification [10], time series classification [11–13], energy consumption forecasting [14, 15], feature extraction [16] and batch bioprocesses [17, 18].

However, the using of random input and reservoir matrix in an ESN often leads to problems: the best input and hidden layers weights usually need a good design principle or extensive experience, as the recurrent and input layers weights are randomly created before the model creation and a few hyper-parameters of the reservoir which can significantly affect its performance need to be optimized [19, 20]. Recent researches have made lots of efforts to design good recurrent matrices and algorithms to enhance the input layer weights, e.g., Ma et al. [21] employed a linear topology reservoir ground on a given short-term memory for adjusting the edge of chaos. An ESN with an improved topology had been investigated through a smooth composite reservoir activation function for sequence forecasting by Li et al. [22]. In particular, Daniel et al. [23] proposed a Nonlinear Vector Autoregressive (NVAR) type reservoir which does not require random inputs and reservoir matrices marks the future direction of ESN.

It has been shown in [23] that mathematically RC and NVAR can be fully equivalent. In NVAR, no input weights as well as reservoir weights are needed: the characteristic representation vector of NVAR consists of k time-delayed representations of the dynamic input datasets. Thus, the classical reservoir is equated by a combination of linear and nonlinear operations on these observations, namely, the reservoir can be implicit in a designed NVAR. Notably, it is actually much easier to design an NVAR than an optimal reservoir for a specific dynamic systems owing to few parameters design of NVAR. In consequence, the study of NVAR-based ESN can effectively reduce the parameter design of ESN models, increase the training speed and interpretability of ESNs.

Lately, the communities highlighted the network structure of ESN and how to represent more multiple time-scale temporary information by ESNs [24, 25]. Multi-layer ESNs (MLESN) that sequentially connect each fixed reservoir externally with other reservoirs allow for a long term memory, and this spontaneous MLESNs allow an abundant multiple time-scales representation of the transient information [26]. Overall, MLESNs have many potential advantages, which enable ESNs to achieve better performance and more multi-timescale transient information, and MLESNs will be an important future direction for the ESN development. However, there are still several disadvantages of MLESNs comparing with single-layer ESNs, such as more hyper-parameters to be optimized and initialized, slower training speed, etc.

Based on the aforementioned theories and algorithms, we utilized the novel RC equivalent framework proposed in [23] to develop a MLESN model with NVAR (NVAR-MLESN) algorithm for sequential data forecasting. Initially, the input sequence is directly converted as the output of the first NVAR layer which does not require the weights of the recurrent and input layers; Furthermore, the output of the first or bottom

NVAR module is treated as the input of the next NVAR module, and so on; Finally, the output layer of the NVAR-MLESN is still kept as a linear output layer as the traditional ESNs.

The rest of this study is structured as listed below: Sect. 2 gives a concise roundup of traditional ESN and MLESN. Section 3 introduces in detail the NVAR algorithm and the establishment process of NVAR-MLESN, which follows the time series forecasting simulation outcomes introduced by Sect. 4, along with a discussion. An abbreviate conclusion part is addressed in Sect. 5.

2 Echo State Network

2.1 Traditional Echo State Network

The idea of a traditional ESN is defined as follows: W_u^s indicate the weights from K input units $U(t)$ to the N reservoir neurons $S(t)$, W_s^s represents an N dimensional sparse recurrent weights matrix, W_s^o indicates an L dimensional readout weights connection matrix, the output datasets are denoted by $O(t)$. The framework of a traditional ESN is displayed in Fig. 1. The supervised ESN training method is obtained by renovating the recurrent layer state $S(t)$ and the final actual output $O(t)$ at time step t as shown in the following equations:

$$S(t) = f(W_u^s \cdot U(t) + W_s^s \cdot S(t-1) + W_o^s O^T(t-1)) \tag{1}$$

$$O(t) = f^{\text{out}}(S^T(t) \cdot W_s^o) \tag{2}$$

where f represents the activation function of the recurrent layer (commonly a hyperbolic tangent function described as $tanh(.)$), f^{out} denotes the activation function of the readout layer (typically a linear function described as $lin(.)$), W_o^s represents the feedback weights.

W_s^s, W_o^s and W_u^s are randomly initialized weights which generated through a uniform distribution number set. Meanwhile, the aforementioned weights are kept unchanged in the training and testing procedure. Only W_s^o requires to be tuned by the supervised linear regression algorithm. In order to accomplish a property of echo state, W_s^s is typically scaled as,

$$W_s^s \leftarrow \frac{\alpha}{|\lambda_{\text{max}}|} W_s^s \tag{3}$$

where $|\lambda_{\text{max}}|$ represents the spectral radius of recurrent weights and the range of scaling hyperparameter is $0 < \alpha < 1$

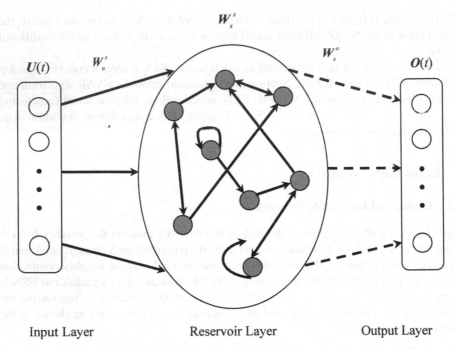

Fig. 1. Architecture of traditional ESN framework. Black lines are fixed and untrainable synaptic connections. Dashed lines represent the trainable synaptic connections of the output layer.

Once the training initialization is complete, the training process of tradition ESN is technically based on supervised calculation of the corresponding recurrent layer states which are ensembled in matrix S as the following equation,

$$S = \begin{bmatrix} s^T(1) \\ s^T(2) \\ \vdots \\ s^T(m) \end{bmatrix} \tag{4}$$

meanwhile the matching target readouts are converged in matrix O,

$$O = \begin{bmatrix} o(1) \\ o(2) \\ \vdots \\ o(m) \end{bmatrix} \tag{5}$$

where m denotes the length of training datasets. In consequence, the practical of a elementary linear regression method (e.g., least square method or ridge regression algorithm) to obtain the output weights W_s^o based on matrix S and O,

$$S \cdot W_s^o = O \tag{6}$$

the traditional method is to utilize the least squares algorithm,

$$W_s^o = \arg \min_w \| Sw - O \|^2 \tag{7}$$

where $\| \cdot \|$ represents the Euclidean norm. Then W_s^o can be computed in one step utilizing the Moore-Penrose pseudo inverse as shown as follows,

$$W_s^o = \tilde{S}O = (S^T S)^{-1} S^T O \tag{8}$$

where $\tilde{S}$ represents the generalized inverse of matrix S.

2.2 Multi-layer Echo State Network

The classical single-layer reservoir lacks the ability to extract the feature information of multiple time scales for the transient features of different time-scale time series. Therefore, numerous different hierarchical ESNs [27–29] have been addressed for different forecasting and classification tasks based on the powerful learning capability and multi-scale transient information stacking property of MLESN. The structure of traditional MLESN is explained as Fig. 2. An MLESN consists of an input layer $U(t)$, n multiple recurrent layers, and is connected to the readout layer simultaneously $O(t)$. $[W_u^1, W_u^2, ..., W_u^n]$ indicate the connection weights of n reservoirs respectively. $[W_u^1, W_u^2, ..., W_u^n]$ represent the internal weights of n reservoirs respectively.

For the MLESN, the output state update equation of ith reservoir $S_i(t)$ and the model readout matrix $O(t)$ at time step t are given as shown as the following equations:

$$S_i(t) = f(W_u^i \cdot S_{i-1}(t) + W_s^i \cdot S_i(t-1)) \tag{9}$$

$$O(t) = f^{\text{out}}(S_n^T(t) \cdot W_s^o) \tag{10}$$

3 MLESN with Nonlinear Vector Autoregression Reservoir

3.1 Nonlinear Vector Autoregression Reservoir

It has been shown in literature [30, 31] that a reservoir with a linear activation function combined with a nonlinear autoregressive feature vector can be equivalent to a powerful universal function approximator. Based on the above deduction, Daniel et al. [23] proposed the NVAR theory which can mathematically identical to traditional RC. In the proposed NVAR algorithm. There is no requirement for input weights and reservoir at all in the NVAR, and the NVAR feature vector consists of only linear and nonlinear combinations of k time-delayed observations as displayed in Fig. 3.

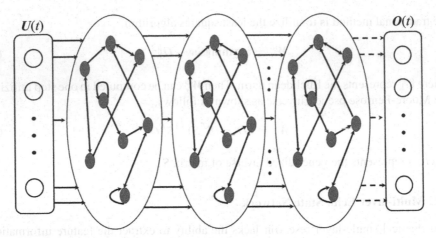

Fig. 2. The structure of traditional MLESN.

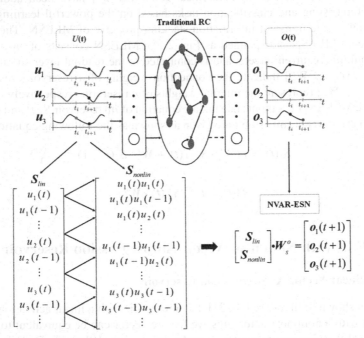

Fig. 3. Equivalent process of a conventional RC and NVAR.

It has been demonstrated that the NVAR is easier to design than traditional reservoir, and the advantages of NVAR are shown in three challenging benchmark areas [23]: (1) predicting the short term dynamic of a time series dataset; (2) reconstructing the attractors and the long term dynamics of a chaotic signal. (3) Predicting the unseen behavior of a dynamic system. As shown in Fig. 3, the reservoir output of a traditional ESN is equivalent to a combination of a linear vector autoregression matrix S_{lin} $(m \times k)$

and a d-th order nonlinear vector autoregression matrix S_{nonlin} as shown as follows,

$$S_{\text{lin}} = \begin{bmatrix} u_1(t) & u_1(t-1) & ... & u_1(t-k) \\ u_2(t) & u_2(t-1) & ... & u_2(t-k) \\ \vdots & \vdots & \vdots & \vdots \\ u_m(t) & u_m(t-1) & ... & u_m(t-k) \end{bmatrix} \tag{11}$$

$$S_{\text{nonlin}} = \begin{bmatrix} u_1(t)u_1(t)...u_1(t) & u_1(t)u_1(t)...u_1(t-1) & ... & u_1(t)u_1(t)...u_1(t-k) \\ u_1(t)u_2(t)...u_2(t) & u_1(t)u_2(t)...u_2(t-1) & ... & u_1(t)u_2(t)...u_1(t-k) \\ \vdots & \vdots & \vdots & \vdots \\ u_1(t)u_2(t)...u_d(t) & u_1(t)u_2(t)...u_d(t-1) & ... & u_1(t)u_2(t)...u_d(t-k) \end{bmatrix} \tag{12}$$

Then, the output matrix of traditional reservoir can be equated to a combination of S_{lin} and S_{nonlin}. In this case, the linear regression process can be defined as,

$$\begin{bmatrix} S_{lin} \\ S_{nonlin} \end{bmatrix} \cdot W_s^o = S_{total} \cdot W_s^o = O \tag{13}$$

Thus, the readout weights W_s^o in NVAR-ESN can be computed through the following equation,

$$W_s^o = (S_{total}^T S_{total})^{-1} S_{total}^T O \tag{14}$$

3.2 Multi-layer Nonlinear Vector Autoregression Reservoir

Although NVAR-ESN can effectively solve the problems such as too many parameters of traditional ESN, dilemma in constructing reservoir. Single-layer NVAR-ESN may still not meet the requirement of feature extraction performance for multi-scale time series data. Therefore, considering the multi-layer mechanism of MLESN and the advantages of NVAR (few parameters to be optimized and interpretability), a NVAR-MLESN with multiple NVAR in series configuration is employed in this paper. The framework of the addressed NVAR-MLESN is explained in Fig. 4.

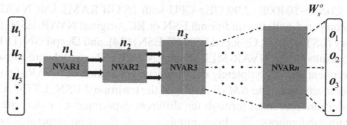

Fig. 4. Structure of NVAR-MLESN. (n_1, n_2, n_3, ...) denote the output dimension of (NVAR1, NVAR2, NVAR3,...) respectively.

The NVAR-MLESN comprises an input layer, n multiple NVAR layers (NVAR1, NVAR2,..., NVARn) which consists of a linear and nonlinear combination of the k time-delayed previous inputs, and an readout layer. The input units number denoted as K, the length of the training set is represented as m, and the output neurons are L. The output dimension $(n_1, n_2, n_3, ...)$ of every NVAR is determined by the time-delayed parameter k and the polynomial order d based on Eqs. (11)–(13) as shown in the following equation,

$$n_{i+1} = \frac{\sum_{i=1}^{n_i} (i+1)i}{2} + d \ (n_1 = k) \tag{15}$$

The idea of NVAR-MLESN in this paper is presented as follows. First, the linear vector autoregression matrix S_{lin} is explained by Eq. (11). Thus, the readout of NVAR1 $S_{\text{nonlin}}(\text{NVAR1})$ is explained by Eq. (12) and the first $S_{\text{nonlin}}(\text{NVAR1})$ is treated as the input of NVAR2 and so on. Finally, the last output state of NVARn, S_{total} is concatenated by S_{lin} and S_{nonlin} as shown in Eq. (13), and the readout weights of NVAR-MLESN can be computed by Eq. (14). Generally, the performance of the ESN-based models is evaluated by the Normalized Mean Square Error ($NMSE$),

$$NMSE = \sum_{t=1}^{m} \frac{(o(t) - \hat{o}(t))^2}{m\sigma^2} \tag{16}$$

where the real output is described as $\hat{o}(t)$ and the wanted readout is represented as $o(t)$, σ^2 is the variance of the actual output, and m is the amount length of $o(t)$.

4   Time Series Prediction Simulation Outcomes

A widespread simulating evaluation of the employed NVAR-MLESN network ground on NVAR theory is proposed as follows, applying several sequential tasks which are generally utilized in the community of RC [32–36]: Sunspot time series dataset, Santa Fe Laser sequence and the synthetic Mackey-Glass (MG) sequential dataset. In the next sections, each benchmark for forecasting is split into three portions in order to train, validate and test the prediction model, respectively. The validating portion is utilized to create the initial parameters k and d. The hyperparameters that make the NVAR-MLESN obtain the minimum validating loss are utilized as the last hyperparameter. k is selected from an integer range of 1–10 with an interval of 1. d is selected from an integer range of 2–5 with an interval of 1. All the forecasting simulation experiments are operated on Intel-based Core i5-10400F (2.90 GHz CPU with 16 GB RAM). Our NVAR-MLESN network is compared with the traditional ESN or RC, original NVAR-ESN, long short-term memory (LSTM) [37], Grouped ESN (G-ESN) [24], and Deep ESN (D-ESN) [26]. The $NMSE$ results of the NVAR-MLESN and NVAR-ESN models are presented only by one experiment due to completely deterministic generation of NVAR if parameters k and d are set in advance. The $NMSE$ results of the traditional ESN, LSTM, G-ESN and D-ESN models are presented through ten different experiments, as they are generated with a certain randomness. The layer number of all the deep structures used in this paper is set to 2 ($n = 2$) because the model structure will become more complex when n is greater than 3 according to Eq. (15). n_1 and n_2 are not set in advance because it is calculated directly from k and d. Part of the parameter settings are displayed in Table 1.

Table 1. Parameter values for the selected NVAR-MLESN of different tasks.

Task	k	d	n	n_1	n_2	Data length	Training length	Validation length	Testing length
Sunspot	8	2	2	36	666	2,899	1,000	1,000	899
Laser	7	2	2	28	406	10,000	7,000	1,000	2,000
MG	6	2	2	21	231	5,000	2,000	1,000	2,000

4.1 Experimental Data Presentation

The Sunspot sequential dataset is a time-dependent real-world benchmark. This benchmark is usually proposed to test the one-step prediction performance of the ESN. The Santa Fe laser dataset is a real-world complex periodicity laser sequence which is generally addressed as a measure to evaluate the one-step performance of different sequential forecasting models. The laser benchmark gained through taking of far-infrared laser samples is generally used to validate the one-step forecasting performance. The laser simulation experiment is a near-chaotic sequence, which includes a lot of roundoff noise and variable time scales, which makes the laser task extremely complex. The MG synthetic benchmark is a traditional task based on a chaotic attractor for the one-step or mult-step forecasting model which has been frequently utilized to ESN predicting models in the community. The MG dataset is obtained from a time-delay differential function as shown as follows:

$$\frac{\partial o(t)}{\partial t} = \frac{0.2o(t - \alpha)}{1 + o(t - \alpha)^{10}} - 0.1o(t) \tag{17}$$

4.2 Model Performance and Comparison

The three time series tasks are employed in order to forecast the next sequential point (one-step ahead prediction). The Fitting regression plots over a chosen length 300 obtained by NVAR-MLESN for the above time series tasks introduced in Sect. 4.1 is shown in Fig. 5. The *NMSE* results of NVAR-MLESN and aforementioned other forecasting models for the three sequential tasks are shown in Tables 2, 3 and 4.

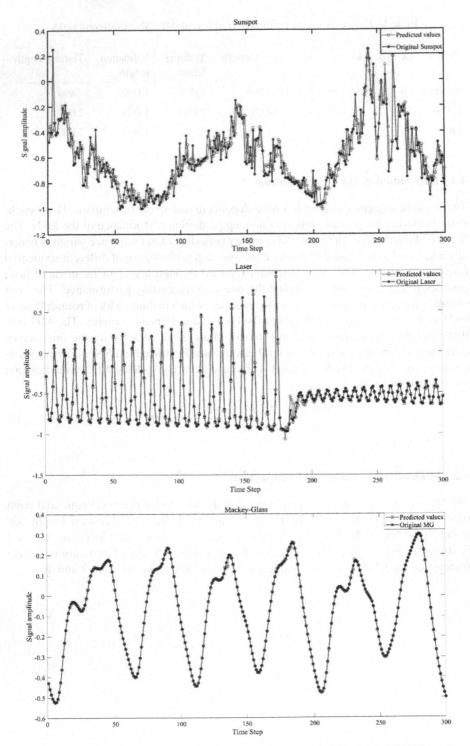

Fig. 5. Output and error performance of RUP-AE-SCRN and SCRN for the NARMA task.

4.3 Discussion

The experimental results clearly show that NVAR-MLESN significantly improves the performance of ESN, also continues to improve the performance of NVAR-ESN. As we can see from Tables 2, 3 and 4, NVAR-MLESN increases the computing time of NVAR-ESN, however the computation time of NVAR-MLESN is still comparable to ESN and less than that of G-ESN and D-ESN. The NVAR-MLESN model performs better than other models in all the one-step ahead prediction experiments.

Table 2. The testing *NMSE* results of NVAR-MLESN and other forecasting models for the Sunspot benchmark.

Model	N	Testing *NMSE*	Training time (s)
ESN	500	0.1569	0.541
LSTM	20	0.1606	3.417
G-ESN	$[n_1 = 300, n_2 = 300]$	0.1325	2.071
D-ESN	$[n_1 = 200, n_2 = 200]$	0.1425	1.988
NVAR-ESN	44	0.0982	0.447
NVAR-MLESN	$[n_1 = 36, n_2 = 666]$	**0.0898**	0.605

Table 3. The testing *NMSE* results of NVAR-MLESN and other forecasting models for the Laser benchmark.

Model	N	Testing *NMSE*	Training time (s)
ESN	500	0.0167	0.747
LSTM	20	0.0178	5.312
G-ESN	$[n_1 = 300, n_2 = 300]$	0.0110	2.976
D-ESN	$[n_1 = 200, n_2 = 200]$	0.0083	2.376
NVAR-ESN	35	0.0101	0.545
NVAR-MLESN	$[n_1 = 28, n_2 = 406]$	**0.0071**	0.806

Table 4. The testing *NMSE* results of NVAR-MLESN and other forecasting models for the MG benchmark.

Model	N	Testing *NMSE*	Training time (s)
ESN	500	4.566e−3	0.662
LSTM	20	3.046e−3	4.203
G-ESN	$[n_1 = 300, n_2 = 300]$	3.012e−3	2.564
D-ESN	$[n_1 = 200, n_2 = 200]$	2.918e−3	2.009
NVAR-ESN	27	2.803e−3	0.478
NVAR-MLESN	$[n_1 = 21, n_2 = 231]$	**2.658e−4**	0.714

5 Conclusion

A novel NVAR-MLESN prediction network is addressed in this study to extend the applicability of NVAR-ENS and increase the feature extraction properties of NVAR-ESN for multi-timescale time series data, through the NVAR theory. Three frequently utilized sequential tasks are proposed to evaluated the merits of the NVAR-MLESN. The comprehensive simulation results demonstrate that the proposed NVAR-MLESN can significantly allow performance of the NVAR-ESN and original Deep ESNs improved. Meanwhile, NVAR-MLESN does not require connection weights between each NVAR layer and hidden layer recurrent weights in the establishment process, which significantly reduces the amount of parameter optimization and initialization difficulty of traditional Deep ESN or MLESN.

Acknowledgments. The authors gratefully acknowledge the support of the following foundations: National Natural Science Foundation of China (61603343 and 62173309), Key Research Projects of Henan Higher Education Institutions (22A413009).

References

1. Tanaka, G., Yamane, T., Héroux, J.B., et al.: Recent advances in physical reservoir computing: a review. Neural Netw. **115**, 100–123 (2019)
2. Lukoševičius, M., Jaeger, H.: Reservoir computing approaches to recurrent neural network training. Comput. Sci. Rev. **3**(3), 127–149 (2009)
3. Lukoševičius, M., Jaeger, H., Schrauwen, B.: Reservoir computing trends. KI-Künstliche Intelligenz **26**(4), 365–371 (2012)
4. Verstraeten, D., Schrauwen, B., d'Haene, M., et al.: An experimental unification of reservoir computing methods. Neural Netw. **20**(3), 391–403 (2007)
5. Jaeger, H., Haas, H.: Harnessing nonlinearity: predicting chaotic systems and saving energy in wireless communication. Science **304**, 78–80 (2004)
6. Jaeger, H.: The "echo state" approach to analysing and training recurrent neural networks. Technology GMD Technical Report 148, German National Research Center for Information, Germany (2001)

7. Li, D., Han, M., Wang, J.: Chaotic time series prediction based on a novel robust echo state network. IEEE Trans. Neural Netw. Learn. Syst. 23(5), 787–799 (2012)
8. Skowronski, M.D., Harris, J.G.: Automatic speech recognition using a predictive echo state network classifier. Neural Netw. 20(3), 414–423 (2007)
9. Trentin, E., Scherer, S., Schwenker, F.: Emotion recognition fromspeechsignals via a probabilistic echo-state network. Pattern Recogn. Lett. 66, 4–12 (2015)
10. Ishu, K., van Der Zant, T., Becanovic, V., et al.: Identification of motion with echo state network. In: MTS/IEEE Techno-Ocean 2004 (IEEE Cat. No. 04CH37600), pp. 1205–1210. IEEE (2004)
11. Wang, L., Wang, Z., Liu, S.: An effective multivariate time series classification approach using echo state network and adaptive differential evolution algorithm. Expert Syst. Appl. 43, 237–249 (2016)
12. Ma, Q., Shen, L., Chen, W., et al.: Functional echo state network for time series classification. Inf. Sci. 373, 1–20 (2016)
13. Tanisaro, P., Heidemann, G.: Time series classification using time warping invariant echo state networks. In: The 15th IEEE International Conference on Machine Learning and Applications, pp. 831–836. IEEE (2016)
14. Hu, H., Wang, L., Lv, S.X.: Forecasting energy consumption and wind power generation using deep echo state network. Renew. Energy 154, 598–613 (2020)
15. Hu, H., Wang, L., Peng, L., et al.: Effective energy consumption forecasting using enhanced bagged echo state network. Energy 193, 116778 (2020)
16. Sun, L., Jin, B., Yang, H., et al.: Unsupervised EEG feature extraction based on echo state network. Inf. Sci. 475, 1–17 (2019)
17. Wang, H., Ni, C., Yan, X.: Optimizing the echo state network based on mutual information for modeling fed-batch bioprocesses. Neurocomputing 225, 111–118 (2017)
18. Wang, H., Yan, X.: Reservoir computing with sensitivity analysis input scaling regulation and redundant unit pruning for modeling fed-batch bioprocesses. Ind. Eng. Chem. Res. 53(16), 6789–6797 (2014)
19. Yperman, J., Becker, T.: Bayesian optimization of hyper-parameters in reservoir computing. arXiv preprint arXiv:1611.05193 (2016)
20. Thiede, L.A., Parlitz, U.: Gradient based hyperparameter optimization in echo state networks. Neural Netw. 115, 23–29 (2019)
21. Ma, Q., Chen, W., Wei, J., et al.: Direct model of memory properties and the linear reservoir topologies in echo state networks. Appl. Soft Comput. 22, 622–628 (2014)
22. Li, X., Bi, F., Yang, X., et al.: An echo state network with improved topology for time series prediction. IEEE Sens. J. 22, 5869–5878 (2022)
23. Gauthier, D.J., Bollt, E., Griffith, A., et al.: Next generation reservoir computing. Nat. Commun. 12(1), 1–8 (2021)
24. Gallicchio, C., Micheli, A., Pedrelli, L.: Deep reservoir computing: a critical experimental analysis. Neurocomputing 268, 87–99 (2017)
25. Gallicchio, C., Micheli, A.: Deep echo state network (DeepESN): a brief survey. arXiv preprint arXiv:1712.04323 (2017)
26. Gallicchio, C., Micheli, A., Pedrelli, L.: Design of deep echo state networks. Neural Netw. 108, 33–47 (2018)
27. Li, X., Zhang, W., Ding, Q.: Deep learning-based remaining useful life estimation of bearings using multi-scale feature extraction. Reliab. Eng. Syst. Saf. 182, 208–218 (2019)
28. Chouikhi, N., Ammar, B., Alimi, A.M.: Genesis of basic and multi-layer echo state network recurrent autoencoders for efficient data representations. arXiv preprint arXiv:1804.08996 (2018)
29. McDermott, P.L., Wikle, C.K.: Deep echo state networks with uncertainty quantification for spatio-temporal forecasting. Environmetrics 30(3), e2553 (2019)

30. Gonon, L., Ortega, J.P.: Reservoir computing universality with stochastic inputs. IEEE Trans. Neural Netw. Learn. Syst. **31**(1), 100–112 (2019)
31. Hart, A.G., Hook, J.L., Dawes, J.H.P.: Echo state networks trained by Tikhonov least squares are L2 (μ) approximators of ergodic dynamical systems. Physica D **421**, 132882 (2021)
32. Wang, H., Yan, X.: Optimizing the echo state network with a binary particle swarm optimization algorithm. Knowl. Based Syst. **86**, 182–193 (2015)
33. Rodan, A., Tino, P.: Minimum complexity echo state network. IEEE Trans. Neural Netw. **22**, 131–144 (2011)
34. Hu, R., Tang, Z.-R., Song, X., Luo, J., Wu, E.Q., Chang, S.: Ensemble echo network with deep architecture for time-series modeling. Neural Comput. Appl. **33**(10), 4997–5010 (2020). https://doi.org/10.1007/s00521-020-05286-8
35. Wang, L., Su, Z., Qiao, J., Yang, C.: Design of sparse Bayesian echo state network for time series prediction. Neural Comput. Appl. **33**(12), 7089–7102 (2020). https://doi.org/10.1007/s00521-020-05477-3
36. Yang, C., Qiao, J., Wang, L., Zhu, X.: Dynamical regularized echo state network for time series prediction. Neural Comput. Appl. **31**(10), 6781–6794 (2018). https://doi.org/10.1007/s00521-018-3488-z
37. Ding, Y., Zhu, Y., Feng, J., et al.: Interpretable spatio-temporal attention LSTM model for flood forecasting. Neurocomputing **403**, 348–359 (2020)

Observer-Based Adaptive Security Control for Network Control Systems Under TDS Actuator Attacks

Liang Wang and Ping Zhao(✉)

School of Information Science and Engineering, Shandong Normal University,
Jinan 250358, People's Republic of China
zhaoping@amss.ac.cn

Abstract. Networked control systems (NCS) have been used widely in many practical fields. The communication connection in NCS allows agents to communicate with each other quickly, which can quickly respond to abrupt changes in the system. NCS depend on computers and multi-purpose networks for operation, rendering them vulnerable to attacks, especially cyber attacks. This paper focuses on time delay switch (TDS) attacks. A method based on Lyapunov theory is proposed to detect and estimate TDS attack in real time, so as to detect the unstable impact of TDS attack and recover from it. Simulation results illustrate the effectiveness of the proposed TDS attack estimator and security control strategy.

Keywords: Networked control systems · Adaptive control design · Time-delay switch attack · Luenberger observer

1 Introduction

With the rapid development of Internet technologies and novel control strategies, networked control system (NCS) as a new discipline and technical field has emerged over recent years. Not only does the emergence of NCS conforms to the development trend of modern technology, but also reflects the development trend of the intersection and integration of theoretical knowledge and applications of various disciplines based on information science. Therefore, NCS is widely used in practical fields, such as national defense, telecommunications, electrical power systems and other fields [1]. But it is worth noting that more and more security problems have followed. The confidentiality, integrity, and availability (CIA) of data exchanged in NCS can be vulnerable to malicious network attacks due to the open network connections among controllers, sensors, actuators [2]. There is no doubt that this threat is mainly initiated by malicious attackers in cyber-space in order to obtain huge economic benefits or disrupt human life. Consequently, the security of NCS has gradually become a research hotspot in the field of international control.

© The Author(s), under exclusive license to Springer Nature Singapore Pte Ltd. 2022
H. Zhang et al. (Eds.): NCAA 2022, CCIS 1637, pp. 503–514, 2022.
https://doi.org/10.1007/978-981-19-6142-7_38

The data in NCS needs to be transmitted through the networks which are extremely vulnerable to intrusion by attackers. The most known cyber attacks are denial of service (DoS). DoS can disable a service by interference or depletion of system resources or the access to system information [3]. And beyond that, there are false data injection (FDI) which can intentionally manipulate the injection of wrong data or change transmission strategy by hijacking physical devices or network channels [4], and replay attack which makes effective data transmission repeated or delayed maliciously and deceptively [5], and a newly found attack, the time-delay switch (TDS) [6].

Generally speaking, time delay is very common in control systems. The existence of time delay usually affects the stability of control systems. When the adversary injects random delay switching attack into the NCS, there may be worse delays resulting in reduced system efficiency and even leads to system instability. This paper mainly studies TDS attacks on NCS [7]. TDS attack is made by inserting time delays into communication channels which will result in a delay in the transmission of measurement signals from the device output to the controller. As is known to all, NCS depends on timeliness. The controller needs to update the measured signal in real time to generate the control signal, but TDS may cause serious damage [8]. The NCS transmits the sensor measurements from agents to the centralized control unit through the communication channel. At this time, the injection of delay will lead to the instability of the NCS. Hence, it is essential to design an adaptive security control scheme to resist TDS attack.

The stability of time-delay electrical power systems has been considered in many researches [9,10]. The influence of time delay on power system was expounded in [11,12]. In [13], Milano and Anghel have presented the effect of delays on small-signal angle stability of power systems. Although many literatures have shown that TDS attack can lead to NCS instability [14], only a few studies can detect the real TDS attack and studying the compensation controller of TDS attack by designing a security control. The effects of TDS of power system was further analyzed in [15], and proposed a generic analytical framework to get the effective derive conditions of TDS. Sargolzaei and Yen proposed n adaptive control algorithm to mitigate and estimate TDS attacks introduced into measurement signals in [8]. This work focused on detecting and mitigating TDS attacks timely and compensating the time delay injected by hacker. In [16], Tan has developed a neural network (NN) approach as a tool for estimating a time delay in industrial communication systems with nonlinear dynamics, but the stability of this controller has not been investigated. A built-in attack characterization scheme based on neural networks for TDS attacks was developed to estimate the delay value in [17]. NN algorithm needs off-line training and can not detect TDS attack in real time. In [18], a scheme of robust load frequency control was presented for power system by Xiahou, Liu and Wu to cope with the random time-delay attack. But the system is not efficient due to its robustness to potential faults, failures, and attacks.

This paper will focus on the problem that NCS suffers from TDS attacks and causes system instability. We supposed that a hacker inject TDS attack into control channel and affect the state of the observer. To mitigate the TDS affects, system controller must be redesigned to be capable of estimating the delay caused by the TDS attack. This paper will introduce a new, easily effective method to address time-delay attacks. The proposed method utilizes a state feedback controller, an observer and a time-delay estimator. The contribution of the paper is its proposal of a novel TDS attack detection technique and the design of an observer-based adaptive security control for NCS under TDS actuator attacks. The proposed algorithm is able to estimate and compensate for TDS attacks in real time with low computational complexity compared with learning-based methods.

This paper is organized as follows: Sect. 2 presents a dynamic model of NCS under TDS attack. In Sect. 3, a state feedback controller with a Luenberger observer will be proposed. In Sect. 4, under this controller, the stability for NCS with TDS attack will be analyzed. Section 5 proposes an output feedback controller design and analyze the corresponding stability for NCS with TDS attack and delay detection method. In Sect. 6, simulations are provided to demonstrate our results. Conclusions are given in Sect. 7.

2   Dynamic Model of NCS

NCS is a control system implemented in the network environment. It is a full distributed, networked real-time feedback control system. NCS realizes the data transmission between devices by connecting sensors, controllers and actuators through communication networks. The control signal of the controller and the measurement signal of the sensor need to be transmitted through the network. In this section, we present the NCS under TDS attack. Firstly, we describe the dynamic model of a single NCS and then extend it to the NCS dynamic model of multiple plants. In order to present the NCS dynamics of the interconnected NCS of N agents, the i-th agent of NCS is given by:

$$\begin{cases} \dot{x}_i(t) = A_i x_i(t) + B_i u_i(t), \\ y_i(t) = C_i x_i(t), \end{cases} \tag{1}$$

where $x_i(t)$, $u_i(t)$ and $y_i(t)$ are the state, control input and output of the i-th agent of the NCS respectively.

$$x_i(t) = \left[x_{i,1}(t) \; x_{i,2}(t) \; \cdots \; x_{i,n_{xi}} \right]^T, \tag{2}$$

$$u_i(t) = \left[u_{i,1}(t) \; u_{i,2}(t) \; \cdots \; u_{i,n_{ui}} \right]^T, \tag{3}$$

$$y_i(t) = \left[y_{i,1}(t) \; y_{i,2}(t) \; \cdots \; y_{i,n_{yi}} \right]^T, \tag{4}$$

where n_{xi}, n_{ui}, and n_{yi} are the dimensions of them, respectively, for the i-th agent. Matrices A_i, B_i and C_i are constant matrices with suitable dimensions. Let $x(t) = [x_1^T(t), x_2^T(t), \cdots, x_N^T(t)]^T$, $u(t) = [u_1^T(t), u_2^T(t), \cdots, u_N^T(t)]^T$ and $y(t) = [y_1^T(t), y_2^T(t), \cdots, y_N^T(t)]^T$. Then, extending a single dynamic NCS model

(1) to the NCS model with multiple plants which can be described as

$$\begin{cases} \dot{x}(t) = Ax(t) + Bu(t), \\ y(t) = Cx(t), \end{cases} \tag{5}$$

where system matrices

$$A - \begin{bmatrix} A_1 & 0 & \cdots & 0 \\ 0 & A_2 & \cdots & 0 \\ \vdots & \vdots & \ddots & \vdots \\ 0 & 0 & \cdots & A_N \end{bmatrix}, B = diag\begin{bmatrix} B_1 & B_2 & \cdots & B_N \end{bmatrix} \text{ and } C = diag\begin{bmatrix} C_1 & C_2 & \cdots & C_N \end{bmatrix}. \tag{6}$$

3   State Feedback Control Design

3.1   Observer and Controller Design Under Normal Conditions

NCS is prone to cyberattacks because NCS realizes the control and feedback data packets transmission between agents and controllers through the network. The control signal will not transmit correctly through the actuator line under cyberattacks. The wrong control signal will endanger the performance and safe operation of the whole system and cause losses. To solve this problem, we design a Luenberger observer to make the error between the system measurement and estimated output converge to zero:

$$\begin{cases} \dot{\hat{x}}(t) = A\hat{x}(t) + B\hat{u}(t) + L(y(t) - \hat{y}(t)), \\ \hat{y}(t) = C\hat{x}(t), \end{cases} \tag{7}$$

where the vectors $\hat{x}$, $\hat{u}$, $\hat{y}$ are the state, input and output of the observer, respectively. The Luenberger gain L is the observer gain that multiplies the error between the output of the system $y = Cx$ and the estimated value from the observer $\hat{y} = C\hat{x}$.

The class of controllers considered for the system (5) can be state feedback, and the controller of an NCS is considered to be an optimal controller, which is described as:

$$u(t) = -K\hat{x}(t). \tag{8}$$

Meanwhile, we can get the input of the observer is:

$$\hat{u}(t) = -K\hat{x}(t). \tag{9}$$

3.2   Observer and Controller Design Under TDS Attack

In this subsection, we considered that the control signal sent by the control center is attacked by TDS in the process of network transmission. Then the state of the control channel after TDS attack can be described as:

$$u(t) = \begin{cases} u(t)_{Normal} = -K\hat{x}(t) & t \leq t_a, \\ u(t)_{Attack} = -K\hat{x}(t - \tau) & t \geq t_a \geq \tau, \end{cases} \tag{10}$$

where τ is an unknown time delay injected into the communication channel by the attacker. t_a is the starting time of the attacker's TDS attack.

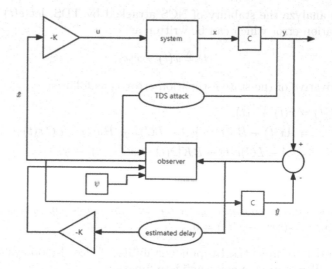

Fig. 1. Block diagram of the resilient observer-based controller.

Since communication channels where the sensor signal is transmitted to the control center and the control signal is transmitted to the power station actuator are more susceptible than control channels. TDS attack is that the attacker injects delay into the above communication channel. Then, we set a compensator signal ψ to counteract the impact of TDS attack as shown in Fig. 1.

$$\begin{cases} \dot{\hat{x}}(t) = A\hat{x}(t) + B\hat{u}(t) + L(y(t) - \hat{y}(t)) + \psi, \\ \hat{y}(t) = C\hat{x}(t), \end{cases} \tag{11}$$

The compensator signal ψ will be designed based on the subsequent stability analysis. In order to obtain state $x(t)$, here we set C as a full rank matrix.
Substituting the $y = Cx$ and $\hat{y} = C\hat{x}$ in (11), we can get:

$$\begin{cases} \dot{\hat{x}}(t) = A\hat{x}(t) + B\hat{u}(t) + LC[x(t) - \hat{x}(t)] + \psi, \\ \hat{y}(t) = C\hat{x}(t). \end{cases} \tag{12}$$

Grouping the elements with $\hat{x}$, the system representation can be described as:

$$\begin{cases} \dot{\hat{x}}(t) = (A - LC)\hat{x}(t) + B\hat{u}(t) + LCx(t) + \psi, \\ \hat{y}(t) = C\hat{x}(t). \end{cases} \tag{13}$$

The system (5) attacked by TDS with resilient controller can be described as:

$$\dot{x} = Ax(t) - BK\hat{x}(t - \tau). \tag{14}$$

4 Stability for NCS with TDS Attack and Delay Detection Method

In order to analyze the stability of NCS attacked by TDS, let $e(t)$ denote the state estimation error which can be written as:

$$e(t) \triangleq x(t) - \hat{x}(t). \tag{15}$$

The derivative of the state estimation error is as follows:

$$
\begin{aligned}
\dot{e}(t) &= \dot{x}(t) - \dot{\hat{x}}(t) \\
&= Ax(t) + Bu(t) - [(A - LC)\hat{x} + B\hat{u}(t) + LCx(t) + \psi] \\
&= (A - LC)e(t) + BK[\hat{x}(t) - \hat{x}(t - \tau)] - \psi.
\end{aligned}
\tag{16}
$$

Theorem 1. *Consider the NCS (5) with TDS attack where C is a full rank matrix. Then, with the controller given by (10), the close-loop system is Lyapunov stable and $\lim_{t \to \infty}(x(t) - \hat{x}(t)) = 0$.*

Proof. In order to prove the Lyapunov stability of closed-loop system (16), we choose the following Lyapunov candidate function

$$V = \frac{1}{2}e^T e + \frac{\alpha}{2}\tilde{\tau}^T \tilde{\tau}, \tag{17}$$

where α is a positive gain for characterizing the system to make it possible to run the computation of the estimated delay value, $\tilde{x} \triangleq x - \hat{x}$ is the state estimation error, and $\tilde{\tau} \triangleq \tau - \hat{\tau}$ is the delay estimation error.

Take the derivative of the Lyapunov function as follows:

$$
\begin{aligned}
\dot{V} &= e^T \dot{e} + \alpha \tilde{\tau}^T \dot{\tilde{\tau}} \\
&= e^T \{(A - LC)e + BK[\hat{x}(t) - \hat{x}(t - \tau)] - \psi\} + \alpha \tilde{\tau}^T (\dot{\tau} - \dot{\hat{\tau}}).
\end{aligned}
\tag{18}
$$

Consider a constant delay in the channel, even if it happens over short periods of time, the derivative of the delay is going to be null ($\dot{\tau} = 0$). Using a Taylor series, the delayed signal is modeled approximately up to the first derivative term as $\hat{x}(t - \tau) = \hat{x} - \dot{\hat{x}}\tau$.

$$
\begin{aligned}
\dot{V} &= e^T \{(A - LC)e + BK[\hat{x}(t) - \hat{x}(t) + \dot{\hat{x}}\tau] - \psi\} - \alpha \tilde{\tau}^T \dot{\hat{\tau}} \\
&= e^T [(A - LC)e + BK\dot{\hat{x}}\tau - \psi_2] - \alpha \tilde{\tau}^T \dot{\hat{\tau}}.
\end{aligned}
\tag{19}
$$

Substituting $\tau = \hat{\tau} + \tilde{\tau}$ into (19), we obtain:

$$\dot{V} = e^T [(A - LC)e + BK\dot{\hat{x}}(\hat{\tau} + \tilde{\tau}) - \psi] - \alpha \tilde{\tau}^T \dot{\hat{\tau}}. \tag{20}$$

In order to find the estimated delay $\hat{\tau}$, we set $e^T BKC\dot{\hat{x}}\tilde{\tau} = \alpha \tilde{\tau}^T \dot{\hat{\tau}}$. After calculation, we can get:

$$\dot{\hat{\tau}} = e^T \frac{BK}{\alpha}\dot{\hat{x}}. \tag{21}$$

For the purpose of meeting the stability condition of Lyapunov theory, we set:

$$\psi = BK\dot{\hat{x}}\hat{\tau}. \tag{22}$$

The Lyapunov theory is applied to guarantee that the system remains stable at different operating points. The Luenberger gain L must satisfy the following equation:

$$e^T(A - LC)e \leq 0. \tag{23}$$

Note that,

$$\ddot{V} = 2e^T(A - LC)\{(A - LC)e + BK[\hat{x}(t) - \hat{x}(t - \tau)] - BK\dot{\hat{x}}\hat{\tau}\} \tag{24}$$

is bounded for all $t > 0$ since e is bounded for all $t \geq 0$. Thus, $\dot{V}$ is uniformly continuous in t. Following Barbalat's lemma [10], $\lim_{t\to\infty} \dot{V} = 0$, and hence, $\lim_{t\to\infty} e(t) = 0$.

Remark 1. Note that in (22), the form of ψ can be achieved. There was a similar method presented in [20], and the feasibility of this method has been proved.

5   Output Feedback Control Design

5.1   Controller Design

In this section, we will present the design of output feedback control for system (5). Considered the TDS attack on control signal, the control channel subject to TDS attack is described as follows:

$$u(t) = -K\hat{y}(t - \tau), \tag{25}$$

and the input of the observer is

$$\hat{u}(t) = -K\hat{y}(t). \tag{26}$$

5.2   Observer Design

A Luenberger observer is designed by (13) such that the error between the system measurement and estimated output converges to zero given. And the error is defined as that in (15).

Take the derivative of the state estimation error as follows:

$$\dot{e}(t) = (A - L_2C)e(t) + BKC[\hat{x}(t) - \hat{x}(t - \tau)] - \psi_2. \tag{27}$$

Theorem 2. *Consider the NCS (5) with TDS attack. Then, with the controller given by (25), the close-loop system is Lyapunov stable and $\lim_{t\to\infty}(x - \hat{x}) = 0$.*

Proof. In order to prove the Lyapunov stability of closed-loop system (27), we choose the following Lyapunov candidate function

$$V = \frac{1}{2}e^T e + \frac{\alpha}{2}\tilde{\tau}^T \tilde{\tau}. \tag{28}$$

Take the derivative of the Lyapunov function as follows:

$$
\begin{aligned}
\dot{V} &= e^T \dot{e} + \alpha \tilde{\tau}^T \dot{\tilde{\tau}} \\
&= e^T \{(A - L_2 C)e + BKC[\hat{x}(t) - \hat{x}(t - \tau)] - \psi_2\} + \alpha \tilde{\tau}^T (\dot{\tau} - \dot{\hat{\tau}}) \\
&= e^T [(A - L_2 C)e + BKC\dot{\hat{x}}\tau - \psi_2] - \alpha \tilde{\tau}^T \dot{\hat{\tau}} \\
&= e^T [(A - L_2 C)e + BKC\dot{\hat{x}}(\hat{\tau} + \tilde{\tau}) - \psi] - \alpha \tilde{\tau}^T \dot{\hat{\tau}}.
\end{aligned}
\tag{29}
$$

In order to find the estimated delay $\hat{\tau}$, we set $e^T BKC\dot{\hat{x}}\tilde{\tau} = \alpha \tilde{\tau}^T \dot{\hat{\tau}}$. After calculation, we can get:

$$\dot{\hat{\tau}} = e^T \frac{BKC}{\alpha} \dot{\hat{x}}. \tag{30}$$

For the purpose of meeting the stability condition of Lyapunov theory system, we set:

$$\psi = BKC\dot{\hat{x}}\hat{\tau}. \tag{31}$$

The Lyapunov theory is applied to guarantee that the system remains stable at different operating points. The Luenberger gain L must satisfy the following equation:

$$e^T (A - LC)e \leq 0. \tag{32}$$

Same as (24), $\ddot{V}$ is bounded for all $t \geq 0$ since e is bounded for all $t \geq 0$. Thus, $\dot{V}$ is uniformly continuous in t. Following Barbalat's lemma [19], $\lim_{t\to\infty} \dot{V} = 0$, and hence, $\lim_{t\to\infty} e(t) = 0$.

6   Numerical Simulation

To evaluate and illustrate the performance of the proposed secure control design, a numerical example was introduced. Simulations also verify the feasibility and effectiveness of the proposed control scheme. Specifically, the system parameters for simulations are as follows

$$
\begin{bmatrix} \dot{x}_1(t) \\ \dot{x}_2(t) \end{bmatrix} = \begin{bmatrix} -1 & -2 \\ 3 & -6 \end{bmatrix} \begin{bmatrix} x_1(t) \\ x_2(t) \end{bmatrix} + \begin{bmatrix} 0 \\ 1 \end{bmatrix} u(t), \quad \begin{bmatrix} x_1(0) \\ x_2(0) \end{bmatrix} = \begin{bmatrix} 2 \\ -1 \end{bmatrix}, \quad t \geq 0,
$$

Use the linear quadratic regulator design in the "lqrd" continuous cost function (MATLAB 2019a) to generate optimal control laws for the system to operate properly. Then the optimal controller gain is

$$K = \begin{bmatrix} -0.50 & 0.92 \end{bmatrix},$$

the gain matrices of the observer (13) is

$$L = \begin{bmatrix} 3 \\ -6 \end{bmatrix},$$

the compensator signal ψ is as follows:

$$\psi = \begin{bmatrix} 0 & 0 \\ -0.50 & 0.92 \end{bmatrix} \begin{bmatrix} \dot{\hat{x}}_1 \\ \dot{\hat{x}}_2 \end{bmatrix} \hat{\tau},$$

and the unstable linear dynamical system is given by

$$\begin{bmatrix} \dot{x}_1(t) \\ \dot{x}_2(t) \end{bmatrix} = \begin{bmatrix} -1 & -2 \\ 3 & -6 \end{bmatrix} \begin{bmatrix} x_1(t) \\ x_2(t) \end{bmatrix} - \begin{bmatrix} 0 & 0 \\ -0.50 & 0.92 \end{bmatrix} \begin{bmatrix} \hat{x}_1(t-\tau) \\ \hat{x}_2(t-\tau) \end{bmatrix}, \begin{bmatrix} x_1(0) \\ x_2(0) \end{bmatrix} = \begin{bmatrix} 2 \\ -1 \end{bmatrix}, t \geq 0,$$

The derivative of estimated delay $\hat{\tau}$ satisfies

$$\dot{\hat{\tau}} = \begin{bmatrix} (x_1 - \hat{x}_1)^T & (x_2 - \hat{x}_2)^T \end{bmatrix} \begin{bmatrix} 0 & 0 \\ -0.50 & 0.92 \end{bmatrix} \begin{bmatrix} \dot{\hat{x}}_1 \\ \dot{\hat{x}}_2 \end{bmatrix}.$$

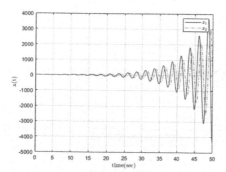

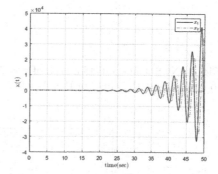

Fig. 2. The NCS under a TDS attack of $\tau = 0.23$s.

Fig. 3. The NCS under a TDS attack of $\tau = 0.24$s.

6.1 Vulnerability Analysis

According to Fig. 2 and Fig. 3, when running the NCS in an optimal control situation with no compensation algorithm to correct the TDS attack, the system becomes unstable. Both attacks were started at $t_0 = 0$ s, with the different intensity of the attack $\tau = 0.23$ s and $\tau = 0.24$ s. It can be observed from the figures that the TDS attack of $\tau = 0.23$ s and $\tau = 0.24$ s affected the state of the system, inducing the system instability, and the system could not work normally under this attack. A resilient controller need to be designed to detect and compensate the impact of a TDS attack.

6.2   TDS Attack Detection and State Estimaion

The same TDS attack as that shown in Fig. 1 was simulated, but the system was
equipped with the proposed secure controller. As Fig. 3 shows, the adverse effect
of the TDS attack was improved by the proposed state estimation mechanism.
The simulation was repeated with different TDS attacks inserted into the NCS
shown in Fig. 4 and Fig. 5 (Fig. 6).

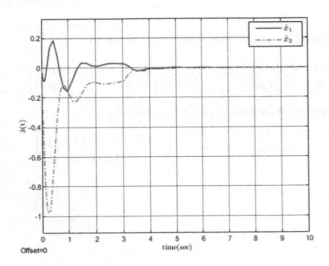

Fig. 4. NCS under a TDS attack of $\tau = 0.23$ s with the Luenberger observer operating
to compensate the effect of the attack

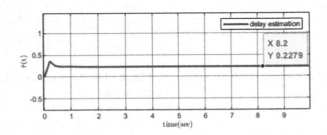

Fig. 5. Delay estimation with the TDS attack of $\tau = 0.23$ s.

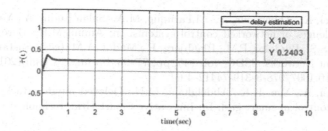

Fig. 6. Delay estimation with the TDS attack of $\tau = 0.24$ s.

7 Conclusion

In this article, two observer-based adaptive security control schemes was proposed to stabilize the system under TDS attack. The compensation signal is defined based on the Lyapunov theory and a Luenberger observer. Specifically, estimating the delay caused by TDS attacks, we prove the effectiveness of the proposed adaptive security control scheme, which guarantees the asymptotic stability of closed-loop dynamic systems under TDS attacks. Unlike other traditional techniques, the design of the controller and estimator is based on Lyapunov stability analysis. When the TDS attacks take place, it is possible to detect and estimate the delay caused by the attack. The proposed approach was evaluated through a case study in which an NCS was monitored and controlled. Future work will enhance the proposed method to investigate time-variable TDS attacks.

References

1. Hespanha, J.P., Naghshtabrizi, P., Xu, Y.: A survey of recent results in networked control systems. Proc. IEEE **95**(1), 138–162 (2007). https://doi.org/10.1109/JPROC.2006.887288
2. Pang, Z.H., Liu, G.P., Zhou, D., Hou, F., Sun, D.: Two-channel false data injection attacks against output tracking control of networked systems. IEEE Trans. Industr. Electron. **63**(5), 3242–3251 (2016)
3. Zhang, C.L., Yang, G.H., Lu, A.Y.: Resilient observer-based control for cyber-physical systems under denial-of-service attacks. Inf. Sci. **545**, 102–117 (2021). https://doi.org/10.1016/j.ins.2020.07.070. https://www.sciencedirect.com/science/article/pii/S0020025520307519
4. Khalghani, M.R., Solanki, J., Solanki, S.K., Khooban, M.H., Sargolzaei, A.: Resilient frequency control design for microgrids under false data injection. IEEE Trans. Industr. Electron. **68**(3), 2151–2162 (2020). https://doi.org/10.1109/TIE.2020.2975494
5. Hosseinzadeh, M., Sinopoli, B., Garone, E.: Feasibility and detection of replay attack in networked constrained cyber-physical systems. In: 2019 57th Annual Allerton Conference on Communication, Control, and Computing (Allerton) (2019)

6. Sargolzaei, A., Abbaspour, A., Al Faruque, M.A., Salah Eddin, A., Yen, K.: Security challenges of networked control systems. In: Amini, M.H., Boroojeni, K.G., Iyengar, S.S., Pardalos, P.M., Blaabjerg, F., Madni, A.M. (eds.) Sustainable Interdependent Networks. SSDC, vol. 145, pp. 77–95. Springer, Cham (2018). https://doi.org/10.1007/978-3-319-74412-4_6

7. Sargolzaei, A., Yen, K.K., Abdelghani, M.N.: Delayed inputs attack on load frequency control in smart grid. In: Innovative Smart Grid Technologies Conference (2014)

8. Sargolzaei, A., Yen, K.K., Abdelghani, M.N.: Preventing time-delay switch attack on load frequency control in distributed power systems. IEEE Trans. Smart Grid 7(2), 1176–1185 (2016)

9. Schenato, L.: Optimal estimation in networked control systems subject to random delay and packet drop. IEEE Trans. Autom. Control 53(5), 1311–1317 (2008)

10. Zhang, C.-K., Jiang, L., Wu, Q.H., He, Y.: Delay-dependent robust load frequency control for time delay power systems. IEEE Trans. Power Syst. 28(3), 2192–2201 (2013)

11. Jiang, Q., Zou, Z., Cao, Y.: Wide-area TCSC controller design in consideration of feedback signals' time delays. In: Power Engineering Society General Meeting, 2005. IEEE (2005)

12. Kamwa, I., Grondin, R., Hebert, Y.: Wide-area measurement based stabilizing control of large power systems-a decentralized/hierarchical approach. IEEE Trans. Power Syst. 16(1), 136–153 (2001)

13. Milano, F., Anghel, M.: Impact of time delays on power system stability. IEEE Trans. Circuits Syst. I Regul. Pap. 59(4), 889–900 (2012). https://doi.org/10.1109/TCSI.2011.2169744

14. Sargolzaei, A., Yen, K.K., Abdelghani, M.N.: Time-delay switch attack on load frequency control in smart grid (2013)

15. Wang, J.K., Peng, C.: Analysis of time delay attacks against power grid stability. In: Workshop on Cyber-Physical Security and Resilience in Smart Grids, pp. 67–72 (2017)

16. Tan, Y.: Time-varying time-delay estimation for nonlinear systems using neural networks. Int. J. Appl. Math. Comput. Sci. 14(1), 63–68 (2004)

17. Lou, X., et al.: Learning-based time delay attack characterization for cyber-physical systems. In: 2019 IEEE International Conference on Communications, Control, and Computing Technologies for Smart Grids (SmartGridComm), pp. 1–6 (2019). https://doi.org/10.1109/SmartGridComm.2019.8909732

18. Xiahou, K.S., Liu, Y., Wu, Q.H.: Robust load frequency control of power systems against random time-delay attacks. IEEE Trans. Smart Grid 12(1), 909–911 (2021). https://doi.org/10.1109/TSG.2020.3018635

19. Haddad, W.M., Chellaboina, V.: Nonlinear Dynamical Systems and Control. Princeton University Press, Princeton (2010)

20. Victorio, M., Sargolzaei, A., Khalghani, M.R.: A secure control design for networked control systems with linear dynamics under a time-delay switch attack. Electronics 10(3), 322 (2021)

Feature Selection for High-Dimensional Data Based on a Multi-objective Particle Swarm Optimization with Self-adjusting Strategy Pool

Yingyu Peng[1] , Ruiqi Wang[2] , Dandan Yu[3]([✉]) , and Yu Zhou[2]

[1] College of Mathematics and Statistics, Shenzhen University, Shenzhen, China
pengyingyu2018@email.szu.edu.cn
[2] College of Computer Science and Software Engineering, Shenzhen University, Shenzhen, China
yu.zhou@szu.edu.cn
[3] Information Center The First Affiliated Hospital of Dalian Medical University, Dalian, China
dayiyudandan@sina.com

Abstract. The development of data collection increases the dimensionality of the data in many fields, which might even cause the curse of dimensionality, arising a challenge to many existing feature selection. This paper proposes a method based on a feature prioritization approach and a multi-objective particle swarm optimization with self-adjusting strategy pool (MOPSO-SaSP). The proposed algorithm adopts a backward filter approach based on Approximate Markov blankets as a pre-processing step which prioritizes features into predominant features and subordinate features to reduce the search space and lessen the randomness of the initialization step. Afterwards, an adaptive mechanism is applied that enables each particle to adaptively choose a strategy from the strategy pool according to the latest information collected. The strategy pool is constructed through a series of experiments on training data. Different strategies specialized in different searching methods. Experimental results on eight benchmark micro-array gene datasets reveal that the feature prioritization approach can deal features in an efficient way and the adaptive mechanism is a powerful searching method to achieve a competitive classification accuracy and reduce the number of features.

Keywords: Feature selection · MOPSO · Strategy pool · Approximate Markov Blanket

1 Instruction

Feature selection (FS) is frequently used as a pre-processing step for classification issues. As the difficulty of studying objects increases, high-dimensional data becomes common, and feature selection methods are facing the challenge of

H. Zhang et al. (Eds.): NCAA 2022, CCIS 1637, pp. 515–529, 2022.
https://doi.org/10.1007/978-981-19-6142-7_39

larger search space [1]. Problems related to feature selection processes are always NP-hard [2]. FS is definitely a multi-objective problem because the objectives of maximizing the classification accuracy and minimizing the size of feature subset are in conflict. Notably, compared to other EC-based multi-objective feature selection methods, multi-objective PSO methods have less computational costs and faster convergence speed [3].

Pursuing the two goals - maximum approximation to optima and avoiding the problem of stagnation in local optima - is the top priority of most PSO algorithm. However, the above two goals are usually operated separately in existing methods, for example, a feature selection method based on a subset similarity distance measure named MOPSO-SiD proposed by Nguyen et al. which improves the approximation to optima one of the two goals [4], but due to the lack of good ability to improve the exploration ability of the population, it is easy to fall into local optima when the population iterates to the later stage. Take another example, a feature selection method based on crowding, mutation and dominance named CMDPSO which yields a good set of feature subsets but there is still a lot of potential to explore and improve the approximation to optima [5]. Many methods like these make the two goals independent of each other and ignores the dynamic relationship of them. As the dynamic relationship is hard to record or control during iterations and these two goals have varying priorities at searching process, a self-adaptive mechanism is needed. Moreover, generally there is only a single updating formula in each PSO algorithm that used to generate new candidate solutions and usually these formulas can perform well when used to achieve only one of the two goals. Considering simply making several parameters of a single updating formula vary adaptively is not enough for large-scale feature selection problem, we managed to combine several formulas that each of them has the advantage in term of one of the two goals. In this paper, in order to achieve both the two goals simultaneously through updating formulas during searching process, we present a Multi-Objective Particle Swarm Optimization with Self-adjusting Strategy Pool (MOPSO-SaSP). The main contributions are listed below.

- A fast feature prioritization approach based on Approximate Markov Blankets is applied to analysis the correlations between categories and features as well as correlations between features and features. As a result, it significantly shrinks the search space for the subsequent optimization process and prioritizes the features aiming to increase the likelihood of meaningful features being chosen.
- A self-adjusting strategy pool is designed to self-adaptively adjust the choice of strategy allocated for each particle from a designed strategy pool, according to the information collected during the searching process. Here the strategy means the updating formula. This method does not simply adapt particle swarm to the dynamic search process by adjusting some isolated parameters whose effect is sometimes subtle, instead, it straightforward gives different updating formulas for different particles at different iterations by calculating the empirical probabilities matrix of the strategy pool.

The paper's outline is presented as follows. Section 2 introduces related works of the proposed method. Section 3 illustrates the proposed feature selection method in detail. Section 4 display the experimental results on 8 microarray benchmark data sets. Finally, Sect. 5 gives the conclusion and future work.

2 Related Works

2.1 PSO-Based Feature Selection

Among the EC methods, PSO-based algorithms have been popularly adopted for feature selection problems. Chuang et al. [6] developed an improved binary PSO method to maintain the classification performance while reducing the size of selected features. In the PSO iterations, the *gbest* will be forced to be nullified if no better solution occur in 3 consecutive generations. Xue et al. [7] proposed a PSO-based multi-objective feature selection algorithm to solve the classification problem with the expectation of achieving a Pareto front of nondominated solutions. In [8], the author proposed a binary PSO with a two-level particle cooperation strategy, which used ReliefF to form the initialized population in the first level and used bare-bone (BBPSO) under the MOEA/D framework to search for Pareto optimal solutions in the second level. In [9], a multi-objective optimization framework for discretization-based feature selection based on a flexible cut-point strategy was developed and incorporated into PSO (FCPSO), which generated a rather competitive result. In recent years, adding adaptive mechanism into PSO methods has been considered an effective way to solve various optimization problems including feature selection. For instance, Rani et al. [10] conducted a chaotic local search for PSO and proposed a chaotic adaptive particle swarm optimization (CSAPSO) algorithm that dynamically adjusts the velocity. It's worth mentioning that the majority of these variant algorithms used only single fixed formula for generating new solutions.

2.2 Strategy Pool

Recently, there are some PSO algorithm started to use multiple strategies. For example, Zhao et al. [3] proposed a multiple strategies interactive learning mechanism based PSO (DPPSO), in which each particle generates two intermediate particles from two strategies and selects a better one as its new position. In term of more than two strategies, Xue et al. [11] proposed a self-adaptive particle swarm optimization (SaPSO) algorithm which applies a strategy pool containing 5 strategies that each of them has different characteristics. These strategies are selected from updating formulas of various existing PSO algorithms by a series of experiment. This method has wonderful performance in feature selection problem that less than 1000 dimensions, but when the scale is extended to more than 1000 dimensions, it reaches a premature convergence and fails to jump out of local optima. In this article, inspired by the idea of designing strategy pool and aiming to overcome the problem of stagnation in local optima, we proposed a method based on a self-adjusting strategy pool combining a feature prioritization approach.

3   The Proposed Method

The overall flow of the proposed method is shown in Fig. 1. At first, a feature prioritization approach based on Approximate Markov Blanket (AMB) is applied to divide relevant features into predominant features and subordinate features. Then, with the self-adjusting strategy pool, MOPSO-SaSP is applied to gain the Pareto optimal solutions. At last, the candidate solution with the highest classification accuracy is taken as the selected feature subset.

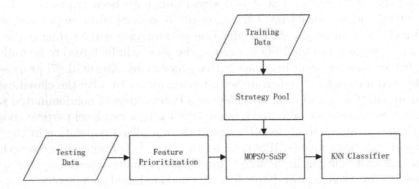

Fig. 1. Process of the MOPSO-SaSP algorithm.

3.1   AMB Feature Prioritization Approach

A method of feature prioritization approach based on Approximate Markov Blanket is added before initialization, which is a filter method to preprocess the original data set and analyse relevance and redundancy among features [12]. We use *entropy* as a measure of uncertainty in random variables. The entropy of the random variable X is defined as

$$H\left(X\right) = -\sum_{i} P\left(x_i\right) \log_2\left(P\left(x_i\right)\right) \tag{1}$$

When another random variable Y is determined, the entropy of the random variable X is defined as

$$H\left(X|Y\right) = -\sum_{j} P\left(y_j\right) \sum_{i} P\left(x_i|y_j\right) \log_2\left(P\left(x_i|y_j\right)\right) \tag{2}$$

The amount by which the information entropy of X decreases after variable Y is observed is called the *information gain* [13], which is defined as

$$IG\left(X|Y\right) = H\left(X\right) - H\left(X|Y\right). \tag{3}$$

The *symmetrical uncertainty (SU)* [14] is a normalized information gain, defined as

$$SU\left(X,Y\right) = 2\left[\frac{IG\left(X|Y\right)}{H\left(X\right) + H\left(Y\right)}\right] \tag{4}$$

The correlation between features F_i and classes C is denoted by $SU(i,c)$, and the correlation between any two features F_i and F_j is denoted by $SU(i,j)$. For two relevant features F_i and F_j $(i \neq j)$, if $SU_{i,c} \geq SU_{j,c}$ and $SU_{i,j} \geq SU_{j,c}$, F_i is considered to form an approximate Markov blanket for F_j. By analyzing the approximate Markov blanket relationship, we divide features into predominant features and subordinate features. For feature F_i, it is a predominant feature if there is no feature that can form approximate Markov blanket for it among all features. The pseudocode of the feature prioritization approach is presented in Algorithm 1.

ALGORITHM 1: Pseudo code of the AMB feature prioritization approach

---

input: $S_0(F_1, F_2, \ldots, F_N,$ C) // a training data set

  δ // a predefined threshold

1: **for** $i \leq N_{Pop}$

2:   calculate $SU_{i,c}$ for F_i;

3:   **if** $SU_{i,c} > \delta$

4:   append F_i to S_0;

5:   **end**;

6:   $F_j = getFirstElement(S_0)$;

7:   **while** $F_j \neq NULL$

8:   $F_i = getNextElement(S_0, F_j)$

9:     **if** $F_i \neq NULL$

10:     **if** $SU_{i,j} \geq SU_{i,c}$

11:       remove F_i from S_0;

12:       append F_i to S_{F_j};

13:     **end if**

14:     $Fi = getNextElement(S_0, F_i)$;

15:   **end if**

16:   $F_j = getNextElement(S_0, F_j)$;

17:   **end while** ;

18: $S_{Pre} = S_0$;

19: **end**;

output: $S_{Pre}\left(F_1', F_2', \ldots, F_m'\right)$// a predominant feature subset

  $S_{F_1'}, S_{F_2'}, \ldots, S_{F_1'}$ // subordinate feature subsets of predominant features

---

3.2 Initialization

In the particle initialization phrase, all predominant features are selected first, and then randomly select s relevant features to add to the feature subset. The number s is determined based on the sample, defined as

$$s = \begin{cases} max\{\frac{N_{rela}-N_{PE}}{N_{pop}}, 0.3 \times N_{PE}\}, N_{PE} < 100 \\ \frac{N_{rela}-N_{PE}}{N_{pop}}, N_{PE} \geq 100 \end{cases} \qquad (5)$$

where N_{rela}, N_{PE} and N_{pop} are the number of relevant features, the number of predominant features and the population size, respectively. When randomly selecting relevant features into the subset, if the predominant feature is sampled, it is replaced by its first three subordinate features. If the selected predominant feature has no subordinate features, it is removed directly. If the selected feature is a subordinate feature, it is directly added to the subset.

3.3   Self-adjusting Strategy Pool

To overcome the limitations of the single update formulation of general PSO, we construct a self-adjusting strategy pool containing multiple strategies to satisfy two major goals simultaneously: approaching the optima and enhancing the diversity of solutions. The algorithm can adaptively adjust the weight of each strategy by analyzing the information collected during the search process.

To construct the strategy pool for the MOPSO-SaSP algorithm, we firstly select five famous update formulas as candidate strategies which perform well in term of maximum approximation to optima and another five that do well in avoiding the problem of stagnation in local optima. These 10 single-strategy PSO algorithms with different strategy are run 20 times and their classification accuracy are recorded. Two strategies of algorithm which result in the highest accuracy in the two classes are added to the strategy pool firstly. Then other strategies are added into the strategy pool in descending order of recorded accuracy one by one and record the result respectively. If the classification accuracy improves after adding a certain strategy, then it would be added in the final strategy pool, and vice versa. Based on the results, four strategies are finally selected. They are presented as follows.

(1) The standard PSO [15]:

$$v_{id}^{t+1} = w * v_{id}^t + c_1 * r_1 * \left(p_{id} - x_{id}^t\right) + c_2 * r_2 * \left(p_{gd} - x_{id}^t\right) \tag{6}$$

$$x_{id}^{t+1} = x_{id}^t + v_{id}^{t+1} \tag{7}$$

where w is an inertia weight usually set to 0.7298. c_1 and c_2 are acceleration constants both set to 1.49618. r_1 and r_2 are two random values uniformly distributed in [0,1].

(2) The strategy with learning factors that vary according to iterations is selected:

$$v_{id}^{t+1} = w * v_{id}^t + 2 * (0.7 - \mu) * c_1 * P_{besti} + 2 * (0.3 + \mu) * c_2 * G_{best} \tag{8}$$

$\mu = \frac{0.4*t}{t_{max}}$, t_{max} is the maximum number of iterations.

(3) An improved CLPSO velocity updating strategy from Ref. [16] is applied:

$$v_{id}^{t+1} = w * v_{id}^t + 0.5 * c_1 * r_1 * \left(pbest_{f_i(d)} - x_{id}^t + p_{gd} - x_{id}^t\right) \tag{9}$$

where $f_i = [f_i(1), f_i(2), ..., f_i(D)]$ defines whose personal best should be used by the current particle.

(4) A different updating strategy from Ref. [17] is selected in our article:

$$x_{id}^{t+1} = r_1 * x_{id}^t + r_2 * p_{id} + r_3 * \left(x_{ad}^t - x_{bd}^t\right) \tag{10}$$

The purpose of setting the self-adjusting strategy pool mechanism is to generate the probabilities for each strategy based on their previous performance, and select an appropriate update strategy for each particle according to these probabilities. If the particle can find a better solution using a certain strategy in recent generations, this strategy is likely to be used more often in future generations. In our article, the probability of the ith strategy being selected is denoted by P_{stra_i}. The initial probability of each strategy is equal and represented as $P_{stra_i} = \frac{1}{N_{stra}}$ and N_{stra} is the number of candidate strategies. In each iteration, the particle selects a strategy randomly according to the probability of the strategy to generate a new solution. We need to record whether a selected strategy successfully generate a better new solution or fail to do this. This information is recorded by the binary matrices $SFlag_{N_{pop} \times N_{stra}}$ and $FFlag_{N_{pop} \times N_{stra}}$ ($i = 1, 2, \ldots, N_{pop}, j = 1, 2, \ldots, N_{stra}$, where N_{pop} is the number of particles and N_{stra} is the number of the candidate strategies). In the first generation,

$$SFlag = \begin{pmatrix} 0 \cdots 0 \\ \vdots \ddots \vdots \\ 0 \cdots 0 \end{pmatrix}_{N_{pop} \times N_{stra}} , \quad FFlag = \begin{pmatrix} 0 \cdots 0 \\ \vdots \ddots \vdots \\ 0 \cdots 0 \end{pmatrix}_{N_{pop} \times N_{stra}} \tag{11}$$

Assume that the ith particle generates a new candidate solution using the jth strategy that dominates the previous generation's solution, which means that this strategy is successful and sets $SFlag_{i,j}$ to 1. Otherwise, the jth strategy is a failure and $FFlag_{i,j}$ is set to 1. After each iteration, all elements in the matrices $SFlag_{i,j}$ and $FFlag_{i,j}$ will be reinitialized to 0.

We set the strategy success count matrix S and the failure count matrix F. When initializing the matrix,

$$S = \begin{pmatrix} 0 \cdots 0 \\ \vdots \ddots \vdots \\ 0 \cdots 0 \end{pmatrix}_{LP \times N_{stra}} , \quad F = \begin{pmatrix} 0 \cdots 0 \\ \vdots \ddots \vdots \\ 0 \cdots 0 \end{pmatrix}_{LP \times N_{stra}} \tag{12}$$

After the evolutionary process of the current generation is finished, update the corresponding rows in the matrices S and F, which is given as follows:

$$S_{i,j} = \sum_{k=1}^{N_{pop}} SFlag_{k,j} \tag{13}$$

$$F_{i,j} = \sum_{k=1}^{N_{pop}} FFlag_{k,j} \tag{14}$$

After the evolutionary process has already repeated for LP generations, the matrices S and F are initialized by Eq. (12). Besides, the probabilities of each strategy are recalculated. The one for the jth strategy is calculated as follows:

$$S_j^1 = \sum_{k=1}^{LP} S_{k,j} \tag{15}$$

$$S_j^2 = \begin{cases} \varepsilon & , S_j^1 = 0 \\ S_j^1 & , otherwise \end{cases} \tag{16}$$

$$S_j^3 = S_j^1 / \left(S_j^2 + \sum_{k=1}^{LP} F_{k,j} \right) \tag{17}$$

$$p_j = S_j^3 / \sum_{j=1}^{N_{stra}} S_j^3 \tag{18}$$

where the value ε is set to 0.0001 to avoid division by zero. These equations are used to recalculate the probability for the strategies in the strategy pool based on their performance during the LP generations. The pseudo-code of the self-adjusting strategy pool mechanism is shown as Algorithm 2.

3.4   Proposed Algorithm: MOPSO-SaSP

We apply the feature prioritization approach and self-adjusting strategy pool mechanism mentioned in the previous subsection to the multi-objective particle swarm optimization and propose an algorithm called MOPSO-SaSP. In our article, two objectives are formulated as an MOP to be minimized which are balanced classification error rate, and the average SU. They are defined as follows:

$$balanced\ calssification\ error rate = \frac{1}{c} \sum_{i=1}^{c} \frac{FP_i}{|S_i|} \tag{19}$$

$$average SU = \frac{\sum_{i \neq j} SU_{i,j}}{N_f} \tag{20}$$

where c denotes the number of the classes, FP_i is the number of samples in class i that are predicted incorrectly, $|S_i|$ is the number of samples in each class, N_f is the number of feature pairs in the subset of selected features and is used to reflect the redundancy among features.

In our method, $gbest$ is updated according to crowding distance. In every generation of the evolutionary process, the non-dominated solution with the smallest crowding distance is selected as $gbest$ from the external archive [18]. Additionally, the $gbest$ mutation operator is added to the algorithm in order to prevent the particles from suffering the problem of stagnation in local optimal. When the $gbest$ of the population is not updated for more than n times, this $gbest$

ALGORITHM 2: Pseudo code of the self-adjusting strategy pool algorithm

---

1: **while**   iter $\leq$ MaxIter $\times N_{Pop}$

2: **for**   i $\leq N_{Pop}$

3: Select one strategy from the strategy pool for x_i based on$\{P_1, P_2, ..., P_{N_{Stra}}\}$. Suppose the selected strategy is the j th strategy. Generate a new particle x_i^{new} by this strategy, and calculate its fitness value;

4: **if** x_i^{new} is better than x_i

5:    $nsFlag_{i,j} = 1$;

6:    **if** x_i^{new} is better than $pbest_i$ **then**

7:       Update $pbest_i$ with x_i^{new};

8:       **if** x_i^{new} is better than $gbest$ **then**

9:          Update $gbest$ with x_i^{new};

10: **else**

11: $nfFlag_{i,j} = 1$;

12: i = i + 1;

13: Replace x_i with x_i^{new};

14: iter = iter + 1;

15: k = iter − flagiter;

16: Replace the kth row of $S_{LP \times N_{Stra}}$ and $F_{LP \times N_{Stra}}$ with the sum of all the rows in $nsFlag$ and $nfFlag$ respectively;

18: Initial $nsFlag$ and $nfFlag$ as Eq. (11);

19: **if** $k = LP$

20: flagiter = iter;

21: Update $\{p1, p2, ..., p_{N_{Stra}}\}$ based on $S_{LP \times N_{Stra}}$ and $F_{LP \times N_{Stra}}$ as Eqs. (15)–(18);

22: Initial $S_{LP \times N_{Stra}}$ and $F_{LP \times N_{Stra}}$ as Eq. (12);

23: Output: $gbest$;

---

is considered as a local optimal solution and has no reference, so the position of this $gbest$ is assigned a random value. In this way, the particle will explore randomly towards other regions to search for more solutions. Since classification accuracy is the most important factor in the feature selection problem, we give priority to this objective in selecting solutions. In our article, we select the solution with the highest classification accuracy from Pareto set as the final feature subset.

4   Experimental Results and Analysis

4.1   Datasets

In order to demonstrate the effectiveness of the proposed MOPSO-SaSP, eight benchmark real-world gene datasets [19] are applied in the experiments. Table 1 shows the detailed description of these datasets including the number of features, samples and classes.

Table 1. Information of the datasets

Dataset	Features	Samples	Class
SRBCT	2308	83	4
Leukemia 1	5327	72	3
DLBCL	5469	77	2
9Tumors	5726	60	9
Brain Tumor1	5920	90	5
Leukemia 2	11225	72	3
11Tumors	12533	174	11
Lung Cancer	5920	90	5

4.2 Compared Algorithms and Parameter Setting

To evaluate the improvement and performance of MOPSO-SaSP, we compare the classification performance with the standard PSO and several FS algorithms based on PSO (including PPSO, VLPSO, VLPSO-LS, SOP-COPSO and LFSDC). Table 2 shows the balanced accuracy (BA) of compared algorithms. The proposed MOPSO-SaSP can perform better than any other in 5 data sets.

Besides, to verify the effectiveness the feature prioritization method and the self-adjusting strategy pool, two algorithms without these two method respectively were applied for comparison. Furthermore, to verify the necessity of multi-objective frame, we also design a PSO-SaSP algorithm for comparison.

10-fold cross-validation was used to divide the data into training and testing sets. The kNN algorithm with $k = 1$ was applied to evaluate the classification accuracy. The parameter settings used in the experiment were present in Table 2. In order to make full use of computational resources and improve the quality of the solutions, we also set the following parameter to improve the particle swarm. After every 3 iterations, the two objective function values of each particle are summed and sorted in descending order. To make the worst particles jump out of the bad results, we reinitialize the top 40% of the particles.

Table 2. Parameter setting of different method

Algorithm	Parameter	Setting
PSO, PPSO, VLPSO VLPSO-LS	Population size	Feature size/20 (≤ 300, ≥ 150)
	Maximum iteration	100
	Threshold	0.6
MOPSO-SaSP	Population size	100
	Maximum iteration	80
	Threshold	0.6
	LP	20

Table 3. Balanced accuracy of compared algorithms

Dataset	PSO	PPSO	VLPSO	VLPSO-LS	SOP-COPSO	LFSDC	MOPSO-SaSP
SRBCT	89.51	95.78	99.67	**99.75**	99.01	99.29	96.73
DLBCL	83.67	86.22	86.51	96.13	89.93	91.55	**99.85**
9Tumors	42.72	59.28	55.11	56.78	50.68	57.85	**67.37**
11Tumors	71.81	76.83	80.81	82.81	79.85	88.48	**88.83**
Leukemia 1	80.60	94.37	93.31	93.75	91.36	93.98	**98.34**
Leukemia 2	89.83	96.74	91.56	95.39	93.97	94.39	**99.90**
Lung Cancer	78.77	79.38	89.47	90.17	84.30	**93.24**	92.04
Brain Tumor1	73.73	74.40	71.19	75.54	74.02	**89.24**	85.50

4.3   Performance of Feature Prioritization

To demonstrate the effectiveness of the feature prioritization method, we designed a set of controlled experiments using random initialization instead of the feature prioritization method. In the random initialization method, all the parameters are set in the same way as those in the evolutionary process in our proposed MOPSO-SaSP. As shown in Table 4, the algorithm with AMB feature prioritization perform better for all data sets though the computation time are slightly longer for 7 data sets. We can conclude that AMB feature prioritization can efficiently achieve high degree of dimension reduction and enhance or maintain predictive accuracy with selected features.

Table 4. Results of comparison with the method with random initialization

Dataset	BA		Size		Time(s)	
	Random initialization	MOPSO-SaSP	Random initialization	MOPSO-SaSP	Random initialization	MOPSO-SaSP
SRBCT	95.55%	96.73%	54.1	8.0	880.1	882.7
DLBCL	99.08%	99.85%	59.4	8.1	209.5	218.8
9Tumors	59.82%	67.37%	371.8	50.9	1301.2	1314.1
11Tumors	83.13%	88.83%	1066.9	322.9	2605.8	2333.1
Leukemia 1	97.34%	98.34%	79.8	8.3	611.4	621.9
Leukemia 2	98.05%	99.90%	86.9	17.6	801.9	807.2
Lung Cancer	90.55%	92.04%	190.8	36.7	1502.1	1518.2
Brain Tumor1	81.23%	85.50%	144	13	1379.5	1395.5

4.4   Performance of Self-adjusting Strategy Pool

To investigate the effectiveness of the self-adjusting strategy pool mechanism, we compare the proposed algorithm with the MOPSO algorithm using a single update strategy and the performance results are given in Table 5. To further analyze the rationality of the strategy pool, the changing curves of the strategy success probabilities are plotted as shown as follows. Because of limited space, Fig. 2 shows the final success probability of each strategy for each of the 10 runs of the algorithm on 3 testing sets. In this graph, the x axis represents the success probability, the y axis represents the number of runs for each strategy. It can be seen that each strategy performs differently on the same test set for each run, and the suitability of the strategy varies depending on the initialization and search process. Each strategy has a different probability of success in each run. All the probabilities change during six evolutionary processes. This phenomenon indicates that the performance of the strategies changes at different evolutionary runs. The applicability of the strategy is variable and the idea of the dynamic strategy is reasonable. The self-adjusting pool mechanism can select the most suitable strategies at different numbers of runs and evolutionary stages. Only the SRBCT test set is shown here, the results are similar for the other datasets. Figure 3 shows the average success rate of each strategy for the algorithm running on the eight datasets. From this figure, it can be concluded that the suitability of each strategy on different datasets is different. This self-adjusting mechanism can select appropriate update strategies in the process of feature selection for different datasets.

Table 5. Results of comparison with single strategy method

Dataset	BA		Size		Time(s)	
	Single strategy	MOPSO-SaSP	Single strategy	MOPSO-SaSP	Single strategy	MOPSO-SaSP
SRBCT	94.01%	96.73%	18.5	8.0	751.2	882.7
DLBCL	96.14%	99.85%	21.8	8.1	255.1	218.8
9Tumors	52.21%	67.37%	56.2	50.9	910.9	1314.1
11Tumors	84.59%	88.83%	331.7	322.9	2605.8	2333.1
Leukemia 1	96.34%	98.34%	15	8.3	611.4	621.9
Leukemia 2	97.06%	99.90%	18.9	17.6	801.9	807.2
Lung Cancer	90.01%	92.04%	190.8	36.7	1502.1	1518.2
Brain Tumor1	78.92%	85.50%	45.3	13	1001.8	1395.5

4.5 Performance of Multi-objective Frame

Though we chose one of the two objectives-classification accuracy-as our priority to select *gbest* in the multi-objective frame, it doesn't mean the other objective is meaningless. We modified the MOPSO-SaSP algorithm into a single-objective method named PSO-SaSP as a comparative experiment. Compared with the single-objective-based self-adjusting strategy pool method, our proposed multi-objective mechanism can better solve the large-scale multi-objective feature selection problems. In the single-objective method, the classification accuracy

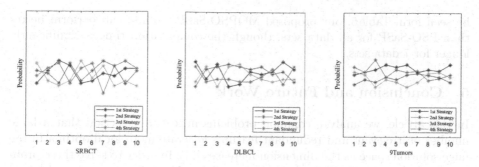

Fig. 2. Changing curves of strategies success probabilities on 10 runs of 3 data sets.

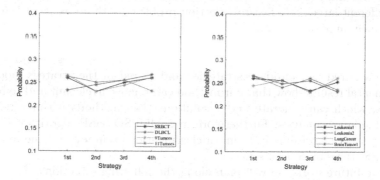

Fig. 3. Changing curves of strategies average success probabilities on 8 data sets.

Table 6. Results of comparison with single-objective algorithm

Dataset	BA		Size		Time(s)	
	PSO-SaSP	MOPSO-SaSP	PSO-SaSP	MOPSO-SaSP	PSO-SaSP	MOPSO-SaSP
SRBCT	94.98%	96.73%	16.5	8.0	763.2	882.7
DLBCL	98.43%	99.85%	15.7	8.1	254.5	218.8
9Tumors	61.82%	67.37%	46.7	50.9	913.1	1314.1
11Tumors	83.79%	88.83%	351.7	322.9	1349.4	2333.1
Leukemia 1	96.23%	98.34%	14.9	8.3	589.3	621.9
Leukemia 2	99.43%	99.90%	18.3	17.6	1415.4	807.2
Lung Cancer	90.98%	92.04%	50.4	36.7	1201.8	1518.2
Brain Tumor1	83.23%	85.50%	28.5	13	1084.5	1395.5

is used as the only objective function, and the single solution with the highest classification accuracy is selected as the new *gbest* during the evolutionary process. As The single-objective PSO has only one best solution, the diversity of solutions does not work well compared to the multi-objective PSO [20]. As can be seen form Table 5, our proposed MOPSO-SaSP method can perform better than PSO-SaSP for all data sets, though the computation time is significantly longer for 7 data sets.

5   Conclusion and Future Work

In this article, we analyze common problems in feature selection that a large number of irrelevant and redundant features generate more local optima in the huge solution space as the dimension increases [21]. In order to solve these problems, we propose an AMB feature prioritization and initialization approach based on relevance and redundancy analysis. In addition, it may be low-performance to solve different large-scale feature selection problems with an existing PSO algorithm since large-scale feature selection problems with different datasets may have different properties [22]. We design a self-adjusting strategy pool algorithm bases on PSO and the experiments are performed on 10 strategies and 4 strategies are selected as candidate strategies and stored in the strategy pool. The experimental results show that our method can remove irrelevant and redundant features, which can generate better solutions than methods without relevance and redundancy analysis. Furthermore, the MOPSO-SaSP algorithm is better than the other algorithms in terms of classification accuracy and the number of selected features on 8 data sets.

In the future work, we will plan along the following directions. The process of designing the strategy pool is tedious. Moreover, there are many evolutionary strategies that have been used so far, and more new strategies will be studied in the future. More complete experiments should take all these strategies into account, but the computational complexity of the algorithm will be increased. We will develop a more efficient method to design a strategy pool to solve large-scale feature selection problems. Furthermore, except for the two objectives involved in this paper, a distance metric might be also considered as an objective [23]. In addition, we will implement more advanced EC based feature selection methods and additional effort is needed to experiment our method on other datasets to improve the performance of MOPSO-SaSP.

References

1. Xue, B., Zhang, M., Browne, W.N., Yao, X.: A survey on evolutionary computation approaches to feature selection. IEEE Trans. Evol. Comput. **20**(4), 606–626 (2016)
2. Kira, K., Rendell, L.A.: The feature selection problem: traditional methods and a new algorithm. In: Proceedings of the Tenth National Conference on Artificial Intelligence, pp. 129–134 (1992)
3. Zhao, X., Liu, Z.: Hybrid particle swarm optimization with differential and perturbation. J. Front. Comput. Sci. Technol. **8**(2), 218 (2014)
4. Nguyen, H.B., Xue, B., Zhang, M.: A subset similarity guided method for multiobjective feature selection. In: Ray, T., Sarker, R., Li, X. (eds.) ACALCI 2016. LNCS (LNAI), vol. 9592, pp. 298–310. Springer, Cham (2016). https://doi.org/10.1007/978-3-319-28270-1_25

5. Xue, B., Cervante, L., Shang, L., Borowne, W.N., Zhang, M.J.: Multi-objective evolutionary algorithms for filter based feature selection in classification. Int. J. Artif. Intell. Tools **22**(4), 1–34 (2013)
6. Chuang, L.-Y., Chang, H.-W., Tu, C.-J., Yang, C.-H.: Improved binary PSO for feature selection using gene expression data. Comput. Biol. Chem. **32**(1), 29–38 (2008). https://doi.org/10.1016/j.compbiolchem.2007.09.005
7. Xue, B., Zhang, M., Browne, W.N.: Particle swarm optimization for feature selection in classification: a multi-objective approach. IEEE Trans. Cybern. **43**(6), 1656–1671 (2013). https://doi.org/10.1109/TSMCB.2012.27469
8. Zhou, Y., Kang, J., Guo, H.: Many-objective optimization of feature selection based on two-level particle cooperation. Inf. Sci. **532**, 91–109 (2020). https://doi.org/10.1016/j.ins.202
9. Zhou, Y., Kang, J., Kwong, S., Wang, X., Zhang, Q.: An evolutionary multi-objective optimization framework of discretization-based feature selection for classification. Swarm Evol. Comput. **60**, 100770 (2021). https://doi.org/10.1016/j.swevo.20. ISSN 2210-6502
10. Rani, C., Kothari, D.P.: Dynamic economic emission dispatch problem with valve-point effect. In: International Conference on Emerging Trends in Electrical Engineering and Energy Management. IEEE (2013)
11. Xue, Y., Xue, B., Zhang, M.: Self-adaptive particle swarm optimization for large-scale feature selection in classification. ACM Trans. Knowl. Discov. Data **13**(5), 27. Article 50 (2019). https://doi.org/10.1145/3340848
12. Yu, L., Liu, H.: Efficient feature selection via analysis of relvance and redundancy. J. Mach. Learn. Res. **5**(12), 1205–1224 (2004)
13. Quinlan, J.R.: Discovering rules by induction from large collections of examples. Expert Syst. Micro Electron. Age (1979)
14. Press, W.H., Teukolsky, S.A., Vetterling, W.T., Flannery, B.P.: Numerical Recipes in C. Cambridge University Press, Cambridge (1988)
15. Kennedy, J., Eberhart, R.: Particle swarm optimization. In: Proceedings IEEE International Conference Neural Networks, vol. 4. pp. 1942–1948 (1995)
16. Wang, Y., Li, B., Weise, T., Wang, J.Y., Yuan, B., Tian, Q.J.: Self-adaptive learning based particle swarm optimization. Inf. Sci. **181**(20), 4515–4538 (2011)
17. Wang, H., Sun, H., Li, C.H., Rahnamayan, S., Pan, J.S.: Diversity enhanced particle swarm optimization with neighborhood search. Inf. Sci. **223**(2013), 119–135 (2013)
18. Li, X.: A non-dominated sorting particle swarm optimizer for multiobjective optimization. In: Proceedings Annual GECCO, pp. 37–48 (2003)
19. Tran, B., Xue, B., Zhang, M.: A new representation in PSO for discretization-based feature selection. IEEE Trans. Cybern. **48**(6), 1733–1746 (2018). https://doi.org/10.1109/TCYB.2017.2714145
20. Cheng, S., Shi, Y., Qin, Q.: Population diversity of particle swarm optimizer solving single and multi-objective problems. Int. J. Swarm Intell. Res. **3**(4), 23–60 (2012)
21. Tran, B., Xue, B., Zhang, M.: Variable-length particle swarm optimisation for feature selection on high-dimensional classification. IEEE Trans. Evol. Comput. **23**(3), 473–487 (2018). https://doi.org/10.1109/TEVC.2018.2869405.1-1
22. Mohamad, M.S., Omatu, S., Deris, S., Yoshioka, M.: A modified binary particle swarm optimization for selecting the small subset of informative genes from gene expression data. IEEE Trans. Inf. Technol. Biomed. **15**(6), 813–822 (2011)
23. Zhou, Y., Zhang, W., Kang, J., Zhang, X., Wang, X.: A problem-specific non-dominated sorting genetic algorithm for supervised feature selection. Inf. Sci. **547**, 841–859 (2021). ISSN 0020-0255

Data-Driven Recommendation Model with Meta-learning Autoencoder for Algorithm Selection

Xianghua Chu[1,2], Yongsheng Pang[1], Jiayun Wang[1], Yuqiu Guo[3], Yuanju Qu[1], and Yangpeng Wang[1(✉)]

[1] College of Management, Shenzhen University, Shenzhen, China
{x.chu,wangyangpeng}@szu.edu.cn
[2] Institute of Big Data Intelligent Management and Decision, Shenzhen University, Shenzhen, China
[3] The Mount School, York, UK

Abstract. To improve the efficiency of problem-solving for complex optimization problems, meta-learning was applied in algorithm selection to choose the most appropriate algorithm recently. However, the common meta-learners are feature-sensitive, where the selection and extraction of meta-features impact the quality of algorithm recommendations. In this study, we propose a data-driven recommendation model to implement the intelligent algorithm selection based on deep meta-features. A new kind of supervised stacked Autoencoder, named meta-learning Autoencoder, has been designed to process the deep meta-feature which is suitable both for instance-based and model-based meta-learners. To evaluate the performance of the proposed model, experiments have been conducted on some benchmark problems. Experimental results show that the recommendation accuracy of the model achieves nearly 100% in the seen problems and more than 80% in the unseen problems.

Keywords: Recommendation model · Meta-learning · Algorithm selection · Supervised Autoencoder · Stacked Autoencoder

1 Introduction

As the complexity of real-world problems increases greatly, many algorithms have been designed. Meanwhile, according to the 'No Free Lunch' theorem [1], no single algorithm can perform well in all problems, which makes algorithm selection necessary for the accuracy and efficiency of problem-solving. Choosing the appropriate algorithm from thousands of alternatives is a challenging job. Previously, engineers utilized the 'trial-and-error' method to assess multiple algorithms on the target problem and select the best-performing algorithm after spending a significant amount of time and computational resources. Then, engineers tend to choose algorithms based on their personal experience. This is a kind of 'Intelligence of Human', which is usually not the ideal answer owing to the limitations of humans themselves.

With the successful application of meta-learning in many fields, researchers have recently started to apply it to algorithm selection. Meta-learning is a kind of 'learn to learn' approach, where the meta-learner extracts the deep meta-knowledge behind the data to guide the base learner to better accomplish the task. This can be considered as a kind of 'Intelligence of Machine'. In the previous studies, meta-learning frameworks have all demonstrated high performance in algorithm selection. Cui et al. [2] proposed a recommendation system to select an appropriate machine learning algorithm for surrogate modeling, which can still perform well with a large number of candidates and without sufficient prior knowledge of the problem. Chu et al. [3] advanced an adaptive recommendation model to select population-based algorithms for different optimization problems and the model showed good flexibility and extendibility. Sehta and Thakar [4] used various meta-learner to recommend heuristic algorithms on Capacitated VRP, proving the effectiveness of the meta-learning framework.

However, the traditional meta-learning algorithm recommendation frameworks rely heavily on the selection and quality of meta-features. According to a survey by Khan et al. [5], there are five kinds of generic meta-features proposed to advance the effectiveness of the frameworks. Researchers also need to find more problem-specific meta-features for different optimization problems [6–10]. Despite the promising performance of meta-learning methods, a major shortcoming is that the meta-learner is feature-sensitive, which needs sufficient prior knowledge on the specific problem to select suitable features. A new dilemma, the construction and selection of meta-features, has emerged.

In this study, we propose a data-driven method to process the deep meta-features efficiently and reduce the reliance of meta-learner on specific meta-features. Different from the traditional frameworks, the data-driven method will make full use of the sample data to process more training data for meta-learners. Autoencoder [11] is a powerful unsupervised deep learning model and several Autoencoders usually are stacked to extract deep features from the dataset. To reduce learning errors and generate meta-features that are more relevant to the algorithm recommendation target, a stacked supervised Autoencoder, named Meta-learning Autoencoder (MLAE), is proposed to map the problem feature space to algorithm performance space. Its latent features that reflect both problem-specific and algorithm-performance-specific information serve as the meta-features.

Based on the novel kind of meta-features which are both suitable for model-based and instance-based meta-learners, a Data-driven Recommendation Model (DRM) is designed for algorithm selection. Experiments had been implemented on twenty benchmark functions from the CEC2010 large-scale optimization competition to evaluate the performance of DRM. Results show that DRM achieves a recommendation accuracy close to 100% on the seen problems and about 80% on the unseen problems.

In summary, there are three contributions in this work:

1. MLAE is proposed to process the meta-features that are closely related to both the target problem and algorithm performance.
2. The latent layers of the trained MLAE output the deep meta-features that are different from the traditional statistical and landmark features.
3. A data-driven recommendation model is advanced to improve the previous meta-learning framework of algorithm selection.

The rest of this paper is organized as follows: background about algorithm selection and Autoencoder are presented in Sect. 2. Section 3 describes a detailed description of DRM. Experiments and results are discussed in Sect. 4. Section 5 provides a discussion of the conclusions and future work.

2 Background

2.1 Algorithm Selection

Algorithm selection is the preliminary work for algorithm application, which picks out the appropriate algorithm that performs well in different specific problems. It can be a selection of single best algorithms or a set of high-performing candidates. Rice [12] first proposed a theoretical framework for algorithm selection using the features of problems and the performances of algorithms, as shown in Fig. 1.

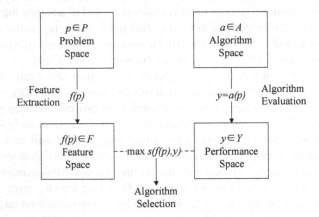

Fig. 1. The framework of Rice's model.

Suppose P is the problem space and there are a lot of problems p belonging to it. By applying a feature extracting method to problem p, we will get the extracted features $f(p)$ which are included in the feature space F. A is the algorithm space that contains all the alternative algorithms and a is one of these algorithms. A corresponding performance space Y of these algorithms can be obtained by evaluating every algorithm a on all problems. In order to select the appropriate algorithm, we will establish an algorithm performance mapping $s(f(p), y)$ and try to find the a which will maximize this mapping.

2.2 Meta-learning

Meta-learning is an advanced machine learning method. A meta-learner will be trained by learning the useful knowledge from the existing problem instances, to guide kinds of basic learners to accomplish the specified classification or regression tasks. Once the meta-learner is trained, it has the ability to "learn to learn" and can quickly learn new tasks based on its own knowledge.

In the meta-learning task, the meta-representation of the problem is of great importance, which will help the meta-learner to fully understand the essence of the problem. In the previous studies, researchers use meta-features to represent different sorts of problems and large numbers of meta-features have been proposed. However, not all features enable the meta-learners to achieve the best prediction performance. How to construct and select useful meta-features is still a problem that needs further research.

2.3 Autoencoder

Autoencoder (AE) [11] is a symmetric unsupervised neural network model, which consists of the input layer, hidden layer, and output layer, and its structure is shown in Fig. 2. Since AE learns by setting the input data as the label and minimizing the reconstruction error, its output layer has the same number of nodes as the input layer. Usually, the number of nodes in its hidden layer is less than the input/output layer, which is also called the bottleneck layer, to achieve the effect of dimensionality reduction and data compression.

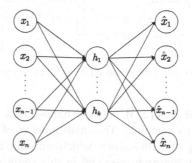

Fig. 2. The structure of Autoencoder.

Since AE has only one hidden layer, a single AE can only extract limited and shallow hidden features of the input data. In order to implement continuous dimensionality reduction and further mining of deep features, researchers stacked several AEs together to build a Stacked Autoencoder [13]. In the training process of Stacked AE, several AEs are pre-trained first and the hidden layer of each AE is taken as the hidden layer of Stacked AE. Then, a fine-tuning will be implemented to adjust the parameters of the model slightly.

In some machine learning studies, researchers use the output of the hidden layers as deep latent features for classification tasks, such as driver identification [14] and breast cancer diagnosis [15]. Due to the remarkable feature-extracting performance of Stacked AE, it is used for the meta-feature extraction in this study. Unlike the traditional statistical meta-features, deep meta-features extracted by Autoencoder are a nonlinear projection of the original input data. This kind of meta-features do not have any realistic physical meaning in every dimension.

3 Data-Driven Recommendation Model

Based on the theoretical framework presented by Rice [12], this study proposes a Data-driven Recommendation Model (DRM) with the structure shown in Fig. 3. The DRM consists of two main components: the Meta-learning module and the Algorithm recommending module. The details of various modules are described as follows.

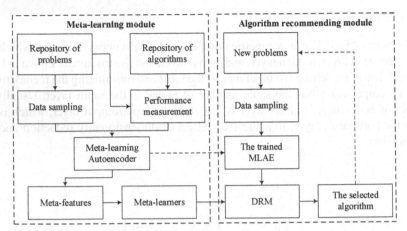

Fig. 3. The framework of the proposed data-driven recommendation model.

In the traditional framework [2, 3], only one meta-feature data can be computed for each problem instance after sampling. The small number of training samples may be detrimental to the training of meta-learners which are some machine learning models. However, the data-driven approach proposed in this study can make full use of the sampled data to generate the same number of meta-features data. This can improve the performance of the meta-learners by increasing the number of training samples. The difference between the traditional framework and this study is shown in Fig. 4.

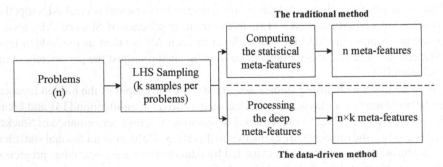

Fig. 4. The difference between the traditional and data-driven frameworks.

3.1 Meta-learning Module

Before the algorithm recommendation, the extraction of meta-features is needed to represent the characteristic of the problems and the performance of the algorithms. This study proposes a stacked supervised Autoencoder, Meta-learning Autoencoder (MLAE), to process the deep meta-features. Then, the deep meta-features will be used to train kinds of meta-learners.

3.1.1 Meta-learning Autoencoder

In this module, MLAE is used to transfer the problem samples and algorithm performance data into meta-features for DRM. Different from the vanilla Stacked Autoencoder, MLAE is composed of some supervised AE [16] which allow for purposeful dimensionality reduction of the input data.

To make the AE better learn the deep features associated with the target output, the output label of the original AE is replaced with the value of the objective function corresponding to the input sample or the performance of the algorithms for different problems. Furthermore, two new kinds of supervised AE (SAE) are obtained, called Obj-based SAE and AP-based SAE respectively, and shown in Fig. 5. Since the objective function value of the sample is only one value, Obj-based SAE is a single-output AE. While the performance of different algorithms are multiple values, AP-based SAE is a multiple-output AE.

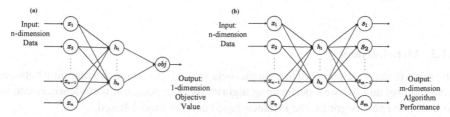

Fig. 5. The structure of the Obj-based SAE (a) and AP-based SAE (b).

The MLAE consists of two parts. The first part consists of several Obj-based SAEs stacked, and the first Obj-based SAE in this part is trained with the original sample data as the input data and the target function value as the output label. After the training of it is completed, the output of the hidden layer is obtained as Obj-based features and input to the subsequent Obj-based SAEs for further dimensionality reduction. The final Obj-based features related to the problem can be obtained through a series of stacked Obj-based SAEs as the intermediate data for the next part.

The second part is similar to the first part and consists of several AP-based SAEs stacked. The first AP-based SAE is trained with the final Obj-based features of the previous part as input and the algorithm performance as output. The final AP-based features that are relevant to both the problem and the algorithm performance can be obtained as the meta-features adopted by DRM through a series of stacked AP-based SAEs.

The structure of MLAE is shown in Fig. 6. Unlike traditional recommendation frameworks where various types of meta-features are computed directly from the data, meta-features used in this paper are the product of dimensionality reduction by MLAE, are closely related to both the target problem and algorithm performance.

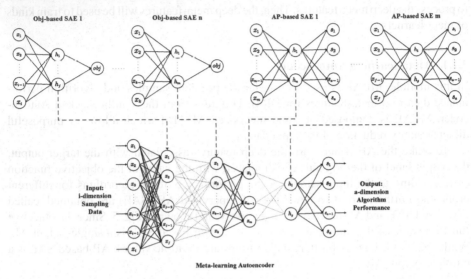

Fig. 6. The training process and structure of Meta-learning Autoencoder.

3.1.2 Meta-learners

The meta-features are prepared to train the meta-learners, to establish a mapping between meta-features and the well-performing algorithms. The common meta-learners can be divided into two categories: the instance-based and the model-based.

The instance-based meta-learner determines the similarity of different instances by computing some distance metrics and predicts the performance of different algorithms in a new instance corresponding to several similar instances, making algorithm recommendations accordingly. K nearest neighbors (KNN) [17] is the most common instance-based meta-learner, which selects the k nearest neighbors based on meta-features by measuring the distance between the new instance and the existing problem instances with a known algorithm performance. The algorithm performance of these k neighbors is then weighted to obtain the predicted algorithm performance of the new instance.

The model-based meta-learner learns the mapping relationship with the meta-features by learning experience in historical algorithms performance. Among the existing studies, Multilayer Perceptron (MLP) and Random Forest (RF) [18] are the two most commonly used model-based meta-learners. MLP is an excellent non-linear fitting approach, which can find the appropriate mapping relationship between meta-features and targets. RF is an ensemble model that generates a certain number of decision tree models by randomly selecting a subset of features and then votes through these single tree models to get the final results.

3.2 Algorithm Recommending Module

After completing the training of the meta-learner, we are ready to perform the meta-recommend task. In order to make an algorithm recommendation for a new problem, the following steps need to be performed:

1. Sample data from the new problems.
2. Use the sampled data to train a series of Obj-based SAEs in the first part of MLAE to obtain the final Obj-based features
3. Input the final Obj-based features to a series of AP-based SAEs already trained in DRM to get the final AP-based features which will be used as meta-features.
4. Input the meta-features into meta-learners to get the prediction performance of each algorithm in the repository of algorithms
5. Select one or several algorithms with better predictive performance for solving the new problems

4 Experiment and Results

For a comprehensive evaluation, several benchmark functions are employed to evaluate the performance of the proposed DRM. The details of this experiment are as follows.

4.1 Experimental Setup

Twenty benchmark functions from the CEC2010 competition on large-scale global optimization are utilized to create the problem repository. Each function is set to 1000 dimensions and 1000 samples are randomly sampled by the Latin hypercube sampling [19]. The Repository of algorithms consists of seven algorithms: DECC-G [20], DECCG-DG [21], CBCC3-DG2 [22], SaNSDE [23], SGCC [24], SLPSO [25] and CSO [26]. The evaluation of algorithms' performance achieves on the PlatEMO [27] platform, and the parameters of all algorithms are kept at default settings.

The MLAE used in the experiments consists of four SAEs stacked with the parameters set as shown in Table 1. In addition, each AE is trained using Adam as the optimizer with a maximum training epoch of 100, and the Early stopping (with the patience of five epochs) strategy is set. As for meta-learners, while the RF model uses the default parameters, the K of the KNN model is set to 2 and the weights of different instances are assigned according to the distance.

To evaluate the algorithm selection performance of DRM, two common metrics are used as evaluation metrics: SRCC [28] and SR. The SRCC is employed to evaluate the agreement between recommended rankings and ideal rankings, which is calculated as follows:

$$\rho = 1 - 6 \times \frac{\sum_{i=1}^{N} D_i^2}{N(N^2 - 1)} \tag{1}$$

where D_i is the distance between the recommended ranking and the ideal ranking of algorithm i; N is the number of algorithms. For this coefficient, the value of 1 means

Table 1. The parameter settings of SAEs in MLAE.

	The nodes of the hidden layer	The activation function of the hidden layer	The nodes of the output layer	The activation function of the output layer
Obj-based SAE 1	500	Sigmoid	1	Sigmoid
Obj-based SAE 2	200			
AP-based SAE 1	100		7	Softmax
AP-based SAE 2	30			

full agreement while − 1 refers to full disagreement. The SR measures the radio of the exact matches between the ideal best algorithm and recommended the most appropriate algorithm on the problems. In the recommendation of algorithms, users are primarily concerned about the recommended best algorithm and tend to use it.

$$SR = \frac{\text{\#number of well-predicted ideal algorithm}}{\text{\#total number of samples from problems}} \tag{2}$$

4.2 Experimental Designs and Results

4.2.1 Experiment on Seen Problems

With the idea of 5-fold cross-validation, 80% of the samples from each of the 20 benchmark functions are taken for the training of DRM, and the remaining 20% of the samples are used for testing. Five experiments are conducted and all samples have ever been used for training and testing. The average result of the 5 experiments is taken as the final performance of DRM, and details are shown in Table 2.

Table 2. The experimental results of different meta-learner on seen problems.

Meta-learners	SRCC	SR
KNN	1.000 ± 0.000	1.000
MLP	0.963 ± 0.050	0.922
RF	0.999 ± 0.003	1.000

The results reveal that all three meta-learners perform admirably. In all experiments, KNN outperforms the competition with 100% SRCC and SR, accurately predicting the ideal algorithm rankings. It indicates that the meta-features processed by MLAE not only can characterize the problem itself, but also effectively reflect the corresponding algorithm performance. While MLP and RF have relatively lower performances, they are still acceptable in practice. It is probably because the model-based meta-learners are more sensitive to tiny differences and misidentify the noise introduced by MLAE.

4.2.2 Experiment on Unseen Problems

With the idea of 'Leave-one-out' cross-validation, nineteen out of twenty benchmark functions are used for training DRM at a time, and the remaining one for testing. Twenty experiments are conducted and the results are shown in Table 3. Since the function used for testing is not seen by DRM during training, the experiment is able to effectively verify the generalization ability of DRM on the new problem.

Table 3. The experimental results of different meta-learner on unseen problems.

Test problems	MLP		KNN		RF	
	SRCC	SR	SRCC	SR	SRCC	SR
F1	0.852 ± 0.262	1.000	0.783 ± 0.398	0.782	0.761 ± 0.437	0.908
F2	0.850 ± 0.267	0.750	0.862 ± 0.279	0.750	0.923 ± 0.131	0.750
F3	0.919 ± 0.147	0.862	0.877 ± 0.271	0.912	0.911 ± 0.191	0.750
F4	0.760 ± 0.417	0.750	0.911 ± 0.155	1.000	0.933 ± 0.117	1.000
F5	0.680 ± 0.556	0.750	0.780 ± 0.382	0.750	0.894 ± 0.194	0.989
F6	0.923 ± 0.135	0.750	0.929 ± 0.124	0.750	0.865 ± 0.235	0.750
F7	0.834 ± 0.288	0.750	0.834 ± 0.289	0.750	0.750 ± 0.433	0.750
F8	0.814 ± 0.323	0.750	0.763 ± 0.420	0.750	0.747 ± 0.438	0.750
F9	0.929 ± 0.129	0.971	0.928 ± 0.143	1.000	0.988 ± 0.030	1.000
F10	0.783 ± 0.377	0.750	0.830 ± 0.294	0.750	0.851 ± 0.258	0.750
F11	0.938 ± 0.107	0.750	0.946 ± 0.093	0.750	0.835 ± 0.289	0.750
F12	0.789 ± 0.366	0.750	0.944 ± 0.103	0.998	0.908 ± 0.162	0.997
F13	0.721 ± 0.483	0.750	0.768 ± 0.402	0.750	0.763 ± 0.412	0.750
F14	0.795 ± 0.360	0.750	0.863 ± 0.307	0.829	0.880 ± 0.214	1.000
F15	0.829 ± 0.297	0.750	0.862 ± 0.249	0.750	0.586 ± 0.721	0.750
F16	0.963 ± 0.072	1.000	0.937 ± 0.112	0.754	0.830 ± 0.297	0.993
F17	0.766 ± 0.407	0.750	0.911 ± 0.154	1.000	0.833 ± 0.346	0.948
F18	0.779 ± 0.400	0.811	0.775 ± 0.391	0.750	0.884 ± 0.292	0.888
F19	0.723 ± 0.479	0.750	0.670 ± 0.572	0.750	0.724 ± 0.517	0.834
F20	0.809 ± 0.335	0.750	0.831 ± 0.298	0.750	0.821 ± 0.330	0.750

For the mean of the average SRCC, three meta-learners achieve a ranking accuracy higher than 80%. We can find that MLP performs a relatively poor accuracy lower than 80% on nine out of twenty problems, while KNN and RF just on six problems. Further analysis on the standard deviation shows there only are nine problems where the standard deviation of 1000 samples is lower than 0.3, while they are higher than 0.4 on six problems. We believe that MLP is more sensitive to sampling bias and noise, which makes MLP raise the variance while reducing bias.

It is worth noting that the standard deviation of SRCC for 1000 samples of each problem is around 0.3. This indicates that when there are fewer samples on the unseen problems, the recommendation effect may be erratic due to less knowledge of new problems learned by DRM. In addition, the SR of all three meta-learners is greater than 0.75 on twenty problems, and the average value is close to 0.8, which demonstrates the promising performance of DRM for the selection of the most appropriate algorithm (Table 4).

Table 4. The experimental results of traditional and data-driven frameworks.

		KNN	MLP	RF
Traditional framework	SRCC	0.791 ± 0.012	0.732 ± 0.051	0.714 ± 0.120
	SR	**0.839 ± 0.078**	**0.828 ± 0.081**	0.778 ± 0.028
Data-driven framework	SRCC	**0.850 ± 0.076**	**0.823 ± 0.079**	**0.834 ± 0.092**
	SR	0.814 ± 0.103	0.795 ± 0.089	**0.853 ± 0.113**

Comparing the experimental results of traditional and data-driven frameworks, it can be found that the data-driven framework is superior in SRCC metrics across the board, while the SR metrics are slightly worse. It shows that the data-driven framework can output better ranking results, but the best performing algorithms sometimes do not appear first in the recommendation results. This may be a result of the training of MLAE with using the ranking of the algorithms' performance, which makes the model better learn the ranking relationships between algorithms. Overall, the data-driven framework proposed in this study is competitive with the traditional framework and even can achieve better performance.

5   Conclusion

Before automated algorithm selection is widely used, people rely on the trial-and-error methods to evaluate the feasibility of various algorithms. It consumes a lot of computational resources and time, which is not necessary. In recent years, many meta-learning frameworks for algorithm selection have been proposed and successfully applied to various practical problems. After reviewing a large number of studies, we found that most meta-learning frameworks rely on problem-specific meta-features. Therefore, this study proposes MLAE to process a novel meta-feature directly from the original sample data. This kind of meta-feature is closely related to the target problem and algorithm performance, but it does not have an actual physical meaning like some statistical features do. Although it is processed by a neural network model, experimental results showed that it is applicable to both instance-based and model-based meta-learners. What's more, experimental results on benchmark functions illustrate that DRM is effective in recommending the most appropriate algorithm for seen problems and performs satisfactorily on unseen problems. Considering the strong scalability and generalization capability of

DRM, future work will focus on improving the generality of this model and trying to apply it to online algorithm recommendations.

Acknowledgement. This work was partially supported by the National Natural Science Foundation of China (Grant No. 71971142), and the Natural Science Foundation of Guangdong Province (No. 2022A1515010278, 2021A1515110595 and 2016A030310067).

References

1. Wolpert, D.H., Macready, W.G.: No free lunch theorems for optimization. IEEE Trans. Evol. Comput. **1**(1), 67–82 (1997)
2. Cui, C., Hu, M., Weir, J.D., Wu, T.: A recommendation system for meta-modeling: a meta-learning based approach. Expert Syst. Appl. **46**, 33–44 (2016)
3. Chu, X., Cai, F., Cui, C., Hu, M., Li, L., Qin, Q.: Adaptive recommendation model using meta-learning for population-based algorithms. Inf. Sci. **476**, 192–210 (2019)
4. Sehta, N., Thakar, U.: A meta-learning approach for algorithm selection for capacitated vehicle routing problems. In: Cyber-Physical, IoT, and Autonomous Systems in Industry 4.0, pp. 255–268 (2021)
5. Khan, I., Zhang, X., Rehman, M., Ali, R.: A literature survey and empirical study of meta-learning for classifier selection. IEEE Access **8**, 10262–10281 (2020)
6. Cui, C., Wu, T., Hu, M., Weir, J.D., Li, X.: Short-term building energy model recommendation system: a meta-learning approach. Appl. Energy **172**, 251–263 (2016)
7. Chu, X., et al.: Meta-feature extraction for multi-objective optimization problems. In: Zhang, H., Yang, Z., Zhang, Z., Wu, Z., Hao, T. (eds.) NCAA 2021. CCIS, vol. 1449, pp. 432–445. Springer, Singapore (2021). https://doi.org/10.1007/978-981-16-5188-5_31
8. Pulatov, D., Kotthof, L.: Utilizing software features for algorithm selection. In COSEAL Workshop, co-located with the 15th ACM/SIGEVO Workshop on Foundations of Genetic Algorithms (2019)
9. Beel, J., Tyrell, B., Bergman, E., Collins, A., Nagoor, S.: Siamese meta-learning and algorithm selection with 'Algorithm-Performance Personas' [Proposal]. arXiv preprint arXiv: 2006.12328 (2020)
10. Tyrrell, B., Bergman, E., Jones, G.J., Beel, J.: Algorithm-performance personas 'for Siamese meta-learning and automated algorithm selection. In 7th ICML Workshop on Automated Machine Learning (2020)
11. LeCun, Y.: Connexionist learning models (1987)
12. Rice, J.R.: The algorithm selection problem. Adv. Comput. **15**, 65–118 (1976)
13. Gehring, J., Miao, Y., Metze, F., Waibel, A.: Extracting deep bottleneck features using stacked auto-encoders. In 2013 IEEE International Conference on Acoustics, Speech and Signal Processing, pp. 3377–3381. IEEE (2013)
14. Chen, J., Wu, Z., Zhang, J.: Driver identification based on hidden feature extraction by using adaptive nonnegativity-constrained autoencoder. Appl. Soft Comput. **74**, 1–9 (2019)
15. Xu, J., Xiang, L., Hang, R., Wu, J.: Stacked Sparse Autoencoder (SSAE) based framework for nuclei patch classification on breast cancer histopathology. In: 2014 IEEE 11th International Symposium on Biomedical Imaging (ISBI), pp. 999–1002. IEEE (2014)
16. Wang, Y., Yang, H., Yuan, X., Shardt, Y.A., Yang, C., Gui, W.: Deep learning for fault-relevant feature extraction and fault classification with stacked supervised auto-encoder. J. Process Control **92**, 79–89 (2020)

17. Goldberger, J., Hinton, G.E., Roweis, S., Salakhutdinov, R.R.: Neighbourhood components analysis. In: Advances in Neural Information Processing Systems, 17 (2004)
18. Breiman, L.: Random forests. Mach. Learn. **45**(1), 5–32 (2001)
19. McKay, M.D., Beckman, R.J., Conover, W.J.: A comparison of three methods for selecting values of input variables in the analysis of output from a computer code. Technometrics **21**, 23,9 (1979)
20. Yang, Z., Tang, K., Yao, X.: Large scale evolutionary optimization using cooperative coevolution. Inf. Sci. **178**(15), 2985–2999 (2008)
21. Omidvar, M.N., Li, X., Mei, Y., Yao, X.: Cooperative co-evolution with differential grouping for large scale optimization. IEEE Trans. Evol. Comput. **18**(3), 378–393 (2013)
22. Omidvar, M.N., Yang, M., Mei, Y., Li, X., Yao, X.: DG2: a faster and more accurate differential grouping for large-scale black-box optimization. IEEE Trans. Evol. Comput. **21**(6), 929–942 (2017)
23. Yang, Z., Tang, K., Yao, X.: Self-adaptive differential evolution with neighborhood search. In: 2008 IEEE Congress on Evolutionary Computation (IEEE World Congress on Computational Intelligence), pp. 1110–1116. IEEE (2008)
24. Liu, W., Zhou, Y., Li, B., Tang, K.: Cooperative co-evolution with soft grouping for large scale global optimization. In: 2019 IEEE Congress on Evolutionary Computation (CEC), pp. 318–325. IEEE (2019)
25. Li, C., Yang, S., Nguyen, T.T.: A self-learning particle swarm optimizer for global optimization problems. IEEE Trans. Syst. Man Cybern. Part B (Cybern.) **42**(3), 627–646 (2011)
26. Cheng, R., Jin, Y.: A competitive swarm optimizer for large scale optimization. IEEE Trans. Cybern. **45**(2), 191–204 (2014)
27. Tian, Y., Cheng, R., Zhang, X., Jin, Y.: PlatEMO: a MATLAB platform for evolutionary multi-objective optimization [educational forum]. IEEE Comput. Intell. Mag. **12**(4), 73–87 (2017)
28. Neave, H., Worthington, P.: Distribution-free tests. Contemp. Sociol. **19**(3), 488 (1990)

Author Index

Wang, Jingyuan   II-234
Wang, Kunlun   II-401
Wang, Lei   I-15
Wang, Liang   I-503
Wang, Ming   II-401
Wang, Qi   I-184, II-80
Wang, Ruiqi   I-515
Wang, Sheng   II-287
Wang, Shuai   I-197, II-150, II-205
Wang, Shuqiang   II-1
Wang, Wanyuan   II-27
Wang, Wenjun   I-394
Wang, Wenkong   II-413, II-427
Wang, Xuhui   II-162
Wang, Yangpeng   I-530
Wang, Yongji   I 154
Wang, Yu   II-494
Wang, Zenghui   I-378, II-302
Wang, Zijian   I-169
Wei, Zichen   I-197, II-150, II-205
Weng, Heng   I-129, II-345
Wong, Leung-Pun   I-115
Wu, Bin   I-406
Wu, Jiali   II-150, II-205
Wu, Kaihan   II-372
Wu, Shiwei   I-286
Wu, Weiwei   II-27
Wu, Zeqiong   II-359
Wu, Zhou   II-302, II-331

Xia, Heng   I-71
Xiang, Xiaohong   I-184, I-353
Xiao, Ke   II-314
Xin, Jianbin   I-489
Xin, Shaofei   I-56
Xin, Zheng   II-80
Xing, Sisi   I-286
Xing, Yu   I-142
Xiong, Tianhua   II-387
Xu, An   II-123
Xu, Baolei   I-274
Xu, Chao   I-286
Xu, Hongke   II-482
Xu, Hongkui   I-1
Xu, Xueyong   II-27
Xu, Yibin   I-434
Xue, Wanli   II-177
Xue, Yangtao   I-235

Yan, Jun   I-129
Yang, Cuili   I-262
Yang, Jing   II-482
Yang, Xiang   II-482
Yang, Yimin   I-316
Yang, Zhaohui   II-387
Yao, Shoufeng   II-109
Ye, Luyao   II-247
Ye, Tingyu   I-197, I-394, II-205
Ye, Xianming   II-302
Yu, Dandan   I-515
Yu, Shiwei   II-40, II-66
Yu, Xinrui   I-247
Yuan, Chengsheng   I-327
Yuan, Jing   I-419, II-123
Yuan, Qi   I-247
Yue, Xiaowen   II-138

Zeng, Kun   I-434
Zeng, Nanli   I-434
Zeng, Qingwei   II-482
Zeng, Shaoxiong   II-302
Zeng, Tao   I-197, I-394
Zhai, Xiangping Bryce   I-44
Zhan, Choujun   I-115, II-359, II-372
Zhang, Gang   I-210
Zhang, Hai   I-197, I-394, II-150, II-205
Zhang, Haijun   I-247, I-378
Zhang, Hao   I-353, II-162, II-314
Zhang, Li   I-235
Zhang, Menghua   II-413, II-427
Zhang, Qiang   II-40, II-66
Zhang, Qisen   II-314
Zhang, Rui   I-142, II-192
Zhang, Shaohong   I-449
Zhang, Tao   II-287
Zhang, Xiao   I-100
Zhang, Xin   II-331
Zhang, Xinyu   I-247
Zhang, Yafei   II-454
Zhang, Yijia   I-30
Zhang, Yong   II-413, II-427
Zhang, Yong-Feng   I-100
Zhang, Yunchu   I-301
Zhang, Zifeng   I-1
Zhang, Zongfeng   I-142
Zhao, Hongnan   II-413
Zhao, Jia   I-197
Zhao, Jin   II-94
Zhao, Jun   I-184